Pipefitting

Level Four

Trainee Guide

Boston Columbus Indianapolis New York San Francisco Amsterdam
Cape Town Dubai London Madrid Milan Munich Paris Montreal Toronto Delhi
Mexico City Sao Paulo Sydney Hong Kong Seoul Singapore Taipei Tokyo

NCCER

President and Chief Executive Officer: Boyd Worsham
Vice President, Innovation and Advancement: Jennifer Wilkerson
Chief Learning Officer: Lisa Strite
Pipefitting Curriculum Project Managers:
Elizabeth Schlaupitz, Lauren Corley
Senior Manager of Projects: Chris Wilson
Senior Production Manager: Erin O'Nora
Technical Writers: Troy Staton, Gary Ferguson, Heidi Saliba
Managing Editor: Jordan Hutchinson
Art Manager: Carrie Pelzer
Digital Content Coordinator: Rachael Downs
Editorial Assistance: Breanna Bean, Daniel Geiger
Production Assistance: Edward Fortman, Joanne Hart

Pearson

Director of Employability Solutions: Kelly Trakalo
Content Producer: Alexandrina B. Wolf
Assistant Content Producer: Alma Dabral
Digital Content Producer: Jose Carchi
Composition: NCCER
Printer/Binder: LSC Communications
Cover Printer: LSC Communications
Text Fonts: Palatino and Univers
Content Technologies: Gnostyx

Credits and acknowledgments for content borrowed from other sources and reproduced, with permission, in this textbook appear at the end of each module.

15 2023

ISBN-13: 978-0-13-748749-3

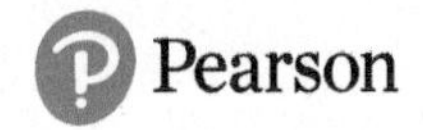

Preface

To the Trainee

There are some who may consider pipefitting synonymous with plumbing, but these are in fact two very distinct crafts. Plumbers install and repair the water, waste disposal, drainage, and gas systems in homes and in commercial and industrial buildings. Pipefitters, on the other hand, install and repair both high- and low-pressure pipe systems used in manufacturing, in the generation of electricity, and in the heating and cooling of buildings.

If you're trying to imagine a setting involving pipefitters, think of large power plants that create and distribute energy throughout the nation; think of manufacturing plants, chemical plants, and piping systems that carry all kinds of liquid, gaseous, and solid materials.

If you're trying to imagine a job in pipefitting, picture yourself in a career that will be highly necessary for the foreseeable future. As the US government reports, the demand for skilled pipefitters continues to outpace the supply of workers trained in this craft. And high demand typically means higher pay, making pipefitters among the highest-paid construction workers in the nation.

While pipefitters and plumbers perform different tasks, the aptitudes involved in these crafts are comparable. Attention to detail, spatial and mechanical abilities, and the ability to work efficiently with the tools of their trade are key. If you think you might have what it takes to work in this high-demand occupation, contact your local NCCER Training Sponsor to see if they offer a training program in this craft, or contact your local union or non-union training programs. You might be the perfect fit.

We wish you success as you embark on your first year of training in the pipefitting craft and hope that you'll continue your training beyond this textbook. There are more than half a million people employed in this work in the United States, and as most of them can tell you, there are many opportunities awaiting those with the skills and desire to move forward in the construction industry.

New with *Pipefitting Level Four*

NCCER is proud to release *Pipefitting* with our latest instructional systems design, linking learning objectives to each module's content. Each module has been updated to reflect feedback from the subject matter experts committee who contributed to the update of curriculum information and provided recommendations for prioritizing competencies required by pipefitters, with portions of the previous content edited for clarity and efficiency. There are additional study questions and updated graphics and photos. In module 08401, several sections related to the associated blueprints were combined for conciseness and ease of reading. The previous edition's module on steam traps (08404-07) has now been incorporated into module 08405 – "In-Line Specialties." Finally, module 46101 – "Fundamentals of Crew Leadership," from the Management craft series, has replaced the previous edition's module titled "Introduction to Supervisory Roles" (08409-07).

Our website, **www.nccer.org**, has information on the latest product releases and training

Your feedback is welcome. You may email your comments to **curriculum@nccer.org** or send general comments and inquiries to **info@nccer.org**.

NCCER Standardized Curricula

NCCER is a not-for-profit 501(c)(3) education foundation established in 1996 by the world's largest and most progressive construction companies and national construction associations. It was founded to address the severe workforce shortage facing the industry and to develop a standardized training process and curricula. Today, NCCER is supported by hundreds of leading construction and maintenance companies, manufacturers, and national associations. The NCCER Standardized Curricula was developed by NCCER in partnership with Pearson, the world's largest educational publisher.

Some features of the NCCER Standardized Curricula are as follows:

- An industry-proven record of success
- Curricula developed by the industry, for the industry
- National standardization providing portability of learned job skills and educational credits
- Compliance with the Office of Apprenticeship requirements for related classroom training (*CFR 29:29*)
- Well-illustrated, up-to-date, and practical information

NCCER also maintains the NCCER Registry, which provides transcripts, certificates, and wallet cards to individuals who have successfully completed a level of training within a craft in NCCER's Curricula. *Training programs must be delivered by an NCCER Accredited Training Sponsor in order to receive these credentials.*

Special Features

In an effort to provide a comprehensive and user-friendly training resource, this curriculum showcases several informative features. Whether you are a visual or hands-on learner, these features are intended to enhance your knowledge of the construction industry as you progress in your training. Some of the features you may find in the curriculum are explained below.

Introduction

This introductory page, found at the beginning of each module, lists the module Objectives, Performance Tasks, and Trade Terms. The Objectives list the knowledge you will acquire after successfully completing the module. The Performance Tasks give you an opportunity to apply your knowledge to real-world tasks. The Trade Terms are industry-specific vocabulary that you will learn as you study this module.

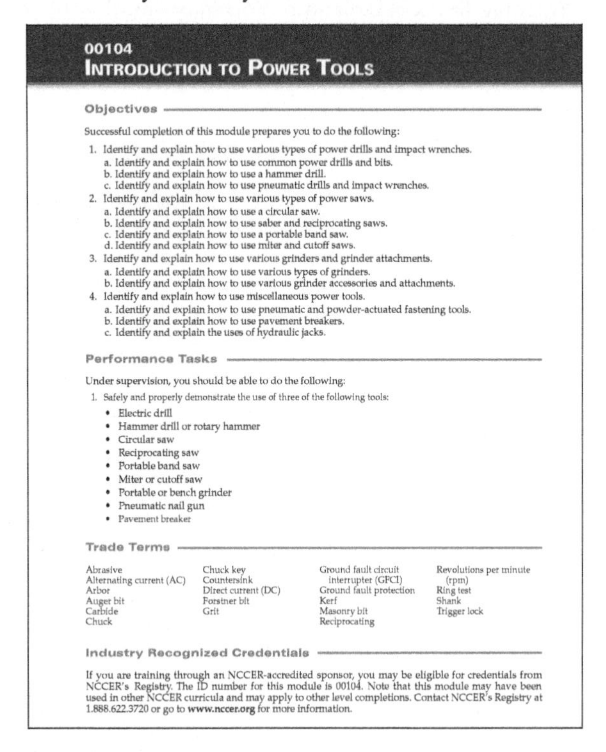

00104
INTRODUCTION TO POWER TOOLS

Objectives

Successful completion of this module prepares you to do the following:

1. Identify and explain how to use various types of power drills and impact wrenches.
   a. Identify and explain how to use common power drills and bits.
   b. Identify and explain how to use a hammer drill.
   c. Identify and explain how to use pneumatic drills and impact wrenches.
2. Identify and explain how to use various types of power saws.
   a. Identify and explain how to use a circular saw.
   b. Identify and explain how to use saber and reciprocating saws.
   c. Identify and explain how to use a portable band saw.
   d. Identify and explain how to use miter and cutoff saws.
3. Identify and explain how to use various grinders and grinder attachments.
   a. Identify and explain how to use various types of grinders.
   b. Identify and explain how to use various grinder accessories and attachments.
4. Identify and explain how to use miscellaneous power tools.
   a. Identify and explain how to use pneumatic and powder-actuated fastening tools.
   b. Identify and explain how to use pavement breakers.
   c. Identify and explain the uses of hydraulic jacks.

Performance Tasks

Under supervision, you should be able to do the following:

1. Safely and properly demonstrate the use of three of the following tools:
   - Electric drill
   - Hammer drill or rotary hammer
   - Circular saw
   - Reciprocating saw
   - Portable band saw
   - Miter or cutoff saw
   - Portable or bench grinder
   - Pneumatic nail gun
   - Pavement breaker

Trade Terms

Abrasive
Alternating current (AC)
Arbor
Auger bit
Carbide
Chuck
Chuck key
Countersink
Direct current (DC)
Forstner bit
Grit
Ground fault circuit interrupter (GFCI)
Ground fault protection
Kerf
Masonry bit
Reciprocating
Revolutions per minute (rpm)
Ring test
Shank
Trigger lock

Industry Recognized Credentials

If you are training through an NCCER-accredited sponsor, you may be eligible for credentials from NCCER's Registry. The ID number for this module is 00104. Note that this module may have been used in other NCCER curricula and may apply to other level completions. Contact NCCER's Registry at 1.888.622.3720 or go to **www.nccer.org** for more information.

Trade Features

Trade features present technical tips and professional practices based on real-life scenarios similar to those you might encounter on the job site.

Bowline Trivia

Some people use this saying to help them remember how to tie a bowline: "The rabbit comes out of his hole, around a tree, and back into the hole."

Figures and Tables

Photographs, drawings, diagrams, and tables are used throughout each module to illustrate important concepts and provide clarity for complex instructions. Text references to figures and tables are emphasized with *italic* type.

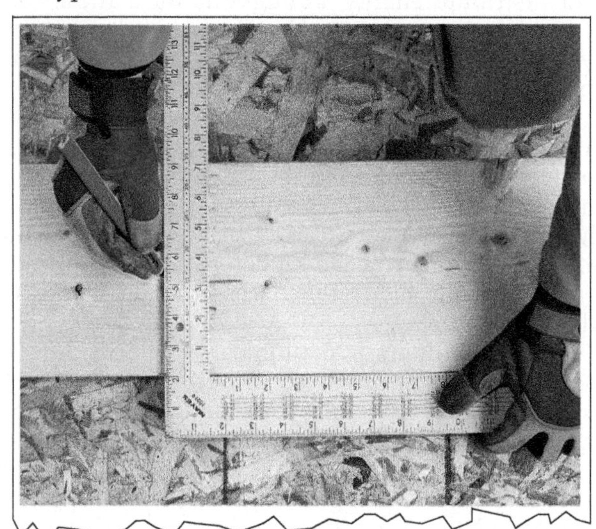

Notes, Cautions, and Warnings

Safety features are set off from the main text in highlighted boxes and categorized according to the potential danger involved. Notes simply provide additional information. Cautions flag a hazardous issue that could cause damage to materials or equipment. Warnings stress a potentially dangerous situation that could result in injury or death to workers.

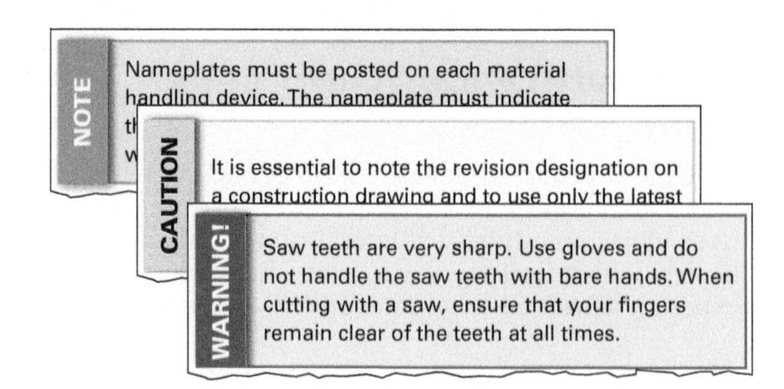
NOTE
Nameplates must be posted on each material
CAUTION
It is essential to note the revision designation on
WARNING!
Saw teeth are very sharp. Use gloves and do not handle the saw teeth with bare hands. When cutting with a saw, ensure that your fingers remain clear of the teeth at all times.

Going Green

Going Green looks at ways to preserve the environment, save energy, and make good choices regarding the health of the planet. Through the introduction of new construction practices and products, you will see how the greening of America has already taken root.

> **Going Green**
>
> **Big Horn Wind Farm**
>
> The Big Horn wind farm uses 133 GE turbines, each rising 119 meters (389 feet) from the ground to the top of the blade tips. The wind farm uses only a

Did You Know?

The Did You Know? features offer hints, tips, and other helpful bits of information.

> **Did You Know?**
>
> **How Blueprints Started**
>
> The process for making blueprints was developed in 1842 by an English astronomer named Sir John F. Herschel. The method involved coating a paper with a special chemical. After the coating dried, an original hand drawing was placed on top of the paper. Both papers were then covered with a piece of glass and set in the sunlight for about an hour. The coated paper was developed much like a photograph. After a cold-water wash, the coated paper turned blue, and the lines of the drawing remained white.

Step-by-Step Instructions

Step-by-Step instructions are used throughout to guide you through technical procedures and tasks from start to finish. These steps show you not only how to perform a task but also how to do it safely and efficiently.

> Perform the following steps to erect this system area scaffold:
>
> *Step 1* Gather and inspect all scaffold equipment for the scaffold arrangement.
>
> *Step 2* Place appropriate mudsills in their approximate locations.
>
> *Step 3* Attach the screw jacks to the mudsills.
>
> *Step 4* Adjust the screw jacks to near their lowest position.
>
> *Step 5* Determine the location of the highest

Trade Terms

Each module presents a list of Trade Terms that are discussed within the text and defined in the Glossary at the end of the module. These terms are denoted in the text with **bold, blue type** upon their first occurence. To make searches for key information easier, a comprehensive Glossary of Trade Terms from all modules is located at the back of this book.

> During a rigging operation, the **load** being lifted or moved must be connected to the apparatus, such as a crane, that will provide the power for movement. The connector—the link between the load and the apparatus—is often a sling made of synthetic, chain, or **wire rope** materials. This section focuses on three types of slings:
>
> - Synthetic slings
> - Alloy steel chain slings
> - Wire rope slings

Review Questions

Review Questions are provided to reinforce the knowledge you have gained. This makes them a useful tool for measuring what you have learned.

> **Review Questions**
>
> 1. Identification tags for slings must include the _____.
>   a. type of protective pads to use
>   b. type of damage sustained during use
>   c. color of the tattle-tail
>   d. manufacturer's name or trademark
> 2. The type of wire rope core that is susceptible to heat damage at relatively low temperatures is the _____.
>   a. fiber core
>   b. strand core
>   c. independent wire rope core
>   d. metallic link supporting core
> 3. Synthetic slings must be inspected _____.
>   a. once every month
>   b. visually at the start of each work week
>   c. before every use
>   d. once wear or damage becomes apparent
> 4. An alloy steel chain sling must be removed from service if there is evidence that _____.
>   a. the sling has been used in different hitch configurations
>   b. replacement links have been used to repair the chain
>   c. the sling has been used for more than one year
>   d. strands in the supporting core have weakened
> 5. A piece of rigging hardware used to couple the end of a wire rope to eye fittings, hooks, or other connections is a(n) _____.
>   a. eyebolt
>   b. hitch
>   c. shackle
>   d. U-bolt
> 6. A lifting clamp is most likely to be used to move loads such as _____.
>   a. steel plates
>   b. piping
>   c.
> 7. Chain hoists are able to lift heavy loads by utilizing a _____.
>   a. rope and pulley system
>   b. rigger's strength
>   c. stationary counterweight
>   d. gear system
> 8. Before attempting to lift a load with a chain hoist, make sure that the _____.
>   a. hoist is secured to a come-along
>   b. load is properly balanced
>   c. tag lines are properly anchored
>   d. tackle is connected to its power source
> 9. A hitch configuration that allows slings to be connected to the same load without using a spreader beam is a _____.
>   a. double-wrap hitch
>   b. choker hitch
>   c. bridle hitch
>   d. basket hitch
> 10. To make the emergency stop signal that is used by riggers, extend both arms _____.
>   a. horizontally with palms down and quickly move both arms back and forth
>   b. directly in front and then move both arms up and down repeatedly
>   c. vertically above the head and wave both arms back and forth
>   d. horizontally with clenched fists and move both arms up and down

NCCER Standardized Curricula

NCCER's training programs comprise more than 80 construction, maintenance, pipeline, and utility areas and include skills assessments, safety training, and management education.

Boilermaking
Cabinetmaking
Carpentry
Concrete Construction
Construction Craft Laborer
Construction Technology
Core: Introduction to Basic Construction Skills
Drywall
Electrical
Electronic Systems Technician
Heating, Ventilating, and Air Conditioning
Heavy Equipment Operations
Heavy Highway Construction
Hydroblasting
Industrial Coating and Lining Application Specialist
Industrial Maintenance Electrical and Instrumentation Technician
Industrial Maintenance Mechanic
Instrumentation
Ironworking
Manufactured Construction Technology
Masonry
Mechanical Insulating
Millwright
Mobile Crane Operations
Painting
Painting, Industrial
Pipefitting
Pipelayer
Plumbing
Reinforcing Ironwork
Rigging
Roofing
Scaffolding
Sheet Metal
Signal Person
Site Layout
Sprinkler Fitting
Tower Crane Operator
Welding

Maritime

Maritime Industry Fundamentals
Maritime Electrical
Maritime Pipefitting
Maritime Structural Fitter
Maritime Welding
Maritime Aluminum Welding

Green/Sustainable Construction

Building Auditor
Fundamentals of Weatherization
Introduction to Weatherization
Sustainable Construction Supervisor
Weatherization Crew Chief
Weatherization Technician
Your Role in the Green Environment

Energy

Alternative Energy
Introduction to the Power Industry
Introduction to Solar Photovoltaics
Power Generation Maintenance Electrician
Power Generation I&C Maintenance Technician
Power Generation Maintenance Mechanic
Power Line Worker
Power Line Worker: Distribution
Power Line Worker: Substation
Power Line Worker: Transmission
Solar Photovoltaic Systems Installer
Wind Energy
Wind Turbine Maintenance Technician

Pipeline

Abnormal Operating Conditions, Control Center
Abnormal Operating Conditions, Field and Gas
Corrosion Control
Electrical and Instrumentation
Field and Control Center Operations
Introduction to the Pipeline Industry
Maintenance
Mechanical

Safety

Fall Protection Orientation
Field Safety
Safety Orientation
Safety Technology

Supplemental Titles

Applied Construction Math
Tools for Success

Management

Construction Workforce Development Professional
Fundamentals of Crew Leadership
Mentoring for Craft Professionals
Project Management
Project Supervision

Spanish Titles

Acabado de concreto: nivel uno (*Concrete Finishing Level One*)
Aislamiento: nivel uno (*Insulating Level One*)
Albañilería: nivel uno (*Masonry Level One*)
Andamios (*Scaffolding*)
Carpintería: Formas para carpintería, nivel tres (*Carpentry: Carpentry Forms, Level Three*)
Currículo básico: habilidades introductorias del oficio (*Core Curriculum: Introductory Craft Skills*)
Electricidad: nivel uno (*Electrical Level One*)
Herrería: nivel uno (*Ironworking Level One*)
Herrería de refuerzo: nivel uno (*Reinforcing Ironwork Level One*)
Instalación de rociadores: nivel uno (*Sprinkler Fitting Level One*)
Instalación de tuberías: nivel uno (*Pipefitting Level One*)
Instrumentación: nivel uno, nivel dos, nivel tres, nivel cuatro (*Instrumentation Levels One through Four*)
Orientación de seguridad (*Safety Orientation*)
Paneles de yeso: nivel uno (*Drywall Level One*)
Seguridad de campo (*Field Safety*)

Acknowledgments

This curriculum was revised as a result of the farsightedness and leadership of the following sponsors:

Applied Trade Solutions
Bechtel
Cianbro Companies
Eastern Constructors Services
Fluor
KBR Services, Inc.
McAllen Careers Institute
S & B Engineers and Constructors, Ltd.
Sundt Construction, Inc.
TIC - The Industrial Company
Turner Industries Group, LLC
Victaulic

This curriculum would not exist were it not for the dedication and unselfish energy of those volunteers who served on the Authoring Team. A sincere thanks is extended to the following:

Arnold Adame, Jr.
Tony Ayotte
Ronnie Balentine
Gerald Bickerstaff
Jack Carbone
Jacob Guzman
John Higdon
Rodney Landry
Myron Laurent
Colin Phelps
Josue Ponce
Shawn Reid
Brian Robinson
Fernando Sanchez
Troy Smith
Michael Stilley
Jody Suchanek

NCCER Partners

NCCER partnering organizations are national associations and organizations that share a common interest in the goals and objectives of NCCER. To learn more about NCCER business partners, go to **www.nccer.org/about-us/partners**.

You can also scan this code using the camera on your phone or mobile device to view these partnering organizations.

Contents

Module One

Advanced Blueprint Reading

Pipefitters need to understand how to read piping and instrumentation drawings (P&IDs), which are also known as blueprints or site plans. These schematic diagrams show process flows, functions, equipment, pipelines, valves, instruments, and controls needed to operate the system. The blueprint package included in the Appendix of this module contains plans that demonstrate how P&ID information appears on the plans and what the information means. (Module ID 08401; 50 hours)

Module Two

Advanced Pipe Fabrication

Pipe fabrication involves the use of either ordinate tables or trigonometry to create fittings and pipe assemblies that suit a process application. Producing ordinates and using them to lay out miters and laterals is important, as are alternative methods for laying out the cuts for laterals, saddles, and mitered turns. In this module, formulas are provided for putting together multiple offsets around obstacles of both equal and unequal spread. (Module ID 08402; 50 hours)

Module Three

Stress Relieving and Aligning

Stress relieving is the process of preheating and post weld heat treatment to keep welds from distorting a pipe assembly. Alignment is the reason for stress relieving, because if the pipe will not fit up accurately to machinery, dynamically balanced pumps will be unbalanced by the distortions of the piping attachments. A skilled pipefitter takes charge of these situations, working with millwrights and others to prevent misalignments and problems which stem from them. (Module ID 08403; 10 hours)

Module Four

In-Line Specialties

Special fittings and instruments used in process piping equipment are known as in-line specialties and described in system documentation as "specials." This category includes steam traps, desuperheaters, bursting discs, strainers, and related equipment. Steam traps protect the lines against water hammer while desuperheaters reduce the temperature of steam. Strainers are used with many types of fluids to keep solids from clogging pipes. Bursting discs provide emergency pressure relief to prevent very high-pressure surges from damaging equipment. Each in-line specialty item serves a specific purpose and must be installed, monitored, and removed by knowledgeable pipefitters who may be working in tandem with other professionals for coordinated activities across the pipe run. (Module ID 08405; 20 hours)

Module Five

Special Piping

Pipefitters must be prepared to assemble small piping and tubing. While these jobs are infrequent, they still require a clear understanding of the skills needed to correctly and safely connect and route pipe on a smaller scale. Some of these skills include brazing, soldering, pipe bending, and installing various fittings made of copper, stainless steel, aluminum, and brass. (Module ID 08406; 25 hours)

Module Six

Hot Taps

When it is necessary to connect to pipes that cannot be shut down or emptied, fluid pressure must be contained to prevent leaks. The method for doing this is referred to as hot tapping, and its procedures will vary according to what is being conveyed through the line. In some situations, it may be possible to temporarily stop the flow while connections are made; this is where line stop plugs, pipe freezing, and pipe plugging become important. Hot tapping is not a regular part of a pipefitter's career, but it is important to understand the environmental factors associated with it as well as ways to safely assist any contractors called on to perform it. (Module ID 08407; 10 hours)

Module Seven

Maintaining Valves

Understanding the function and assembly of valves is essential to the pipefitter's career. While most valves are replaced rather than maintained, it is important to understand the procedures for both. Knowledge on valve maintenance contributes to troubleshooting issues within a pipe run, with the function of valves as a pivotal point. (Module ID 08408; 10 hours)

Module Eight

Fundamentals of Crew Leadership

When a crew is assembled to complete a job, one person is appointed the leader. This person is usually an experienced craft professional who has demonstrated leadership qualities. While having natural leadership qualities helps in becoming an effective leader, it is more true that "leaders are made, not born." Whether you are a crew leader or want to become one, this module will help you learn more about the requirements and skills needed to succeed (Module ID 46101; 22.5 hours)

Glossary

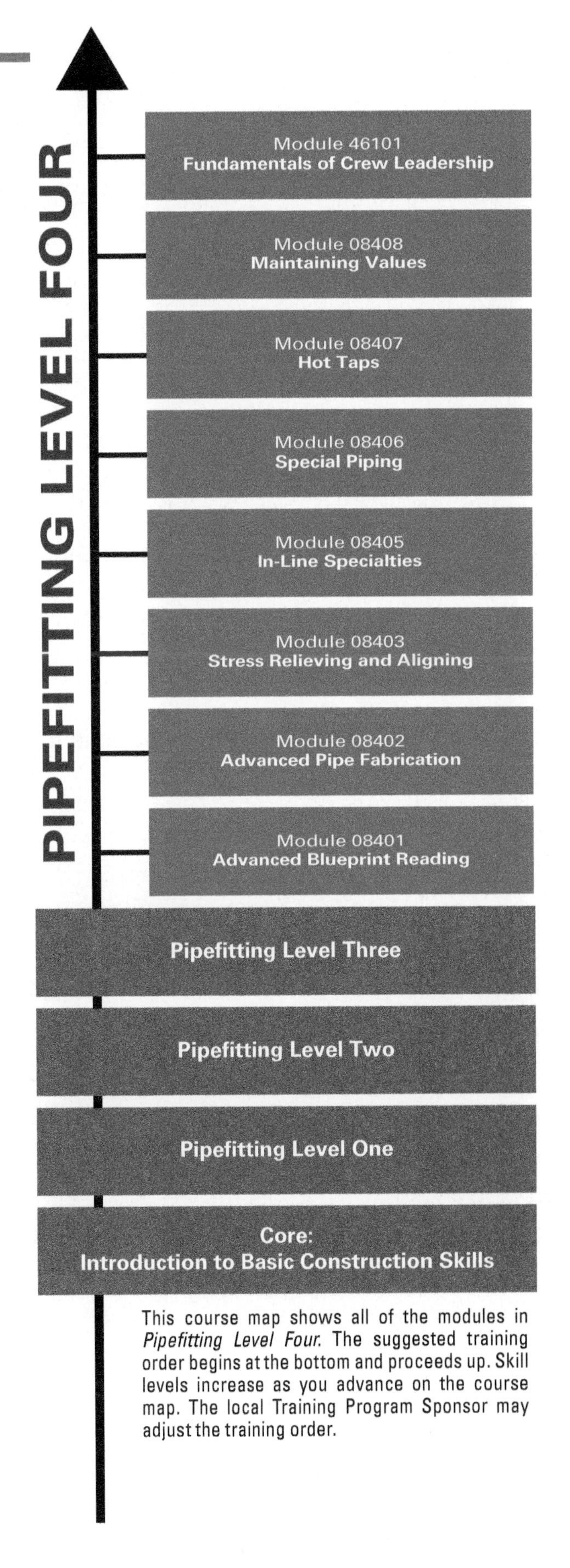

This course map shows all of the modules in *Pipefitting Level Four.* The suggested training order begins at the bottom and proceeds up. Skill levels increase as you advance on the course map. The local Training Program Sponsor may adjust the training order.

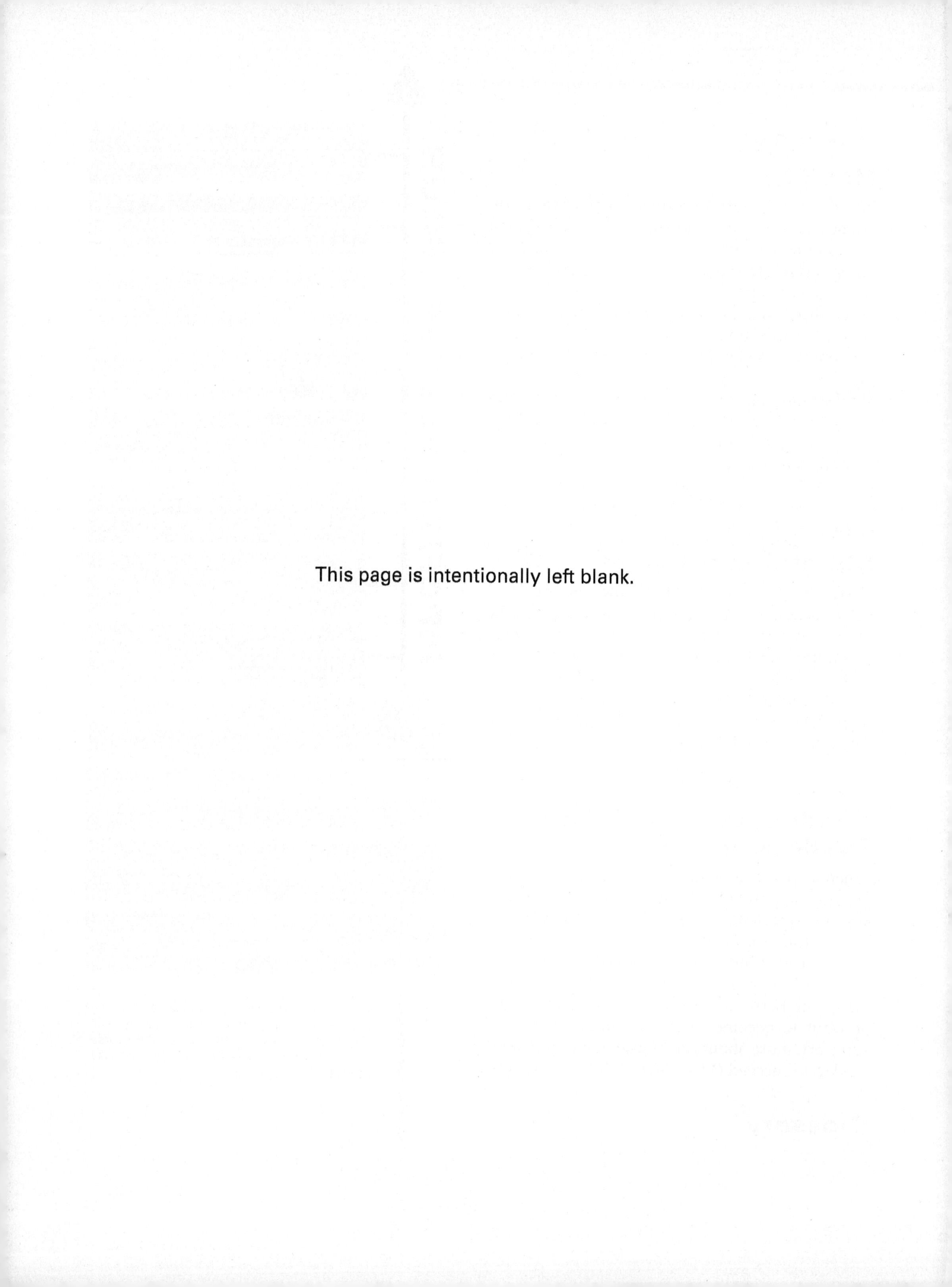

This page is intentionally left blank.

Advanced Blueprint Reading

Overview

Pipefitters need to understand how to read piping and instrumentation drawings (P&IDs), which are also known as *blueprints* or *site plans*. These schematic diagrams show process flows, functions, equipment, pipelines, valves, instruments, and controls needed to operate the system. The blueprint package included in the *Appendix* of this module contains plans that demonstrate how P&ID information appears on the plans and what the information means.

Module 08401

Trainees with successful module completions may be eligible for credentialing through the NCCER Registry. To learn more, go to **www.nccer.org** or contact us at 1.888.622.3720. Our website, **www.nccer.org**, has information on the latest product releases and training.

Your feedback is welcome. You may email your comments to **curriculum@nccer.org**, send general comments and inquiries to **info@nccer.org**, or fill in the User Update form at the back of this module.

This information is general in nature and intended for training purposes only. Actual performance of activities described in this manual requires compliance with all applicable operating, service, maintenance, and safety procedures under the direction of qualified personnel. References in this manual to patented or proprietary devices do not constitute a recommendation of their use.

08401

Advanced Blueprint Reading

Objectives

Successful completion of this module prepares you to do the following:

1. Identify and describe the contents of piping and instrumentation drawings (P&IDs).
   a. Describe how process piping is depicted on P&IDs.
   b. Identify symbols used on P&IDs to represent piping components.
   c. Identify symbols used on P&IDs to represent process equipment.
   d. Identify symbols used on P&IDs to represent instrumentation.
2. Identify and describe the contents of piping arrangement drawings.
   a. Describe how control points are used on drawings to locate work areas horizontally.
   b. Describe how coordinates are used on drawings to define geographic position.
   c. Describe how elevations are used on drawings to locate work areas vertically.
3. Explain how to read and interpret P&IDs, piping arrangement drawings, and isometric drawings.
   a. Explain how to read and interpret P&IDs and piping arrangement drawings.
   b. Explain how to read and interpret isometric drawings.
   c. Explain how to follow a single line from one drawing to another.

Performance Tasks

Under supervision, you should be able to do the following:

1. Calculate the total line length from an ISO.
2. Sketch an ISO from a plan view.

Trade Terms

Dimension
Title block

Industry Recognized Credentials

If you are training through an NCCER-accredited sponsor, you may be eligible for credentials from NCCER's Registry. The ID number for this module is 08401. Note that this module may have been used in other NCCER curricula and may apply to other level completions. Contact NCCER's Registry at 1.888.622.3720 or go to **www.nccer.org** for more information.

Contents

Figures and Tables

This page is intentionally left blank.

Section One

1.0.0 Piping and Instrumentation Drawings

Objective

Identify and describe the contents of piping and instrumentation drawings (P&IDs).

a. Describe how process piping is depicted on P&IDs.
b. Identify symbols used on P&IDs to represent piping components.
c. Identify symbols used on P&IDs to represent process equipment.
d. Identify symbols used on P&IDs to represent instrumentation.

The building blocks of process piping consists of lines, abbreviations, and symbols shown on the P&ID. It may be impossible to memorize all of the symbols included on a P&ID, but pipefitters should be able to read the symbols and abbreviations to gain a clear understanding of the piping's components, equipment, and instrumentation. While symbols and abbreviations may vary depending upon the engineer or contractor who created the plan, a legend is typically included that shows their meaning. If not included on the plan, the legend may appear in a space on the working diagram or on a separate diagram. *Table 1* shows abbreviations commonly used on plans.

1.1.0 Process Piping

Lines, line numbers, and match lines shown on P&IDs depict the entire piping system of a process and provide information about each pipeline. Reading a P&ID is a matter of tracing a system from one location, or piece of equipment, to another and interpreting the symbols and abbreviations on the diagram.

1.1.1 Lines

Lines on a P&ID show flow through a piping system from one place or piece of equipment to another. Each type of line on a P&ID has a special purpose, and the way that a line appears provides the pipefitter with information about that line. *Figure 1* shows types of lines used on P&IDs.

1.1.2 Line Numbers

Line numbers used on P&IDs indicate the pipe size, service of line, piping material specification group number, number of the pipe in a service, and insulation required. In other P&IDs, or other types of drawings of the same system, line numbers are used to identify lines and to transfer pipelines from one drawing to another.

Line numbers enclosed in an oval-shaped symbol with an arrow at one end indicate the direction of flow through the pipeline and should be used as often as necessary to maintain continuity. Line numbers can be shown directly on the line or above the line. *Figure 2* shows how typical line numbers can appear.

In *Figure 2*, 6" indicates the nominal pipe size. In this example, LW is the fluid symbol and represents white liquor. The number 20 indicates the piping material specification, while number 12 indicates the particular line number in the white liquor system. The letter A indicates the insulation group letter from the specifications. Insulation is not required if the letter is absent. Insulation may be presented in a separate symbol.

Since line number appearance varies between engineering companies, always check the legend of the P&ID to determine the system for reading the line numbers. The line number may include the pipe size in the line code.

1.1.3 Match Lines

When an entire system is shown on a single sheet, it is easy to trace. Larger systems may require the use of two or more sheets to show the entire system, and the system must be traced from one P&ID to another using match lines.

A match line (*Figure 3*) is shown at the end of the P&ID to indicate where the system leaves one numbered sheet and starts again on the next numbered sheet. The system is traced to the numbered match line on one sheet and continues at the same number on the match line of the next sheet. The system always continues on the next sheet using the numbering sequence of the P&ID sheets, unless noted on the match line. If the system continues on a P&ID sheet other than the next sheet in sequence, a note appears in a box on the system line at the match line.

Table 1A Abbreviations (1 of 3)

Adapter	ADPT	Detail	DET
Air preheating	A	Diameter	DIA
American Iron and Steel Institute	AISI	Dimension	DIM
American National Standards Institute	ANSI	Discharge	DISCH
American Petroleum Institute	API	Double extra-strong	XX STRG
American Society for Testing and Materials International	ASTM	Drain	DR
American Society of Mechanical Engineers	ASME	Drain funnel	DF
Ash removal water, sluice or jet	AW	Drawing	DWG
Aspiring air	AA	Drip leg or dummy leg	DL
Auxiliary steam	AS	Ductile iron	DI
Bench mark	BM	Dust collector	DC
Beveled	B	Each	EA
Beveled end	BE	Eccentric	ECC
Beveled, both ends	BBE	Elbolet	EOL
Beveled, large end	BLE	Elbow	ELB
Beveled, one end	BOE	Electric resistance weld	ERW
Beveled, small end	BSE	Elevation	EL
Bill of materials	BOM	Equipment	EQUIP
Blind flange	BF	Evaporator vapor	EV
Blowoff	BO	Exhaust steam	E
Bottom of pipe	BOP	Expansion	EXP
Butt weld	BW	Extraction joint	EXP JT
Carbon steel or cold spring	CS	Extraction steam	ES
Cast iron	CI	Fabrication (dimension)	FAB
Ceiling	CLG	Face of flange	FOF
Chain operated	CO	Faced and drilled	F&D
Chemical feed	CF	Factory Mutual	FM
Circulating water	CW	Far side	FS
Cold reheat steam	CR	Feed pump balancing line	FB
Compressed air	CA	Feed pump discharge	FD
Concentric or concrete	CONC	Feed pump recirculating	FR
Condensate	C	Feed pump suction	FS
Condenser air removal	AR	Female	F
Continue, continuation	CONT	Female	FM
Coupling	CPLG	Female pipe thread	FPT

Table 1B Abbreviations (2 of 3)

Field weld	FW	Isometric	ISO
Figure	FIG	Large male	LM
Fillet weld	W	Length	LG
Finish floor	F/F	Long radius	LR
Finish floor	FIN FL	Long tangent	LT
Finish grade	FIN GR	Long weld neck	LWN
Fitting	FTG	Low pressure	LP
Fitting makeup	FMU	Low-pressure drains	DR
Fitting to fitting	FTF	Low-pressure steam	LPS
Flange	FLG	Lubricating oil	LO
Flat face	FF	Main steam	MS
Flat on bottom	FOB	Main system blowouts	BL
Flat on top	FOT	Makeup water	MU
Floor drain	FD	Male	M
Foundation	FDN	Malleable iron	MI
Fuel gas	FG	Manufacturer	MFR
Fuel oil	FO	Manufacturer's Standard Society	MSS
Gauge	GA	National Pipe Thread	NPT
Galvanized	GALV	Nipolet	NOL
Gasket	GSKT	Nipple	NIP
Grating	GRTG	Nominal	NOM
Hanger	HGR	Not to scale	NTS
Hanger rod	HR	Nozzle	NOZ
Hardware	HDW	Outside battery limits	OSBL
Header	HDR	Outside diameter	OD
Heat traced, heat tracing	HT	Outside screw and yoke	OS&Y
Heater drains	HD	Overflow	OF
Heating system	HS	Pipe support	PS
Heating, ventilating, and air conditioning	HVAC	Pipe tap	PT
Hexagon	HEX	Piping and instrumentation diagram	P&ID
Hexagon head	HEX HD	Plain	P
High point	HPT	Plain end	PE
High pressure	HP	Plain, both ends	PBE
Horizontal	HORIZ	Plain, one end	POE
Hot reheat steam	HR	Point of intersection	PI
Hydraulic	HYDR	Point of tangent	PT
Increaser	INCR	Pounds per square inch	PSI
Input/output	I/O	Process flow diagram	PFD
Inside diameter	ID	Purchase order	PO
Insulation	INS	Radius	RAD
Invert elevation	IE	Raised face	RF
Iron pipe size	IPS	Raised face slip-on	RFSO

Table 1C Abbreviations (3 of 3)

Raised face smooth finish	RFSF	Temperature or temporary	TEMP
Raised face weld neck	RFWN	That is	I.E.
Raw water	RW	Thick	THK
Reducer, reducing	RED	Thousand	M
Relief valve	PRV-PSV	Thread, threaded	THRD
Ring-type joint	RTJ	Threaded	T
Rod hanger or right hand	RH	Threaded end	TE
Safety valve vents	SV	Threaded, both ends	TBE
Sanitary	SAN	Threaded, large end	TLE
Saturated steam	SS	Threaded, one end	TOE
Schedule	SCH	Threaded, small end	TSE
Screwed	SCRD	Threadolet	TOL
Seamless	SMLS	Top of pipe or top of platform	TOP
Section	SECT	Top of steel or top of support	TOS
Service and cooling water	SW	Treated water	TW
Sheet	SH	Turbine	TURB
Short radius	SR	Typical	TYP
Slip-on	SO	Underwriters Laboratories	UL
Socket weld	SW	Vacuum	VAC
Sockolet	SOL	Vacuum cleaning	VC
Stainless steel	SS	Vents	V
Standard weight	STD WT	Vitrified tile	VT
Steel	STL	Wall thickness or weight	WT
Suction	SUCT	Weld neck	WN
Superheater drains	SD	Weldolet	WOL
Swage	SWG	Well water	WW
Swaged nipple	SN	Wide flange	WF

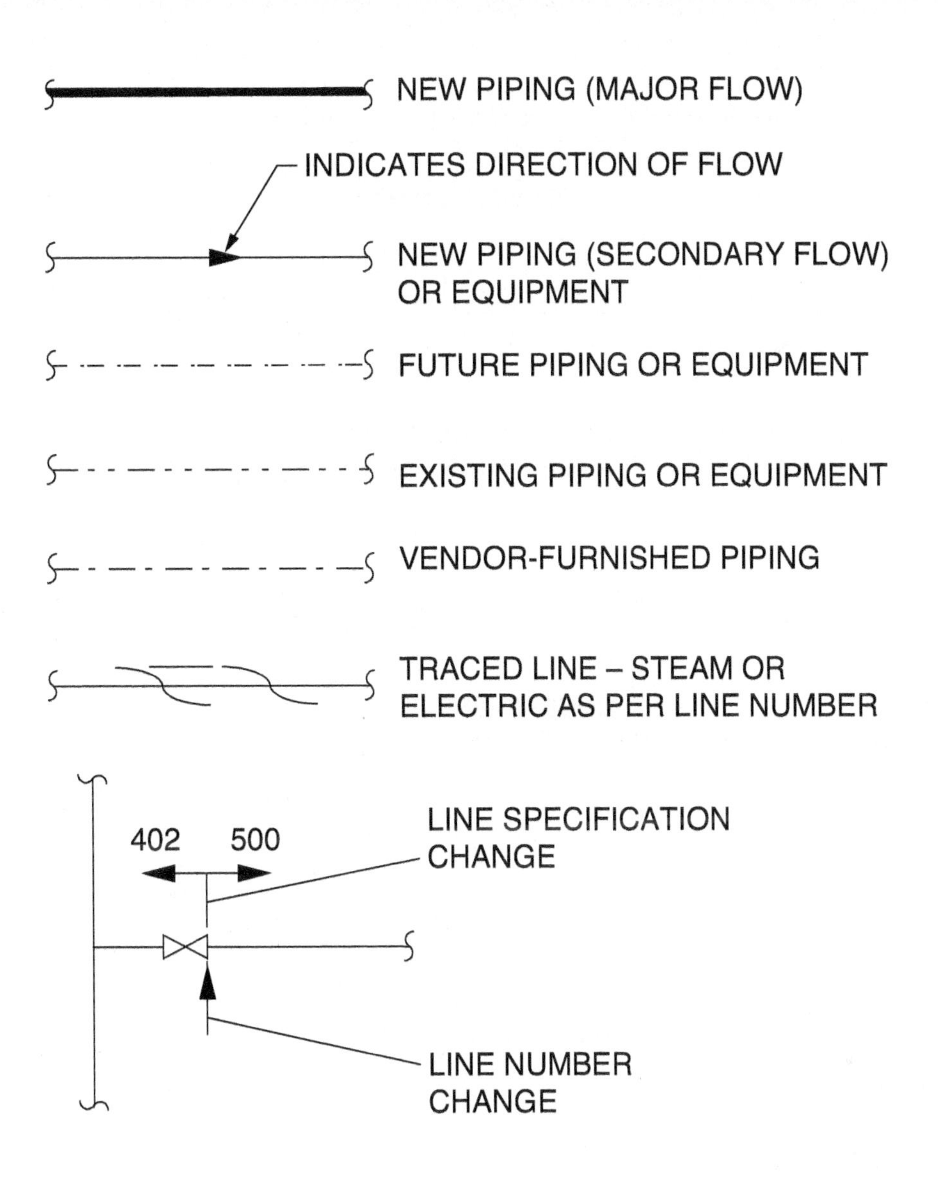

Figure 1 Lines.

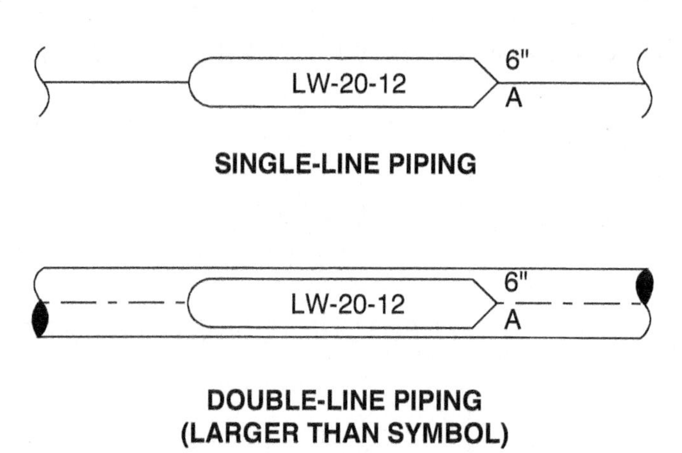

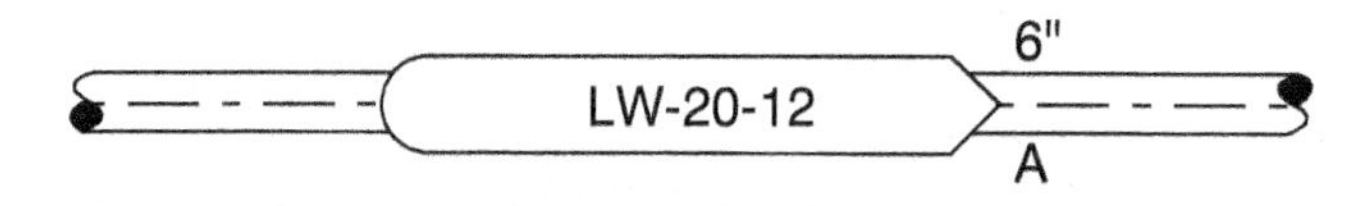

DOUBLE-LINE PIPING
(SMALLER THAN SYMBOL)

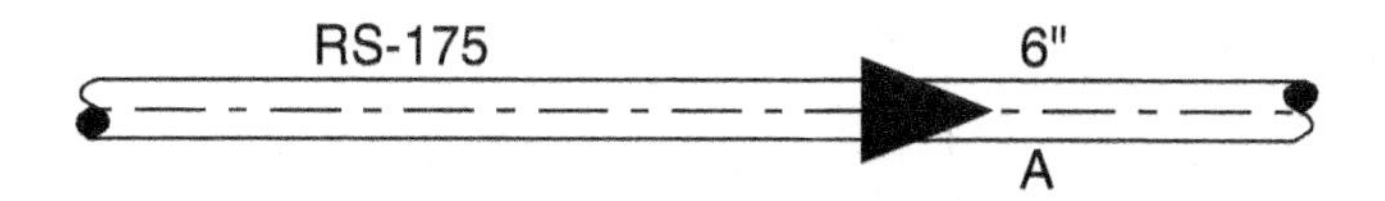

DOUBLE-LINE PIPING
(WITH DIRECTIONAL ARROW)

Figure 2 Line numbers.

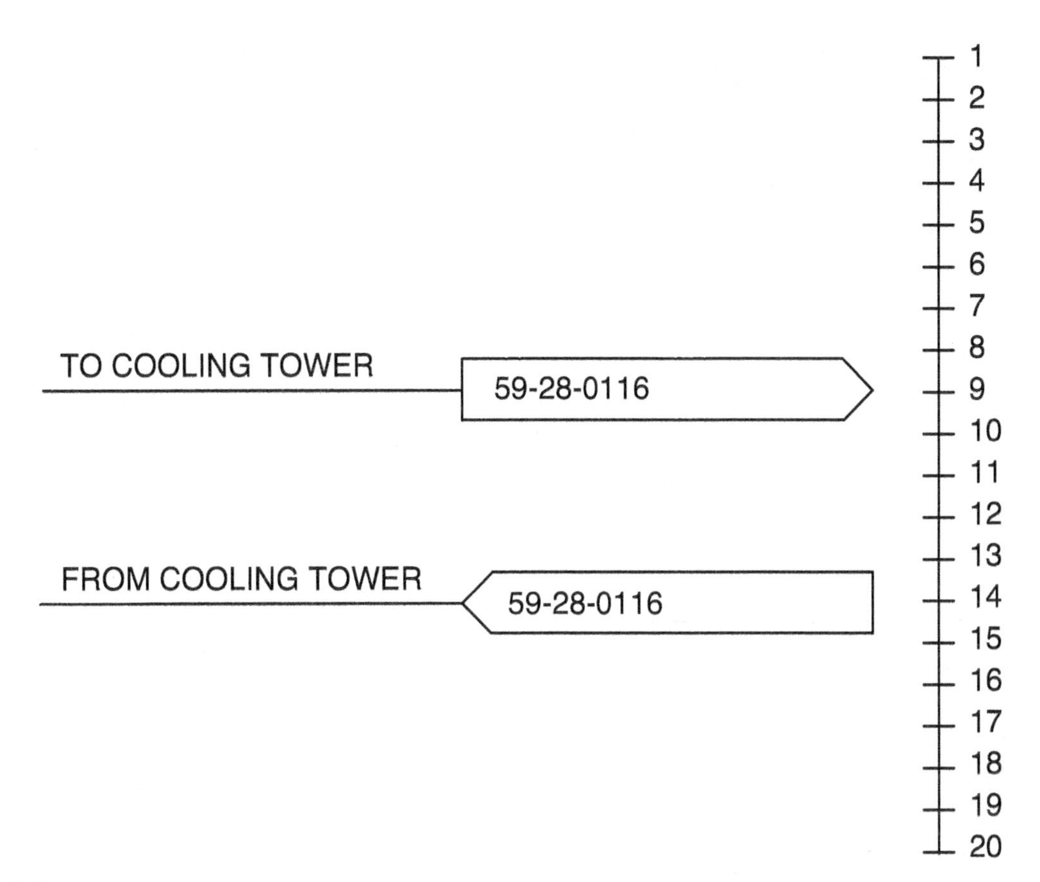

Figure 3 Match lines.

1.2.0 Piping Components

Piping components shown on P&IDs include valves, strainers, expansion joints, clean-outs, flexible connections, dampers, de-superheaters, removable spools, drains, and vents. Symbols and abbreviations are used to describe these components. Always check the legend on the P&ID to determine the meaning of the symbols since the symbols may vary between engineering companies. *Figure 4* shows commonly used piping component symbols.

1.3.0 Process Equipment

Process equipment shown on P&IDs includes pumps, vessels, motors, turbines, condensers, and other major equipment in a system. Plans may also show information about process equipment size, capacity, horsepower, and operating parameters. When the initials VS (vendor's supply) are enclosed by a dashed line, all equipment within the enclosed area is vendor supplied, and the pipefitter is not required to pipe up that area. *Figure 5* shows commonly used equipment symbols, which may vary between engineering companies.

1.4.0 System Instrumentation

System instrumentation shown on P&IDs includes control valve operators as well as flow, pressure, level, temperature, and analytical instruments. *Figure 6* shows typical instrumentation symbols and the meaning of each.

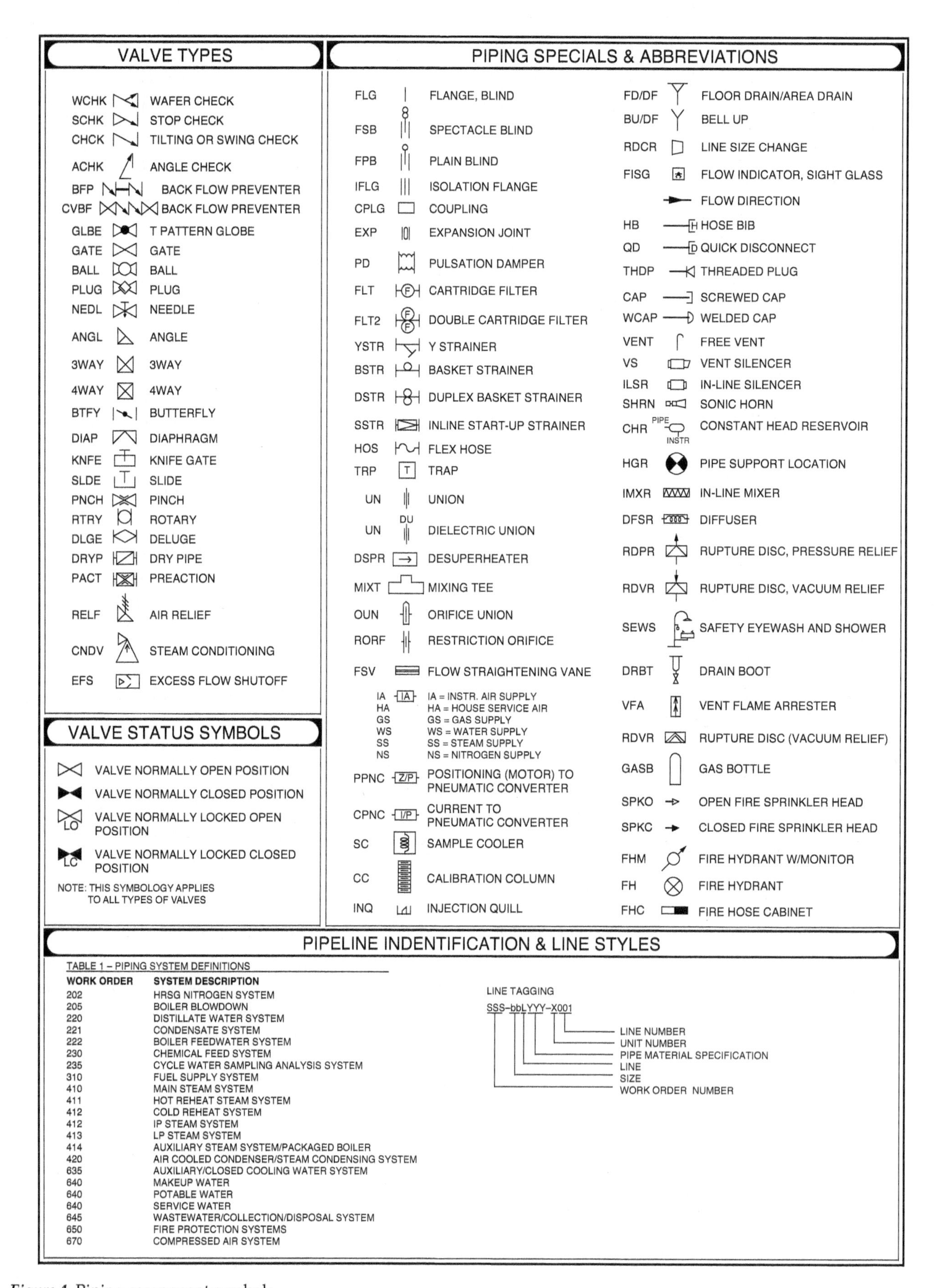

TABLE 1 – PIPING SYSTEM DEFINITIONS

WORK ORDER	SYSTEM DESCRIPTION
202	HRSG NITROGEN SYSTEM
205	BOILER BLOWDOWN
220	DISTILLATE WATER SYSTEM
221	CONDENSATE SYSTEM
222	BOILER FEEDWATER SYSTEM
230	CHEMICAL FEED SYSTEM
235	CYCLE WATER SAMPLING ANALYSIS SYSTEM
310	FUEL SUPPLY SYSTEM
410	MAIN STEAM SYSTEM
411	HOT REHEAT STEAM SYSTEM
412	COLD REHEAT SYSTEM
412	IP STEAM SYSTEM
413	LP STEAM SYSTEM
414	AUXILIARY STEAM SYSTEM/PACKAGED BOILER
420	AIR COOLED CONDENSER/STEAM CONDENSING SYSTEM
635	AUXILIARY/CLOSED COOLING WATER SYSTEM
640	MAKEUP WATER
640	POTABLE WATER
640	SERVICE WATER
645	WASTEWATER/COLLECTION/DISPOSAL SYSTEM
650	FIRE PROTECTION SYSTEMS
670	COMPRESSED AIR SYSTEM

Figure 4 Piping component symbols.

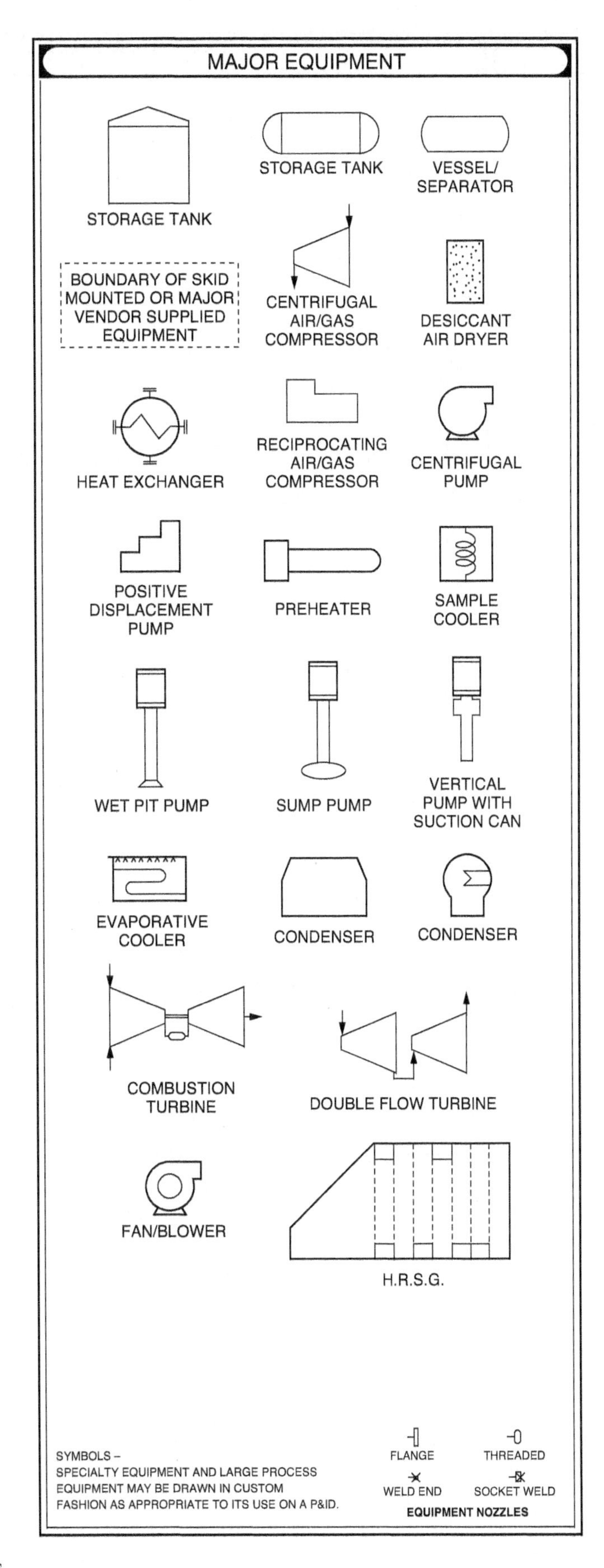

Figure 5 Equipment symbols.

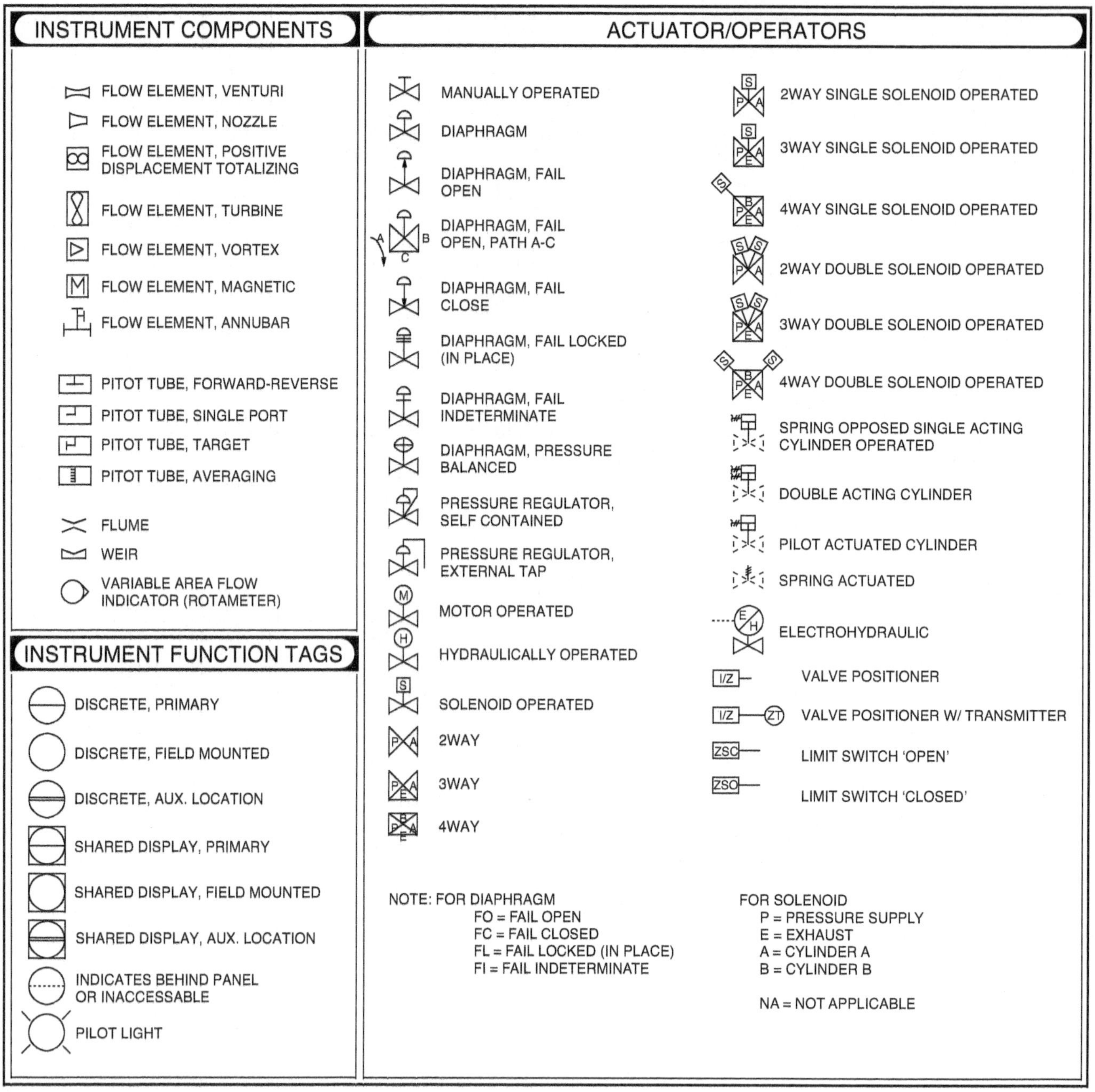

INSTRUMENT COMPONENTS

- FLOW ELEMENT, VENTURI
- FLOW ELEMENT, NOZZLE
- FLOW ELEMENT, POSITIVE DISPLACEMENT TOTALIZING
- FLOW ELEMENT, TURBINE
- FLOW ELEMENT, VORTEX
- M FLOW ELEMENT, MAGNETIC
- FLOW ELEMENT, ANNUBAR
- PITOT TUBE, FORWARD-REVERSE
- PITOT TUBE, SINGLE PORT
- PITOT TUBE, TARGET
- PITOT TUBE, AVERAGING
- FLUME
- WEIR
- VARIABLE AREA FLOW INDICATOR (ROTAMETER)

INSTRUMENT FUNCTION TAGS

- DISCRETE, PRIMARY
- DISCRETE, FIELD MOUNTED
- DISCRETE, AUX. LOCATION
- SHARED DISPLAY, PRIMARY
- SHARED DISPLAY, FIELD MOUNTED
- SHARED DISPLAY, AUX. LOCATION
- INDICATES BEHIND PANEL OR INACCESSABLE
- PILOT LIGHT

ACTUATOR/OPERATORS

- MANUALLY OPERATED
- DIAPHRAGM
- DIAPHRAGM, FAIL OPEN
- A B C DIAPHRAGM, FAIL OPEN, PATH A-C
- DIAPHRAGM, FAIL CLOSE
- DIAPHRAGM, FAIL LOCKED (IN PLACE)
- DIAPHRAGM, FAIL INDETERMINATE
- DIAPHRAGM, PRESSURE BALANCED
- PRESSURE REGULATOR, SELF CONTAINED
- PRESSURE REGULATOR, EXTERNAL TAP
- M MOTOR OPERATED
- H HYDRAULICALLY OPERATED
- S SOLENOID OPERATED
- P A 2WAY
- P A E 3WAY
- B P A E 4WAY
- S P A 2WAY SINGLE SOLENOID OPERATED
- S P A E 3WAY SINGLE SOLENOID OPERATED
- S B P A E 4WAY SINGLE SOLENOID OPERATED
- S S P A 2WAY DOUBLE SOLENOID OPERATED
- S S P A E 3WAY DOUBLE SOLENOID OPERATED
- S S B P A E 4WAY DOUBLE SOLENOID OPERATED
- SPRING OPPOSED SINGLE ACTING CYLINDER OPERATED
- DOUBLE ACTING CYLINDER
- PILOT ACTUATED CYLINDER
- SPRING ACTUATED
- E H ELECTROHYDRAULIC
- I/Z VALVE POSITIONER
- I/Z ZT VALVE POSITIONER W/ TRANSMITTER
- ZSC LIMIT SWITCH 'OPEN'
- ZSO LIMIT SWITCH 'CLOSED'

NOTE: FOR DIAPHRAGM
FO = FAIL OPEN
FC = FAIL CLOSED
FL = FAIL LOCKED (IN PLACE)
FI = FAIL INDETERMINATE

FOR SOLENOID
P = PRESSURE SUPPLY
E = EXHAUST
A = CYLINDER A
B = CYLINDER B

NA = NOT APPLICABLE

Figure 6 Instrumentation symbols.

1.0.0 Section Review

1. Which of the following is *not* true of lines on P&IDs?
   a. Lines indicate the material of the pipe.
   b. Lines show flow through a piping system.
   c. Lines provide information for the pipefitter based on how it appears.
   d. Each line serves a special purpose.

2. To determine the meaning of the symbols on the P&ID, always check the ______.
   a. component list
   b. legend
   c. section
   d. instrumentation table

3. All equipment within an enclosed area is vendor supplied, and the pipefitter is *not* required to pipe up that area unless the initials VS are enclosed by a _____.
   a. solid square
   b. dotted line
   c. circle
   d. dashed line

4. What do control valve operators and analytical, flow, pressure, level, and temperature instruments shown on P&IDs indicate?
   a. Supports
   b. Numbers
   c. Instrumentation
   d. Hardware

SECTION TWO

2.0.0 PIPING ARRANGEMENT DRAWINGS

Objective

Identify and describe the contents of piping arrangement drawings.

a. Describe how control points are used on drawings to locate work areas horizontally.
b. Describe how coordinates are used on drawings to define geographic position.
c. Describe how elevations are used on drawings to locate work areas vertically.

NOTE

Blueprint One (*Appendix*) is a typical site plan that shows the major parts of a power plant, including the steam plant, tanks, heat recovery systems, and various administrative and storage buildings. The site plan is like a map of the site and where the actual plans start. The two objects labeled Unit 1 and Unit 2 are heat recovery steam generators (HRSG). Unit 3 is a different type of generator that also produces power through a steam turbine. Notice the point at the northwest corner of the site, in the middle of the road, at the intersection of two lines. The location of the point is given as N 1675.97; E 231.13. The site and equipment are laid out with a surveyor's total station based on this point and other points surveyed from this point.

Piping arrangement drawings show how equipment is arranged and located in the system. The drawings can be either plan or section views, and equipment shown on the drawings typically corresponds to the basic shape of the item represented.

A plan view shows an object or area from above looking down on the system and depicts the exact locations of the equipment in relation to established reference points, objects, or surfaces. These reference points may be building columns, walls, benchmarks, or other equipment.

Plan views usually show north at the top of the drawing. If the need for space makes it necessary to rotate the drawing, rotate it so that north is to the left side of the sheet. *Figure 7* shows a plan view of an HRSG.

A section, or section view, shows the internal characteristics for part of a system on a cutting plane. The cutting plane may not be continuous; a horizontal cutting plane of a building may be staggered to include specific information. *Blueprint Six* (*Appendix*) shows three section views of headers C, D, E, and G taken from a section view of an HRSG elevation. These views show projections of a plane within the system and they are detailed drawings of a particular part of the plan.

Section views provide visual information not included on elevations or plan views and they often show internal arrangements that may not be obvious on other views. Many symbols are used on drawings to designate section cutting planes. *Figure 8* shows common symbols used for section views.

Piping arrangement drawings use control points, coordinates, and elevations to indicate the exact location of a pipeline, floor penetration, or piece of equipment in a building. While elevations refer to the vertical placement of an object, control points and coordinates refer to the horizontal locations. To identify the location of an object in a building or area, first locate the horizontal location and then establish the elevation.

2.1.0 Control Points

Column line control points are used to locate the work horizontally. The nearest column line on a drawing is shown, and piping dimensions are determined from this column. The column is designated using the grid line system, which designates every column inside a building or structure. On a grid line drawing, an imaginary grid is made using the column lines. Areas within the column lines are known as bays. *Figure 9* shows a grid line drawing.

Numbering and lettering of the system normally begins in the upper left or upper right corner of the system, which corresponds to the north side of the building or structure. Numbers are normally assigned to vertical rows, and letters are assigned to the horizontal rows. In some cases, the letters I, O, and Q and the numbers 1 and 0 are eliminated from the system to avoid confusion. This system allows each beam in the building to be given a reference number or control point; for example, B-3. Building beams are usually numbered on plans so that they are easy to identify. Using column lines allows those familiar with the system to precisely locate any point. When column lines are used with elevations, the locations can be specified three-dimensionally.

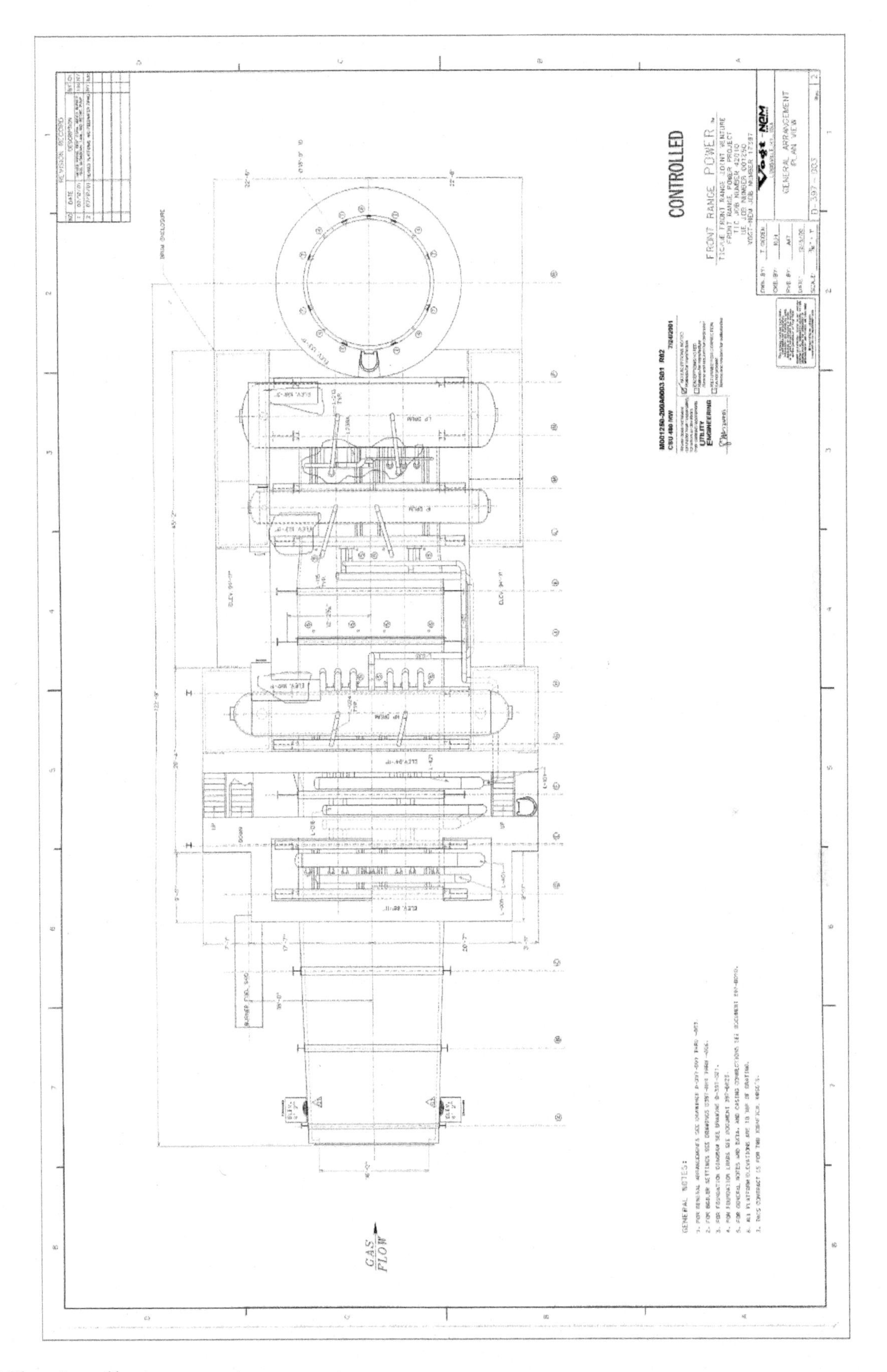

Figure 7 Plan view of heat recovery steam generator.

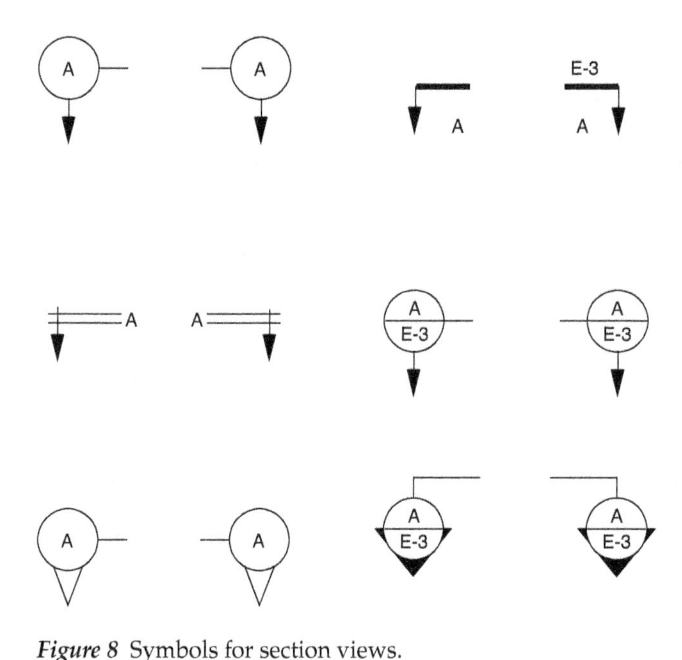

Figure 8 Symbols for section views.

2.2.0 Coordinates

Coordinates define the geographic position of a point, line, piece of equipment, or other object on the plant site. A coordinate is the distance a given point is from a base point, commonly called a benchmark (BM). Coordinates are written as a letter and number combination. The letter designates the direction of the coordinate, normally north or east, in relation to the benchmark; the number designates the distance in feet and inches.

A site plan indicates the position of the originating point or coordinate benchmark. From this point, a grid system is developed and used by all design groups to coordinate their drawings. The north-south axis and east-west axis intersect at this originating point, and every point on the drawing can be referenced by the originating point. For example, a point that is 30' east of the north-south axis through the originating point at N. 1675.97 and E. 231.13 has an east coordinate of E. 261.13. If this same point were 1' north of the east-west axis through the originating point, it would have a north coordinate of N. 1676.97. Combining the two coordinates pinpoint the location as E. 261.13 and N. 1676.97. *Figure 10* shows coordinate origin. Elevation is frequently given with the coordinates.

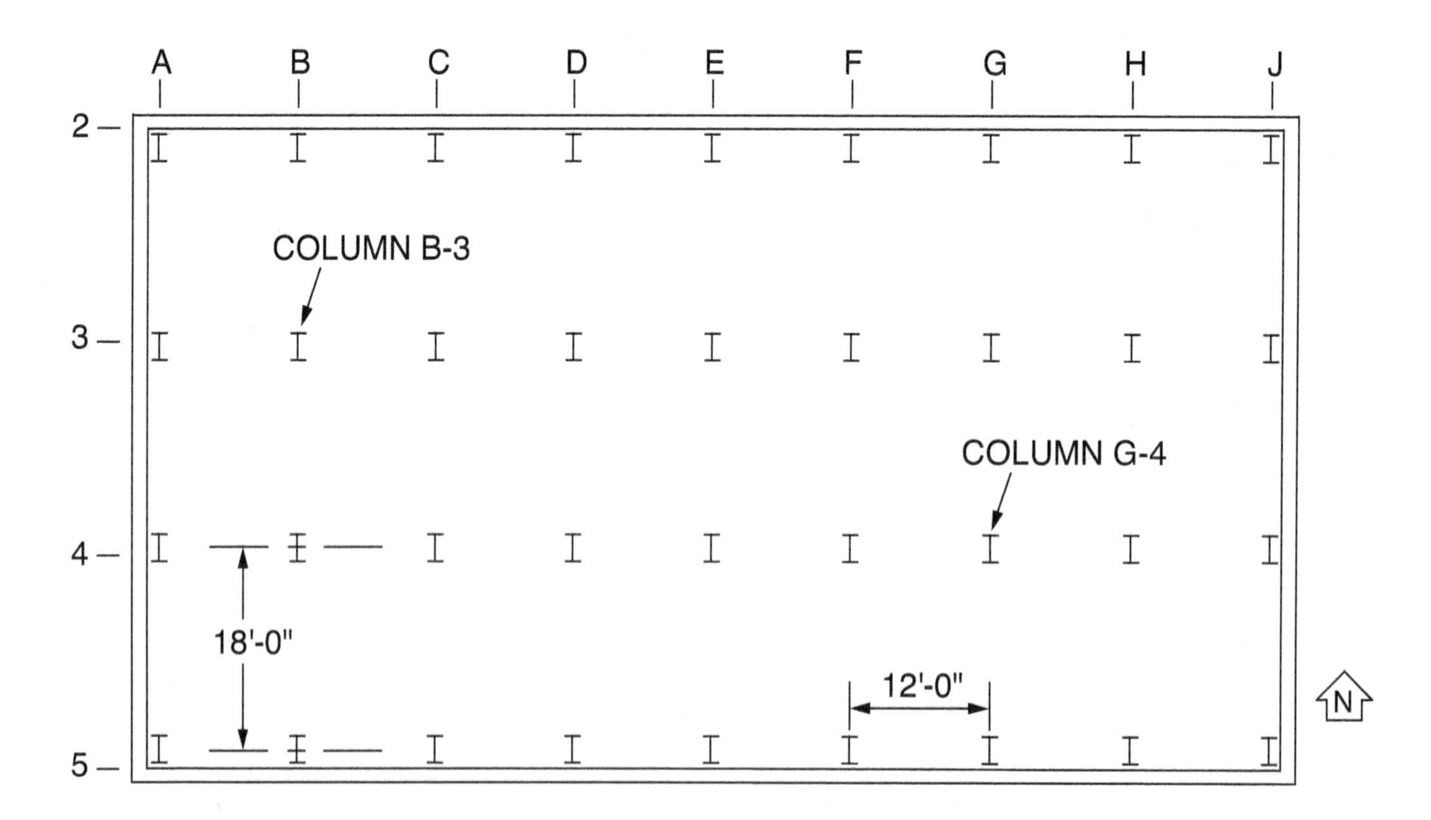

Figure 9 Grid line drawing.

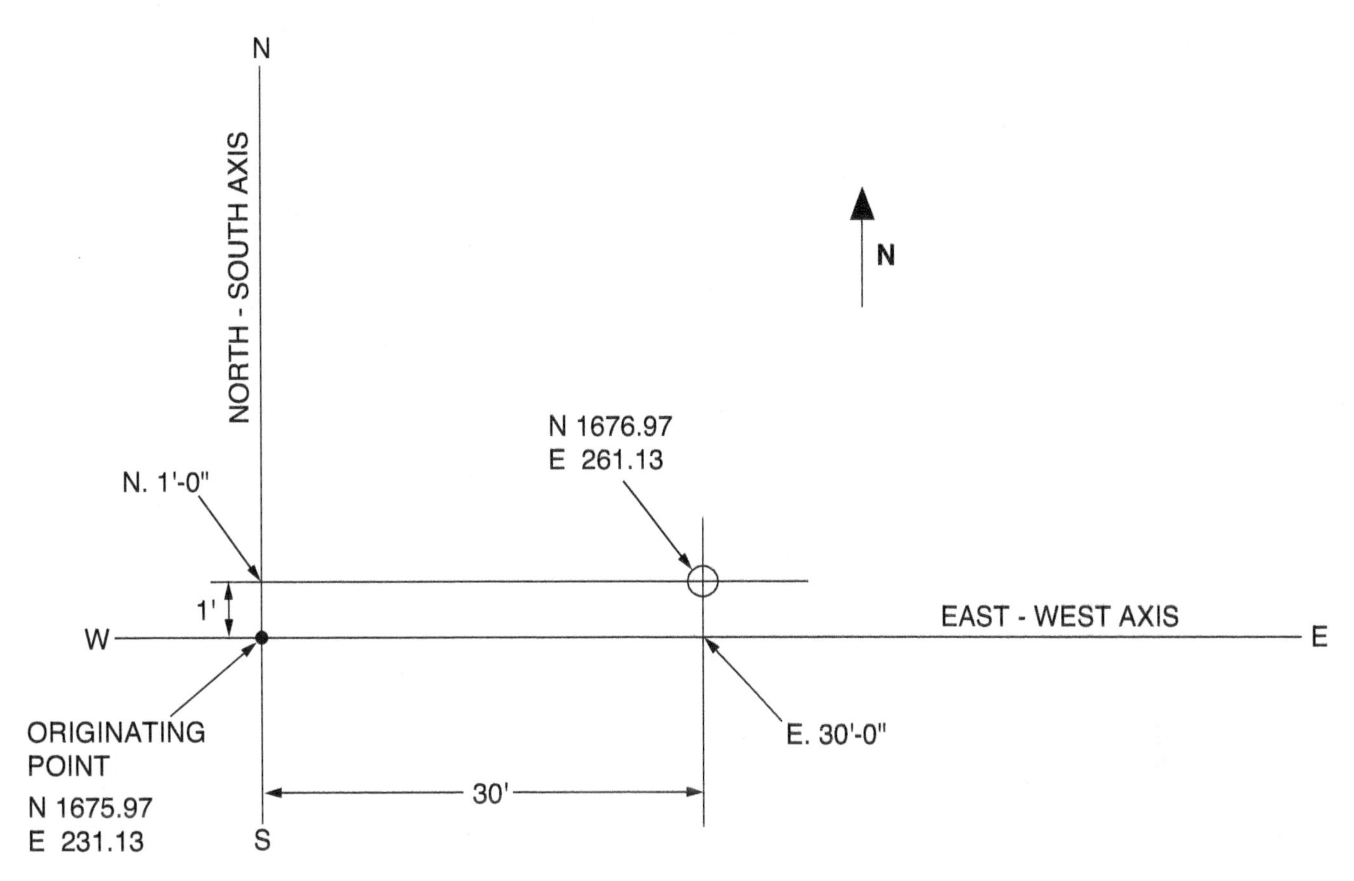

Figure 10 Coordinate origin.

2.3.0 Elevations

Elevations locate piping vertically in relation to other runs of pipe, building floors, or platforms. They may be referenced to the center line (CL) of the pipe, to the bottom of a pipe, to the face of a flange, or to any other convenient point. The drawing must indicate from which point the reference is made and must always be consistent. Plans normally indicate elevations on the structural columns using a reference line. These elevations are written in feet and inches. To determine the exact elevation at which a pipeline must be set, locate the elevation reference point marked by the engineer.

Piping elevations are specified in different ways. Elevation may be provided as an above-the-finished-floor elevation, in which case a measurement is taken from the finished floor to the desired height to establish the equipment elevation. At other times, elevation must be determined from an established elevation benchmark.

An elevation benchmark may be inside or outside the structure, but it is usually located on columns in various places throughout the building. To determine a rough elevation from a benchmark on a column, measure from the benchmark to the floor, using the floor as a measuring point. For more accurate measurements, elevations can be transferred from the benchmark to the bottom of the pipe or face of the flange using a string level, a laser level, a theodolite, or total station. The most precise tool is the laser level.

2.0.0 Section Review

1. Normally numbers are assigned to vertical rows, and letters are assigned to the horizontal rows, when using a(n) _____.
   a. instrumentation drawing
   b. grid line drawing
   c. vertical drawing
   d. site plan drawing

2. Normally north or east in relation to the benchmark, the number designates the distance in feet and inches while the letter designates the direction of the _____.
   a. line
   b. component
   c. flow
   d. coordinate

3. Which of the following is the *most* precise instrument for measuring elevations from benchmark to the bottom of the pipe?
   a. String level
   b. Laser level
   c. Theodolite
   d. Total station

SECTION THREE

3.0.0 READING AND INTERPRETING PIPE DRAWINGS

Objective

Explain how to read and interpret P&IDs, piping arrangement drawings, and isometric drawings.

a. Explain how to read and interpret P&IDs and piping arrangement drawings.
b. Explain how to read and interpret isometric drawings.
c. Explain how to follow a single line from one drawing to another.

Performance Tasks

1. Calculate the total line length from an ISO.
2. Sketch an ISO from a plan view.

Trade Terms

Dimension: A measurement between two points on a drawing.

Title block: A section of an engineering drawing blocked off for pertinent information, such as the title, drawing number, date, scale, material, draftsperson, and tolerances.

The pipefitter's ability to accurately interpret different types of pipe drawings is crucial to a project's safety and success. Three types of drawings typically used by pipefitters include P&IDs, piping arrangement, and isometric drawings.

3.1.0 Reading and Interpreting P&IDs

P&IDs provide an overall control scheme of the piping on a project. They show the types of control mechanisms and how those control mechanisms are used. They show the number of block valves and instruments, identify line numbers and equipment, indicate flow direction, and give other detailed information required on the project. As noted earlier, plan drawings show a view of an object or area from above, looking down on the system, and show the exact locations of the equipment and piping in relation to established reference points, objects, or surfaces. A section view is the result of an imaginary line cut through the system. From this drawing, the height, width, and location of the equipment can be seen. This type of drawing shows, in schematic fashion, the interconnected pipelines of a system. *Blueprint Three* (*Appendix*) is an example of a P&ID that shows a main steam system as it comes off the two HRSG units and connects with the rest of the system.

3.1.1 Symbol and Arrangement Pages

Symbol and general arrangement pages contain information about specific abbreviations, symbols, elevations, and plan views used in the P&ID.

- The mechanical symbol page *Blueprint Four* (*Appendix*) shows symbols used with major equipment, components, tags, and piping.
- The instrumentation symbol page *Blueprint Five* (*Appendix*) shows symbols for instrumentation and for operators and actuators. The categories are: components, functions tags, and line styles for instruments, instrument function abbreviations, and actuators and operators for various equipment.
- General arrangement pages show elevations *Blueprint Six* (*Appendix*) and plan views *Blueprint Seven* (*Appendix*) for units one and two of the HRSGs from the site.

3.2.0 Reading and Interpreting Isometric, Spool, and Vessel Drawings

A pipefitter must have a thorough knowledge of isometric drawings (ISOs) as they are widely used in fabricating and erecting pipe. The biggest advantage that ISOs have over other types of drawings is clarity. In an ISO, the draftsperson can extend a pipeline and show all its components without regard to scale. Therefore, an ISO can show a complete line from one piece of equipment to another or show a complete line as it exists on an orthographic drawing. Additionally, all the information needed to erect and fabricate the line can be clearly shown on an ISO.

3.2.1 Isometric, Spool, and Vessel Drawings

Pipefitters use these types of drawings to gain a clear understanding of the pipe's travel, fabrication requirements, and needed connectivity.

- ISO drawings show the general travel of the pipe. Valves and instruments are also identified on ISOs. *Figure 11* is a non-dimensional isometric of the entire main steam line, providing a general idea of where it all goes.

- Spool drawings are specifically used to fabricate sections for the fitters so that the pieces are manageable and put together in right relationship to other parts of the system. As shown on *Blueprint Three* (*Appendix*), the two, 12" lines 12LKEA1031 and 12LKEA2031 come together and pass through 16" by 12" concentric reducers to form the line 16LKEA3001.
- Vessel or subassembly drawings *Blueprint Eight* (*Appendix*) show the necessary connectivity requirements for the system.

3.3.0 Following a Single Line

Locate *Blueprint Three* (*Appendix*) and find the two, 12" lines from HRSG Unit One and Unit Two. These lines are labeled as 12LKEA1031 and 12LKEA2031. The two lines converge to produce Line 16LKEA3001. Since the steam turbine is within the dashed lines, it is the responsibility of the vendor. The line is shown on the P&ID as if it were a straight line, with attached branch lines and assemblies. It ends at the turbine after being reduced from a 16" line to a 14" line.

Now examine *Blueprint Nine* (*Appendix*), which is a piping general arrangement drawing for a section of the piping running from the HRSG Unit One and Unit Two main steam lines. This shows the junction of 12LKEA-2031 and 16LKEA-3001, and shows 12LKEA-1031 running parallel, before the junction. There is a break, where 16LKEA-3001 runs out of the level shown, and then returns to the level to the left of the page. Several more pages of plan views at different heights follow this sheet. *Figure 12* shows 16LKEA coming from the side and centers over the Steam Turbine. The connection shown on *Figure 13* is on the other end, where it becomes a 14" line connected to the HRSG.

3.3.1 Getting an ISO from a Plan View

When pulling an ISO from a plan view or general arrangement drawing, start from a clear break in the piping and proceed to another clear break. Look up the line in the pipe list (*Figure 14*). The line is neither insulated nor heat traced. In *Figure 15*, the best place to begin or end an ISO is the flange at either end. The location is given, with reference to the benchmark identified in *Blueprint One* (*Appendix*). The center line of the flange face on *Figure 15* is given as East 1733'7½" and North 1583'5⁄16", with the center line elevation being given as 92'6". Since the other end of the run on *Figure 16* is set at the same elevation, and the notes indicate that these are continued as underground piping on another drawing, it is likely that 92'6" is close to the ground, with just enough room for the flange to clear the ground.

The flange is at a 45-degree angle to the ground because there is a 45-degree ell at both ends. The first ell is centered at 2'11¹⁵⁄16" from the flange. In the spool immediately after the ell, there is a 2" line to a pressure control valve followed by a 2-to-1 concentric reducer. It is likely the ISO would show a 2" weldolet on the 12" line, rather than add the branch. However, it might show the line to the 2-to-1 reducer.

The next item on the line is an automated ball valve (ABV). This valve provides control of the flow at this point and is controlled by a pneumatic actuator, which can be activated remotely. According to the valve list on *Figure 14*, this is a 12" by 10" fail closed valve. This line is a fuel line, as shown by the pipe specifications index description for the number 310 (*Figure 17*). The column labeled UE Pipe Specification 3-Digit Code indicates that the FAB in the line label means this is the main fuel gas line that is rated as a 600-pound ANSI pipe class. In the valve list specification, the ABV is referenced to API 6D, which refers to an American Petroleum Institute (API) standard that allows zero leakage for a closed valve.

The next item identified on the line is the support labeled SPS UB24. The specific support will be supplied as a detail, usually from a manufacturers' cutsheet. These are frequently provided as details on a single sheet of the drawings, which keeps details related to hangers and supports in one place.

A spool of 12" pipe continues the line, with flow element 0250 in the line. According to the legend for the P&ID, this is a primary flowrate element, which is known as a flowrate sensor. The end of the spool is flanged and followed by a short spool that has a 12" by 8" concentric reducer welded to the end. The flange bolts to a pressure control valve (PCV), labeled PCV 0251, which is an 8" V-ball PCV. The other side has a 12" by 8" concentric reducer bolted to the flange of the valve. Just before the valve, the spool has a support, and after the valve there are three supports and two PRVs: PRV 0089A and 89B. Another 45-degree vertical ell and a flanged connection end the ISO.

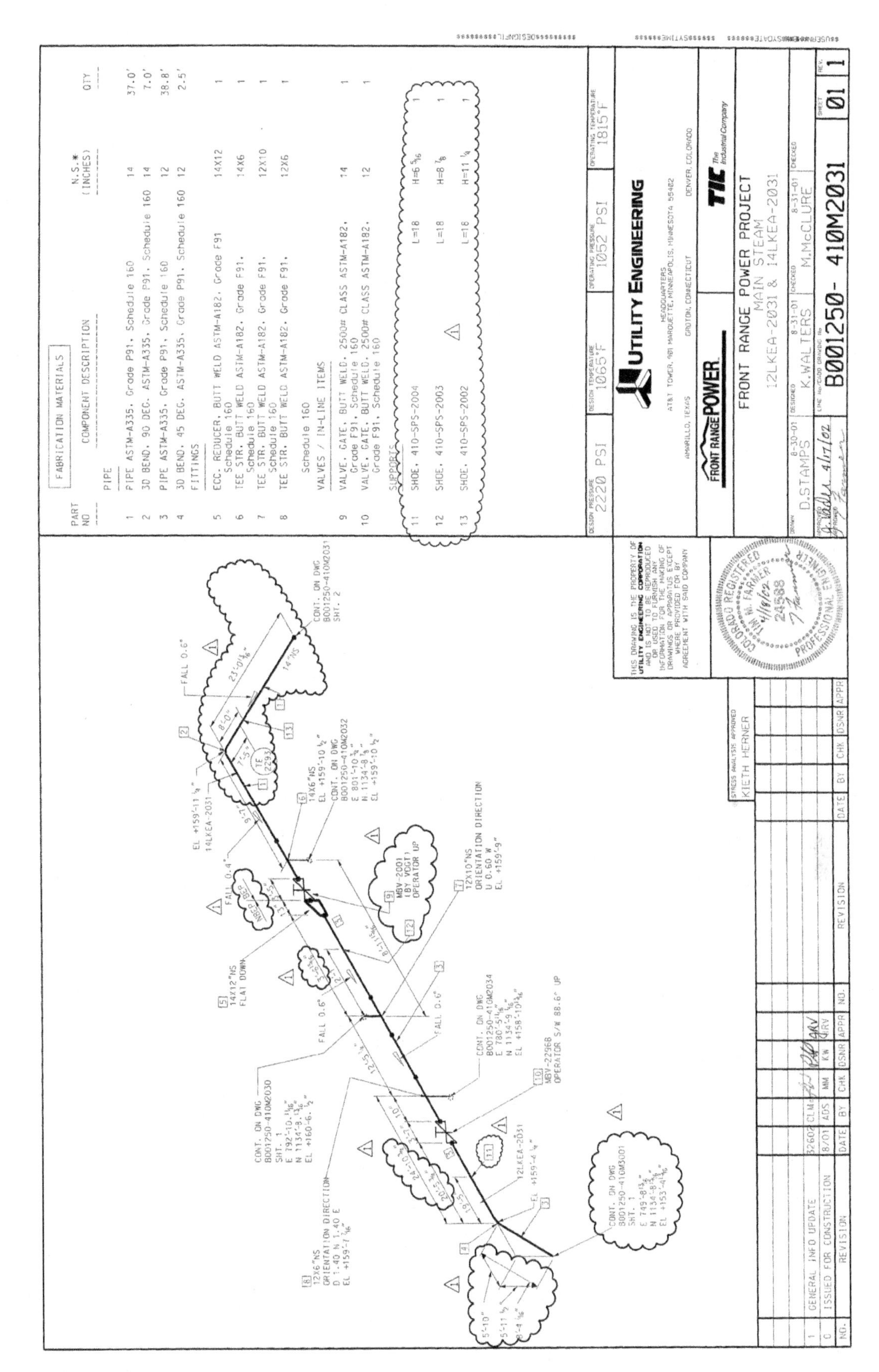

Figure 11 ISO of part of the main steam system.

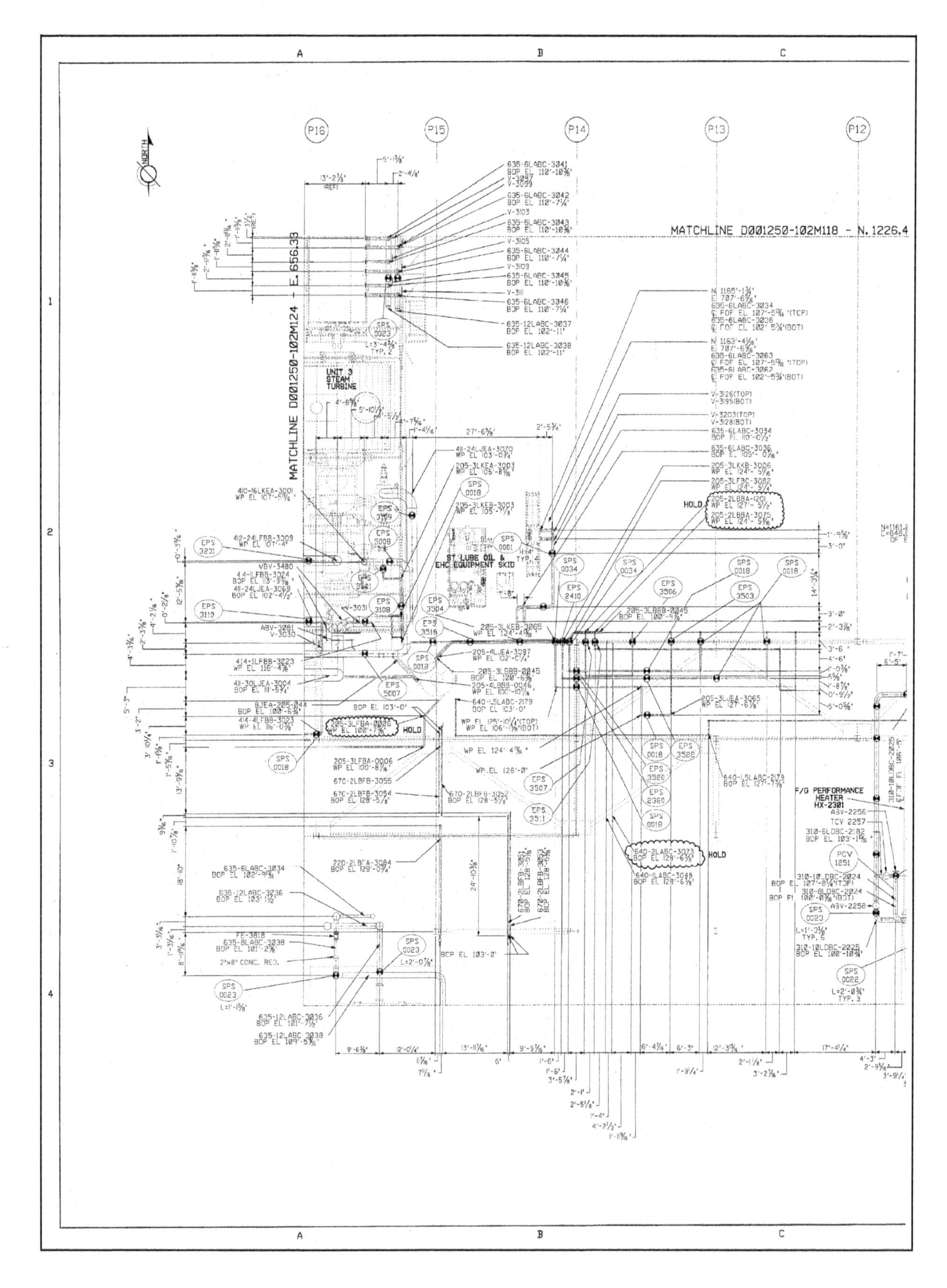

Figure 12 General arrangement piping 2.

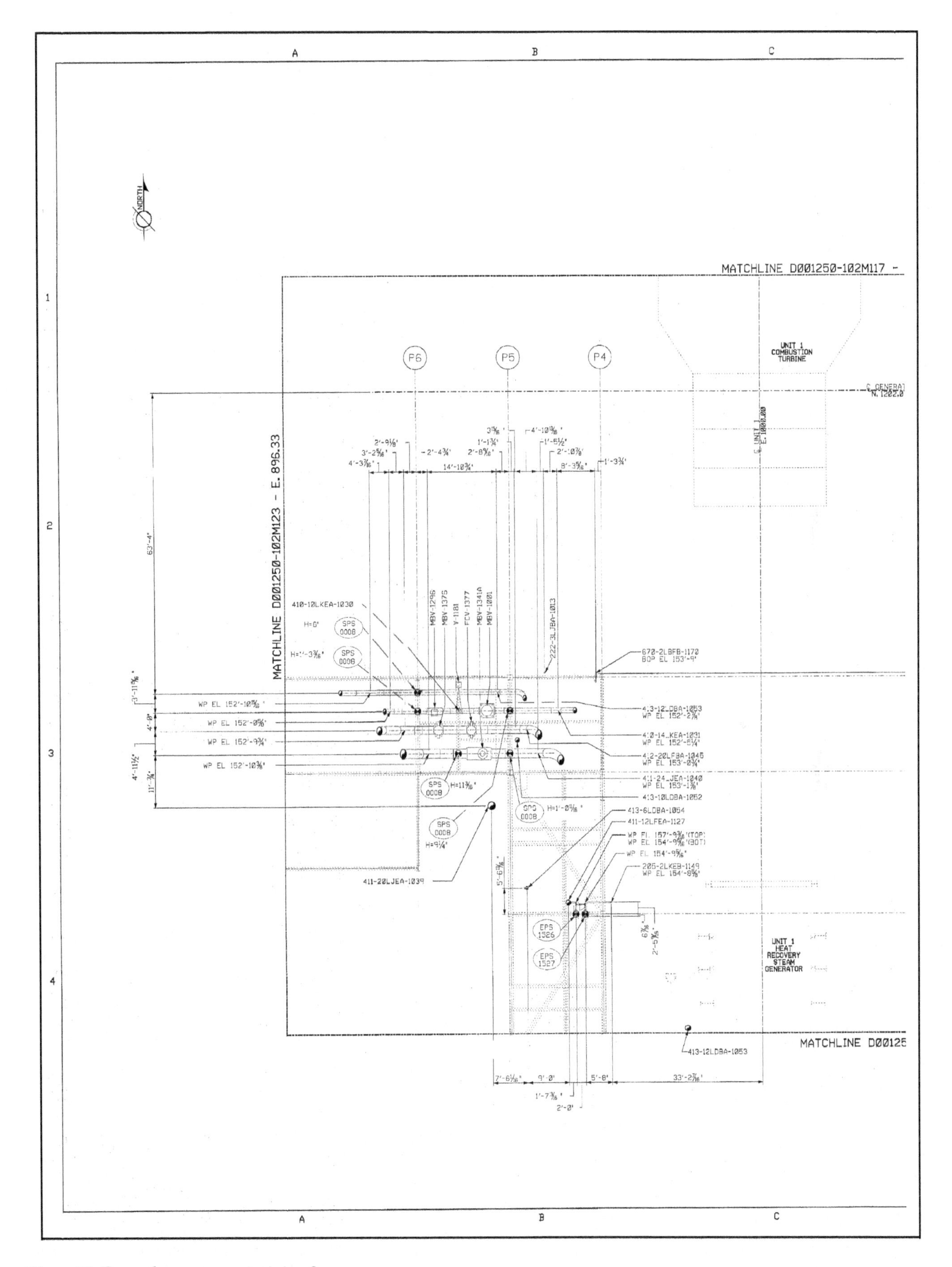

Figure 13 General arrangement piping 3.

Valve and Pipe Lists

Drawing Number	Pipe Tag	Line	Insulation	Size	Heat Trace	
D001250-310M001 S01 R1	12LFBA--0030	30	N/A	12	N/A	

Drawing Number	Line	Tag	Valve Type	Size	Line Type	Remarks
D001250-310M001 S01 R1	310	ABV-0262	BALL	12 ×10	FBA	API-6D, N.O.,Air Fail Closed
D001250-310M001 S01 R1	310	PCV-0251	V-BALL	8	FBA	By I&C (FC)
D001250-310M001 S02 R1	310	PRV-0089A	ANGLE			HOLD, UE bid
D001250-310M001 S02 R1	310	PRV-0089B	ANGLE			HOLD, UE bid

Figure 14 Valve and pipe lists.

Follow the steps below to draw the ISO:

Step 1 Using a piece of ISO paper, start with a 10" flange connection. The center line of the flange should be 92'6", with the line labeled as a continuation from an underground line.

Step 2 On the ISO, identify each component with a small number tag and enter it as an item in the component description list.

Step 3 Ensure the line is at a 45-degree angle to the bend. The run and set are 2'1½". Travel to the center of the bend is 2'11$^{15}/_{16}$".

Step 4 Draw a 45-degree angle and the spool that follows to the ABV. In the middle, indicate the 2" weldolet, then draw a short line to the PCV and another to the 2-to-1 reducer.

Step 5 Draw the 12" ball valve symbol and identify it.

Step 6 Draw the spool and flowmeter that follow. This is a short-flanged spool with an attached direct mass flow meter installed by the instrumentation and control (I&C) division.

The next spool has a 12" by 8" concentric reducer, with the 8" end flanged. A support is shown at the point where the 12" end of the concentric reducer rests. The next element is PCV 0251. Another 12" by 8" concentric reducer brings the size up to 12" again, then a short spool with a pressure relief valve, three more supports, another PRV, then the spool is welded to another 45-degree ell. Last is a spool to the flanged end, with a travel of 4'2$^{13}/_{16}$" from the center of the ell to the flange center line.

3.3.2 From ISO to Spool Sheet

On sheet B001250-410M3001 (*Figure 18*), the line begins with the two converging lines from the ISOs for 12LKEA1031 and 12LKEA2031. Callouts indicate that they are continued from the drawings for those lines.

The cloud shape, or curly line around the end area, indicates that this area has been revised from an earlier set of drawings. The triangle next to the cloud shape has a "1" inside of it. In the revision block, that number indicates that on 3/27/02 a general information update was added.

The direction of the pipe run can be determined by examining changes in the distances, given as dimensions. If the east **dimension** decreases, the pipe is running west. If the north dimension decreases, the pipe is running south. The locations of the endpoints are given as dimensions from a benchmark and elevation. These dimensions are set with a total station using the benchmark, or using marks set from the original benchmark. The spool drawing that matches that beginning is *Figure 19*, which shows the bend and the 12" piping coming into the 16" by 12" eccentric reducing lateral where the two lines merge.

The pipefitter should have a clear understanding of the beginning and end points on the drawing. In this case, the table of materials indicates that the 12" pipe should be assembled from the end of the spool to the bend, and from there to the lateral. The dimensions are given for the fit up to the 12" line, 12LKEA2031, which is from HRSG Unit 2. The system is for high-pressure steam and is made of Schedule 160 P91 chrome. This information is also shown on the ISO for this section, along with the design pressure, operating pressure, and design and operating temperature. These are located on the ISO above the **title block**. The pressure of 1,815 psig and the temperature of 1,052°F indicate this is a superheated steam system. The insulation is described by the symbol on the P&ID as 5.5" of calcium silicate.

Welding instructions are given in the lower right corner of the spool sheet. On these particular drawings, the welder who butt welds the pipe needs to sign or initial each weld they perform.

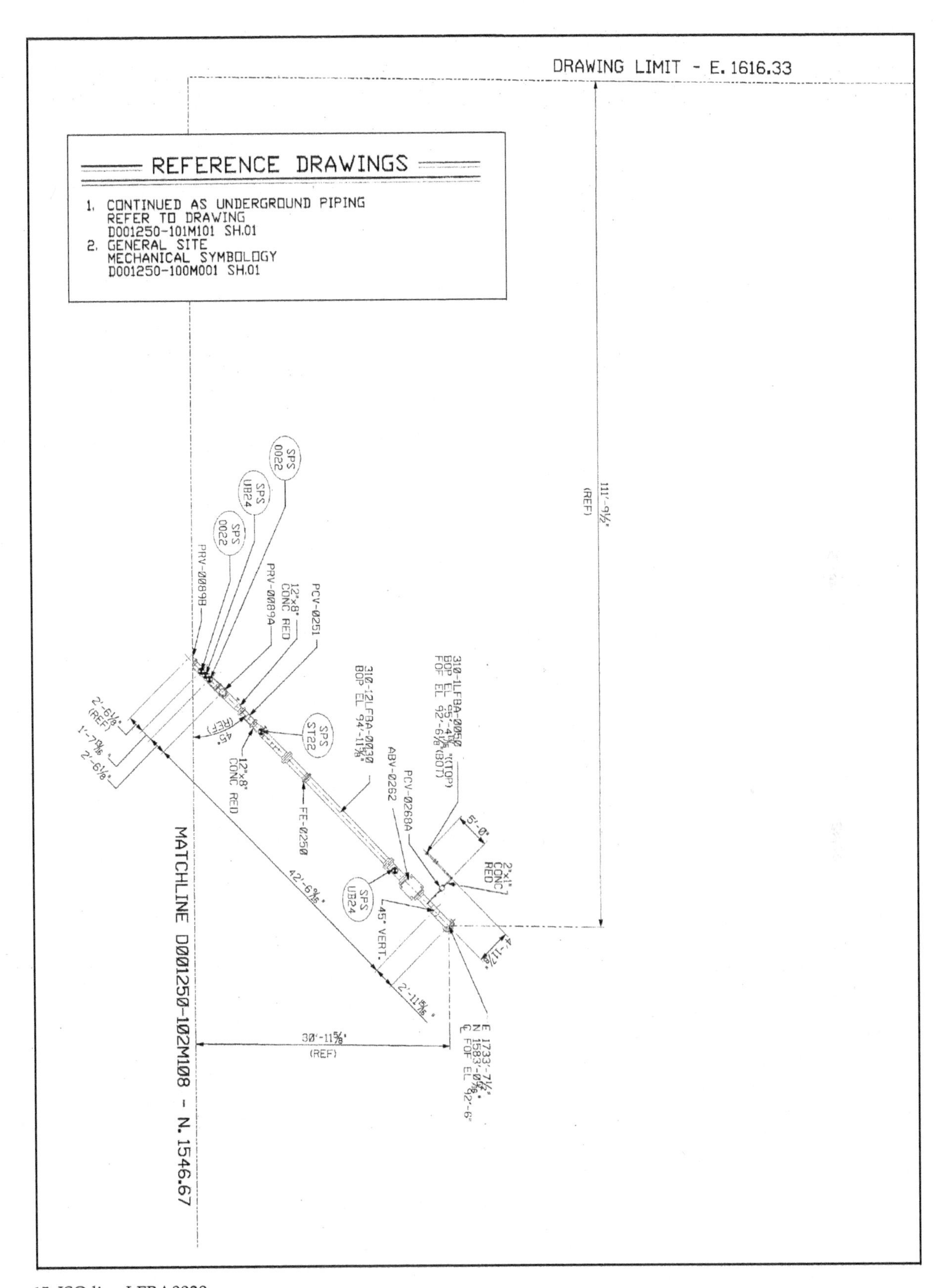

Figure 15 ISO line LFBA0030.

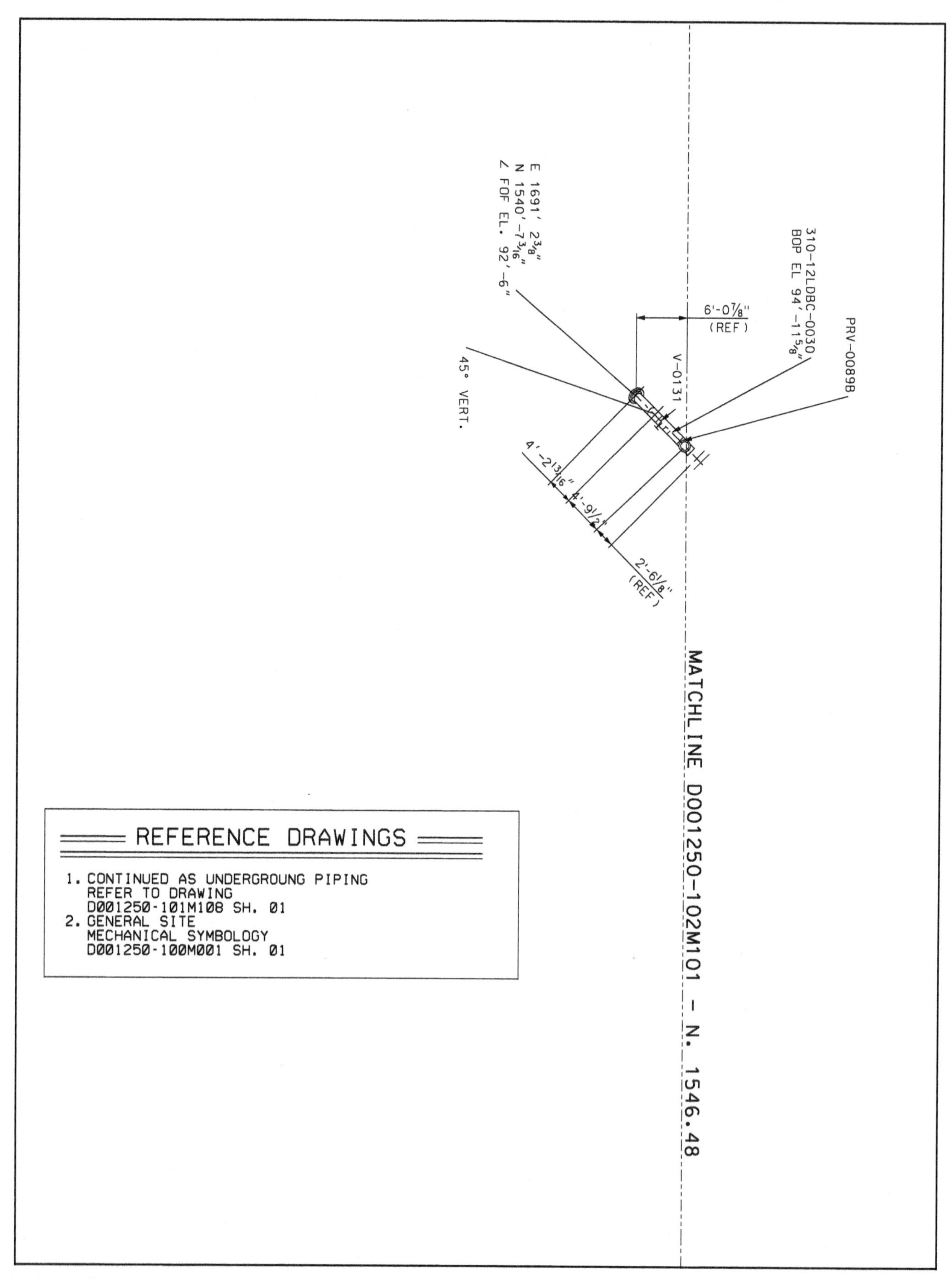

Figure 16 Continuation of ISO line LFBA0030.

COLORADO SPRINGS – FRONT RANGE POWER PROJECT									
Pipe Specification Index									
			Design			Operating			
Work Order #	System	ANSI Pipe Class	Temperature deg F	Pressure psig	Flow lb/hr	Temperature deg F	Pressure psig	Flow lb/hr	UE Pipe Specification 3-Digit Code
205	Boiler Blowdown		660	2,000		625	1,900		
220	Distillate	150	100	150	45,000/225,000	80	110		BFA
221	Condensate Suction	150*	160	25*	1,640,000	130	-5	1,247,823	BBA
221	Condensate, Pump Discharge	300*	160	400*	1,640,000	130	300	1,247,823	DBA
222	Boiler Feedwater Suction	300	335*	200*		310	108		DBA
222	Boiler Feedwater IP Discharge	600*	335*	1,000*		310	700		FBA
222	Boiler Feedwater HP Discharge	1500*	335*	3,000*		315	2,600		JBA
230	Chemical Feed and Sample Panels		1,100	3,000		various	various		TFB/TFD
310	Fuel Gas, Duct Burner	150*	110*	285*		85*	30*		BBC
310	Fuel Gas, Main	300*	110*	550*		85*	460*		DBC/DFA/FBA
410	Main Steam System, Bypass to Cold Reheat	600*	1,065	550*		660	380		
410	Main Steam System, BFP - HP Discharge	1500*	335*	3,000*		315	2,600		JBA
410	HP Steam System	2500*	1,065	2,220		1,052	1,815		KEA
411	Hot Reheat System, Condensate Supply	300*	160	400*		130	300		DBA
411	Hot Reheat System	1500*	1,065	475		1,052	350		JEA/FEA
412	Cold Reheat System	600*	750*	550*		660	395		FBB
413	Low Pressure Steam System, Steam	300	650	150		565	55		DBA
413	Low Pressure Steam System, Condensate Supply	300*	160	400*		130	300		DBA
414	Auxiliary Steam System, Cold Reheat Supply	600*	750*	550*		660	395		FBB
420	Steam Condensing	150*	250	15		121	-10.25		
635	Auxiliary/Closed Cooling Water	150	110*	80*		85*	31*		ABC
640	Service Water	150*	100	150	125,500	80	115	71,000	ABC
640	Softened Water from Nixon Plant	150*	100	150	62,500	80	115	55,000	ABC
645	Waste Water	150*	212	100		60	60		
650	Fire Protection (AG/UG)	175/350	100	175	1,000,000	80	150	500,000	CBA/EDA
670	Compressed Air	150*	150	150		100	125		BFB/BBA

Figure 17 Piping specification index.

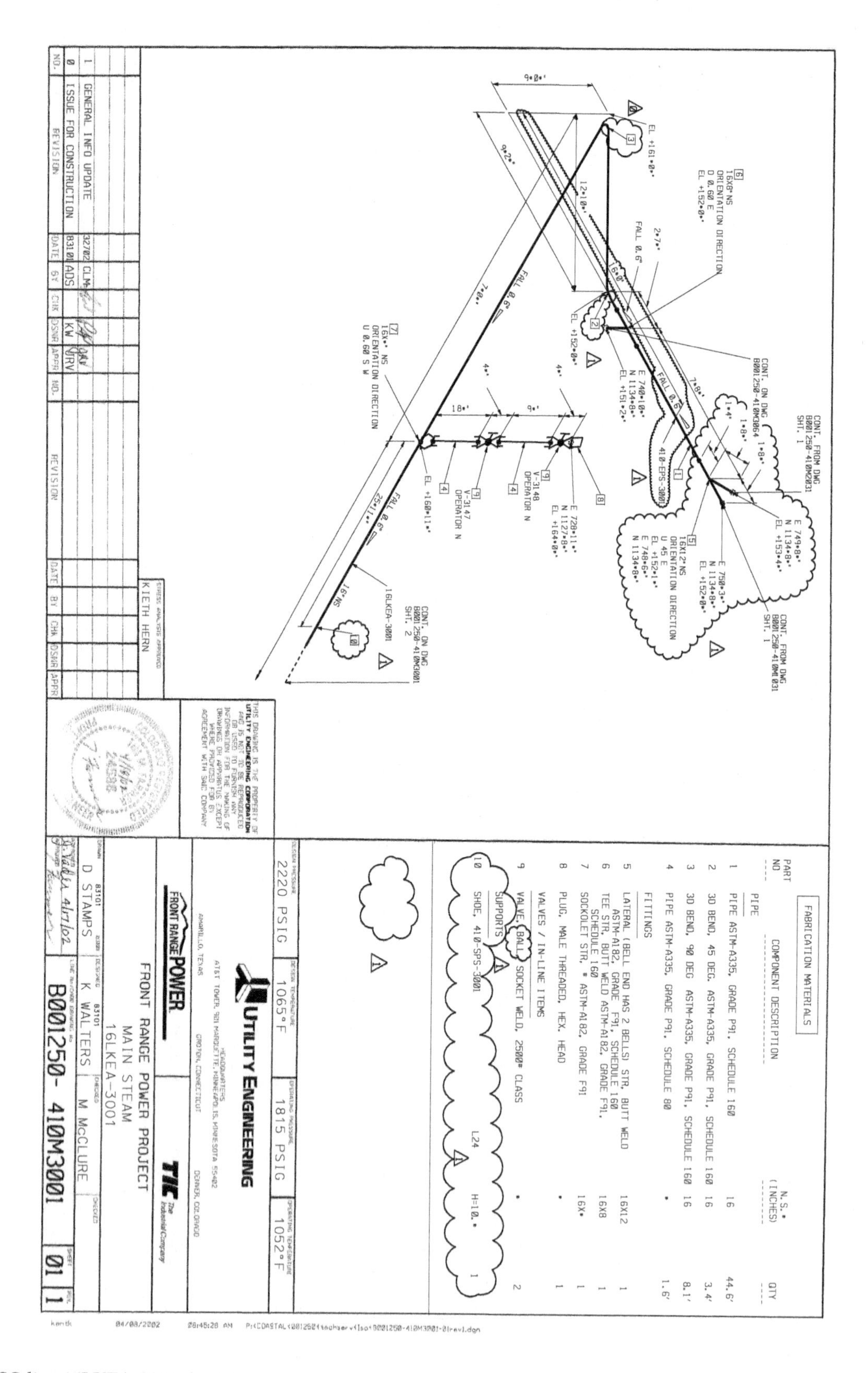

Figure 18 ISO line 16LKEA-3001 sheet 1.

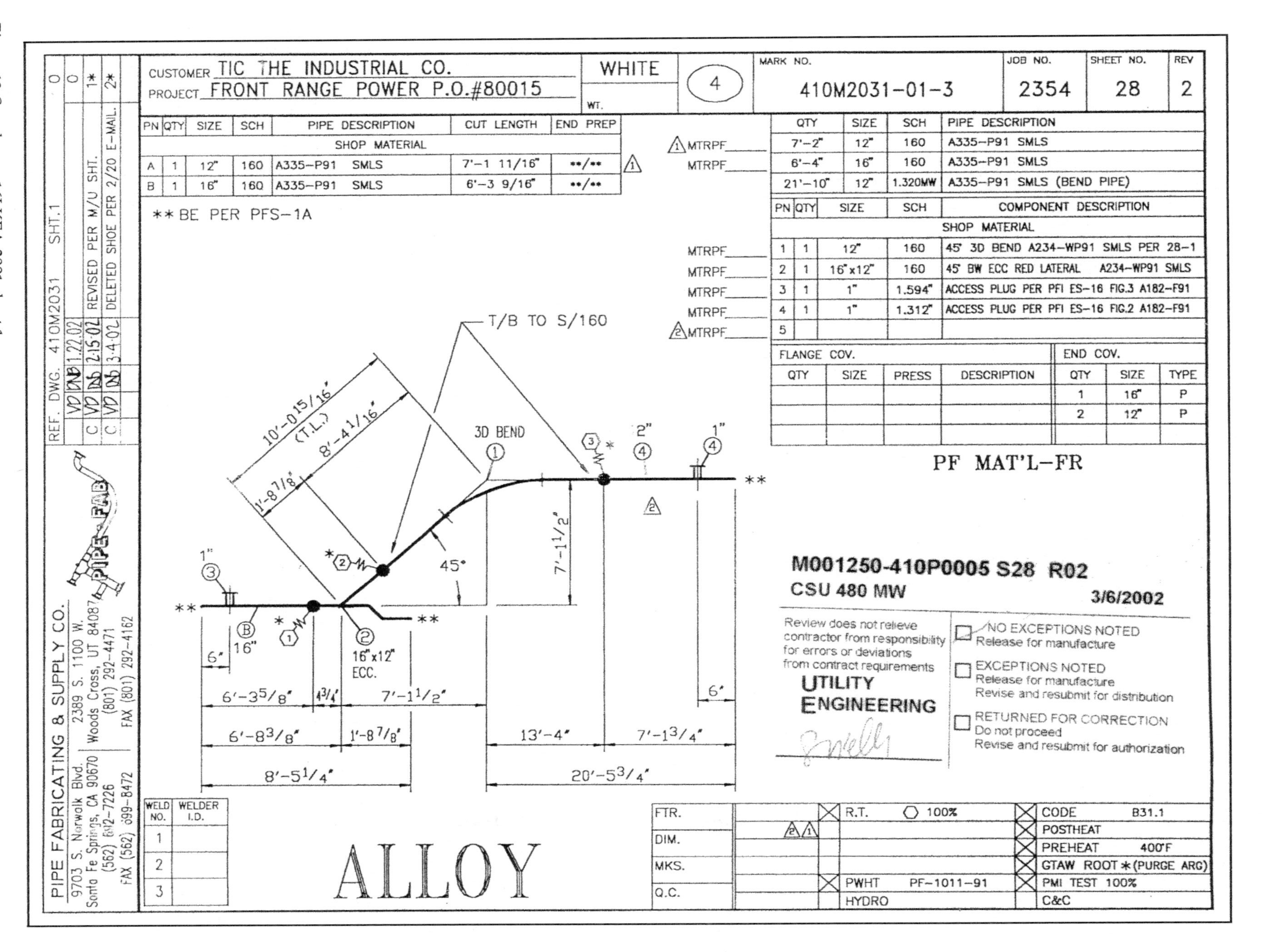

Figure 19 Spool page 16LKEA-3001 sheet 1.

The welds are marked on the spool sheets with a number inside either a hexagon or a diamond that references the signatures. The hexagonal weld symbol is shown as 100 percent X-ray tested, while the diamond is shown as 100 percent magnetic tested.

On the spool sheet, the instructions are given for installing a 1" plug on the run at a point 6" from the end of the spool. These plugs are installed after the weld has passed the x-ray examination, and the spool is shipped with the plug already in place. After the spool is delivered, the pipefitter installs any branch lines shown on the P&ID at that point on the spool.

The next spool (*Figure 20*) has two bends: a 45-degree bend and a 90-degree bend. The 45-degree bend is horizontal and the 90-degree bend is vertical. On the tangent of the 45-degree bend is a 16" by 8" reducing tee, which starts the 2' long, 8" branch. The valves on the ¾" line on the ISO, labeled V-3147 and V-3148, are drawn as T-pattern globe valves, but they are drawn on the P&ID as ball valves and are identified on the ISO component list as ball valves. It is important to clarify such discrepancies before assembling the line, as rework is always be more expensive than doing the job right the first time. Don't assume that the engineer changed the words and didn't change the drawing.

On the spool, the fabricator installs a thermowell 9" from the end of the branch. Additionally, a weldolet is centered 3" from the end. A cap with a ¾" sockolet inside of it is installed on the end. A detail sheet (*Figure 21*) specifies the welding and fit for the fittings. Notice on the ISO that the pipe is to be installed with a 0.6 fall. This slope is incorporated in elevations at points where they are given, but they can also be calculated for each piece as it is installed.

The next spool (*Figure 22*) is also on this ISO. Here, the fabricator in the shop installs a 16" by ¾" sockolet 6" from the beginning of the spool for the vent assembly shown on the ISO. The assembly, which is put together by the fitter, consist of two, ¾" ball valves as isolation valves, with a threaded plug in the end. The revision has also resulted in the addition of a shoe support (item 10).

The next ISO (*Figure 23*) has another shoe support and restrained joint added as a revision. Because of the large difference in temperature in this process, these are sliding supports and variable hangers and supports, which allow expansion and contraction of the pipe to occur without binding. Details from the manufacturer of the supports and hangers are supplied with the drawings. These drawings, also called cut sheets, are supplied with the bid documentation to demonstrate compliance with the specifications for the job.

The spool sheet (*Figure 24*) shows a 1" access plug welded on the straight run. This plug is at a right angle to the axis of the pipe and to the plane of the bend. On the second spool sheet for this ISO, the detail shows a 1" access plug installed at 44.4 degrees from the plane of the bend (*Figure 25*). The spool sheet that follows (*Figure 26*) shows a 45-degree bend with another 1" plug on the far side of the drawing. The next spool sheet (*Figure 27*) has a single bend that is slightly greater than a 90-degree angle, with 1" access plugs 6" from each end of the spool. The next spool (*Figure 28*) from the second ISO is marked 410M3001-02-5 and is a straight run with a 45-degree bend. Again, there is a 1" access plug on the straight end. The last spool (*Figure 29*) for this ISO is a 90-degree ell on one end and an 89.4-degree bend at a right angle to the plane of the elbow, with a straight joint of pipe in between that measures 28$^{11}/_{16}$' long.

The third ISO (*Figure 30*) matches the single spool sheet for it (*Figure 31*). There are three ¾" sockolets for the pressure instruments and a 1" thermowell. There is a 1" plug on the downstream end.

The last ISO of this line (*Figure 32*) is on two spools. The access plugs for purge are on each end of the first spool (*Figure 33*). There is a straight and a 90-degree long radius ell with a straight end. The last spool (*Figure 34*) takes the run to the 16" by 14" reducer, which is the end of the pipe contractor's work. There is another 8" branch that leads to a cap and sockolet, with a thermowell and a 3" weldolet branch on it. The 16" line has a 90-degree ell and then a ¾" sockolet horizontally, a ¾" sockolet at 45 degrees from vertical, and two 1" thermowells at a 45-degree angle from the vertical. Finally, there is the 16" by 14" concentric reducer that ties into the turbine inlet.

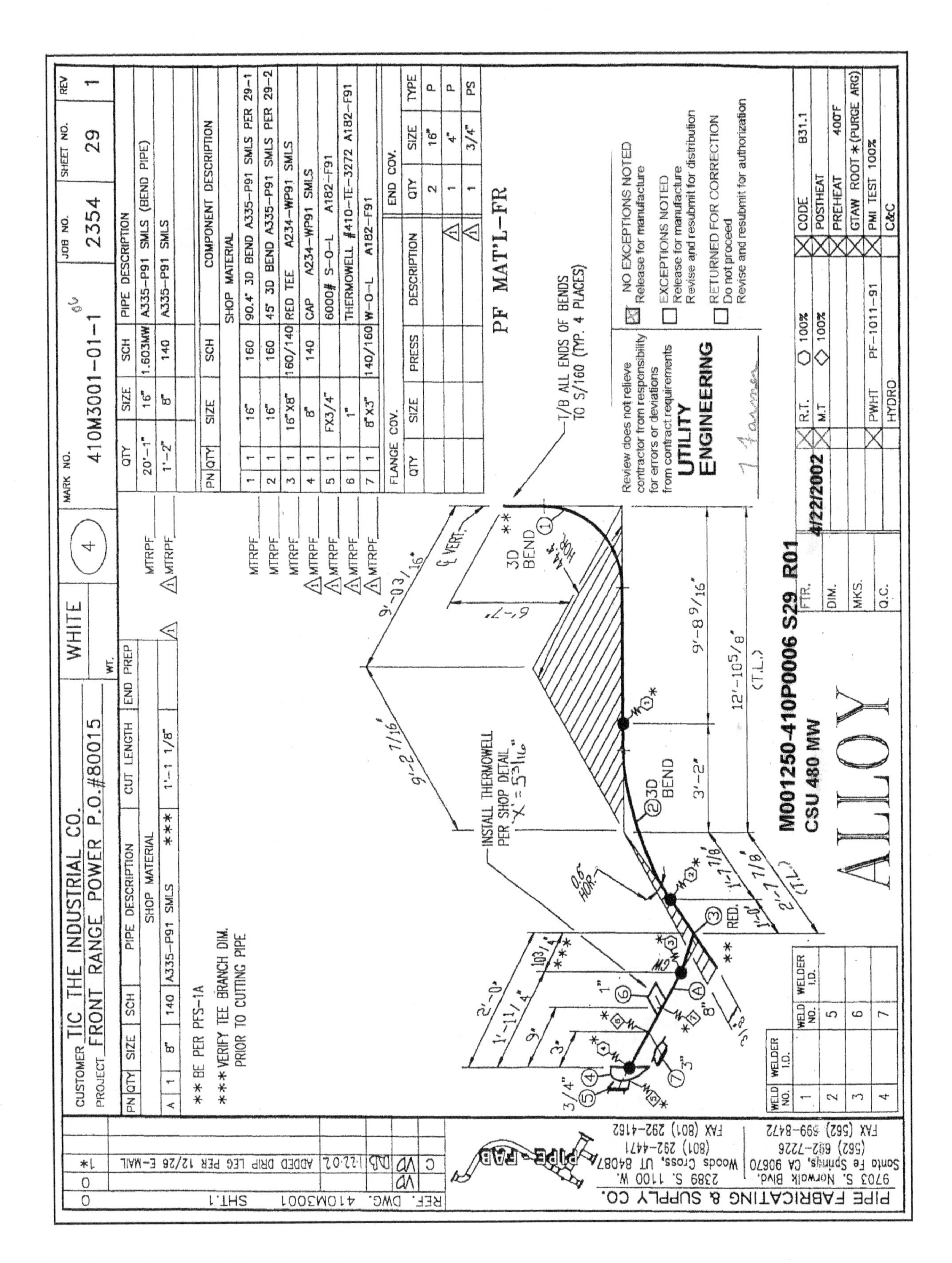

Figure 20 Spool page 16LKEA-3001 sheet 2.

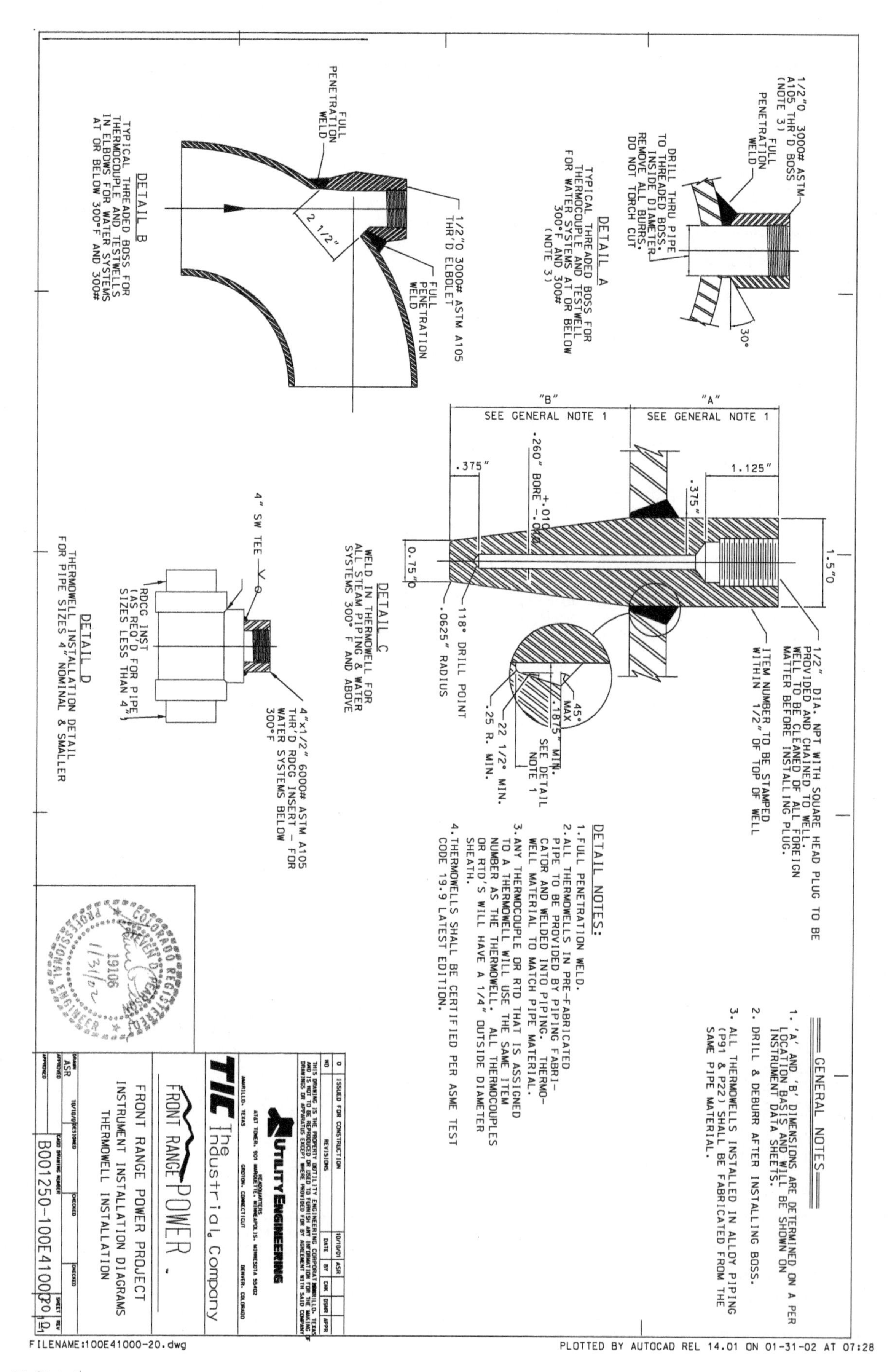

Figure 21 Detail page.

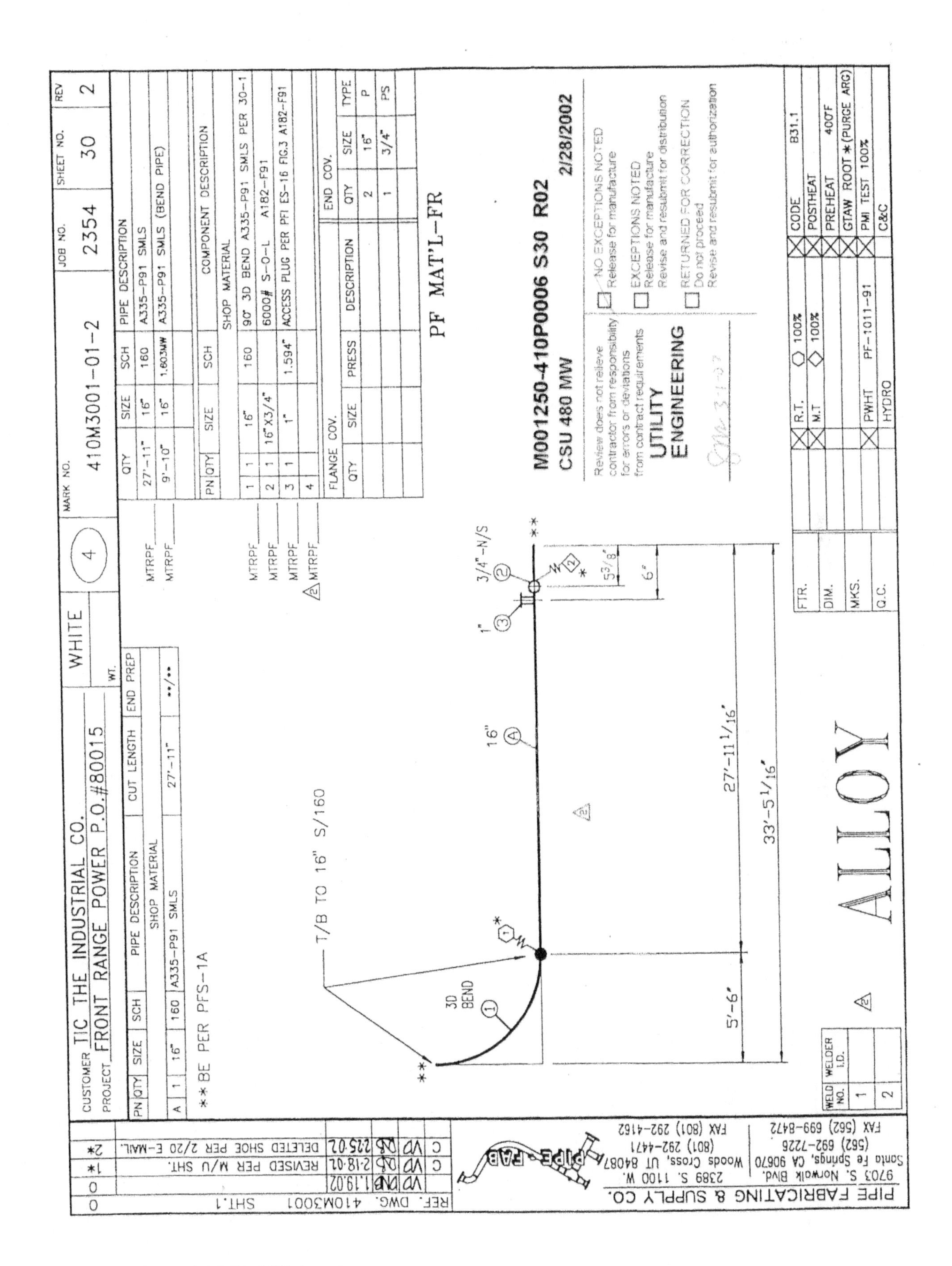

Figure 22 Spool page 16LKEA-3001 sheet 3.

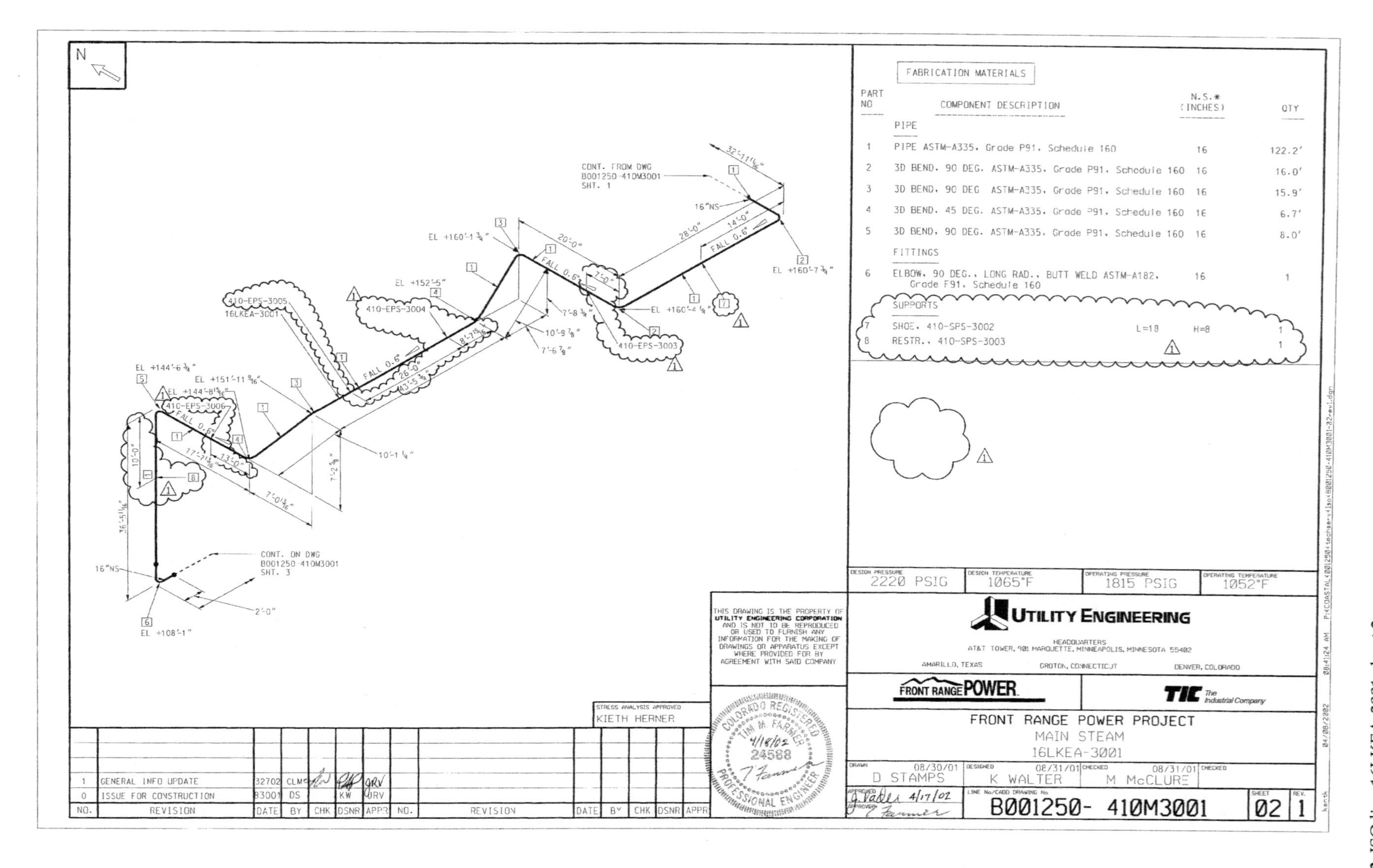

Figure 23 ISO line 16LKEA-3001 sheet 2.

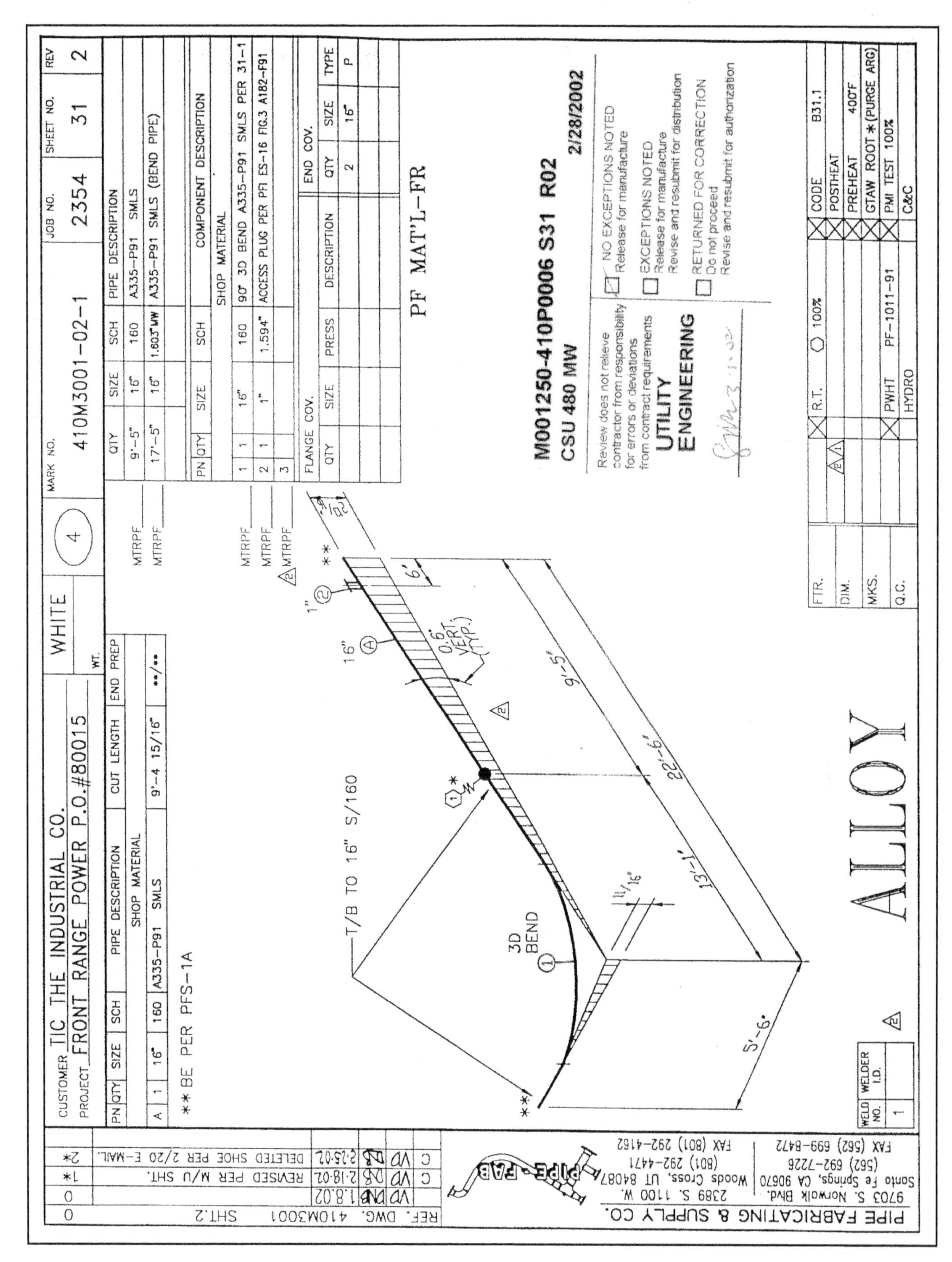

Figure 24 Spool page 16LKEA-3001 sheet 3.

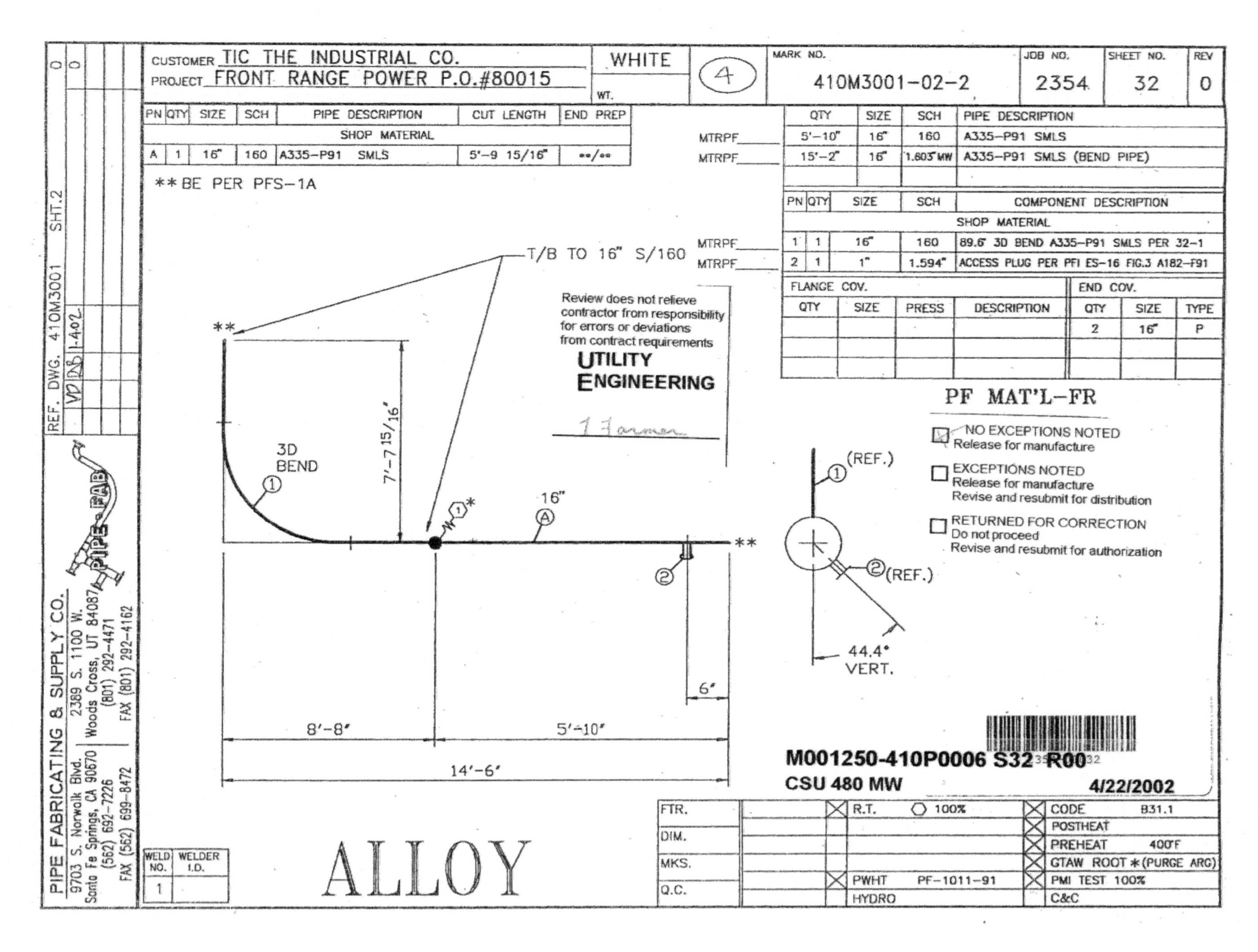

Figure 25 Spool page 16LKEA-3001 sheet 4

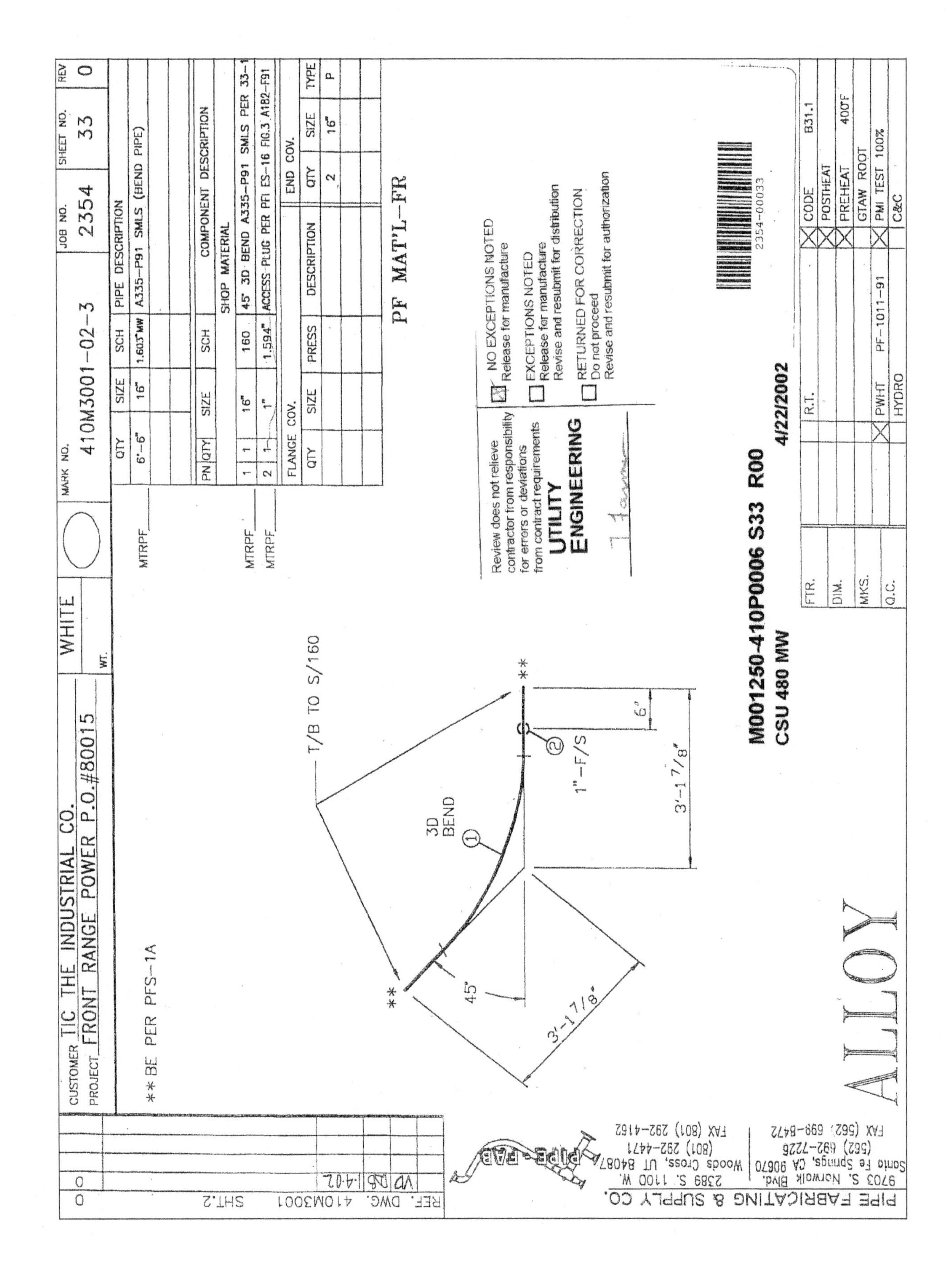

Figure 26 Spool page 16LKEA-3001 sheet 5.

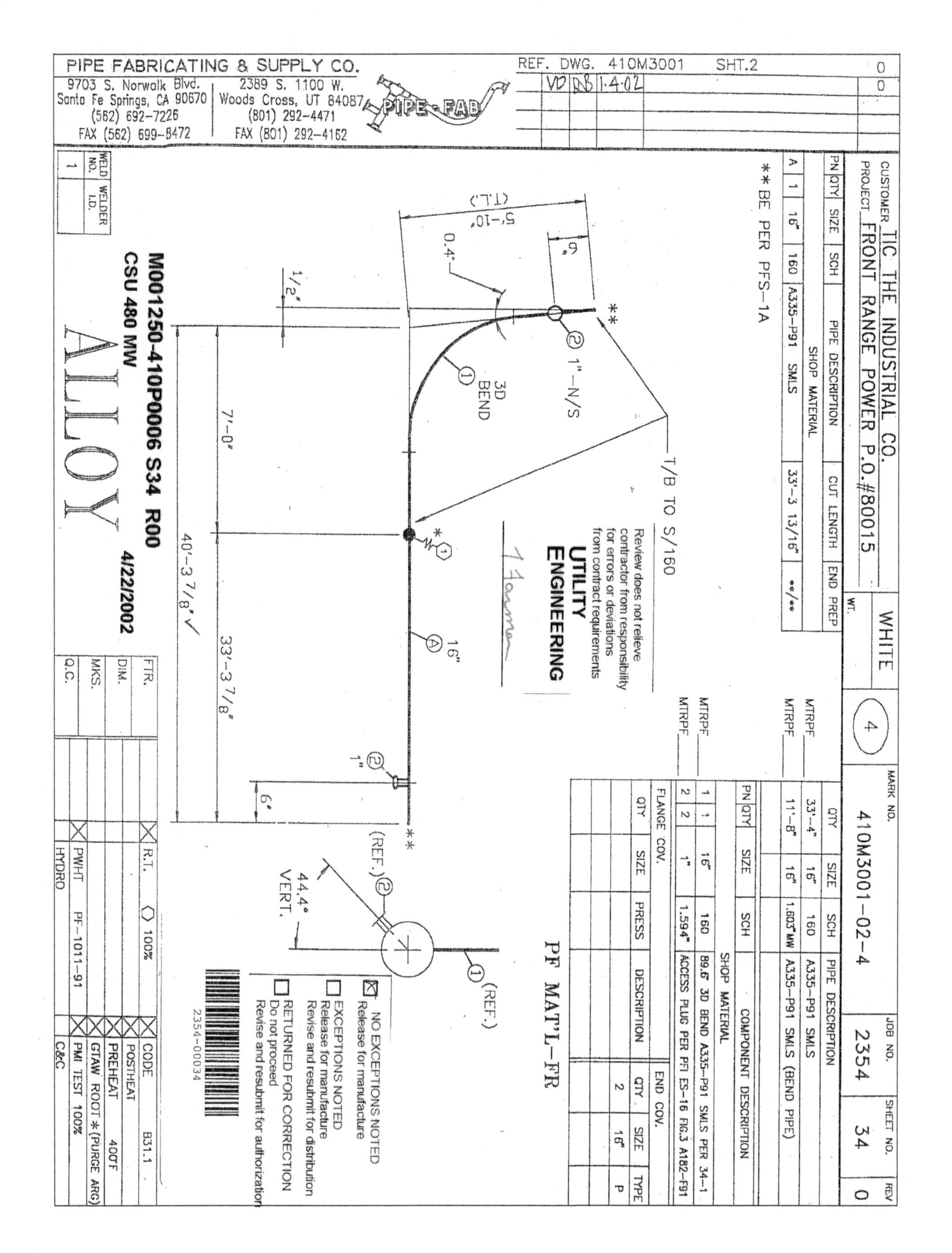

PN	QTY	SIZE	SCH	PIPE DESCRIPTION	CUT LENGTH	END PREP
A	1	16"	160	A335-P91 SMLS	33'-3 13/16"	**/**

QTY	SIZE	SCH	PIPE DESCRIPTION
33'-4"	16"	160	A335-P91 SMLS
11'-8"	16"	1.603" MW	A335-P91 SMLS (BEND PIPE)

PN	QTY	SIZE	SCH	COMPONENT DESCRIPTION
1	1	16"	160	89.6° 3D BEND A335-P91 SMLS PER 34-1
2	2	1"	1.594"	ACCESS PLUG PER PFI ES-16 FIG.3 A182-F91

END COV. QTY	SIZE	TYPE
2	16"	P

Figure 27 Spool page 16LKEA-3001 sheet 6.

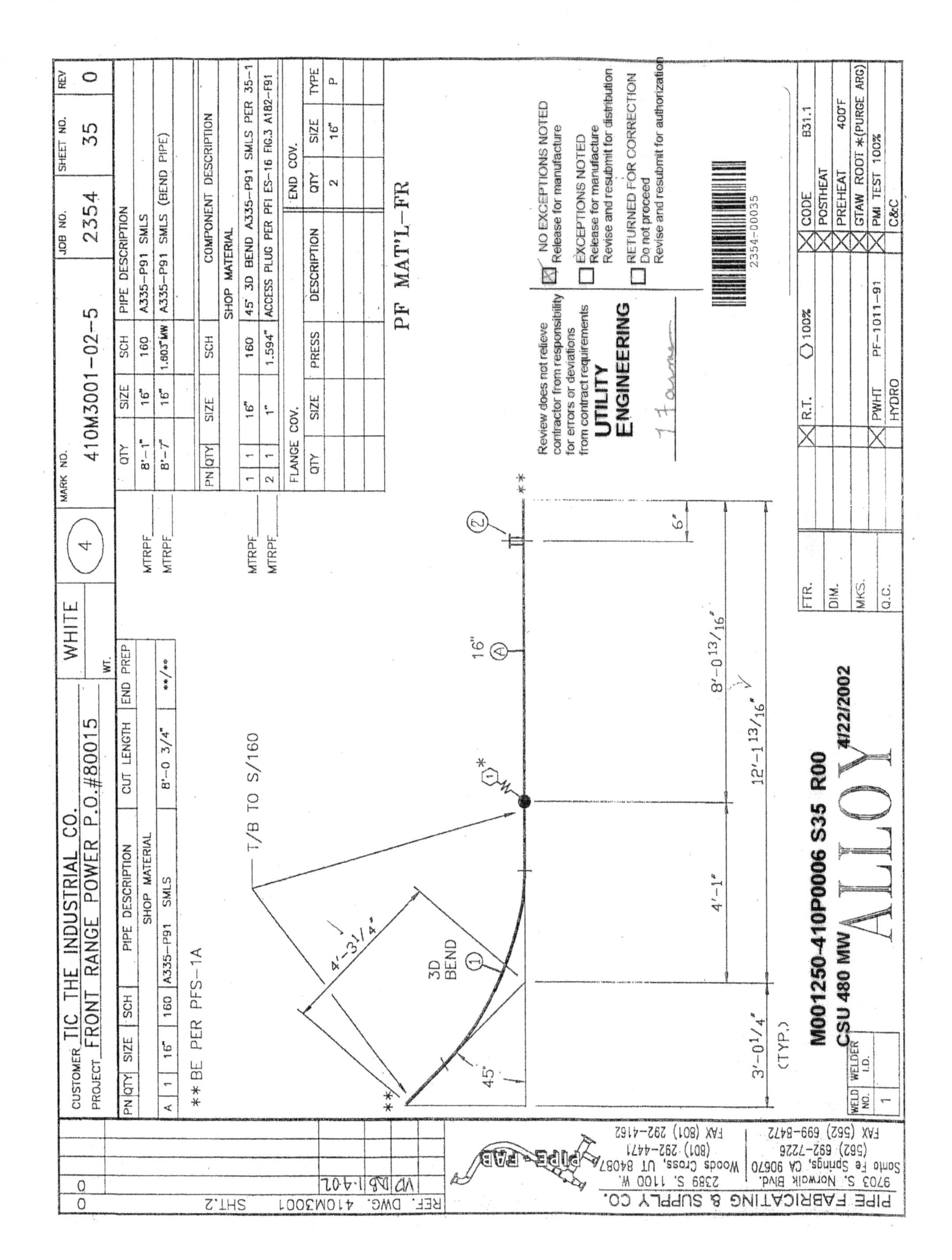

Figure 28 Spool page 16LKEA-3001 sheet 7.

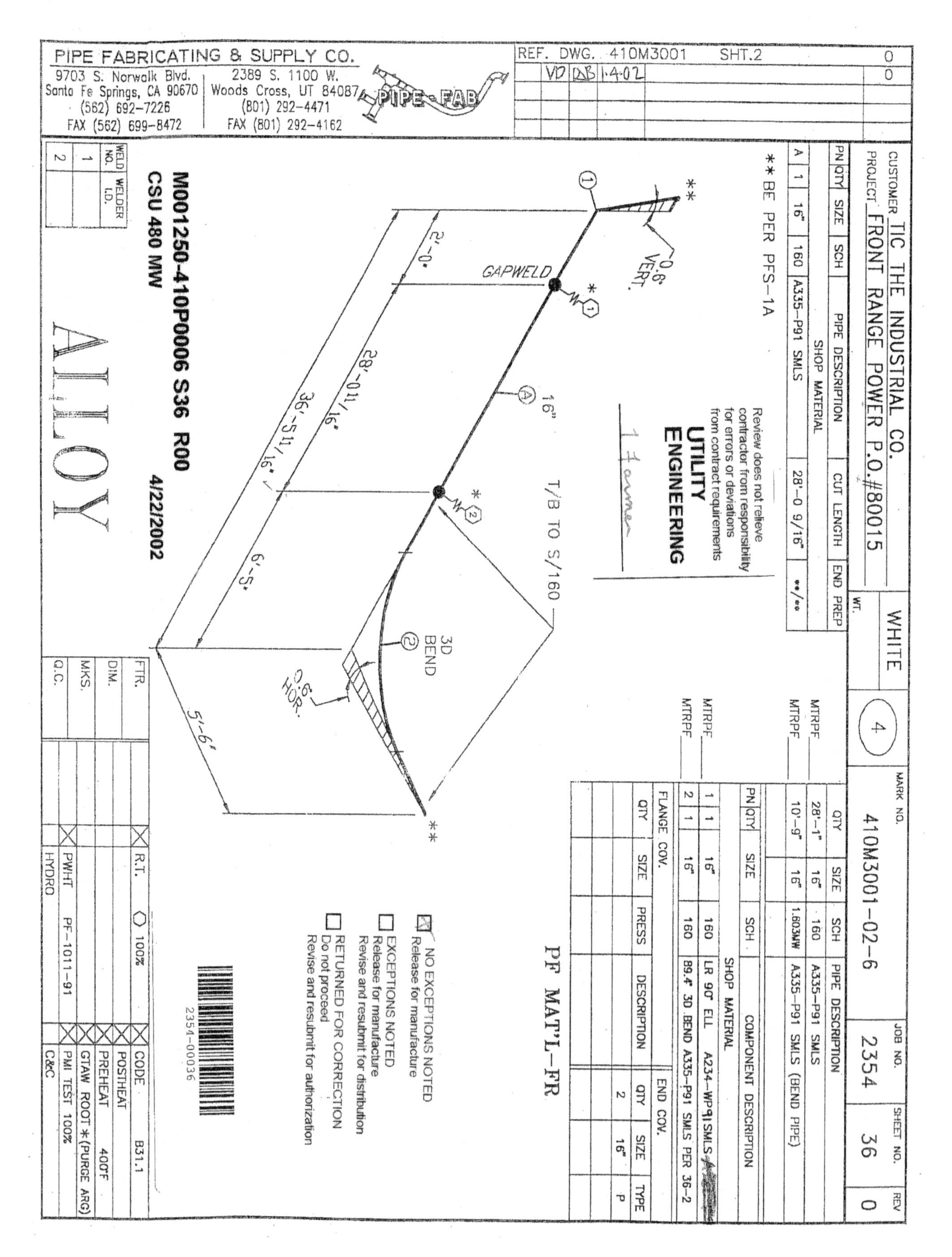

Figure 29 Spool page 16LKEA-3001 sheet 8.

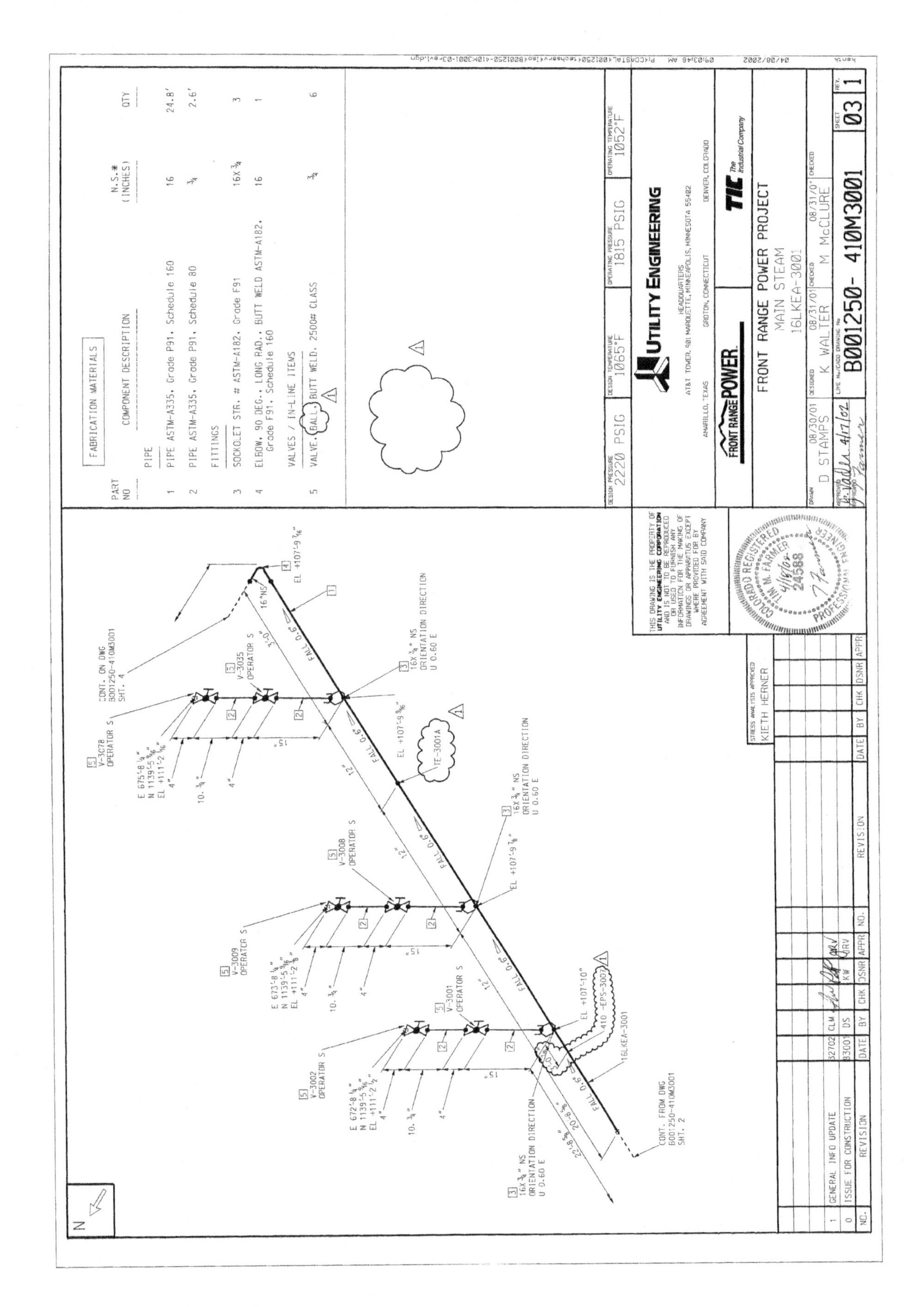

Figure 30 ISO line 16LKEA-3001 sheet 3.

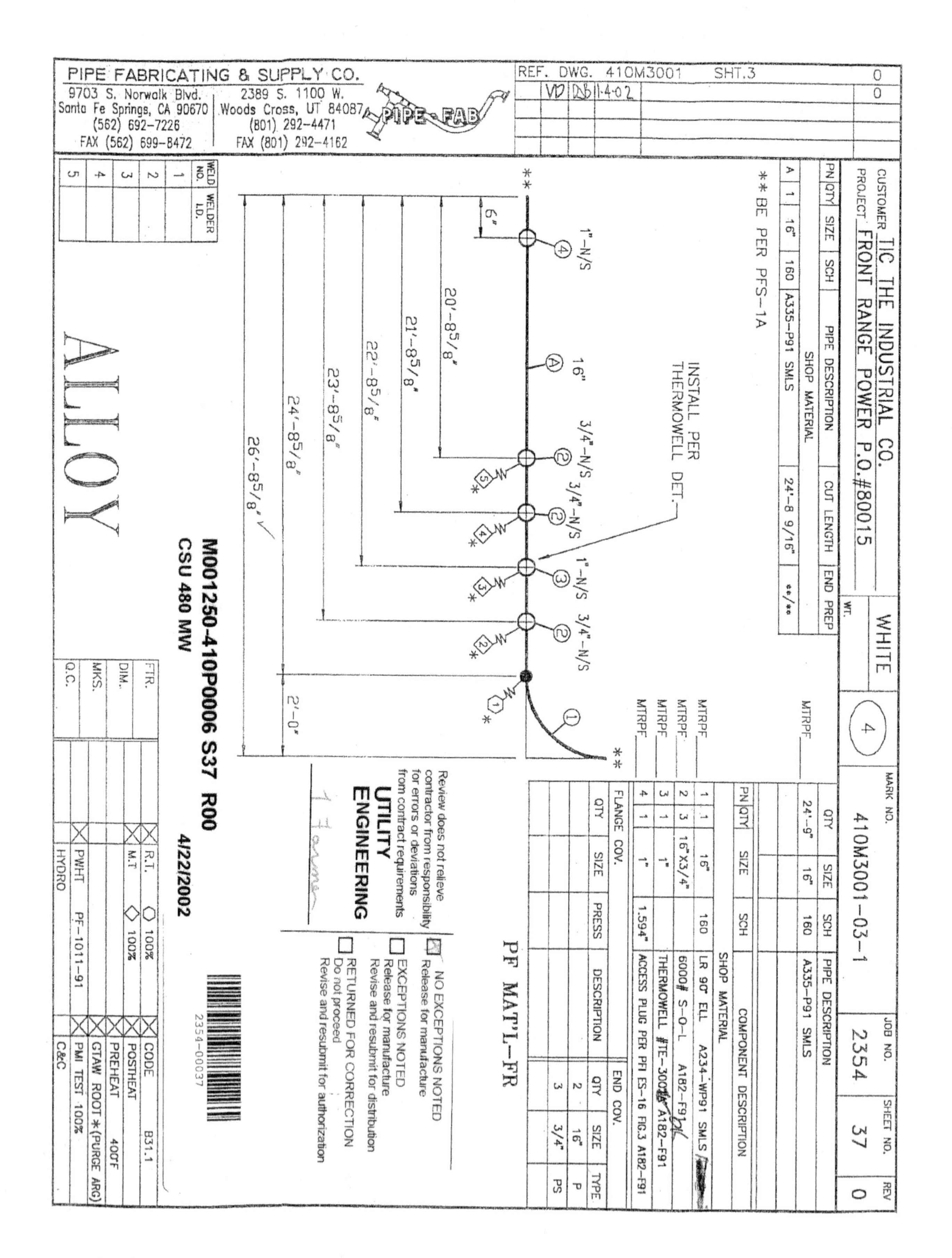

Figure 31 Spool page 16LKEA-3001 sheet 9.

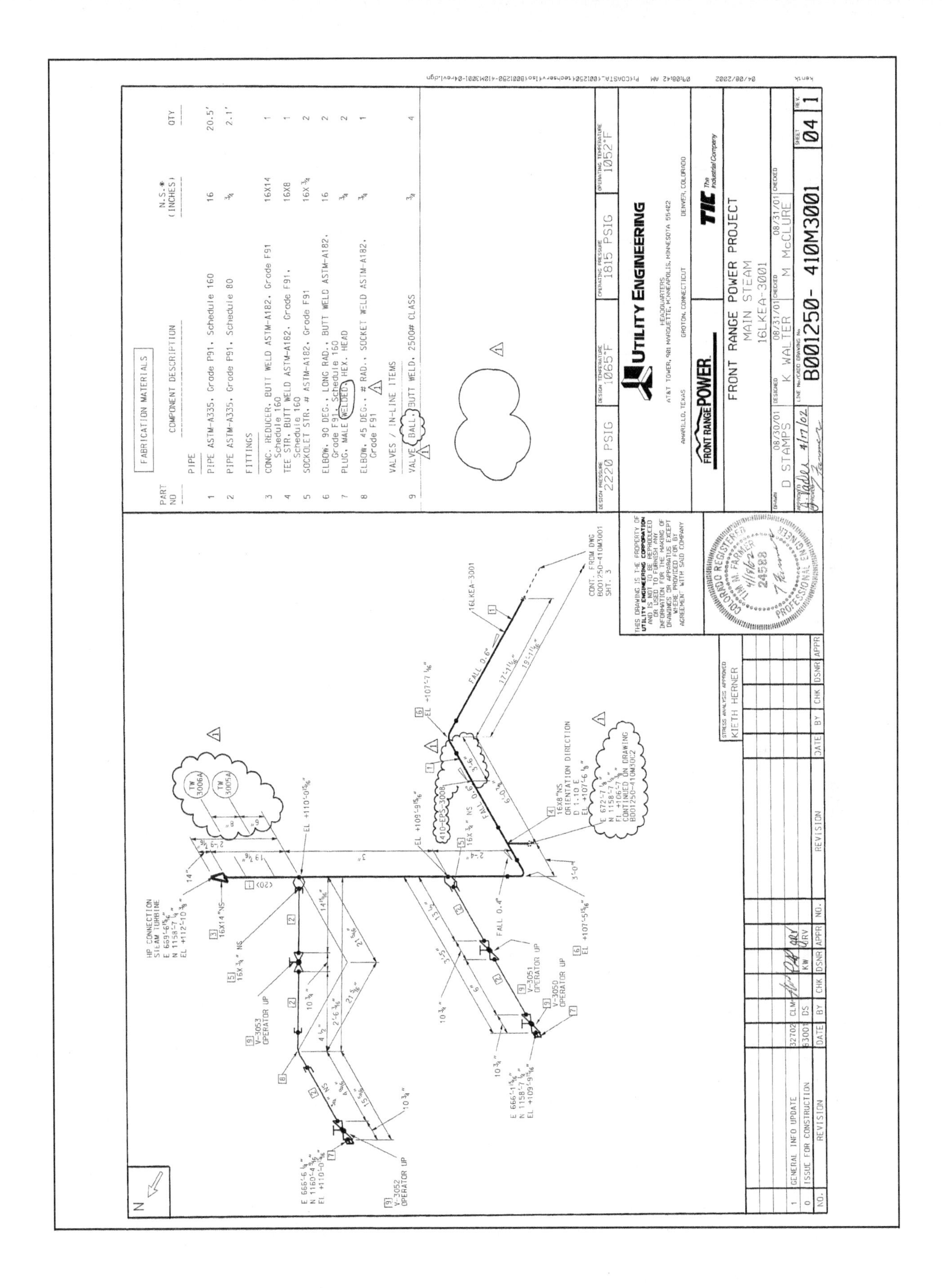

Figure 32 ISO line 16LKEA-3001 sheet 4.

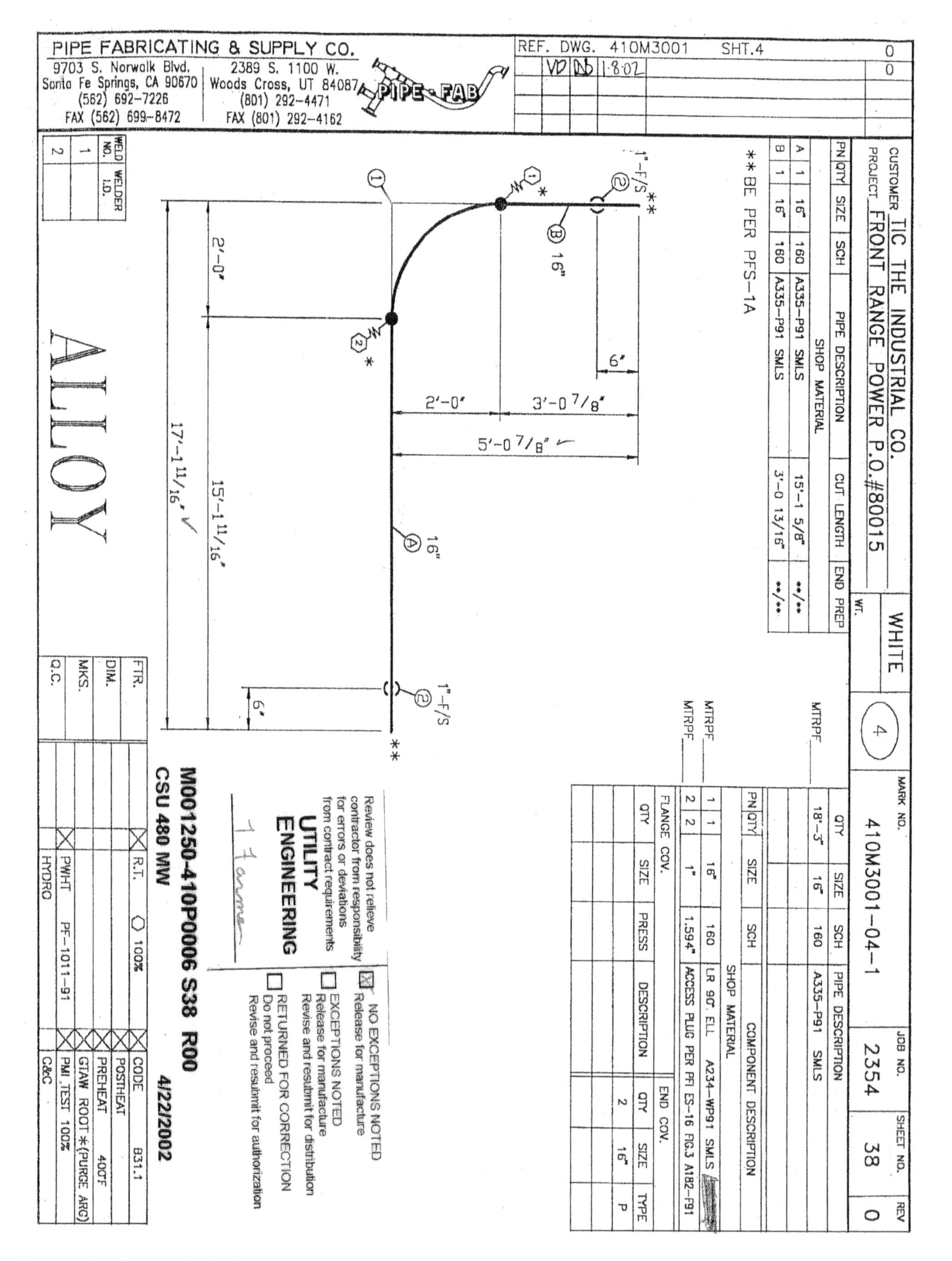

PN	QTY	SIZE	SCH	PIPE DESCRIPTION	CUT LENGTH	END PREP
				SHOP MATERIAL		
A	1	16"	160	A335-P91 SMLS	15'-1 5/8"	**/**
B	1	16"	160	A335-P91 SMLS	3'-0 13/16"	**/**

QTY	SIZE	SCH	PIPE DESCRIPTION
18'-3"	16"	160	A335-P91 SMLS

PN	QTY	SIZE	SCH	COMPONENT DESCRIPTION
				SHOP MATERIAL
1	1	16"	160	LR 90° ELL A234-WP91 SMLS
2	2	1"	1.594"	ACCESS PLUG PER PFI ES-16 FIG.3 A182-F91

WELD NO.	WELDER I.D.
1	
2	

Figure 33 Spool page 16LKEA-3001 sheet 10.

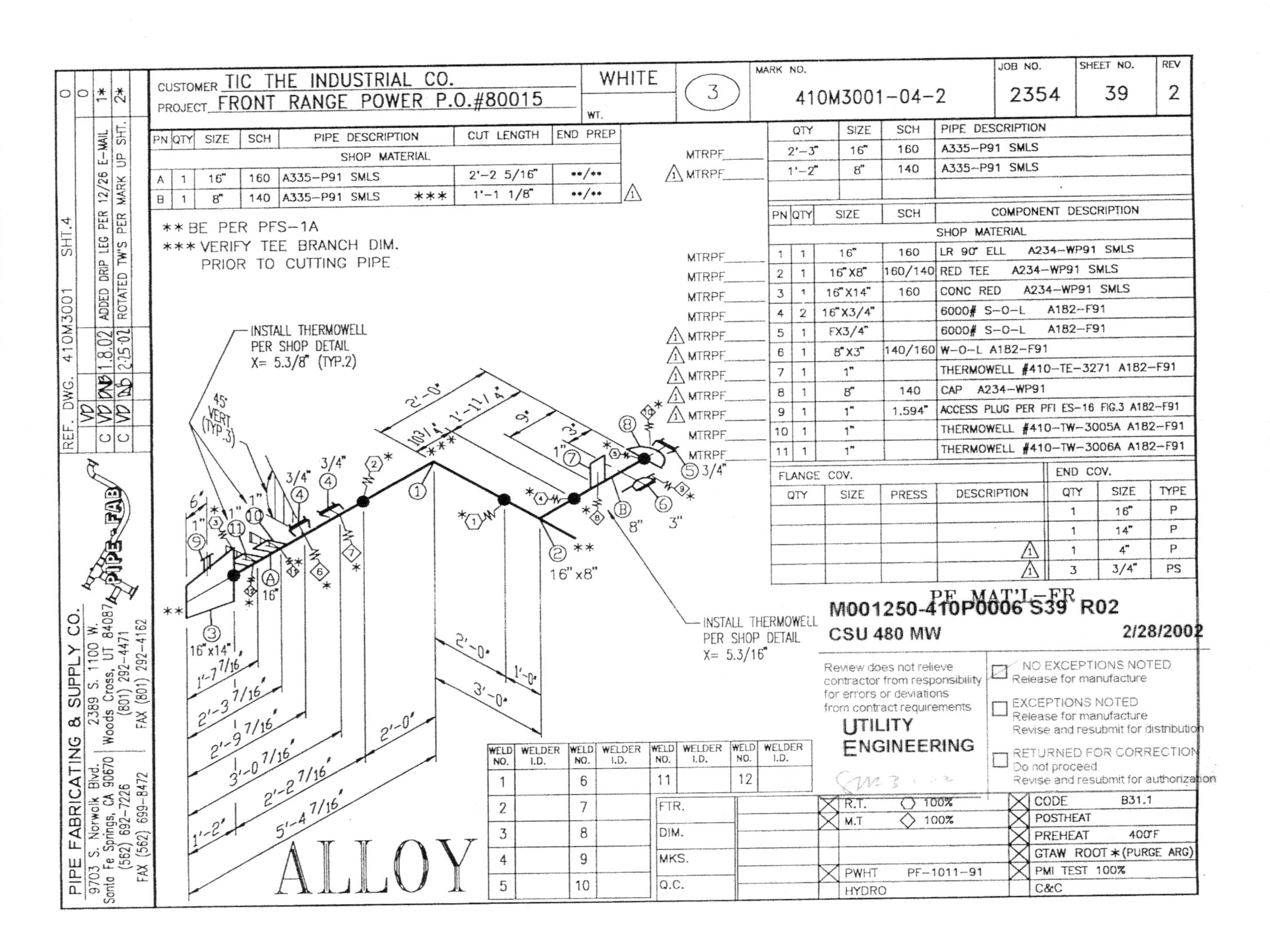

Figure 34 Spool page 16LKEA-3001 sheet 11.

3.3.3 Drawing ISOs

One of the best ways to understand how ISOs are drawn from plan views is to draw one. The ability to make rough isometric sketches in the field is a useful skill that all pipefitters should develop. ISOs are drawn on special paper that has vertical lines and lines projected at 30 degrees.

Before drawing the ISO, try to have a clear understanding of the plan view. Check all dimensions, direction of flow, how far the line drops, its orientation, and how it is organized. Pay close attention to how the ISO is dimensioned. Draw the dimensions accurately and avoid cluttering the ISO with all the information. Understand all the elevations and whether they are Bottom of Pipe (BOP) or center line.

NOTE

Drawing rough elevation and isometric sketches on scrap paper prior to drawing the ISO is a good way to visualize the line.

Follow the steps below to draw an ISO using *Figure 35* as a reference.

Step 1 Find line 221-12LDBA on the plan view.

Step 2 Determine whether the plan shows elevation. In this case, the plan shows 20 vertical feet of elevation, so it requires an elevation drawing to follow the specific lines.

Step 3 Start the ISO by drawing the north arrow first. Orient the line on the ISO exactly as the plan view shows it. Match lines, the line number, and the direction of flow.

Step 4 Draw up a complete bill of material for the ISO by calculating the cut lengths of straight pipe required for each straight section, adding them together, and rounding up to the nearest foot. Be sure to include the following information in the bill of material:

- ISO number
- Line number
- Reference number
- Your name in a box entitled, "Drawn By"

The ISO should enable anyone with a knowledge of piping to fabricate and erect the line.

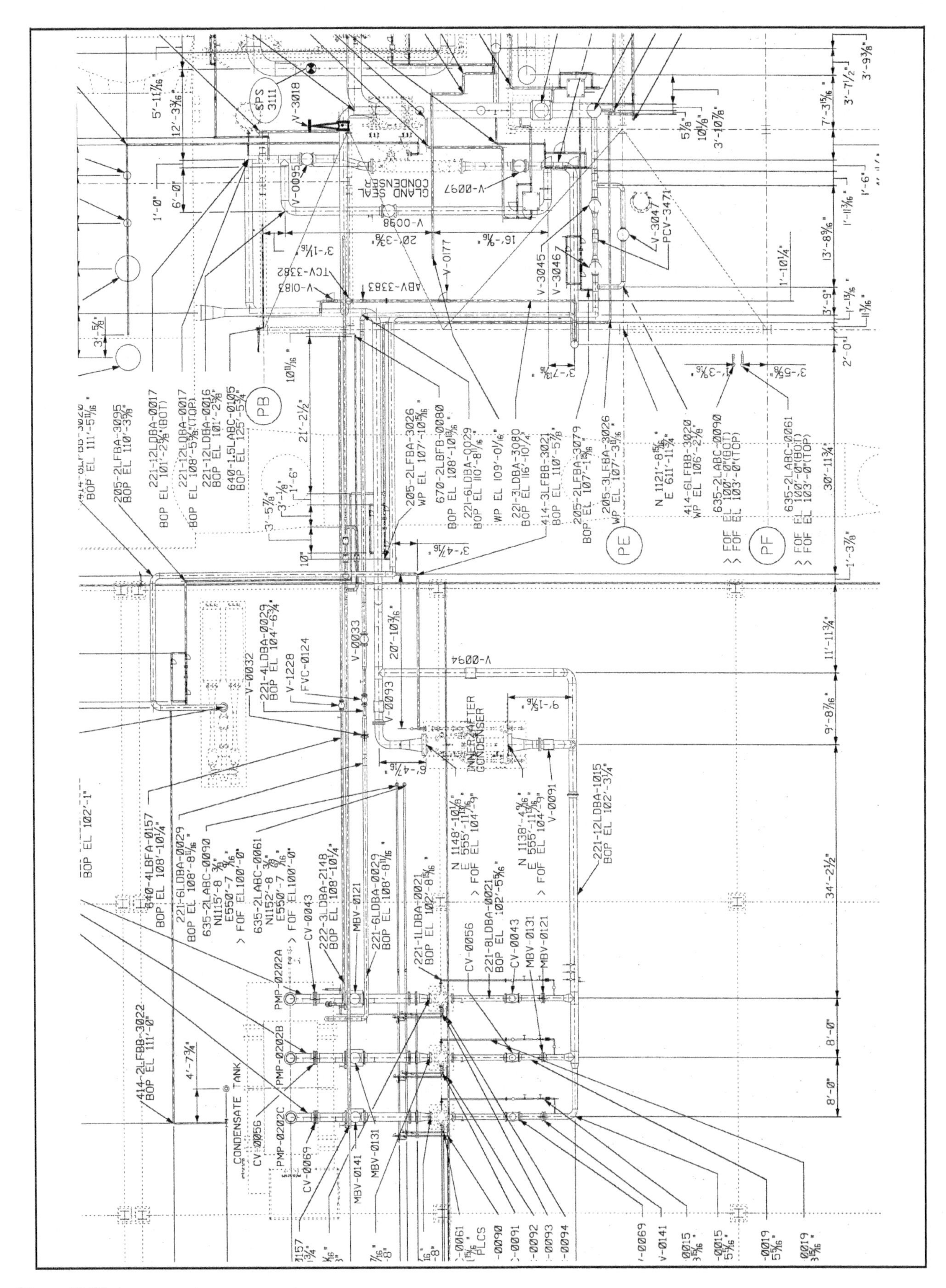

Figure 35 Piping general arrangement.

3.0.0 Section Review

1. A section view is the result of an imaginary line cut through the system. What aspects of the equipment can be seen from this drawing?
   a. Direction, elevation, and size
   b. Number, name, and orientation
   c. Height, width, and location
   d. Elevation, dimension, and direction

2. The biggest advantage that ISOs have over other types of drawings is _____.
   a. clarity
   b. detail
   c. color
   d. length

3. When pulling an ISO from a plan view or general arrangement drawing, start from a clear break in the piping and proceed to _____.
   a. the first bend
   b. another clear break
   c. the first component
   d. the next line

Review Questions

1. What features of a system do P&IDs show?
   a. Piping components
   b. Details
   c. General geographical arrangement
   d. Contents
2. The initials HT stand for _____.
   a. hot tap
   b. heat traced
   c. heating terminal
   d. hot water treatment
3. P&IDs provide information about each pipeline through the use of lines, line numbers, and _____.
   a. detail drawings
   b. coordinates
   c. match lines
   d. elevations
4. When a system is shown on more than one sheet of a P&ID, the continuation is shown in a _____.
   a. legend
   b. cloud
   c. note
   d. match line arrow
5. To show the internal arrangement of an object, engineers use a(n) _____.
   a. plan view
   b. section view
   c. elevation
   d. X-ray view
6. Piping arrangement drawings show location of pipelines by using _____.
   a. coordinates, elevations, and control points
   b. P&IDs
   c. spool drawings
   d. plan views
7. Coordinates are a combination of _____.
   a. columns and floors
   b. numbers and letters
   c. pipe tags and labels
   d. words and numbers

Questions 8–13 refer to *Figure RQ01* on the following page.

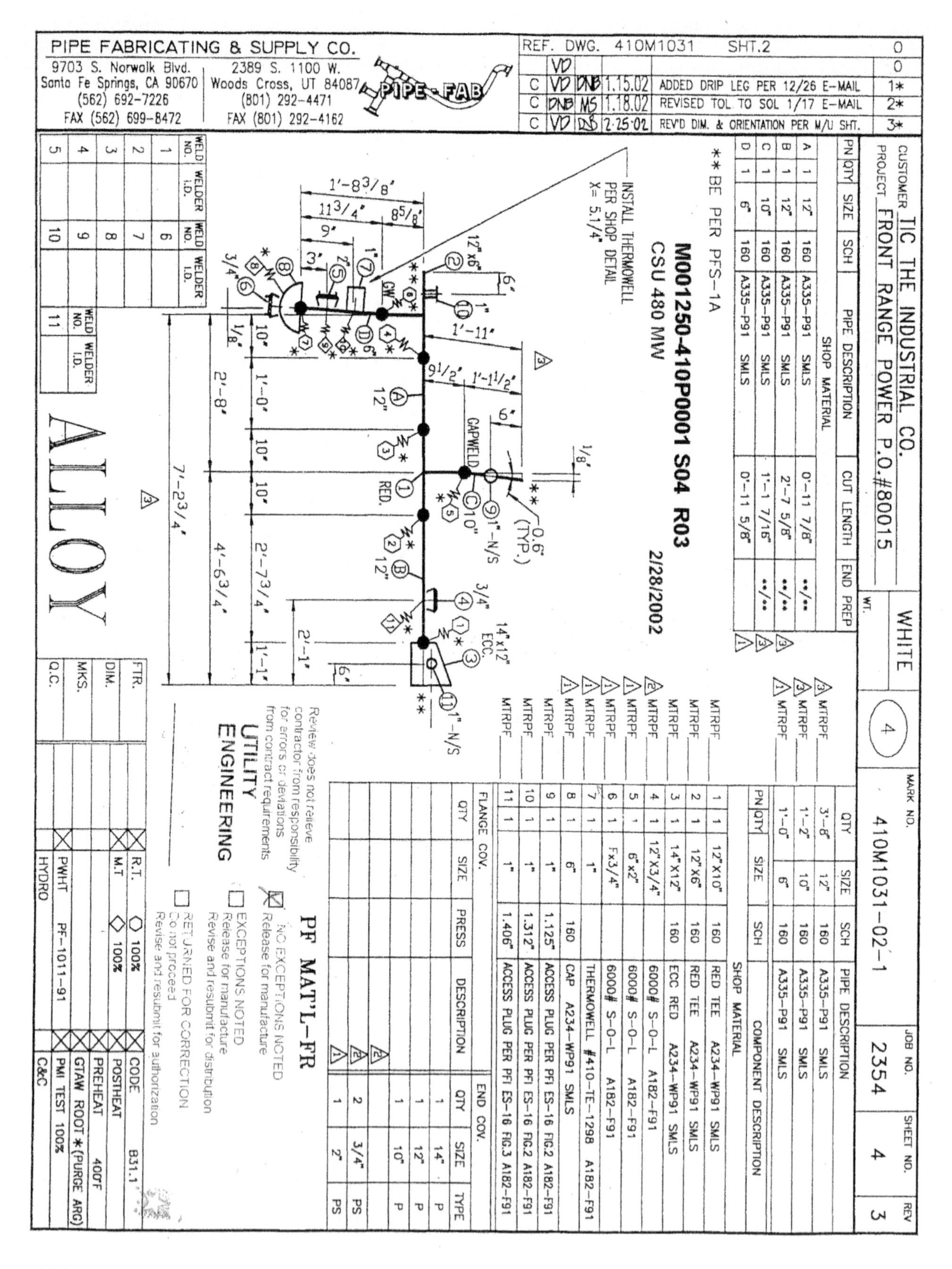

PN	QTY	SIZE	SCH	PIPE DESCRIPTION	CUT LENGTH	END PREP
A	1	12"	160	A335-P91 SMLS	0'-11 7/8"	**/**
B	1	12"	160	A335-P91 SMLS	2'-7 5/8"	**/**
C	1	10"	160	A335-P91 SMLS	1'-1 7/16"	**/**
D	1	6"	160	A335-P91 SMLS	0'-11 5/8"	**/**

QTY	SIZE	SCH	PIPE DESCRIPTION
3'-8"	12"	160	A335-P91 SMLS
1'-2"	10"	160	A335-P91 SMLS
1'-0"	6"	160	A335-P91 SMLS

PN	QTY	SIZE	SCH	COMPONENT DESCRIPTION
1	1	12"x10"	160	RED TEE A234-WP91 SMLS
2	1	12"x6"	160	RED TEE A234-WP91 SMLS
3	1	14"x12"	160	ECC RED A234-WP91 SMLS
4	1	12"x3/4"		6000# S-O-L A182-F91
5	1	6"x2"		6000# S-O-L A182-F91
6	1	Fx3/4"		6000# S-O-L A182-F91
7	1	1"		THERMOWELL #410-TE-1298 A182-F91
8	1	6"	160	CAP A234-WP91 SMLS
9	1	1"	1.125"	ACCESS PLUG PER PFI ES-16 FIG.2 A182-F91
10	1	1"	1.312"	ACCESS PLUG PER PFI ES-16 FIG.2 A182-F91
11	1	1"	1.406"	ACCESS PLUG PER PFI ES-16 FIG.3 A182-F91

QTY	SIZE	TYPE
1	14"	P
1	12"	P
1	10"	P
2	3/4"	PS
1	2"	PS

Figure RQ01

8. The main line on the spool in *Figure RQ01* is _____.
   a. 10" pipe
   b. 12" pipe
   c. 14" pipe
   d. 16" pipe

9. How many radiologically tested welds are there on the spool in *Figure RQ01*?
   a. 5
   b. 6
   c. 7
   d. 11

10. The reducing branches in *Figure RQ01* are _____.
    a. stainless steel pipe
    b. 6" pipe
    c. 6" and 10" pipe
    d. 8" pipe

11. Part number 4 on the drawing in *Figure RQ01* is a _____.
    a. ¾ tee
    b. sensor
    c. weld
    d. 12 × ¾ sockolet

12. Part number 2 on the drawing in *Figure RQ01* is a _____.
    a. 1" access plug
    b. 1" sockolet
    c. 12" pipe
    d. 12 × 6 reducing tee

13. Part number 7 on the drawing in *Figure RQ01* is a _____.
    a. 1" sockolet
    b. 1" thermowell
    c. 6" cap
    d. 6" tee

Questions 14–20 refer to *Figure RQ02* on the following page.

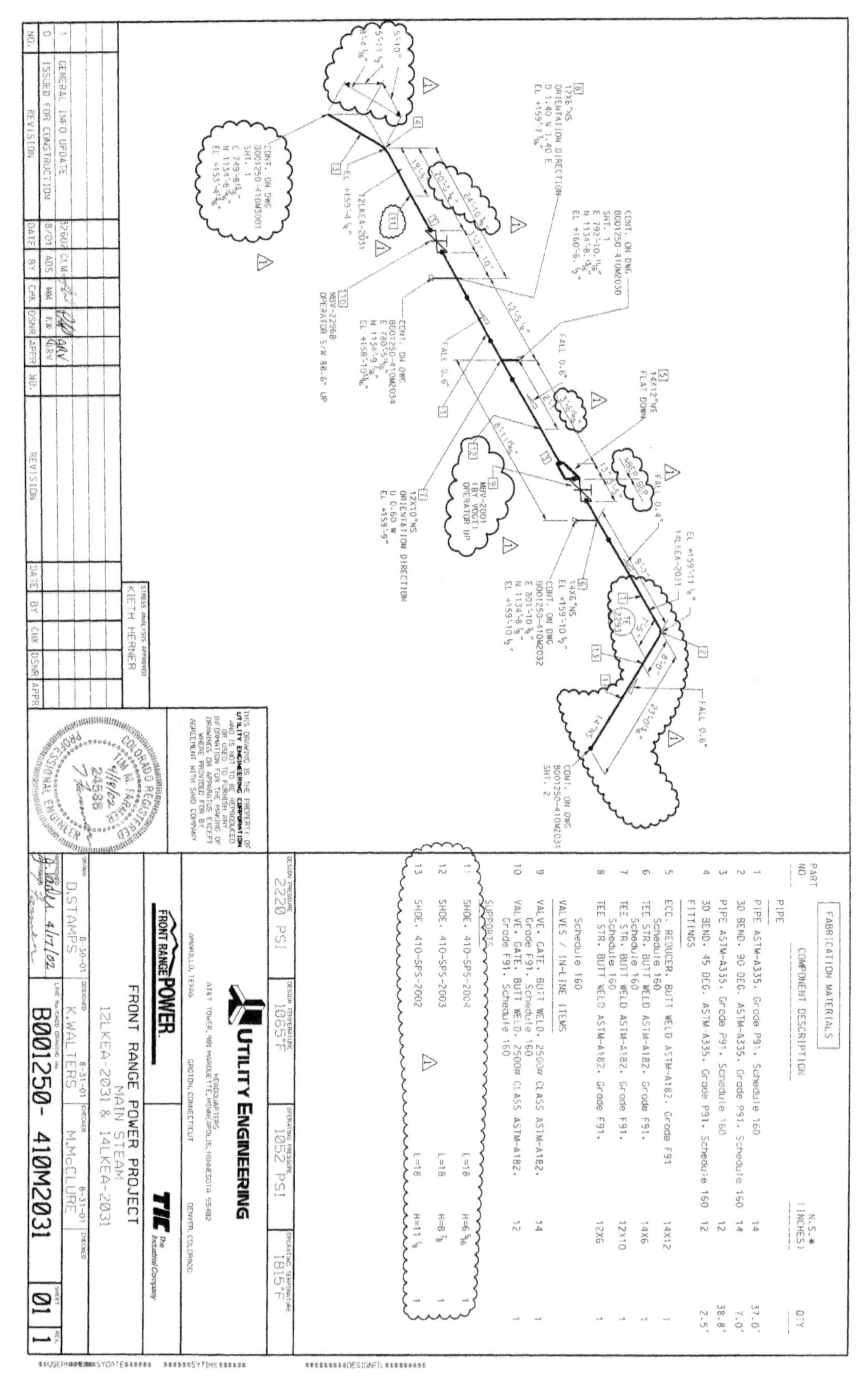
FABRICATION MATERIALS
PART NO
COMPONENT DESCRIPTION
N.S.* (INCHES)
QTY
PIPE
1 PIPE ASTM-A335. Grade P91. Schedule 160 14 37.0'
2 30 BEND. 90 DEG. ASTM-A335. Grade P91. Schedule 160 14 7.0'
3 PIPE ASTM-A335. Grade P91. Schedule 160 12 38.8'
4 3D BEND. 45 DEG. ASTM-A335. Grade P91. Schedule 160 12 2.5'
FITTINGS
5 ECC. REDUCER. BUTT WELD ASTM-A182. Grade F91 Schedule 160 14X12 1
6 TEE STR. BUTT WELD ASTM-A182. Grade F91. Schedule 160 14X6 1
7 TEE STR. BUTT WELD ASTM-A182. Grade F91. Schedule 160 12X10 1
8 TEE STR. BUTT WELD ASTM-A182. Grade F91. Schedule 160 12X6 1
VALVES / IN-LINE ITEMS
9 VALVE. GATE. BUTT WELD. 2500# CLASS ASTM-A182. Grade F91. Schedule 160 14 1
10 VALVE. GATE. BUTT WELD. 2500# CLASS ASTM-A182. Grade F91. Schedule 160 12 1
SUPPORTS
11 SHOE. 410-SPS-2004 L=18 1
12 SHOE. 410-SPS-2003 L=18 1
13 SHOE. 410-SPS-2002 L=18 1
UTILITY ENGINEERING
FRONT RANGE POWER
TIC The Industrial Company
FRONT RANGE POWER PROJECT
MAIN STEAM
12LKEA-2031 & 14LKEA-2031
D.STAMPS
K.WALTERS
M.McCLURE
B001250- 410M2031
2220 PSI
1065°F
1052 PSI
1815°F

Figure RQ02

14. How much 14" pipe is used in the spool in *Figure RQ02*?

    a. 7'
    b. 37'
    c. 38.8'
    d. 44'

15. On what drawing is information on the piping run above this found in *Figure RQ02*?

    a. BOO1250-410M3001 Sheet 1
    b. BOO1250-410M2034
    c. BOO1250-410M2031 Sheet 2
    d. BOO1250-410M2032

16. In *Figure RQ02,* the tee immediately following the eccentric reducer is _____.

    a. 12 × 6
    b. 12 × 8
    c. 12 × 10
    d. 12 × 12

17. According to *Figure RQ02,* the longest part of the pipe run at this stage runs _____.

    a. due east
    b. a little north of west
    c. a little south of west
    d. a little south of east

18. In *Figure RQ02,* the valve next to the eccentric reducer is labeled _____.

    a. flat down
    b. 12LKEA-2031
    c. MBV-2001
    d. MBV-2296B

19. In *Figure RQ02,* the distance from the center of the 14" by 6" tee to the face of valve MBV2001 is _____.

    a. 3'5"
    b. 4'6"
    c. 8'11$^{15}/_{16}$"
    d. 11"

20. In *Figure RQ02,* the cut length of the spool from the end of MBV2296B to the weld at the 45-degree bend is _____.

    a. 18'7"
    b. 19'2$^{1}/_{2}$"
    c. 20'5$^{3}/_{4}$"
    d. 24'10$^{3}/_{4}$"

Blueprint One

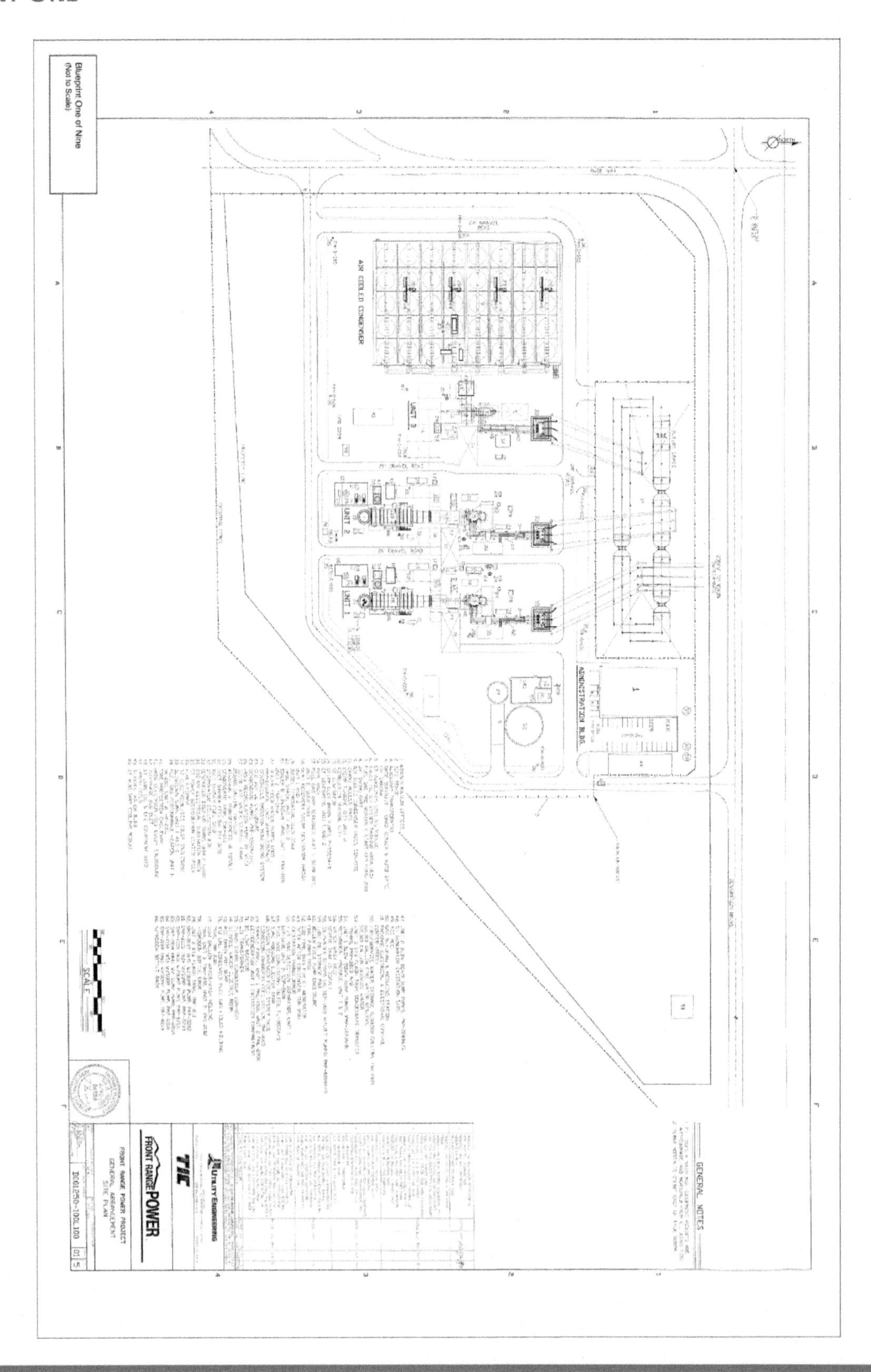

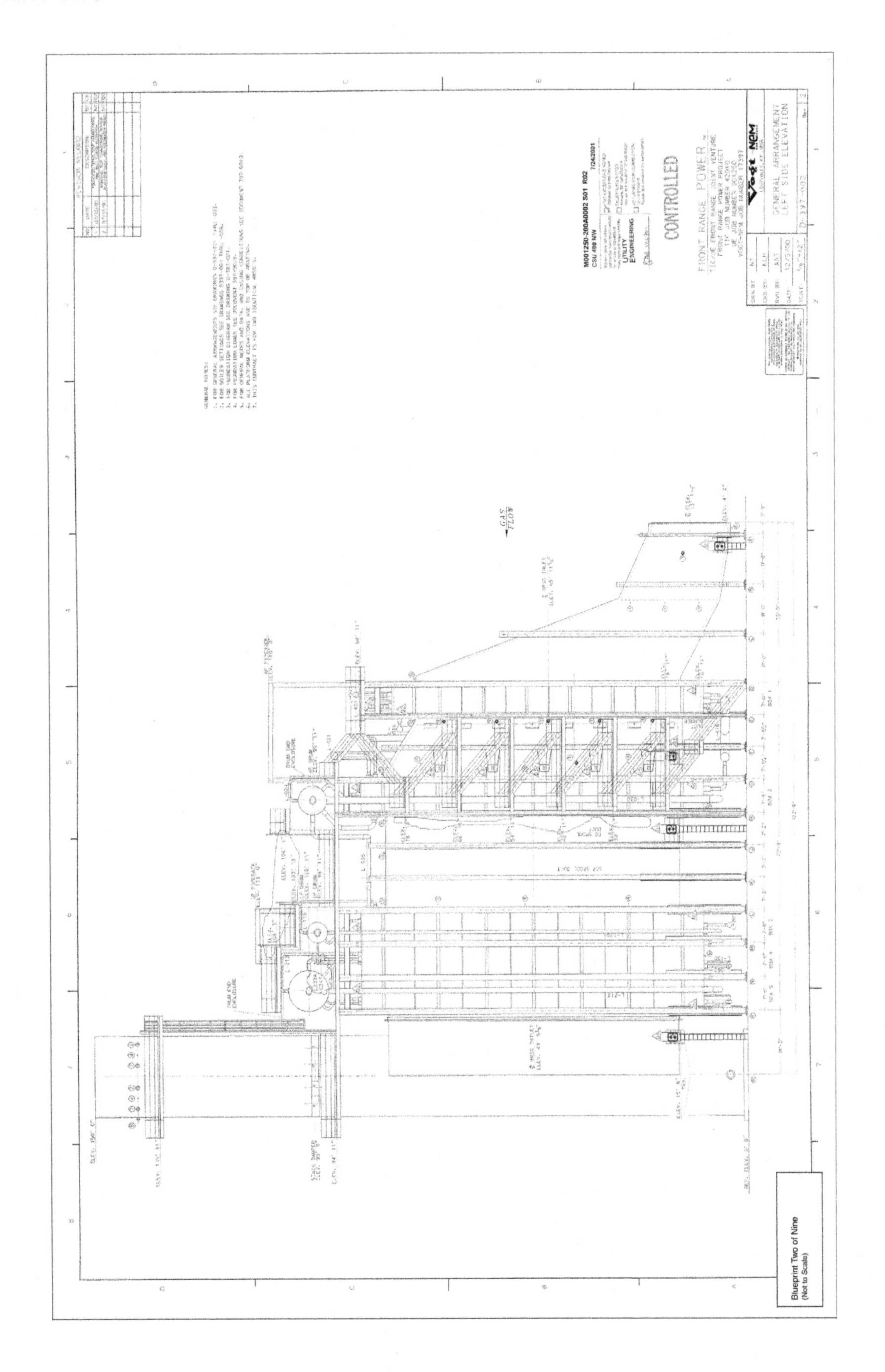

Blueprint Two of Nine
(Not to Scale)

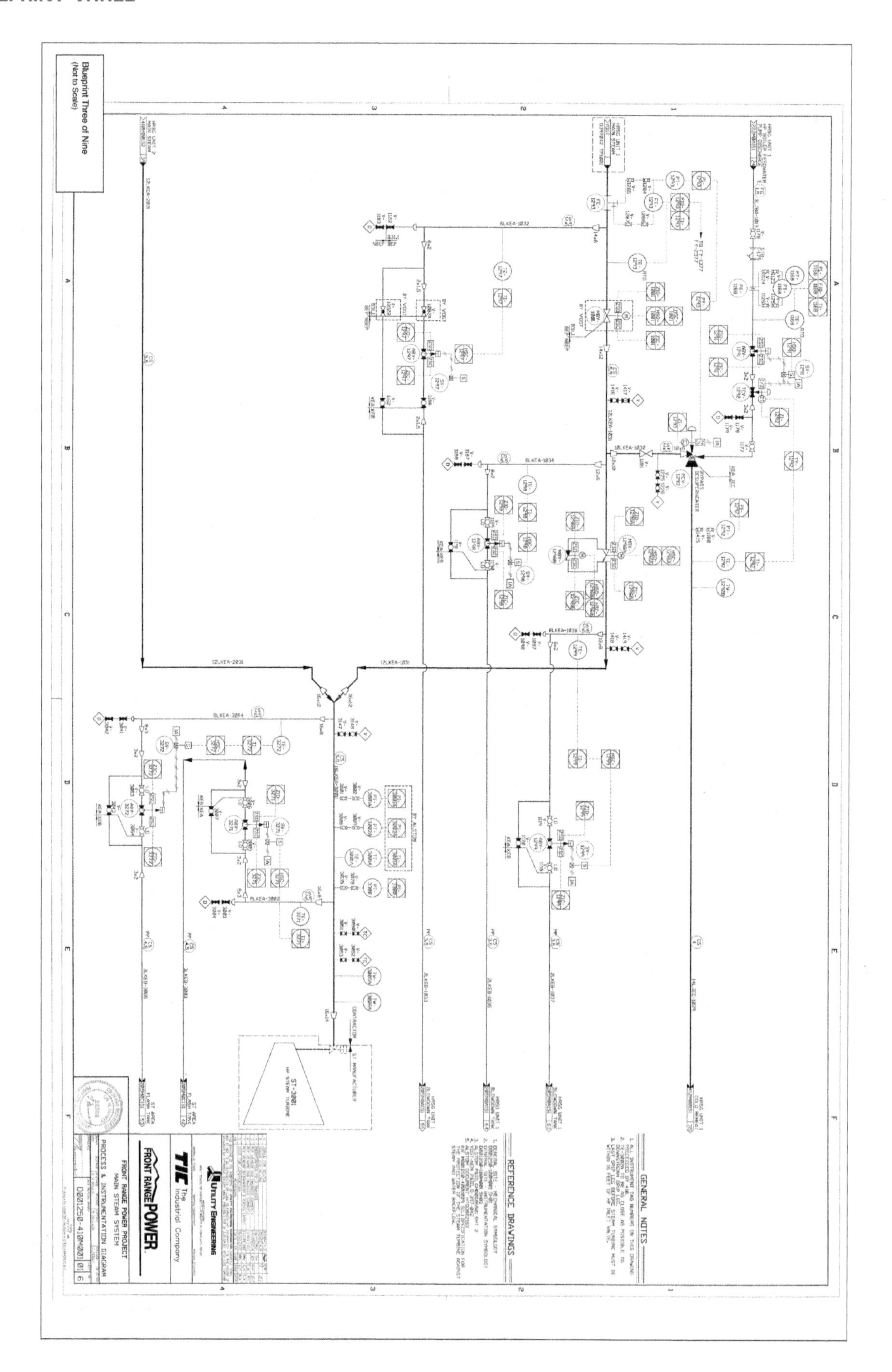

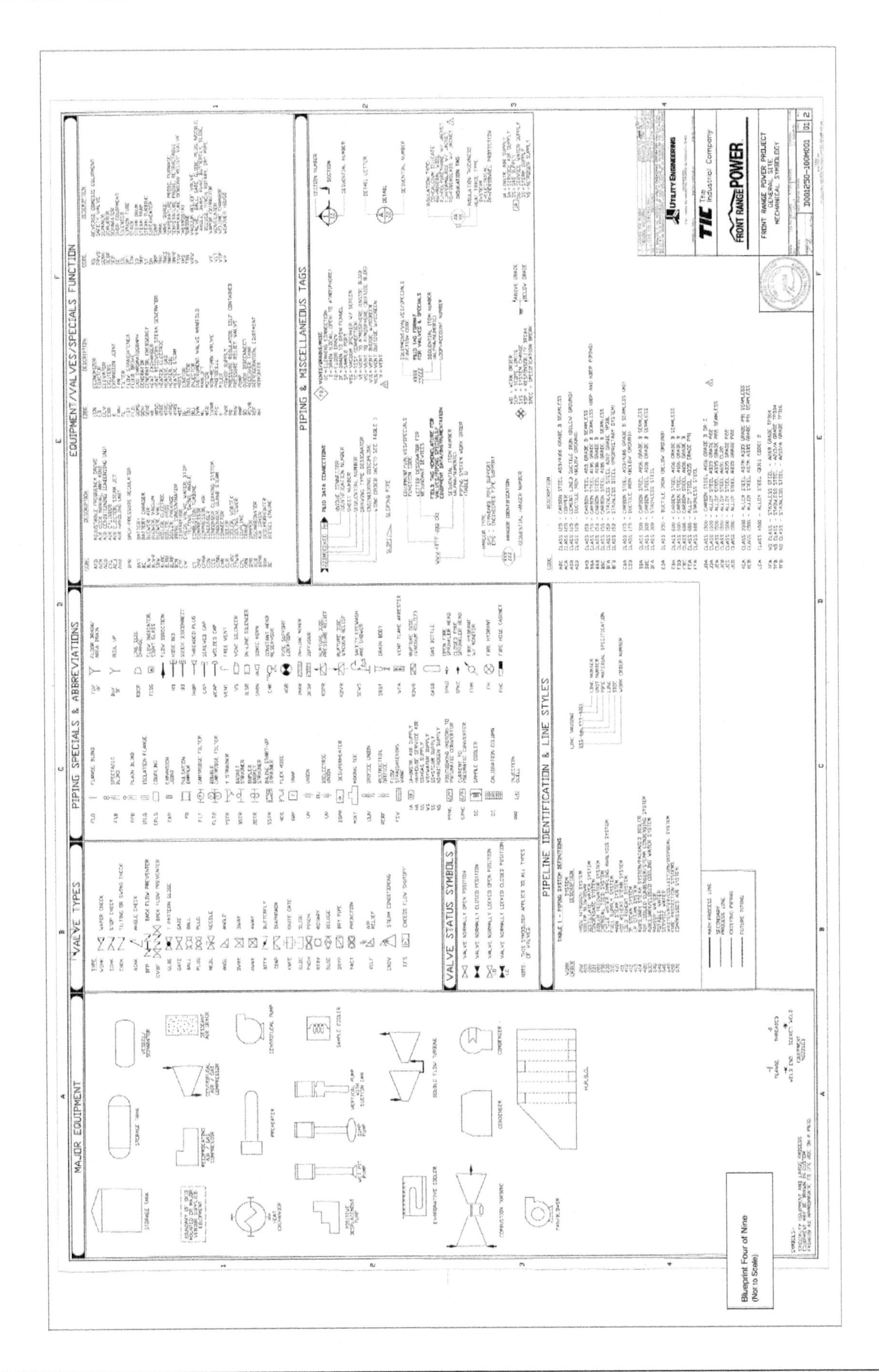
MAJOR EQUIPMENT
VALVE TYPES
PIPING SPECIALS & ABBREVIATIONS
EQUIPMENT/VALVES/SPECIALS FUNCTION
PIPING & MISCELLANEOUS TAGS
VALVE STATUS SYMBOLS
PIPELINE IDENTIFICATION & LINE STYLES
FRONT RANGE POWER
FRONT RANGE POWER PROJECT
GENERAL SITE
MECHANICAL SYMBOLOGY
Blueprint Four of Nine
(Not to Scale)

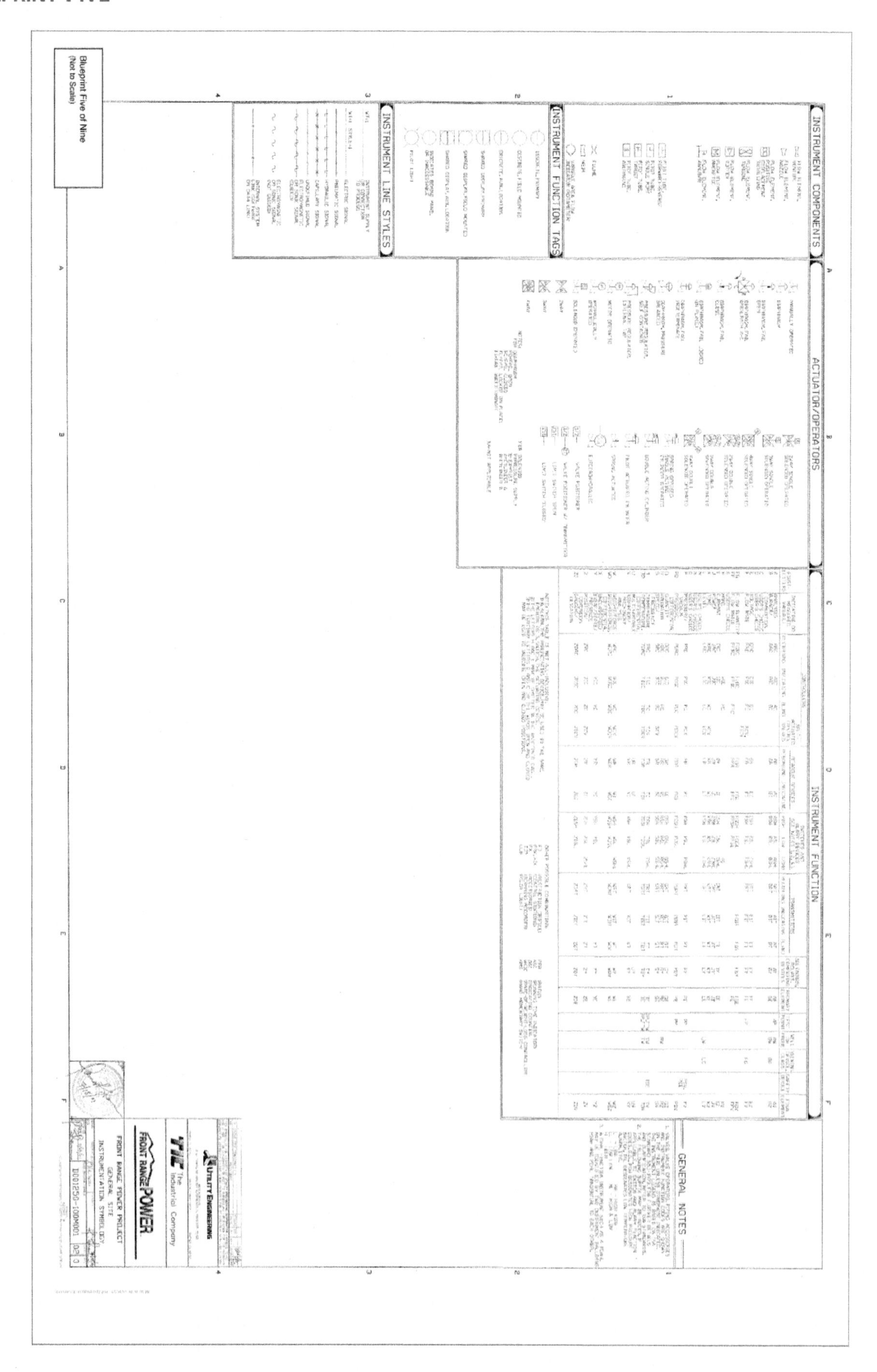
Blueprint Five of Nine
(Not to Scale)
INSTRUMENT COMPONENTS
INSTRUMENT FUNCTION TAGS
INSTRUMENT LINE STYLES
ACTUATOR/OPERATORS
INSTRUMENT FUNCTION
GENERAL NOTES
The Industrial Company
FRONT RANGE POWER

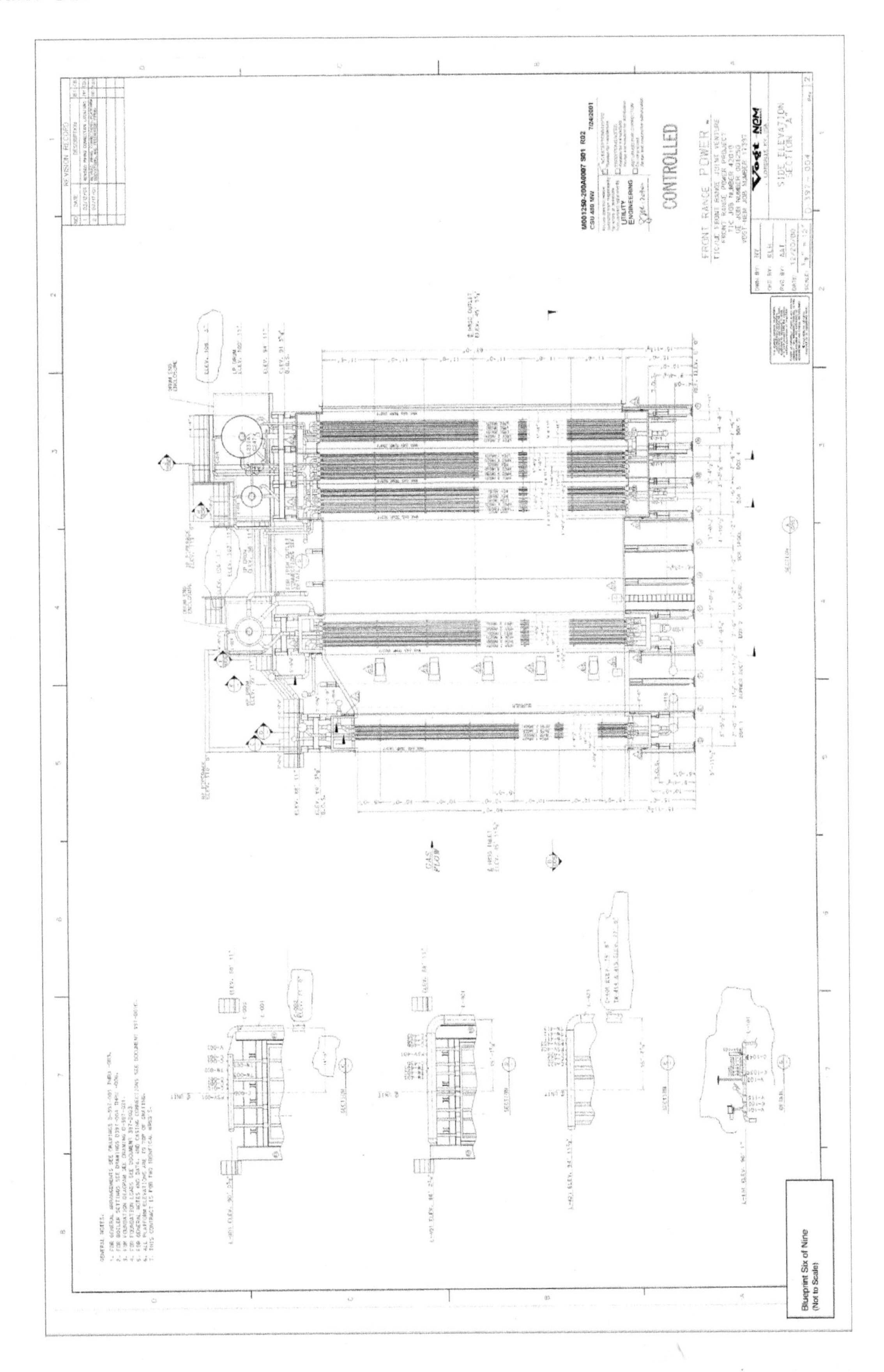
CONTROLLED
FRONT RANGE POWER
SIDE ELEVATION
SECTION "A"
GAS FLOW
Blueprint Six of Nine
(Not to Scale)

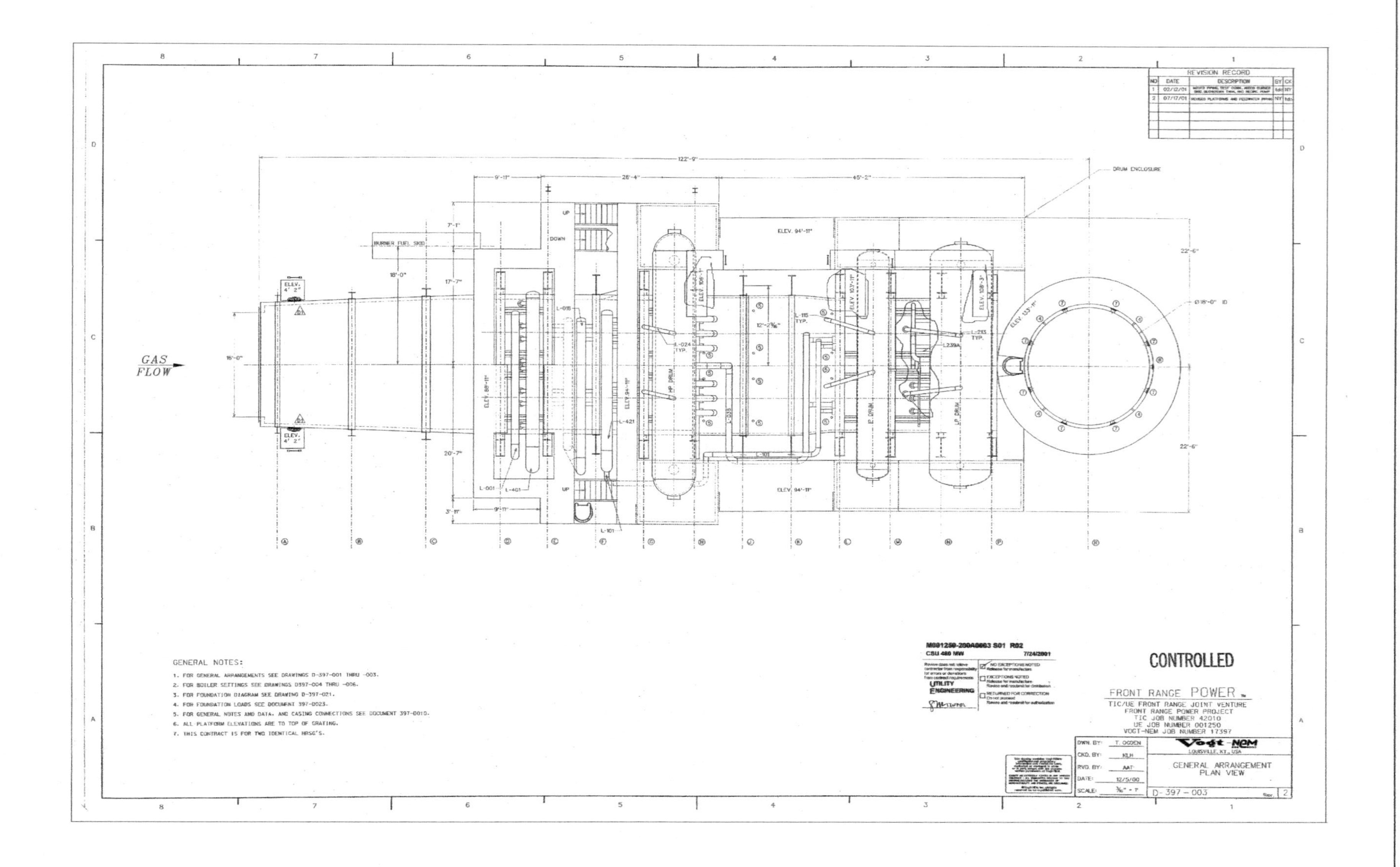
GAS FLOW
GENERAL NOTES:
1. FOR GENERAL ARRANGEMENTS SEE DRAWINGS D-397-001 THRU -003.
2. FOR BOILER SETTINGS SEE DRAWINGS D397-004 THRU -006.
3. FOR FOUNDATION DIAGRAM SEE DRAWING D-397-021.
4. FOR FOUNDATION LOADS SEE DOCUMENT 397-0023.
5. FOR GENERAL NOTES AND DATA, AND CASING CONNECTIONS SEE DOCUMENT 397-0010.
6. ALL PLATFORM ELEVATIONS ARE TO TOP OF GRATING.
7. THIS CONTRACT IS FOR TWO IDENTICAL HRSG'S.
BURNER FUEL SKID
HP DRUM
LP DRUM
DRUM ENCLOSURE
ELEV. 94'-11"
122'-9"
REVISION RECORD
CONTROLLED
FRONT RANGE POWER
TIC/UE FRONT RANGE JOINT VENTURE
FRONT RANGE POWER PROJECT
TIC JOB NUMBER 42010
UE JOB NUMBER 001250
VOGT-NEM JOB NUMBER 17397
LOUISVILLE, KY, USA
GENERAL ARRANGEMENT
PLAN VIEW
D-397-003

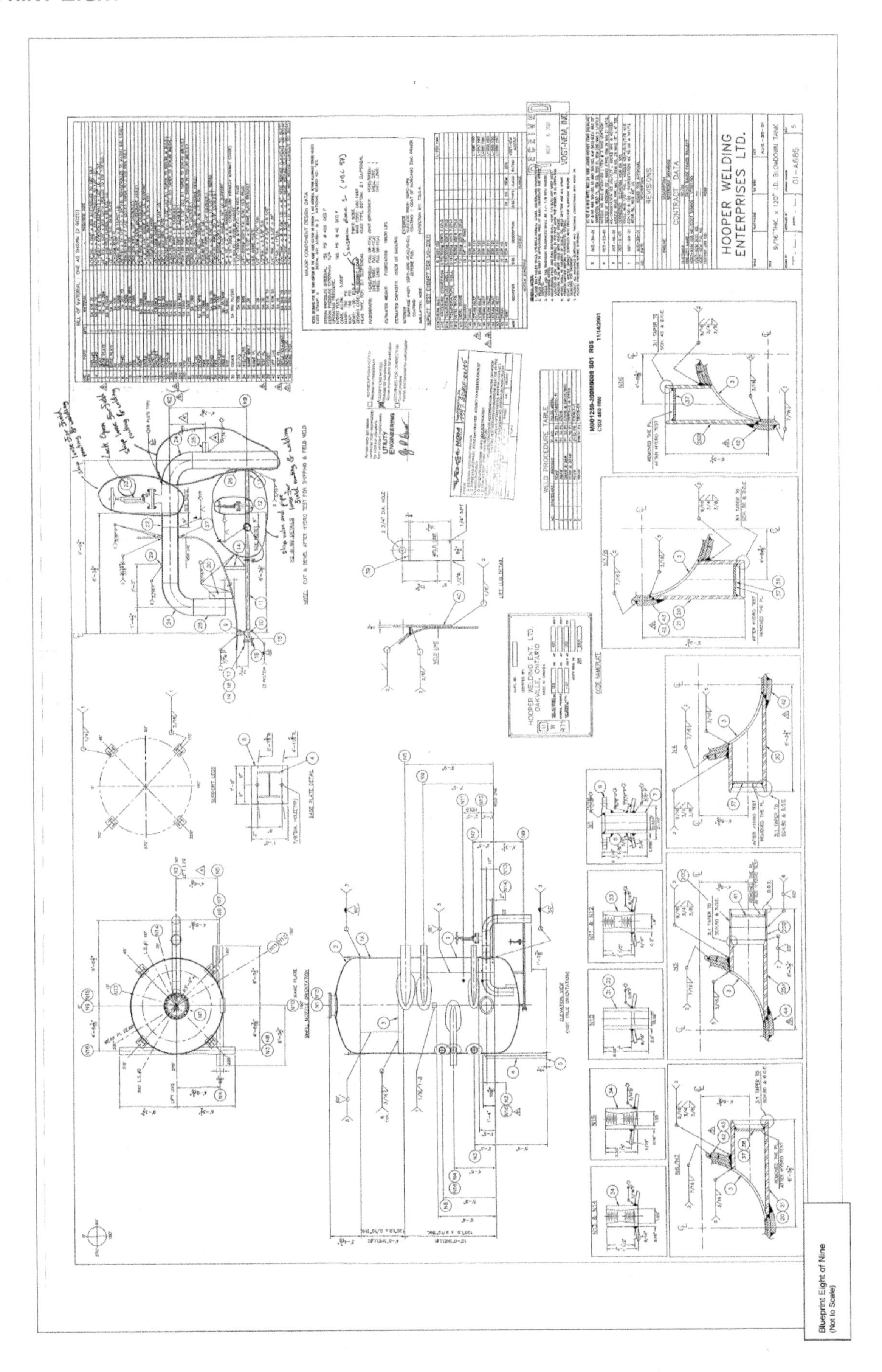

Blueprint Eight of Nine
(Not to Scale)

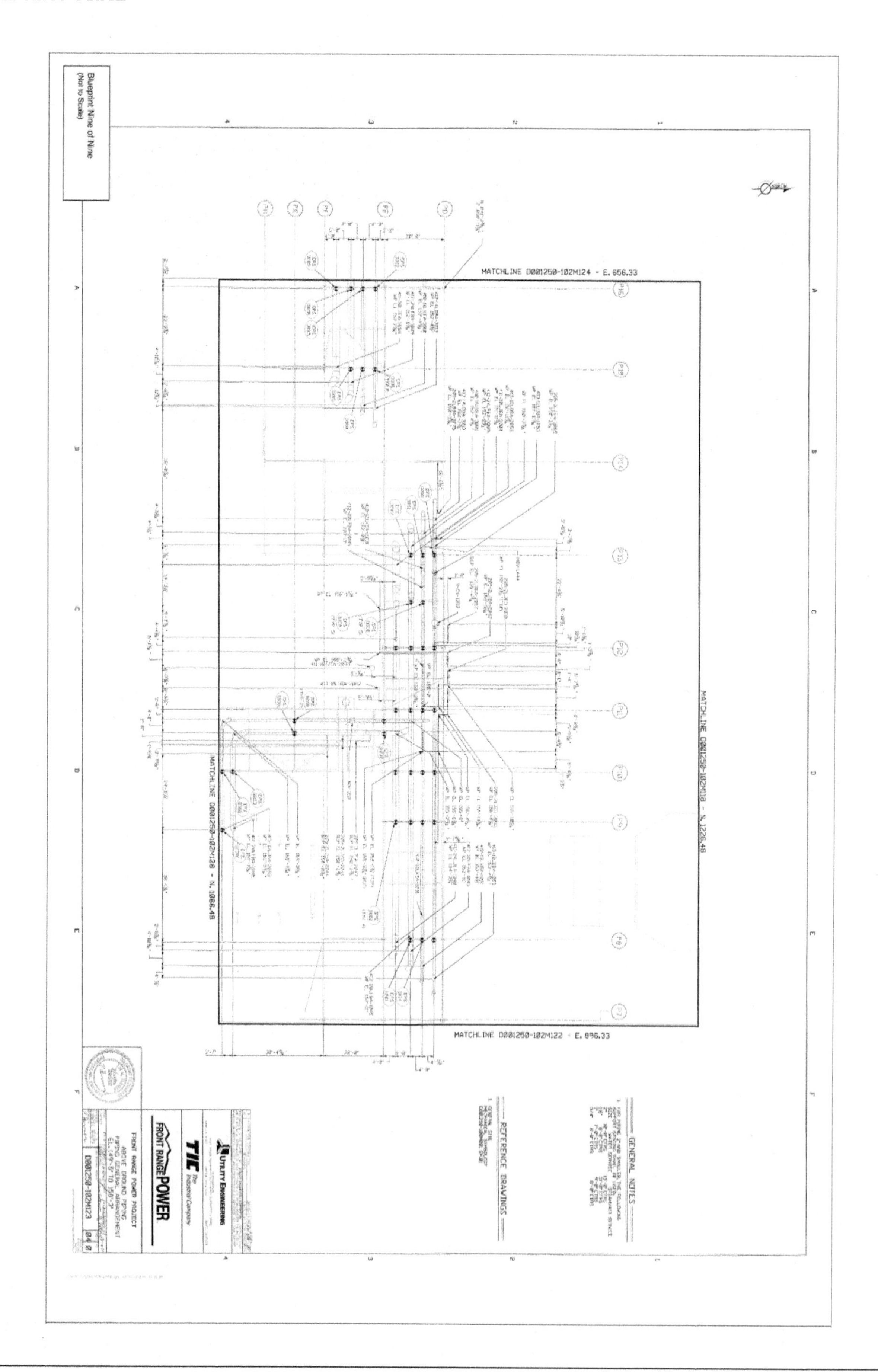
Blueprint Nine of Nine
(Not to Scale)
MATCHLINE D001250-102M124 - E. 656.33
MATCHLINE D001250-102M122 - E. 896.33
FRONT RANGE POWER
GENERAL NOTES
REFERENCE DRAWINGS

Trade Terms Introduced in This Module

Dimension: A measurement between two points on a drawing.

Title block: A section of an engineering drawing blocked off for pertinent information, such as the title, drawing number, date, scale, material, draftsperson, and tolerances.

Additional Resources

This module presents thorough resources for task training. The following reference material is recommended for further study.

Blueprint Reading: Constructions Drawings for the Building Trade. Sam A. A. Kubba, Ph.D. 2009. The McGraw-Hill Companies, Inc.

How to Interpret Piping and Instrumentation Diagrams. 2010. New York, NY: American Institute of Chemical Engineers (AIChE).

Process Piping Drafting, Weaver, Rip; Gulf Publishing Company, Book Division, Houston, TX, 1986.

Figure Credits

Utility Engineering Corp. a Zachry Group Company, Figures 4-7; 11-35; Blueprints One through Eight of *Appendix*

Section Review Answer Key

Section 1.0.0

Answer	Section Reference	Objective
1. a	1.1.1	1a
2. b	1.2.0	1b
3. d	1.3.0	1c
4. c	1.4.0	1d

Section 2.0.0

Answer	Section Reference	Objective
1. b	2.1.0	2a
2. d	2.2.0	2b
3. b	2.3.0	2c

Section 3.0.0

Answer	Section Reference	Objective
1. c	3.1.0	3a
2. a	3.2.0	3b
3. b	3.3.1	3c

This page is intentionally left blank.

NCCER CURRICULA — USER UPDATE

NCCER makes every effort to keep its textbooks up-to-date and free of technical errors. We appreciate your help in this process. If you find an error, a typographical mistake, or an inaccuracy in NCCER's curricula, please fill out this form (or a photocopy), or complete the online form at **www.nccer.org/olf**. Be sure to include the exact module ID number, page number, a detailed description, and your recommended correction. Your input will be brought to the attention of the Authoring Team. Thank you for your assistance.

Instructors – If you have an idea for improving this textbook, or have found that additional materials were necessary to teach this module effectively, please let us know so that we may present your suggestions to the Authoring Team.

NCCER Product Development and Revision

13614 Progress Blvd., Alachua, FL 32615

Email: curriculum@nccer.org

Online: www.nccer.org/olf

❑ Trainee Guide ❑ Lesson Plans ❑ Exam ❑ PowerPoints Other ____________

Craft / Level: Copyright Date:

Module ID Number / Title:

Section Number(s):

Description:

Recommended Correction:

Your Name:

Address:

Email: Phone:

This page is intentionally left blank.

Advanced Pipe Fabrication

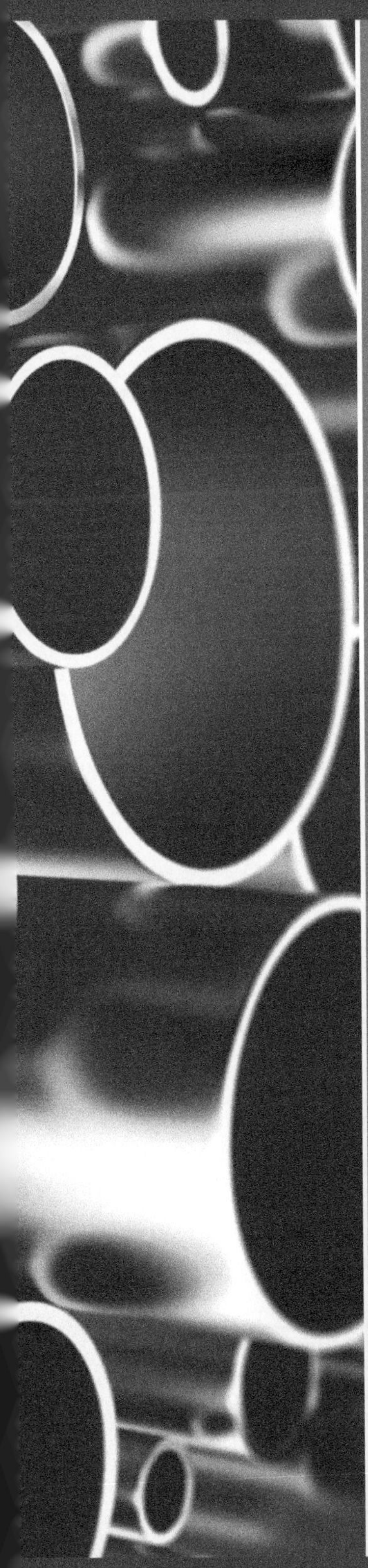

Overview

Pipe fabrication involves the use of either ordinate tables or trigonometry to create fittings and pipe assemblies that suit a process application. Producing ordinates and using them to lay out miters and laterals is important, as are alternative methods for laying out the cuts for laterals, saddles, and mitered turns. In this module, formulas are provided for putting together multiple offsets around obstacles of both equal and unequal spread.

Module 08402

Trainees with successful module completions may be eligible for credentialing through the NCCER Registry. To learn more, go to **www.nccer.org** or contact us at 1.888.622.3720. Our website, **www.nccer.org**, has information on the latest product releases and training.

Your feedback is welcome. You may email your comments to **curriculum@nccer.org**, send general comments and inquiries to **info@nccer.org**, or fill in the User Update form at the back of this module.

This information is general in nature and intended for training purposes only. Actual performance of activities described in this manual requires compliance with all applicable operating, service, maintenance, and safety procedures under the direction of qualified personnel. References in this manual to patented or proprietary devices do not constitute a recommendation of their use.

08402
Advanced Pipe Fabrication

Objectives

Successful completion of this module prepares you to do the following:

1. Explain how to calculate piping offsets.
   a. Explain how to calculate simple offsets.
   b. Explain how to calculate three-line, equal-spread offsets around a vessel.
   c. Explain how to calculate three-line, unequal-spread offsets.
   d. Explain how to lay out and fabricate tank heating coils.
2. Explain how to lay out and fabricate miter turns.
   a. Explain how to lay out ordinate lines and how to use *The Pipe Fitters Blue Book* in doing so.
   b. Explain how to lay out cutback lines and how to use *The Pipe Fitters Blue Book* in doing so.
   c. Explain how lay out mitered turns.
   d. Explain how to lay out and fabricate three-piece, 90-degree mitered turns.
   e. Explain how to lay out and fabricate four-piece, 90-degree mitered turns.
   f. Explain how to lay out the cutback for a wye.
3. Explain how to lay out and fabricate saddle and supports made out of pipe.
   a. Explain how to lay out and fabricate a saddle.
   b. Explain how to lay out and fabricate supports made out of pipe.
4. Explain how to lay out laterals without using references.

Performance Tasks

Under supervision, you should be able to do the following:

1. Calculate a three-line, 45-degree, equal-spread offset.
2. Calculate and lay out a tank coil.
3. Lay out and fabricate a three-piece mitered turn (degree to be determined by instructor).
4. Lay out and fabricate a four-piece, 90-degree, mitered turn.
5. Lay out and fabricate a wye.
6. Using reference charts, lay out and fabricate a 45-degree lateral.
7. Lay out and fabricate a Type 1 pipe support.
8. Lay out a 45-degree lateral by performing geometric layout.
9. Lay out and fabricate a saddle.

Trade Terms

Base line
Branch
Chord
Cutback
Miter
Ordinate lines
Ordinates
Saddle
Segments

Industry Recognized Credentials

If you are training through an NCCER-accredited sponsor, you may be eligible for credentials from NCCER's Registry. The ID number for this module is 08402. Note that this module may have been used in other NCCER curricula and may apply to other level completions. Contact NCCER's Registry at 1.888.622.3720 or go to **www.nccer.org** for more information.

Contents

Figures and Tables

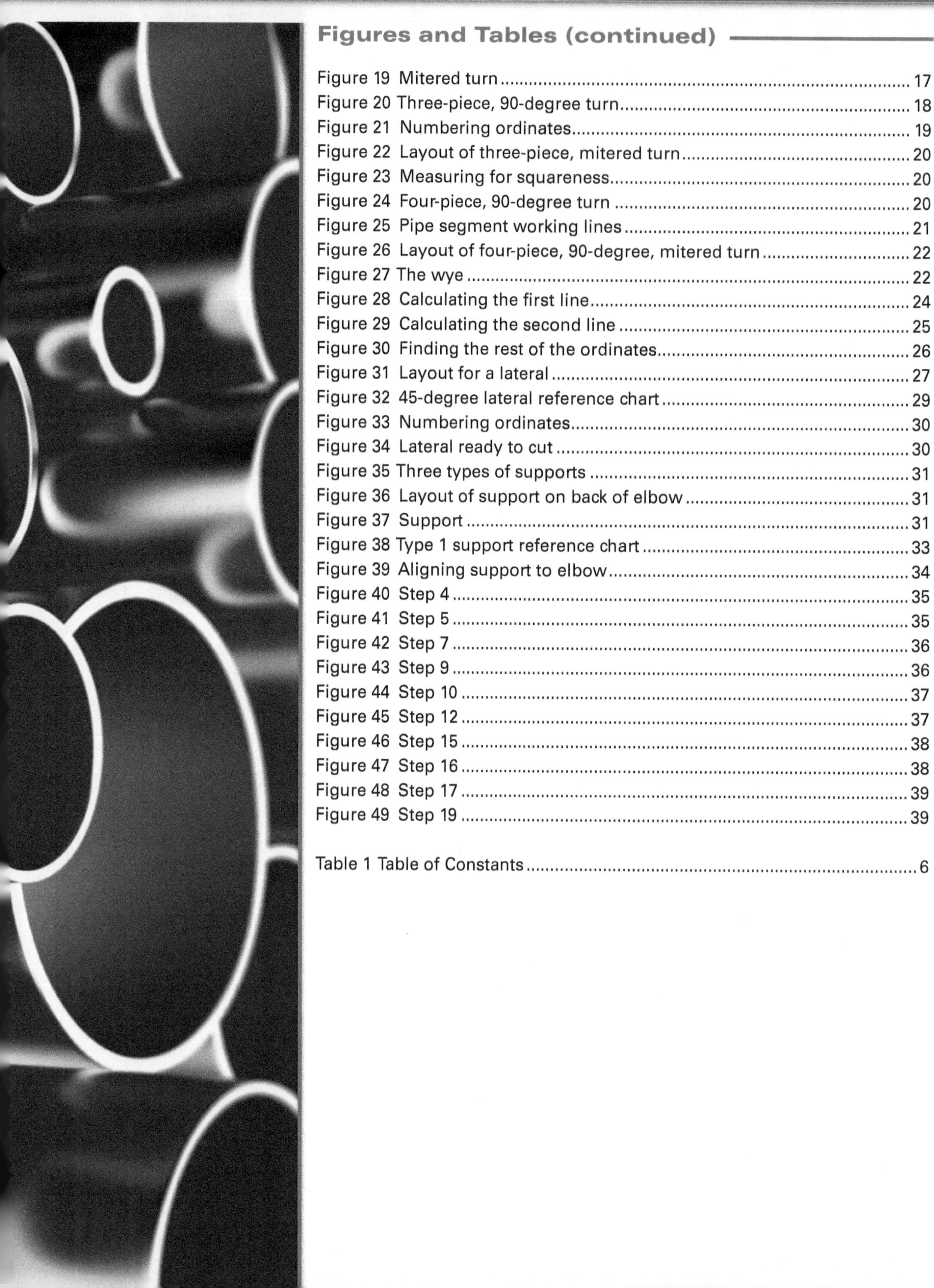

Figures and Tables (continued)

SECTION ONE

1.0.0 CALCULATING PIPING OFFSETS

Objective

Explain how to calculate piping offsets.

a. Explain how to calculate simple offsets.
b. Explain how to calculate three-line, equal-spread offsets around a vessel.
c. Explain how to calculate three-line, unequal-spread offsets.
d. Explain how to lay out and fabricate tank heating coils.

Performance Tasks

1. Calculate a three-line, 45-degree, equal-spread offset.
2. Calculate and lay out a tank coil.

Experienced pipefitters who are skilled with reading blueprints and fabricating piping components are ready to learn advanced concepts in the field. These professionals must be able to create piping systems that bend around objects, rise or drop to meet other lines, and intersect at angles greater than or less than 90 degrees. Although manufactured fittings can be used for most jobs, they can be expensive and difficult to come by. In these cases, advanced pipe fabrication becomes extremely important. The advanced pipe fabricator, in fact, can make the difference between having a problem efficiently solved or having to stop all work until parts may be obtained.

Advanced pipe fabricators are often called on to produce offsets. An offset is a lateral or vertical move that takes the pipe out of its original line to one that is parallel with it. Offsets are used when it is necessary to change the position of a pipeline in order to avoid an obstruction, such as a wall or a tank.

The three basic sides of an offset are the set, run, and travel. The set and run are joined by a 90-degree angle, and the travel connects the span between their end points. The set is the distance, measured center-to-center, that the pipeline is offset. The run is the total linear axial distance required for the offset. The travel is the center-to-center measurement of the offset piping. The angle of the fittings determines the number of degrees that the piping changes direction.

Figure 1 shows the components of an offset.

All piping offsets are based on right triangles, so it is important to understand their parameters. The sum of the three angles of all triangles is 180 degrees. A right triangle is a type of triangle that has a 90-degree angle and two acute angles; acute angles are those with less than 90 degrees. If one of the acute angles of a right triangle is 45 degrees, then the other acute angle must also be 45 degrees (90 + 45 + 45 = 180). If one of the acute angles of a right triangle is 30 degrees, then the other acute angle is 60 degrees (90 + 30 + 60 = 180). *Figure 2* shows two right triangles.

In advanced pipe fabrication, the "solving" of right triangles is a frequent task. To solve a right triangle means to find the length of its unknown sides and the degrees of its unknown angles. The use of right triangle trigonometry will help you calculate the following types of offsets:

- Simple offsets
- Three-line, 45-degree, equal-spread offsets around vessel
- Three-line, 45-degree, unequal-spread offsets
- Tank heating coils

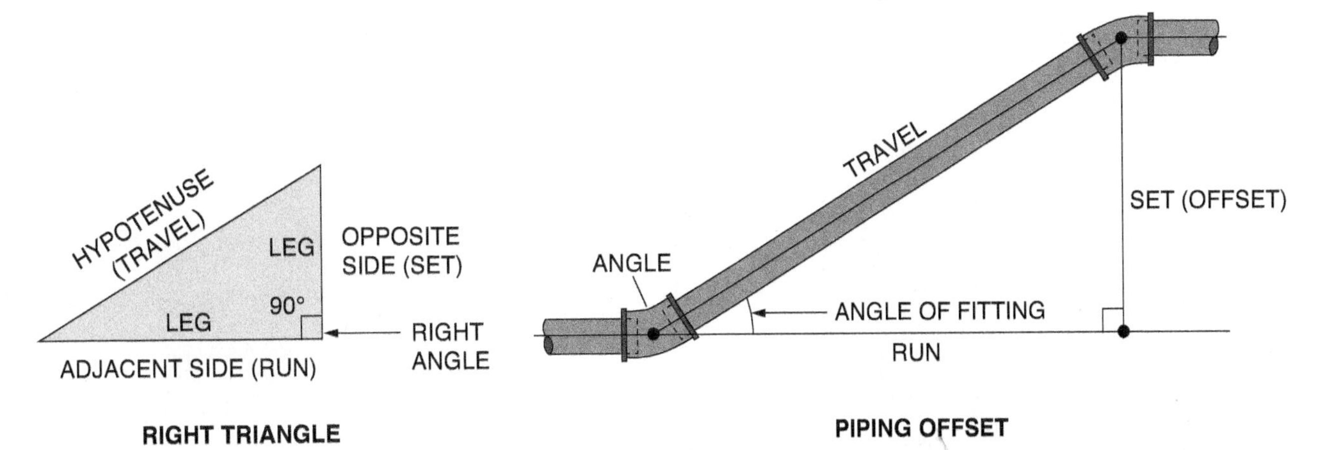

Figure 1 Offset components.

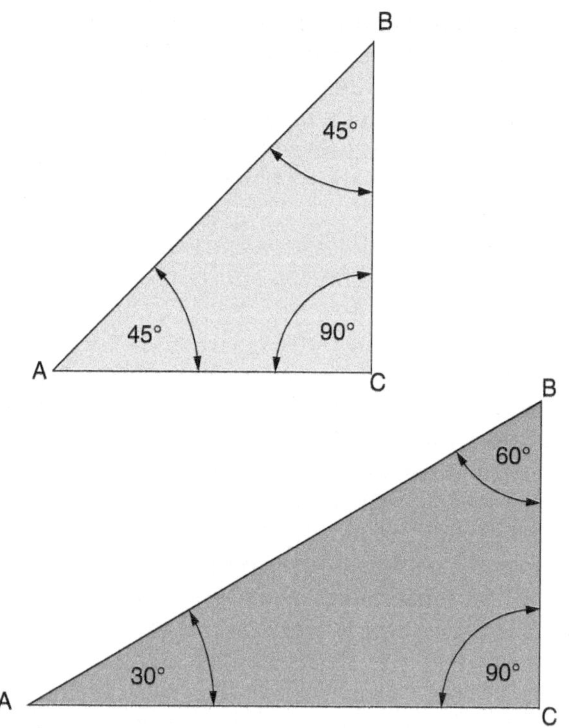

Figure 2 Right triangles.

In order to solve a right triangle, you must either:

1. Know the lengths of two sides, **or**
2. Know the measurements of two angles (including the 90-degree angle) and the length of one side.

The Pythagorean theorem ($a^2 + b^2 = c^2$) underlines this concept, in that knowing two of the components allows you to calculate the third.

Figure 3 shows a right triangle with two sides known. If side *a* is 12, side *c* is 16.96, and side *b* is unknown, the solution is found with the formula $b^2 = c^2 - a^2$ or $b^2 = (16.96)^2 - (12)^2$. This means $b^2 =$ 143.64, so b = 11.99.

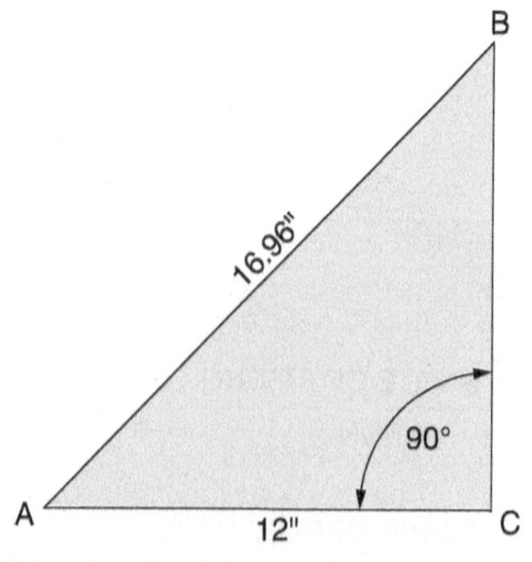

Figure 3 Right triangle with two sides known.

Figure 4 shows a right triangle with two known angles and one known side. Given that angle A = 45 degrees, angle C = 90 degrees, and side b = 12", start with the sine of 45 (12/*c*) to find the length of side *c*:

Step 1 sine 45 = 12/*c*

Step 2 *c* = 12/sin 45

= 12/0.707

= 16.97

Step 3 *a* = *b* since *ABC* is 45-45-90

B = 45 since *ABC* is 45-45-90

1.1.0 Calculating Simple Offsets

To find the length of the travel for a simple 45-degree offset when the set is known, multiply the set by the constant 1.414. This formula, using the constant of 1.414 (which is the square root of 2 and is a non-repeating, continuous decimal), only works on a 45-degree offset. Other trigonometric functions are also used to find unknown sides and angles of piping offsets. The most common functions used to calculate simple offsets are sine, cosine, and tangent. *Figure 5* shows the relationship between trigonometry functions and piping offsets.

Did You Know?

Trigonometry Mnemonic

A mnemonic is a shortened version of a complex idea that helps you remember the details. For the functions of trigonometry, the initialism "SOHCAHTOA" may be helpful.

SOHCAHTOA means:

Sine = **O**pposite/**H**ypotenuse

Cosine = **A**djacent/**H**ypotenuse

Tangent = **O**pposite/**A**djacent

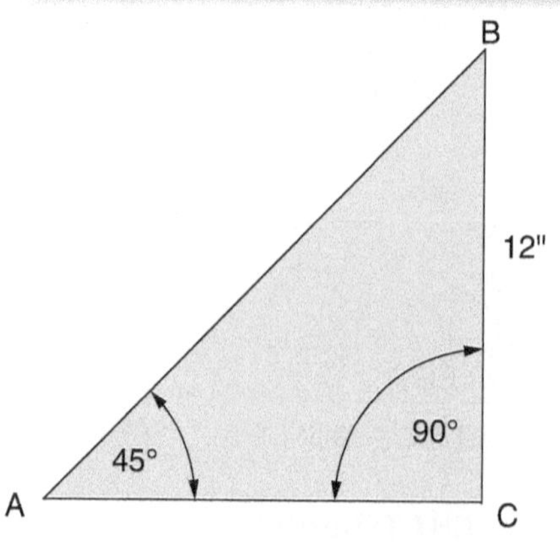

Figure 4 Right triangle with two angles and one known side.

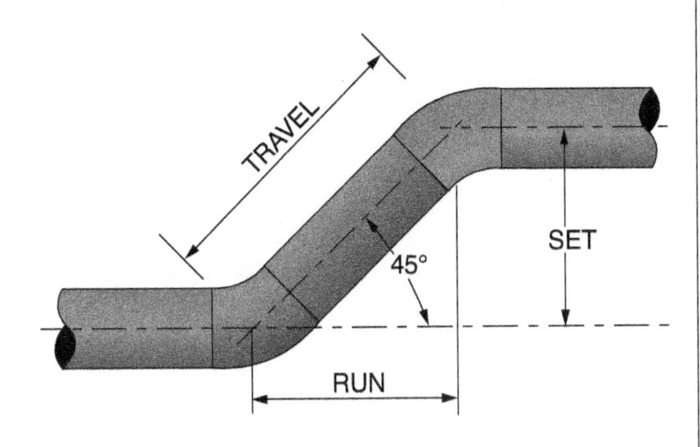

TO DETERMINE ANGLE OF OFFSET WHEN THE LENGTHS OF TWO SIDES ARE KNOWN

SET DIVIDED BY TRAVEL = SINE

RUN DIVIDED BY TRAVEL = COSINE

SET DIVIDED BY RUN = TANGENT

TO DETERMINE LENGTHS OF SIDES WHEN ANGLE AND ONE SIDE ARE KNOWN	ANGLE OF OFFSET				
	60°	45°	30°	22½°	15°
SET = TRAVEL × SINE	0.866	0.707	0.500	0.383	0.259
RUN = TRAVEL × COSINE	0.500	0.707	0.866	0.924	0.966
SET = RUN TANGENT	1.732	1.000	0.577	0.414	0.268

Figure 5 Relationship between trigonometry functions and piping offsets.

For this exercise, assume that the set is 57" and the pipe is a 4", Schedule 40 butt weld piping system.

To find the length of the travel:

Step 1 Multiply the length of the set by 1.414. This gives the length of the travel between the centers of the fittings.

$57 \times 1.414 = 80.598$, or $80\,{}^{5}/_{8}$"

Step 2 Multiply the diameter of the pipe by ⅝" (or 0.625) to obtain the takeout for the 45-degree fittings.

$4 \times 0.625 = 2.5$"

Step 3 Multiply 2.5 by 2 to obtain the takeout for both 45-degree fittings.

$2.5 \times 2 = 5$"

Step 4 Subtract 5" from the length of travel measurement from Step 1.

$80\,{}^{5}/_{8} - 5 = 75\,{}^{5}/_{8}$" length of travel, less takeout for both 45-degree fittings

Step 5 Subtract ¼" from the 75⅝" takeout (for the two ⅛" weld gaps).

$75\,{}^{5}/_{8} - {}^{1}/_{4} = 75\,{}^{3}/_{8}$" total cut length for the travel of the pipes

> **NOTE**
>
> Different welding procedures will require different weld gaps. For calculation purposes, ⅛" is used as an example throughout this module. The proper weld gap should be determined from the materials and proper procedure.

1.2.0 Calculating Three-Line, 45-Degree, Equal-Spread Offsets around a Vessel

Advanced pipe fabricators often have to lay out a series of pipelines in an offset around a vessel and maintain the same distance between each line.

To do this, it is necessary to:

1. Calculate the distance from the center of the vessel to the first line.
2. Calculate the set and travel of that line.
3. Calculate the offsets of the other lines to be routed around the vessel.

Another way to determine the distance from the center of the vessel to the starting point of the offset is to divide the offset angle by 2 and find the tangent of that angle. This can be found in *The Pipe Fitters Blue Book* or by using a calculator. Multiply the tangent by the distance from the center of the vessel to the center of the pipe. For example: a 45-degree offset divided by 2 equals 22½ degrees, and the tangent for 22½ degrees is

0.41421, or 0.4142. Using 14" as the center line of a vessel to the center line of a pipe, the distance from the center of the vessel to the starting point of the 45-degree offset is 14 × 0.4142 = 5.79, or 5¾".

Figure 6 shows a three-line, 45-degree, equal-spread offset.

Refer to *Figure 6* and follow these steps to calculate a three-line, 45-degree, equal-spread offset around a vessel, using 4" pipe:

Step 1 Add the distance from the wall to the side of the vessel, plus the diameter of the vessel, plus the distance from the outside of the vessel to the center of line 1 to obtain the distance of line 1 from the wall before the offset around the vessel (measurement A).

3" + 18" + 5" = 26"

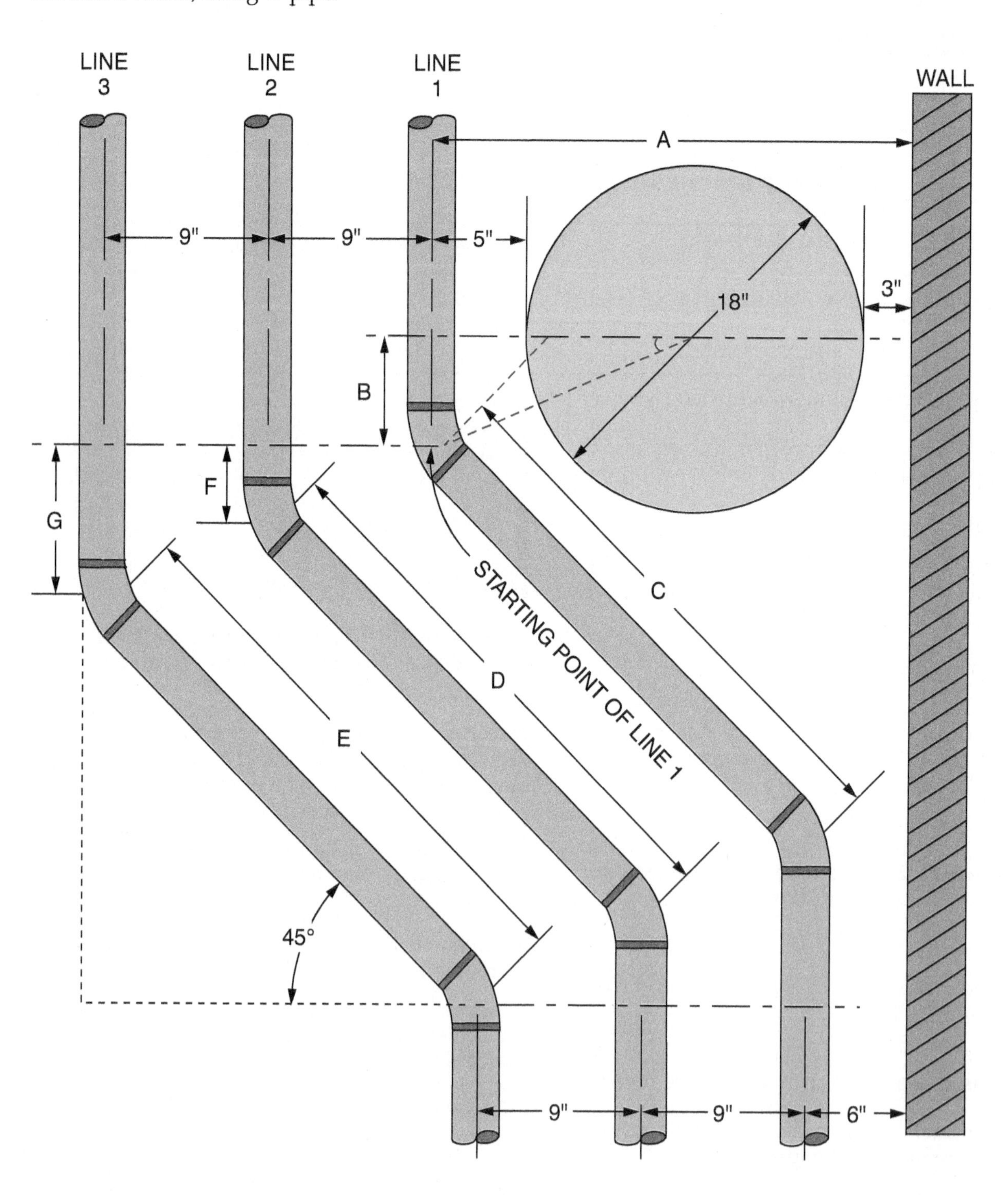

Figure 6 Three-line, 45-degree, equal-spread offset.

Step 2 Subtract the distance from the wall to the center of line 1 after the offset around the vessel from the measurement found in Step 1 to obtain the set of line 1.

$$26'' - 6'' = 20''$$

Step 3 Add the radius of the vessel to the distance from the outside of the vessel to the center of line 1, and multiply by 0.4142 to obtain the distance from the center of the vessel to the starting point of the offset for line 1 (measurement B).

$$9'' + 5'' = 14'' \times 0.4142 = 5.79, \text{ or } 5\ ^{13}/_{16}''$$

Step 4 Divide the set of line 1 by the 45-degree angle sine, 0.7071, to determine the travel of line 1.

$$20'' \div 0.7071 = 28.28, \text{ or } 28\ ^{1}/_{4}''$$

NOTE

The set and travel for all three lines in an equal-spread offset is always the same, but the starting points for the three lines will be different.

Step 5 Multiply the distance between lines 1 and 2 by 0.4142 to obtain the starting point of the offset for line 2 (measurement F). This is the distance past the start of the offset of line 1 that the offset of line 2 will start.

$$9'' \times 0.4142 = 3.727, \text{ or } 3\ ^{3}/_{4}''$$

Step 6 Multiply the distance between line 1 and line 3 by 0.4142 to obtain the starting point of the offset for line 3 (measurement G). This is the distance past the starting point of the offset for line 1 where the offset for line 3 will start.

$$9'' + 9'' \times 0.4142 = 7.455, \text{ or } 7\ ^{7}/_{16}''$$

1.3.0 Calculating Three-Line, 45-Degree, Unequal-Spread Offsets

Calculating unequal-spread offsets is necessary to move piping around an object. Calculating a three-line, 45-degree, unequal-spread offset is similar to calculating a three-line, 45-degree, equal-spread offset except that the travel lengths will be different. *Figure 7* shows a three-line, 45-degree, unequal-spread offset.

NOTE

Measurements must start at line 1. These measurements are used to obtain measurements for lines 2 and 3.

Three-line, 45-degree, unequal-spread offsets can be fabricated using differently-sized pipes. For example, line 1 could be a 2" pipe, line 2 could be a 3" pipe, and line 3 could be a 4" pipe. It is important to use the correct center-to-center measurements when calculating the takeouts for the differently-sized pipes.

1.4.0 Laying Out and Fabricating Tank Heating Coils

Although tank coils (*Figure 8*) may be made on a roll bender as a continuous coil, the pipefitter is sometimes called upon to fabricate tank heating coils. A tank coil is a piping system that is fabricated around the inside walls of a tank and keeps the same radius from the center of the tank. A typical tank coil provides steam or heated liquid to a vessel to keep a material molten.

The pipe for a tank coil must be assembled or erected inside the tank. To determine the length of pipe for a tank coil, the pipefitter must know the radius of the tank and the radius of the pipe coil inside the tank. After these measurements are known, a table of constants can be used to obtain the center-to-center measurements of the tank coil. *Table 1* shows a typical table of constants. The number of pieces for each coil is obtained by dividing the fitting angle into 360 degrees. For example, if the fitting angle is 45 degrees, the pipefitter would divide 360 by 45 to find that 8 pipes per coil are required.

Refer to *Figure 8* and follow these steps to lay out and fabricate a tank coil:

Step 1 Divide the diameter of the tank by 2 to obtain the radius of the tank (measurement A).

$$76'' \div 2 = 38''$$

Step 2 Measure the distance from the outside diameter of the tank to the center line of the coil to obtain measurement B.

10"

Step 3 Subtract the distance from the outside diameter of the tank to the center line of the coil from the radius of the tank to obtain the radius of the coil (measurement R).

$$38'' - 10'' = 28''$$

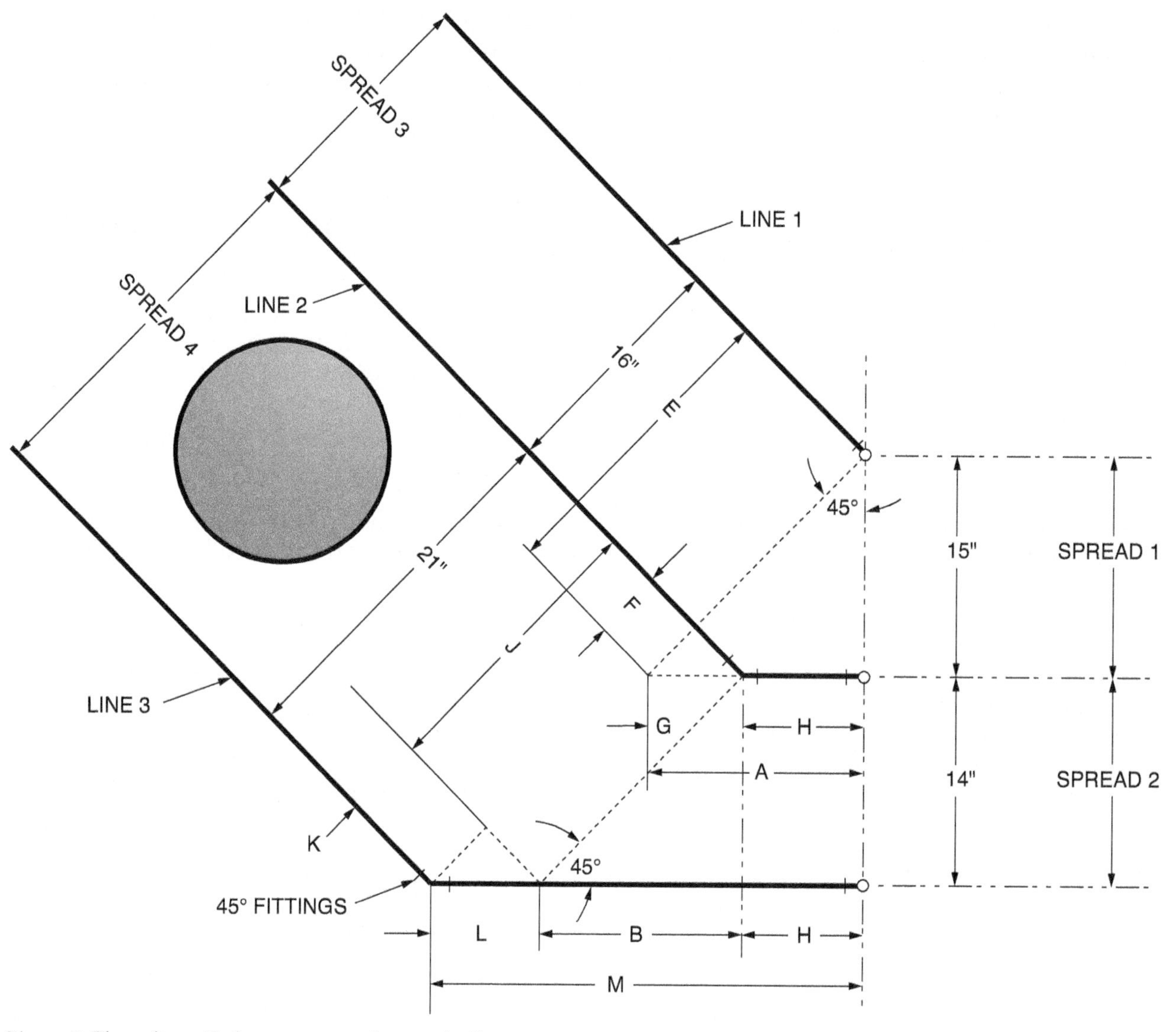

Figure 7 Three-line, 45-degree, unequal-spread offset.

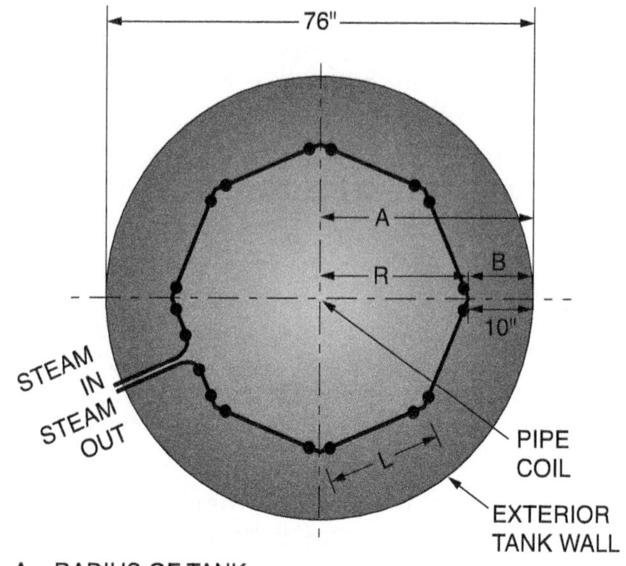

Figure 8 Tank coil.

Table 1 Table of Constants

Angle of Fitting	Number of Pipes per Coil	Constant
90° (A)	4	1.4142
60° (B)	6	1.0000
45° (A)	8	0.7653
30° (B)	12	0.5176
$22\frac{1}{2}$° (B)	16	0.3902
$11\frac{1}{4}$° (B)	32	0.1960
$5\frac{5}{8}$° (B)	64	0.0981
(A) = Factory fitting		
(B) = Filter will fabricate		

Step 4 Multiply the radius of the coil by the constant for a 45-degree fitting to obtain the center-to-center measurement of the pipe travel for the coil (measurement L).

$28" \times 0.7653 = 21.4284$, or $21\ ^{7}/_{16}"$

Instead of the table of constants, you can calculate the center-to-center lengths by multiplying 2 times the radius of the coil by the sine of one half the angle on a calculator. The answer will be more accurate, and more easily obtained. In Step 5, if you are fabricating the coil miters, the only necessary addition to the center-to-center measurement would be the two, $^{1}/_{8}"$ weld gaps.

Step 5 Add the takeout for two 45-degree elbows plus two, $^{1}/_{8}"$ weld gaps to obtain the total takeout for seven of the eight pipes in the coil.

The radius of a 2" pipe,
$1\ ^{1}/_{4}"$ (45-degree takeout) $\times$ 2
$= 2\ ^{1}/_{2}" + (^{1}/_{8}$ [weld gap takeout] $\times$ 2)
$= 2\ ^{1}/_{2} + ^{1}/_{4}$ inch $= 2\ ^{3}/_{4}"$ total takeout

If you were fabricating miters, this would only be the weld gap.

Step 6 Subtract the takeout from the center-to-center measurement of the pipe travel to obtain the cut length for seven of the eight pipes in the coil.

$21.428" - 2.75" = 18.68$ or $18\ ^{11}/_{16}"$

Step 7 Add the takeout for two, 90-degree elbows (3" each) and the takeout for two, $^{1}/_{8}"$ weld gaps to obtain the total takeout for the supply pipe in the coil. You will also have to subtract $2^{1}/_{2}"$ (the radius of the two pipes) plus $^{1}/_{2}"$ to maintain the gap between the inflow and outflow pipes. Otherwise, you would have the two pipes on top of each other.

$3" \times 2 = 6"$
$6" + (^{1}/_{8}" \times 2) = 6\ ^{1}/_{4}"$
$6\ ^{1}/_{4}" + (2\ ^{1}/_{2} + ^{1}/_{2}) = 9\ ^{1}/_{4}"$

Step 8 Subtract the takeout from the center-to-center measurement of the pipe travel to obtain the cut length for the supply pipe in the coil.

$21.428" - 9.25" = 12.178"$, or $12\ ^{3}/_{16}"$

Step 9 Divide the coil supply pipe travel length by 2 to obtain the cut lengths.

$12.178" \div 2 = 6.089$, or $6\ ^{1}/_{16}"$

1.0.0 Section Review

1. The constant 1.414, or the square root of 2, is used to find the length of the _____ for a 45-degree angle.
   a. set
   b. run
   c. travel
   d. tail

2. When laying out pipelines in an offset around a vessel, it's necessary to _____.
   a. calculate for perigee and angularity
   b. maintain the same distance between each line
   c. initiate stop-start patterns across groups of pipelines
   d. square up and use the inverse of the sine and cosine

3. When calculating three-line, 45-degree, unequal-spread offsets, it's important to use the correct _____ measurements when calculating takeouts for differently-sized pipes.
   a. center-to-center
   b. left-to-center
   c. right-to-center
   d. left-to-right

4. The purpose of a tank coil is to _____.
   a. provide structural reinforcement
   b. prevent sludge build-up
   c. create an additional opening for hot taps
   d. provide warmth so that material stays molten

SECTION TWO

2.0.0 FABRICATING MITER TURNS

Objective

Explain how to lay out and fabricate miter turns.

a. Explain how to lay out ordinate lines and how to use *The Pipe Fitters Blue Book* in doing so.
b. Explain how to lay out cutback lines and how to use *The Pipe Fitters Blue Book* in doing so.
c. Explain how lay out mitered turns.
d. Explain how to lay out and fabricate three-piece, 90-degree mitered turns.
e. Explain how to lay out and fabricate four-piece, 90-degree mitered turns.
f. Explain how to lay out the cutback for a wye.

Performance Tasks

3. Lay out and fabricate a three-piece mitered turn (degree to be determined by instructor).
4. Lay out and fabricate a four-piece, 90-degree, mitered turn.
5. Lay out and fabricate a wye.

Trade Terms

Base line: A straight line drawn around a pipe to be used as a measuring point.

Branch: A line that intersects with another line.

Cutback: The point at which a miter fitting is to be cut.

Ordinate lines: Straight lines drawn along the length of the pipe connecting the ordinate marks.

Ordinates: Divisions of segments obtained by dividing the circumference of a pipe into equal parts.

Segments: Parts of a circle that are defined by a chord and the curve of the circumference.

Miter: A specified angle to which a piece of pipe is cut.

Advanced pipefitting craftworkers must know how to cut a piece of pipe to a specified angle, called a **miter**, so that the piping may bend around, rise to, drop to, or intersect with other pipes. Before mitering a piece of pipe or fabricating mitered turns, it is essential to first understand how to lay out **ordinate lines** and **cutback** lines on a piece of pipe

2.1.0 Laying out Ordinate Lines

In pipe fabrication, the circumference of a pipe is divided into 4, 8, 16, or more **segments** depending on the size of the pipe and the complexity of the intersection. At each of these division points, a straight line is drawn along the surface of the pipe. These lines are called ordinate lines and serve as guides for cutting miters and marking contours. Certain distances are marked off on the ordinate lines to form the angle at which the pipe needs to be cut.

To lay out ordinate lines:

Step 1 Obtain a strip of paper with straight edges long enough to wrap around the outside of the pipe $1\frac{1}{2}$ times.

Step 2 Wrap the strip of paper around the outside of the pipe and mark the point at which the strip of paper overlaps (*Figure 9*).

Step 3 Make a square cut on the strip of paper at the mark.

Step 4 Determine the number of **ordinates** needed for the fabrication:

- Divide $\frac{1}{2}$" to 3" pipe into 4 ordinates;
- 4" to 10" pipe into 8 ordinates; and
- 12" pipe and larger into 16 ordinates.

Step 5 Fold the paper enough times to obtain the correct number of ordinates (*Figure 10* shows ordinate folds):

- Fold the paper once for 2 ordinates
- Fold the paper twice for 4 ordinates
- Fold the paper 3 times for 8 ordinates
- Fold the paper 4 times for 16 ordinates

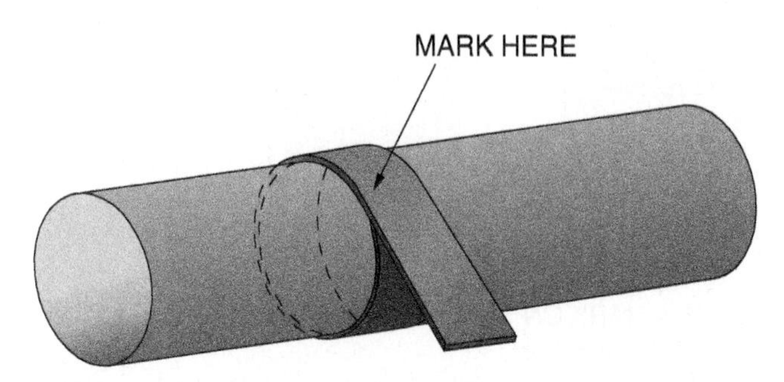

Figure 9 Strip of paper around a pipe.

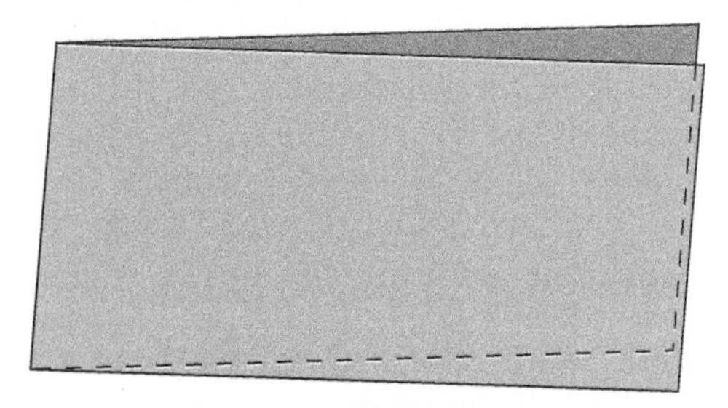
ONE FOLD CREATES
TWO ORDINATES.

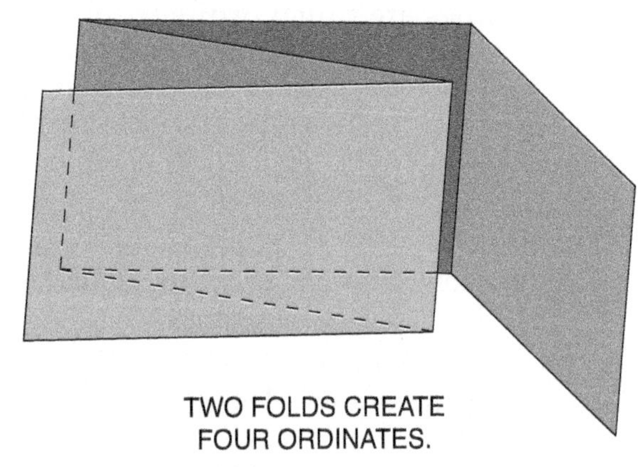
TWO FOLDS CREATE
FOUR ORDINATES.

Figure 10 Ordinate folds.

Step 6 Cut off a small piece at the corner of each fold, making a notch to provide an ordinate template.

Step 7 Wrap the paper around the pipe squarely.

Step 8 Make a mark on the pipe at each fold on the ordinate template.

Step 9 Draw a straight line along the length of pipe at each of the ordinate marks, as shown in *Figure 11*.

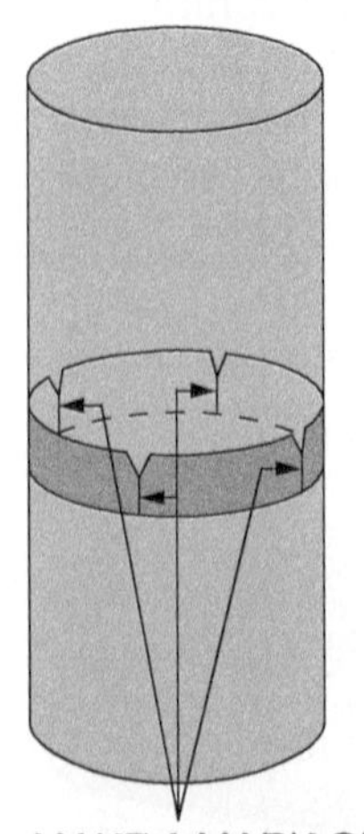
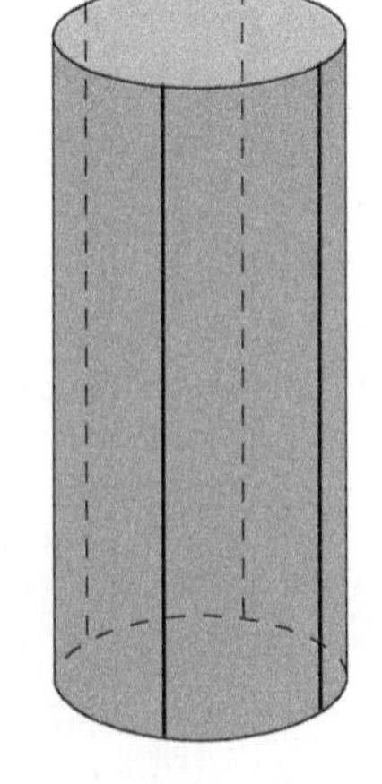
DRAW A
STRAIGHT
LINE THROUGH
EACH MARK
ALONG THE
AXIS OF
THE PIPE.
MAKE A MARK ON
THE PIPE AT EACH
OF THE FOLDS.

Figure 11 Ordinate marks.

A piece of angle iron is a good device to use to make straight ordinate marks on the outside of the pipe, as shown in *Figure 12*.

2.2.0 Laying Out Cutback Lines

The pipefitter must mark the cutback on the ordinate lines to provide the cut line for the pipe fitting. Cutback refers to a cut made back from a straight line marked around a pipe where the ordinate lines are marked. *The Pipe Fitters Blue Book* and other pipefitting handbooks give the cutback measurements for common miter angles. Always check a handbook before manually calculating the cutback measurements.

To lay out cutback lines:

Step 1 Lay out the correct number of ordinate lines based on the size of pipe being used.

Step 2 Number the ordinate lines, as shown in *Figure 13*.

Step 3 Using the wraparound and soapstone, draw a working line around the pipe at the center of the desired turn. This provides a reference point for measuring and marking the necessary cutback lines.

Step 4 Multiply the outside diameter (OD) of the pipe by the tangent of the degree of the miter to be cut. Then, divide by 2 to obtain the cutback distance for the given miter.

Step 5 Measure (to the left of the working line) the cutback distance along ordinate 2 at the top of the pipe. Then, mark the pipe, as shown in *Figure 14*.

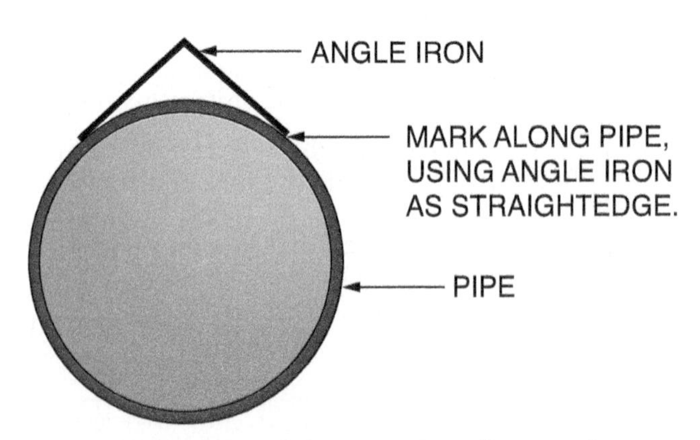
ANGLE IRON
MARK ALONG PIPE,
USING ANGLE IRON
AS STRAIGHTEDGE.
PIPE

Figure 12 Using an angle iron to make ordinate marks.

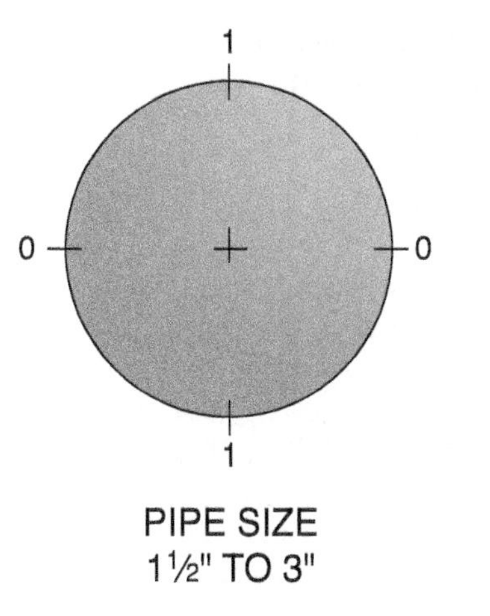

PIPE SIZE
4" TO 10"

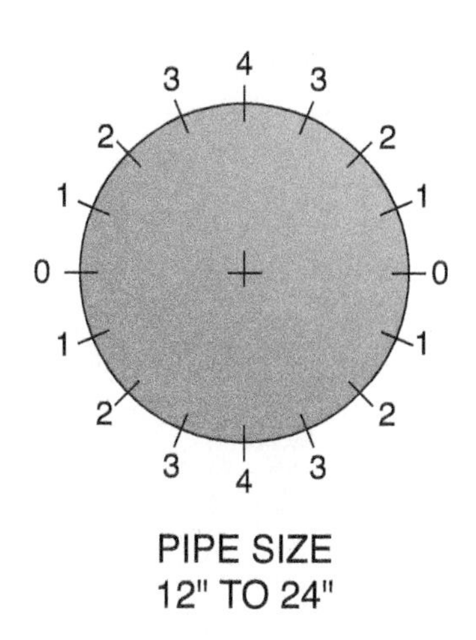

Figure 13 Numbering ordinates.

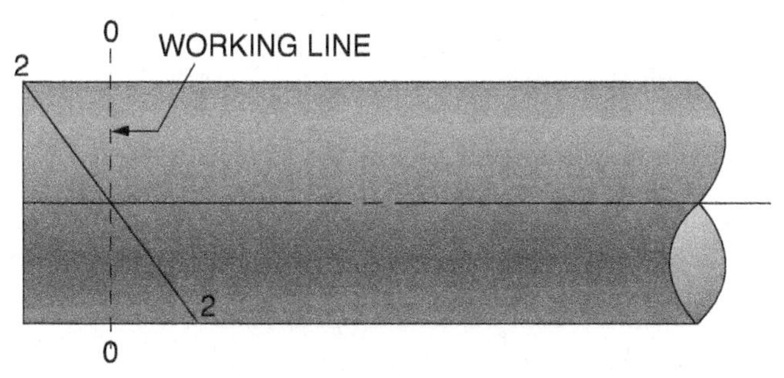

Figure 14 Conventional method for marking ordinates.

Step 6 Measure (to the right of the working line) the cutback distance along ordinate 2 at the bottom of the pipe and mark the pipe. If more ordinate cutbacks are required, either calculate the cutback lengths by trigonometry or use the given formulas and tables in the pipefitters' handbooks to obtain the cutback measurements. *Figure 15, Figure 16*, and *Figure 17* show common cutback measurements for various pipe sizes and miter cuts.

- To determine the ordinate points for a cutback line, either refer to the tables in *The Pipe Fitters Blue Book* or use trigonometry. It's best to know how to use both because the tables in the blue book do not give every possible miter angle. This becomes important when a pipefitter needs to fabricate a very large pipe: trigonometric calculations allow more ordinate points for the longer line involved.
- Two calculations are required to obtain the dimensions for the layout. First is the length of the initial, longest line from the working line to the place where the miter first strikes the surface of the pipe (refer to Step 4).
- The second involves the use of right angle trigonometry. The two sides that define a right angle are the working line and the axis of the pipe. Divide 90 degrees by the number of points that are needed, and this will reveal how far apart, in degrees, the angle will be from the intersection of the first cutback line to the ends of the other ordinates. The first, or number 1 line in a 4-point quadrant will be $22\frac{1}{2}$ degrees from the vertical working line. Since the straight line from the working line to the ordinate point is the opposite side and the long line to that point is the hypotenuse, the axial line is found by sine $22\frac{1}{2}$ times the axial cutback.
- Each line thereafter adds $22\frac{1}{2}$ degrees to the calculation. The second line is sine 45 degrees times the longest line; the third line is the sine of $67\frac{1}{2}$ degrees times the longest line.
- Had this been a large diameter pipe, such as a 36" or 48" pipe, accuracy of the cutline layout might have been increased by laying out more ordinates. In such a case, it would have been possible to divide the quadrant into six equal parts, then to have divided the 90 degrees by 6, giving the multipliers as sine 15 degrees, sine 30 degrees, sine 45 degrees, sine 60 degrees, and sine 75 degrees. The procedure is the same.

Step 7 Line up a wraparound on the pipe at the ordinate cutback marks on the right side of the working line (see *Figure 18*).

Step 8 Connect these marks using soapstone.

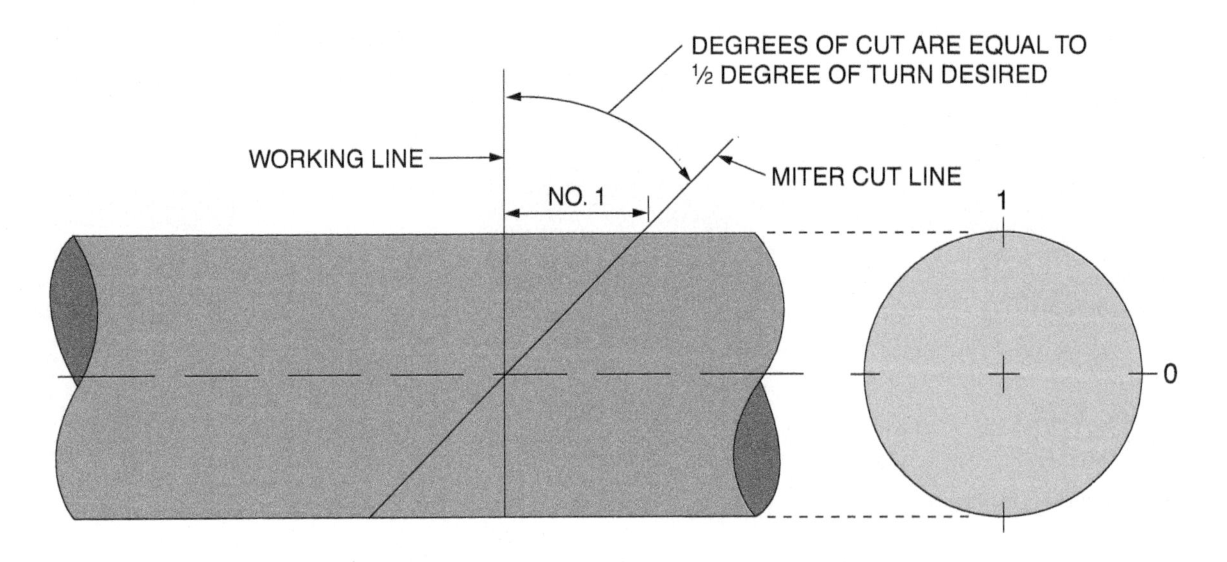

MITER CUTS FOR 1 1/2" THROUGH 3" WITH PIPE DIVIDED INTO 4 ORDINATES.
CUTBACK LINE NO. 1 DIMENSION EQUALS
TANGENT OF CUT × OD OF PIPE DIVIDED BY 2.

1 1/2" THROUGH 3" MITER CUTS – PIPE QUARTERED

7 1/2° CUT FOR 15° TURN		22 1/2° CUT FOR 45° TURN	
SIZE	**NO. 1**	**SIZE**	**NO. 1**
1 1/2	1/8	1 1/2	3/8
2	1/8	2	1/2
2 1/2	3/16	2 1/2	9/16
3	3/16	3	3/4
9° CUT FOR 18° TURN		**30° CUT FOR 60° TURN**	
SIZE	**NO. 1**	**SIZE**	**NO. 1**
1 1/2	1/8	1 1/2	1/2
2	3/16	2	11/16
2 1/2	1/4	2 1/2	13/16
3	1/4	3	1
11 1/4° CUT FOR 22 1/2° TURN		**45° CUT FOR 90° TURN**	
SIZE	**NO. 1**	**SIZE**	**NO. 1**
1 1/2	3/16	1 1/2	1 5/16
2	1/4	2	1 3/16
2 1/2	1/4	2 1/2	1 7/16
3	5/16	3	1 3/4
15° CUT FOR 30° TURN			
SIZE	**NO. 1**		
1 1/2	1/4		
2	5/16		
2 1/2	3/8		
3	7/16		

Figure 15 Ordinate cutbacks for 1 1/2" through 3" pipe.

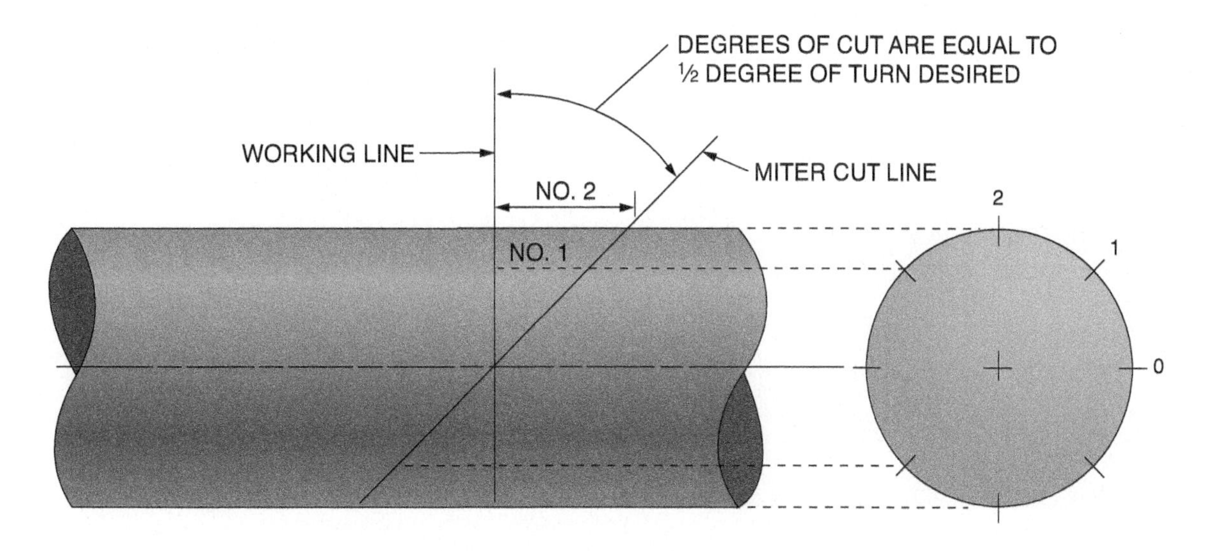

MITER CUTS FOR 4" THROUGH 10" WITH PIPE DIVIDED INTO 8 ORDINATES.
CUTBACK LINE NO. 2 DIMENSION EQUALS
TANGENT OF CUT × OD OF PIPE DIVIDED BY 2.
CUTBACK LINE NO. 1 DIMENSION EQUALS DIMENSION NO. 2 × 0.7071

4" THROUGH 10" MITER CUTS – PIPE IN EIGHTHS

$7\frac{1}{2}$° CUT FOR 15° TURN			$22\frac{1}{2}$° CUT FOR 45° TURN		
SIZE	**NO. 1**	**NO. 2**	**SIZE**	**NO. 1**	**NO. 2**
4	$\frac{3}{16}$	$\frac{1}{4}$	4	$\frac{11}{16}$	$\frac{15}{16}$
6	$\frac{5}{16}$	$\frac{7}{16}$	6	1	$1\frac{3}{8}$
8	$\frac{3}{8}$	$\frac{9}{16}$	8	$1\frac{1}{4}$	$1\frac{3}{4}$
10	$\frac{1}{2}$	$\frac{11}{16}$	10	$1\frac{9}{16}$	$2\frac{3}{16}$
9° CUT FOR 18° TURN			**30° CUT FOR 60° TURN**		
SIZE	**NO. 1**	**NO. 2**	**SIZE**	**NO. 1**	**NO. 2**
4	$\frac{1}{4}$	$\frac{3}{8}$	4	$\frac{15}{16}$	$1\frac{5}{16}$
6	$\frac{5}{16}$	$\frac{1}{2}$	6	$1\frac{5}{16}$	$1\frac{7}{8}$
8	$\frac{1}{2}$	$\frac{11}{16}$	8	$1\frac{3}{4}$	$2\frac{1}{2}$
10	$\frac{5}{8}$	$\frac{7}{8}$	10	$2\frac{3}{16}$	$3\frac{1}{16}$
$11\frac{1}{4}$° CUT FOR $22\frac{1}{2}$° TURN			**45° CUT FOR 90° TURN**		
SIZE	**NO. 1**	**NO. 2**	**SIZE**	**NO. 1**	**NO. 2**
4	$\frac{5}{16}$	$\frac{7}{16}$	4	$1\frac{9}{16}$	$2\frac{1}{4}$
6	$\frac{7}{16}$	$\frac{5}{8}$	6	$2\frac{3}{8}$	$3\frac{5}{16}$
8	$\frac{5}{8}$	$\frac{7}{8}$	8	$3\frac{1}{16}$	$4\frac{5}{16}$
10	$\frac{3}{4}$	$1\frac{1}{16}$	10	$3\frac{13}{16}$	$5\frac{3}{8}$
15° CUT FOR 30° TURN					
SIZE	**NO. 1**	**NO. 2**			
4	$\frac{3}{8}$	$\frac{9}{16}$			
6	$\frac{5}{8}$	$\frac{7}{8}$			
8	$\frac{13}{16}$	$1\frac{1}{8}$			
10	1	$1\frac{7}{16}$			

Figure 16 Ordinate cutbacks for 4" through 10" pipe.

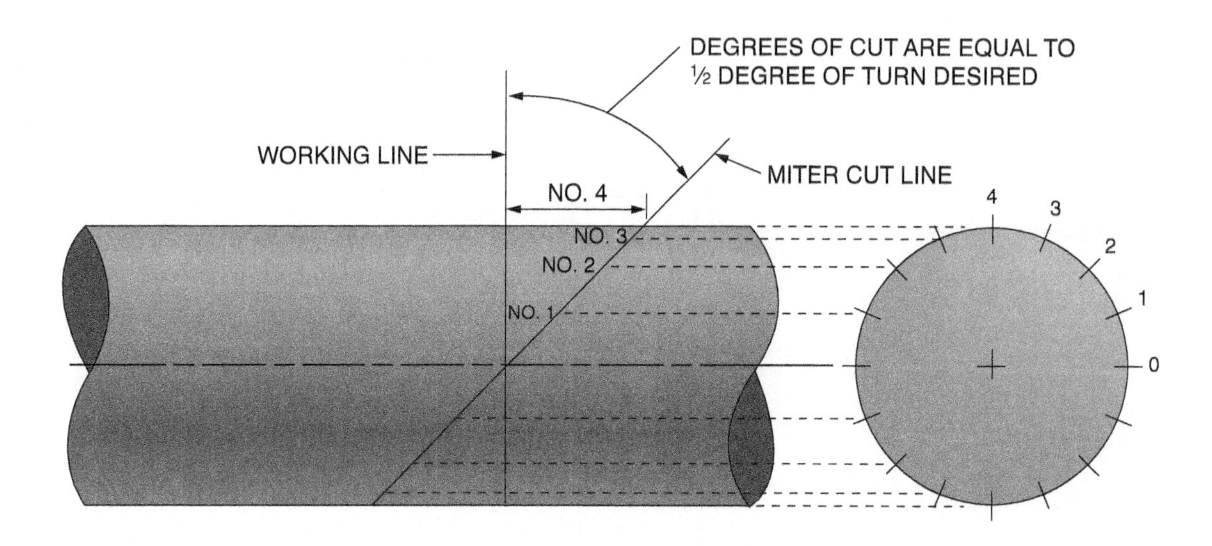

MITER CUTS FOR 12" THROUGH 24" WITH PIPE DIVIDED INTO 16 ORDINATES.
CUTBACK LINE NO. 4 DIMENSION EQUALS
TANGENT OF CUT × OD OF PIPE DIVIDED BY 2.
CUTBACK LINE NO. 3 DIMENSION EQUALS DIMENSION NO. 4 × 0.9239
CUTBACK LINE NO. 2 DIMENSION EQUALS DIMENSION NO. 4 × 0.7071
CUTBACK LINE NO. 1 DIMENSION EQUALS DIMENSION NO. 4 × 0.3827

12" THROUGH 24" MITER CUTS
MARK PIPE IN SIXTEENTHS

7½° CUT FOR 15° TURN				
SIZE	**NO. 1**	**NO. 2**	**NO. 3**	**NO. 4**
12	5/16	9/16	3/4	13/16
14	3/8	5/8	7/8	15/16
16	7/16	3/4	1	1 1/16
18	7/16	13/16	1 1/16	1 3/16
20	1/2	15/16	1 3/16	1 5/16
24	5/8	1 1/8	1 7/16	1 9/16
9° CUT FOR 18° TURN				
SIZE	**NO. 1**	**NO. 2**	**NO. 3**	**NO. 4**
12	3/8	11/16	15/16	1
14	7/16	13/16	1	1 1/8
16	1/2	7/8	1 3/16	1 1/4
18	9/16	1	1 5/16	1 7/16
20	5/8	1/8	1 7/16	1 9/16
24	3/4	1 5/16	1 3/4	1 7/8
11½° CUT FOR 22½° TURN				
SIZE	**NO. 1**	**NO. 2**	**NO. 3**	**NO. 4**
12	1/2	7/8	1 3/16	1 1/4
14	1/2	1	1 5/16	1 3/8
16	5/8	1 1/8	1 7/16	1 9/16
18	11/16	1 1/4	1 11/16	1 12/16
20	3/4	1 3/8	1 13/16	2
24	15/16	1 11/16	2 3/16	2 3/8

12" THROUGH 24" MITER CUTS
MARK PIPE IN SIXTEENTHS

15° CUT FOR 30° TURN				
SIZE	**NO. 1**	**NO. 2**	**NO. 3**	**NO. 4**
12	5/8	1 3/16	1 9/16	1 11/16
14	3/4	1 5/16	1 3/4	1 7/8
16	3/16	1 1/2	2	2 1/8
18	15/16	1 11/16	2 1/4	2 3/8
20	1	1 7/8	2 1/2	2 11/16
24	1 1/4	2 1/4	3	3 3/16
22½° CUT FOR 45° TURN				
SIZE	**NO. 1**	**NO. 2**	**NO. 3**	**NO. 4**
12	1	1 7/8	2 7/16	2 5/8
14	1 1/8	2 1/16	2 11/16	2 7/8
16	1 1/4	2 5/16	3 1/16	3 5/16
18	1 7/16	2 5/8	3 7/16	3 3/4
20	1 9/16	2 15/16	3 13/16	4 1/8
24	1 7/8	3 1/2	4 5/8	5
30° CUT FOR 60° TURN				
SIZE	**NO. 1**	**NO. 2**	**NO. 3**	**NO. 4**
12	1 3/8	2 5/8	3 3/8	3 11/16
14	1 9/16	2 7/8	3 3/4	4 1/16
16	1 3/4	3 1/4	4 1/4	4 5/8
18	2	3 11/16	4 13/16	5 3/16
20	2 3/16	4 1/16	5 5/16	5 3/4
24	2 5/8	4 7/8	6 3/8	6 15/16

Figure 17 Ordinate cutbacks for 12" through 24" pipe.

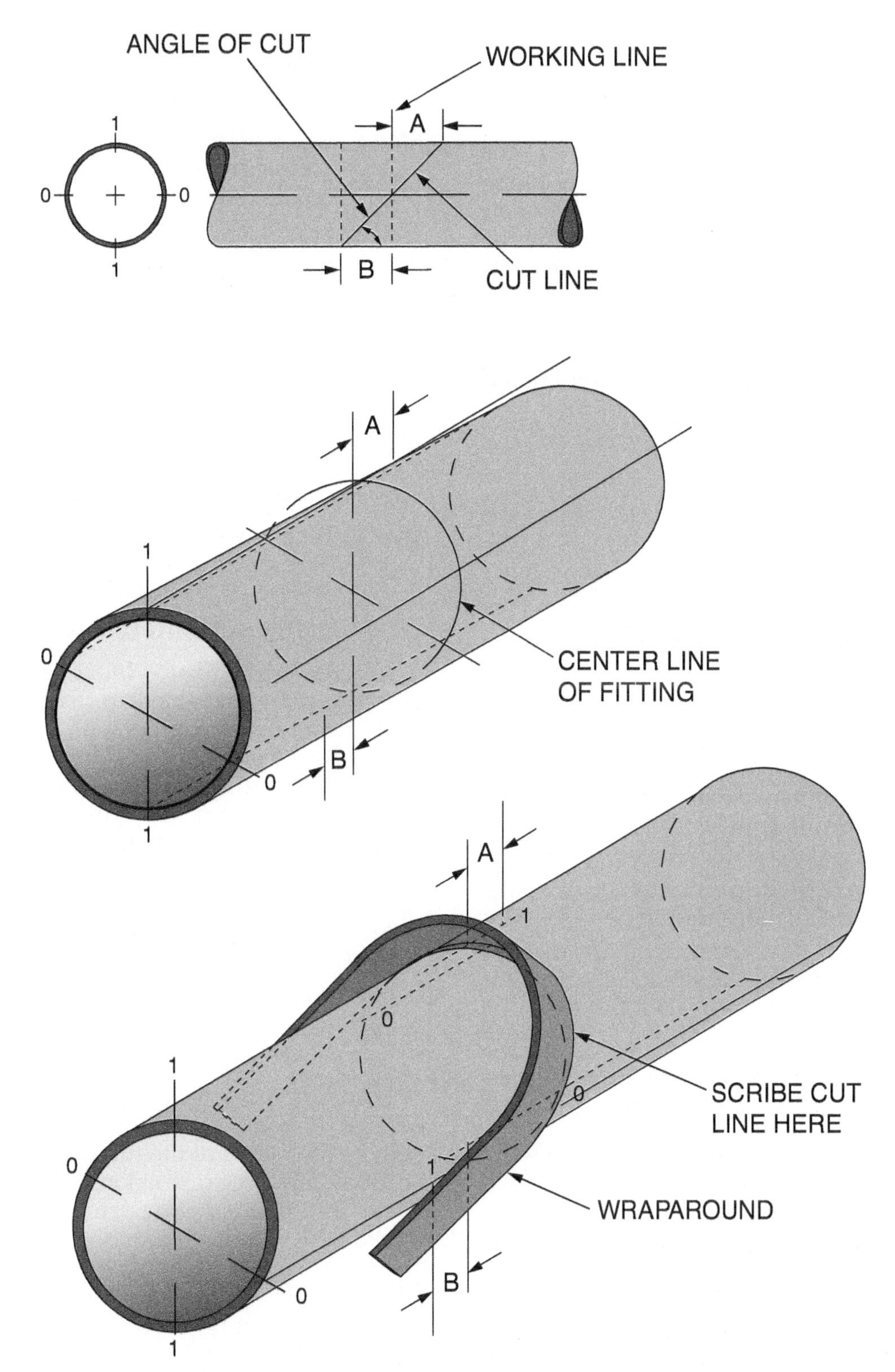

Figure 18 Laying out cutback marks with wraparound.

NOTE

Remember that work will begin on the right side of the working line. After that, work will be focused on the left side of the working line.

Step 9 Line up a wraparound on the pipe at the ordinate cutback marks on the left side of the working line.

Step 10 Connect these marks using soapstone.

Step 11 Secure the pipe in a vise and cut along the line, using a cutting torch. It is important to rotate the pipe to ensure that the torch angle is always at a 90-degree angle with the pipe.

Step 12 Check the miter cut for the correct angle.

Step 13 Bevel the edges of the cut after the cut has been made.

2.2.1 Mitering Exercise

Practice cutting a miter using the 13-step procedure outlined in 2.2.0.

Follow these steps to cut a $22^1/_2$-degree miter on 6" pipe:

Step 1 Support a 6" pipe with two jack stands.

Step 2 Lay out eight ordinate lines on the pipe.

Step 3 Use a wraparound to mark a working line in the center of the desired turn.

Step 4 Find the dimension of cutback line number 2 by using the given formula or checking in a pipefitter's handbook.

- For a $22\frac{1}{2}$-degree miter on 6" pipe, your calculations should be as follows: Pipe OD times tangent of $22\frac{1}{2}$ degrees divided by 2, or (6.625 × 0.4142) ÷ 2, or 2.744 ÷ 2, which equals 1.372" or $1\frac{3}{8}$".

Step 5 Measure and mark the cutback along the number 2 ordinate line and to the left and right side of the working line.

Step 6 Find the dimension for ordinate number 1 by using the given formula or checking a pipefitter's handbook.

- For a $22\frac{1}{2}$-degree miter on 6" pipe, your calculations should be as follows: Dimension of ordinate number 2 times 0.7071, or 1.372 × 0.7071 which equals 0.9701", rounded up to 1".

Step 7 Measure and mark the cutback of 1" along the number 1 ordinate line to the left and to the right of the working line.

Step 8 Use a wraparound to connect the points on the number 0, 1, and 2 cutback marks on the right side of the working line.

Step 9 Use a wraparound to connect the points on the number 0, 1, and 2 cutback marks on the left side of the working line.

Step 10 Secure the pipe in a vise and cut along the line using a cutting torch. It is important to rotate the pipe to ensure that the torch angle is always at a 90-degree angle with the pipe.

Step 11 Check the miter cut for the correct angle.

Step 12 Bevel the edges of the cut after the cut has been made.

2.3.0 Laying out Mitered Turns

A mitered turn (*Figure 19*) consists of several pieces of pipe assembled so that they turn on a specific number of degrees with a specific radius. Each mitered piece of pipe known as a segment. When building a miter, the first consideration should be the radius. In a typical setting, the pipefitter would build the miter similar to the radius of a standard 90-degree elbow for the size of pipe being used. Normally, the number of segments used in 90-degree mitered turns, according to the W.V. Graves *The Pipe Fitters Blue Book*, is as follows:

- *Pipe smaller than 6"* – 2 to 3 segments
- *Pipe 6 to 10"* – 3 to 4 segments
- *Pipe 12" and over* – 4 to 7 segments

There are other factors used to determine the number of segments to use in a mitered turn, such as the radius of the turn, engineering specifications, and flow restrictions. The fewer the segments in a mitered turn, the greater the restriction to flow.

To fabricate a mitered turn, determine the following:

- OD of pipe
- Radius of turn
- Number of degrees in each miter cut
- Center-to-center length of each piece in turn
- Length of end pieces for given radius

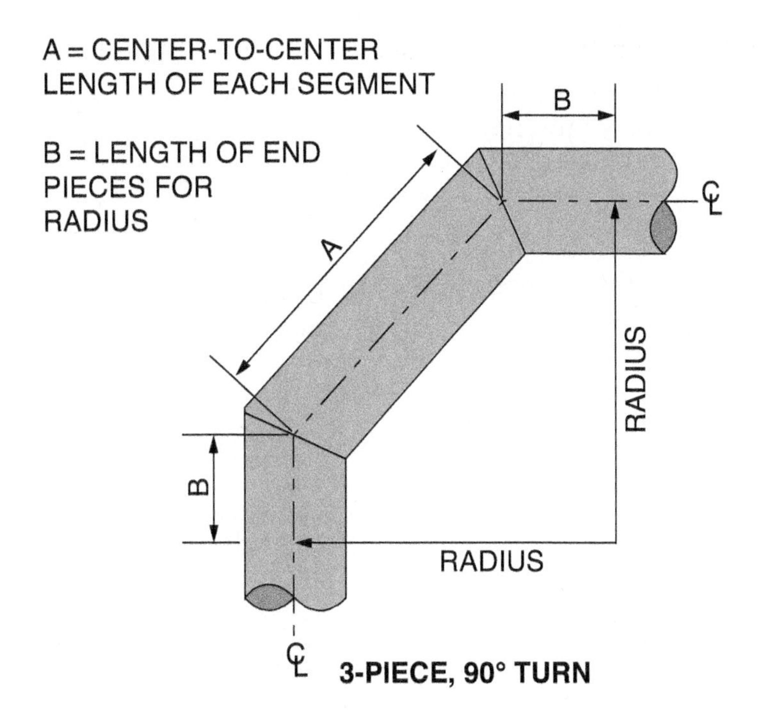

DEGREES IN EACH MITER = NUMBER OF DEGREES IN TURN / 2 × NUMBER OF WELDS

LENGTH OF B = RADIUS × TANGENT OF MITER (ANGLE CUT)
LENGTH OF A = LENGTH B × 2

Figure 19 Mitered turn.

In most cases, the radius of the turn is given. When it is not, it can be measured in the field at the location of the pipe run.

To find the number of degrees in each miter cut of a turn, use the following formula:

$$\frac{\text{Number of degrees in turn}}{2 \times \text{number of welds}}$$

The reason the number of welds is used in this formula is that it is always the total number of pieces in the turn minus 1.

Example: Calculate the number of degrees per miter in a three-piece, 90-degree elbow (two welds).

$$90 \text{ degrees} \div (2 \times 2) = 22\,^{1}/_{2} \text{ degrees per miter}$$

2.3.1 Finding Center-to-End Length

In order for there to be a smooth transition between the segments in a mitered elbow, two things must be consistent: the number of degrees in each miter and the center-to-end length of each segment.

To find the center-to-end length of each segment in a mitered turn, use the following formula:

$$B = \text{radius} \times \text{tangent of miter}$$

Example: The center-to-end length of a three-piece, mitered, 90-degree elbow with a 12" radius is found as follows:

$$B = \text{radius} \times \text{tangent of miter}$$
$$B = 12 \times 0.4142$$
$$B = 4.97, \text{ or } 4\,^{15}/_{16}"$$

2.3.2 Finding Center-to-Center Length

The one center-to-center length for a given radius is calculated by multiplying the center-to-end length by two. Therefore, the formula is as follows:

$$A = B \times 2$$

Using the previous example of a three-piece, mitered, 90-degree elbow on a 12" radius with an A dimension of $4^{15}/_{16}$", dimension A is calculated as follows:

$$A = B \times 2$$
$$A = 4\ ^{15}/_{16}" \times 2$$
$$A = 9.94, \text{ or } 9\ ^{15}/_{16}"$$

When laying out and fabricating three- and four-piece, mitered, 90-degree turns, lay out the complete turn on a straight length of pipe before cutting out the segments. This saves time because all the measuring and cutting can be done at once.

2.4.0 Laying out and Fabricating Three-Piece, 90-Degree Mitered Turns

For training purposes, assume that the task is to lay out and fabricate a three-piece, mitered, 90-degree turn on a 12" radius using 8" pipe. *Figure 20* shows the completed turn.

Follow these steps to lay out and fabricate a three-piece, mitered, 90-degree turn:

Step 1 Find the number of degrees in each miter.

$$\frac{\text{Number of degrees in turn}}{2 \times \text{number of welds}} = \frac{90 \text{ degrees}}{2 \times 2} = 22\ ^{1}/_{2} \text{ degrees}$$

Step 2 Find the cutback for a $22^{1}/_{2}$-degree miter on an 8", Schedule 40 pipe.

$$\text{Cutback} = \frac{\text{pipe OD} \times \text{tangent of miter}}{2}$$
$$\text{Cutback} = \frac{(8.625 \times 0.4142)}{2}$$
$$\text{Cutback} = \frac{3\ ^{9}/_{16}"}{2}$$
$$\text{Cutback} = 1.786 \text{ or } 1\ ^{13}/_{16}"$$

Step 3 Find dimension A for a 12" radius.

$$A = 2 \times \text{radius} \times \text{tangent of miter}$$
$$A = 2 \times 12 \times 0.41421$$
$$A = 24 \times 0.41421$$
$$A = 9.94, \text{ or } 9\ ^{15}/_{16}"$$

Step 4 Divide dimension A in half to find dimension B.

$$B = 9\ ^{15}/_{16} \times\ ^{1}/_{2}$$
$$B = \text{approximately } 5"$$

Step 5 Lay out eight ordinates on the pipe, using an angle iron as a straight edge, and number the ordinates, as shown in *Figure 21.*

Step 6 Establish a working line or **base line** about 3" from the end of the pipe using a wraparound.

Step 7 Measure off dimension B (5") from this working line along ordinates 0 and 0.

THREE-PIECE 90-DEGREE TURN: TWO 45° TURNS EQUALS 22½° CUTS

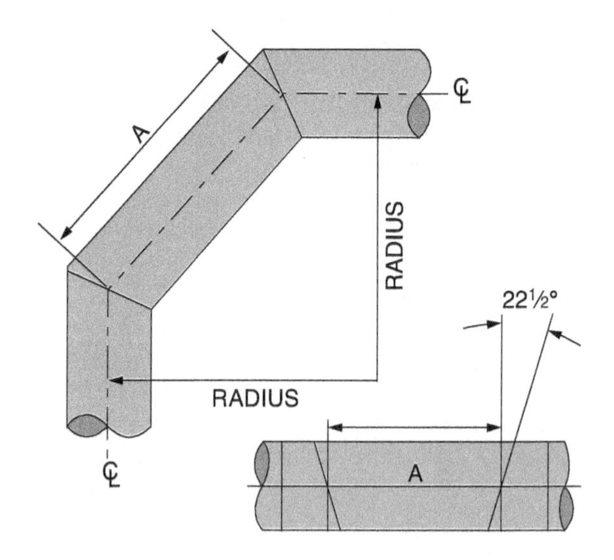

LENGTH A EQUALS RADIUS × 0.4142 × 2

RADIUS (INCHES)	LENGTH A (INCHES)
12	9 15/16
18	14 7/8
24	19 3/4
30	24 7/8
36	29 7/8
42	34 3/4
48	39 3/4

Figure 20 Three-piece, 90-degree turn.

Step 8 Draw a line around the pipe connecting the measurement at ordinates 0 and 0, using a wraparound, to establish a working line to lay out dimension B.

Step 9 Measure off dimension A ($9^{15}/_{16}$") from this working line along ordinates 0 and 0 and draw a line around the pipe at this point. This line becomes the working line used to lay out dimension A.

Step 10 Measure off dimension B (5") from this working line along ordinates 0 and 0.

Step 11 Draw a line around the pipe connecting these points using a wraparound. This line becomes the working line used to lay out dimension B.

Step 12 Lay out the cutback lines on each of the working lines for a $22^1/_2$-degree miter. Remember that each miter slants in the opposite direction from the previous one. For 8" pipe being mitered $22^1/_2$ degrees and divided into 8 ordinates, ordinate number 1 will be $1^1/_4$" from the working line and ordinate number 2 will be $1^3/_4$" from the working line. *Figure 22* shows the layout of a three-piece, mitered turn.

NOTE

Be sure to measure from 0-0 line. If it's not cut square, the lengths will be wrong.

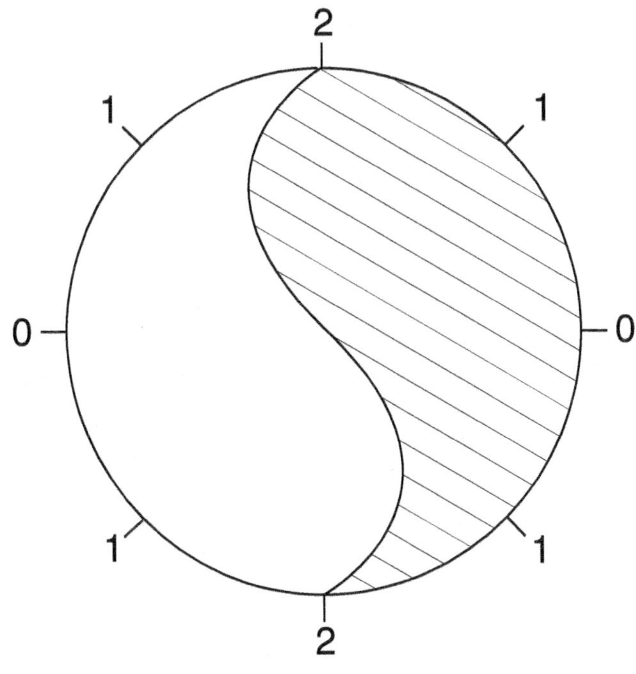

Figure 21 Numbering ordinates.

Step 13 Center-punch marks on either side of ordinates 2 and 2. This helps line up the segments during fit-up.

Step 14 Cut out each segment carefully using an oxyacetylene torch.

Step 15 Dress the edges using a sander or grinder to obtain the proper fit-up.

Step 16 Align the center punch marks and ordinate lines of two segments, and place spacer wires between them. Remember that a certain amount of draw will occur when the segments are tack-welded.

Step 17 Have a qualified welder tack-weld the joint. Adjust the mitered turn as necessary.

Step 18 Align the remaining segment and have a qualified welder tack-weld the joint. Adjust the mitered turn as necessary.

Step 19 Measure the radius to check the squareness of the turn, as shown in *Figure 23*.

2.5.0 Laying out and Fabricating Four-Piece, 90-Degree Mitered Turns

This task calls for laying out and fabricating a four-piece, mitered, 90-degree turn on an 18" radius, using 12" pipe. *Figure 24* shows a finished example.

To lay out and fabricate a four-piece, 90-degree, mitered turn:

Step 1 Find the number of degrees in each miter.

$$\frac{\text{Number of degrees in turn}}{2 \times \text{number of welds}} = \frac{90 \text{ degrees}}{2 \times 3} = 15 \text{ degrees}$$

Step 2 Find the cutback for a 15-degree miter on a 12", standard-weight pipe.

$$\text{Cutback} = \text{pipe OD} \times \text{tangent of miter}$$

$$\text{Cutback} = \frac{(12.75 \times 0.2679)}{2}$$

$$\text{Cutback} = \frac{3.415}{2}$$

$$\text{Cutback} = 1.7078 \text{ or approximately } 1^{11}/_{16}"$$

Step 3 Find dimension A for an 18" radius.

$$A = 2 \times \text{radius} \times \text{tangent of miter}$$
$$A = 2 \times 18 \times 0.2679$$
$$A = 9.64\text{, or approximately } 9\ ^5/_8"$$

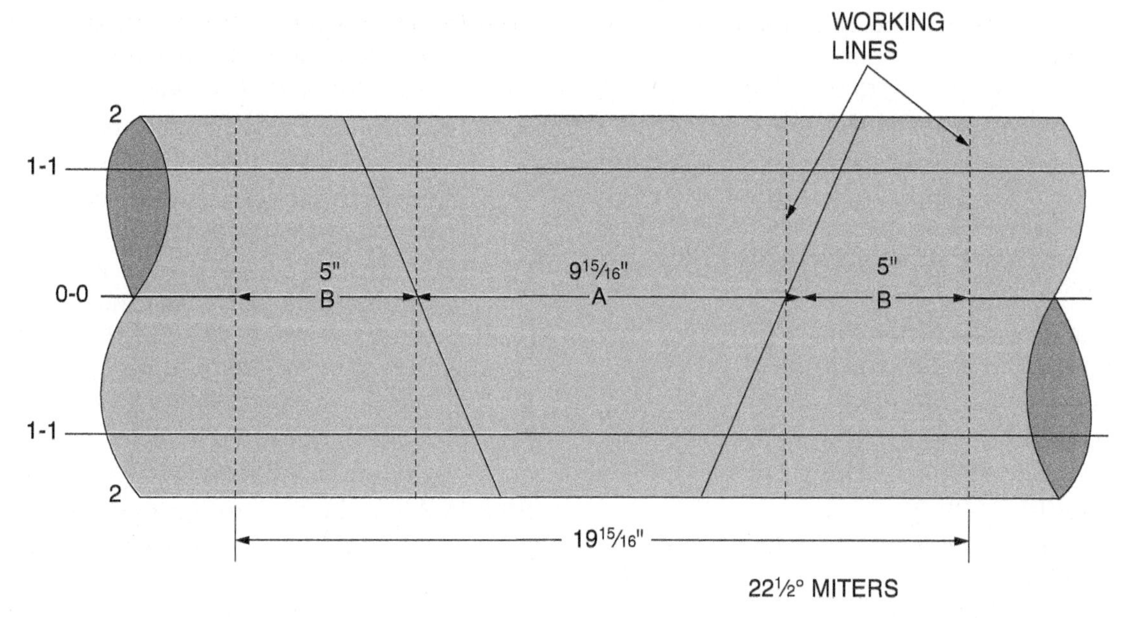

Figure 22 Layout of three-piece, mitered turn.

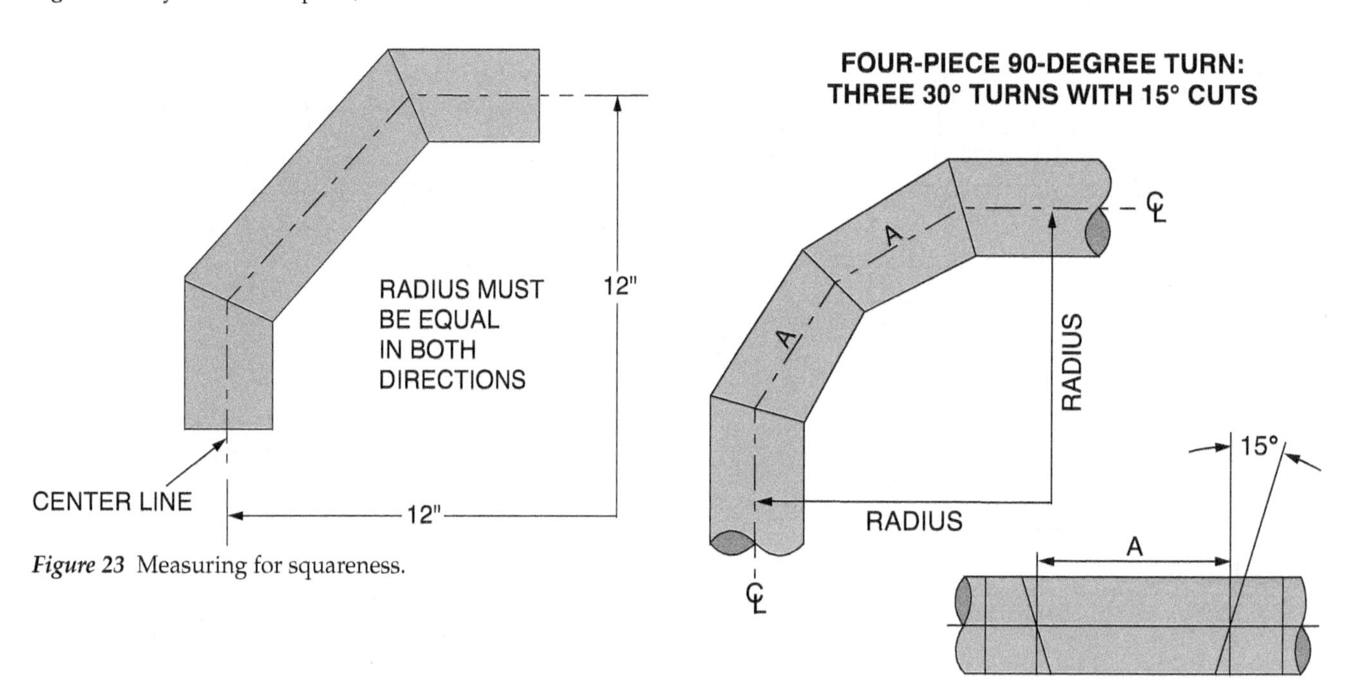

Figure 23 Measuring for squareness.

LENGTH A EQUALS RADIUS × 0.2679 × 2

RADIUS (INCHES)	LENGTH A (INCHES)
24	12⅞
30	16 1/16
36	19 5/16
42	22½
48	25¾
60	32⅛
72	38 9/16
84	45
96	51 7/16

Figure 24 Four-piece, 90-degree turn.

Step 4 Divide dimension *A* by 2 to find dimension *B*.

$$B = \frac{9\,{}^{5}/_{8}}{2}$$

$$B = 4\,{}^{13}/_{16}"$$

Step 5 Lay out and number 16 ordinates on the pipe.

Step 6 Establish a working line about 3" from the end of the pipe using a wraparound.

Step 7 Measure off dimension *B* ($4^{13}/_{16}$") from this working line along ordinates 0 and 0.

Step 8 Draw a line around the pipe along dimension *B*, using a wraparound, to establish a new working line.

Step 9 Measure off dimension *A* ($9^{5}/_{8}$") from this working line along ordinates 0 and 0.

Step 10 Draw a line at this point, using a wraparound, to establish a new working line.

Step 11 Measure off another dimension *A* ($9^{5}/_{8}$") from this working line along ordinates 0 and 0.

Step 12 Draw a line at this point, using a wraparound, to establish the next working line.

Step 13 Measure off another dimension *B* ($4^{13}/_{16}$") from this working line along ordinates 0 and 0.

Step 14 Draw a line around the pipe along dimension *B*, using a wraparound, to establish the final working line. *Figure 25* shows the pipe segment working lines.

Step 15 Lay out the ordinate cutbacks on either side of each of the working lines on ordinate line

NOTE

Remember that each miter slants in the opposite direction from the previous one. For 12" pipe being mitered 15 degrees and divided into 16 ordinates, ordinate number 1 will be $^{5}/_{8}$" from the working line; ordinate number 2 will be $^{13}/_{16}$" from the working line; ordinate number 3 will be $1^{9}/_{16}$" from the working line; and ordinate number 4 will be $1^{11}/_{16}$" from the working line.

Step 16 Use the formulas in the pipefitters' handbooks or calculate the dimensions to mark the cutbacks on the rest of the ordinate lines.

Step 17 Connect the points carefully using a wraparound. *Figure 26* shows the layout of a four-piece, 90-degree, mitered turn.

Step 18 Center-punch marks along either side of ordinates 2 and 2 on each segment.

Step 19 Cut out the segments using an oxyacetylene torch.

Step 20 Bevel the segments and dress the edges, using a sander or grinder.

Step 21 Fit and align the segments, then have a qualified welder tack-weld them.

Step 22 Check the radius of the completed turn.

2.6.0 Mitering a Wye

The procedure for laying out a cutback for a wye is very similar to that for laying out a miter. The working line is first drawn on the branch and the header. The first cutback for the branches is the

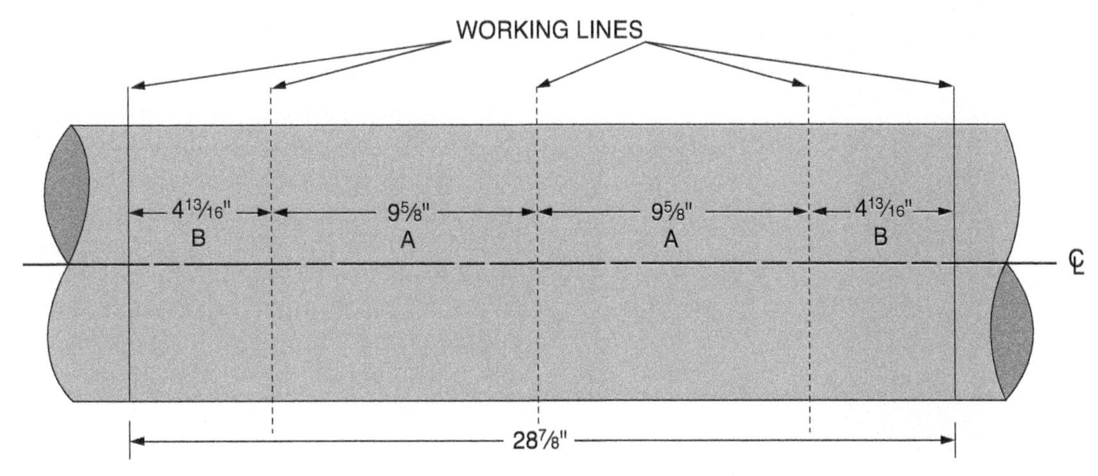

Figure 25 Pipe segment working lines.

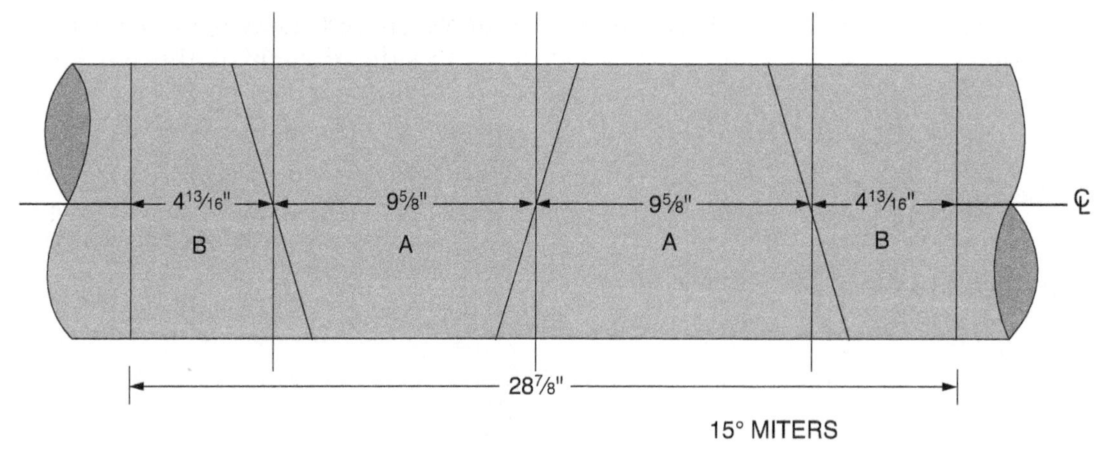

Figure 26 Layout of four-piece, 90-degree, mitered turn.

same as the cutback for the header. The header has two cuts that mirror each other. The branches have a cut that makes the miter with the header and a cut that makes the miter to the other branch (*Figure 27*).

To start, decide where the intersection of the two pipes will be. The best way is to consider the intersection to be at the center of the header, where the miters meet. Set a working line there. Mark the 4, 8, or 16 axial center lines of the header and the branches at 90 degrees around the pipes, just as when creating miter turns. The lines will be where measurements occur for the ordinates of cutbacks.

The first calculation is the miter angle, which will be $\frac{1}{2}$ the intended deviation from the line of the header. The calculation here is the same as that for a miter, except that there is only one weld, so the equation is:

Miter angle = Degree of Bend ÷ 2.

If the wye is to be 60 degrees, the degree of deviation from the axis of the header on each side is 30 degrees, so the miters between the branches and the header are cut at 15 degrees. The first ordinate (the long cut) on a 4" header and branch is:

$$\text{Cutback} = \frac{\text{pipe OD} \times \text{tangent of miter}}{2}$$

$$\text{Cutback} = \frac{(4.5 \times 0.2679)}{2}$$

$$\text{Cutback} = 0.6028\text{, or about } \tfrac{5}{8}\text{"}$$

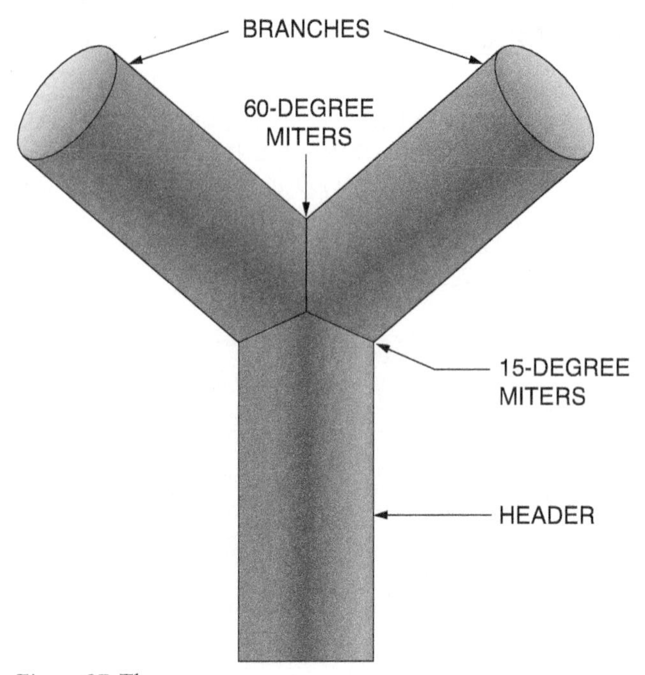

Figure 27 The wye.

The cutback for the branches of the wye is essentially the same, since the calculation for a 15-degree cutback using the OD of the pipe (12.75") = 1.7075", which also rounds off to $1\frac{5}{8}$". Other ordinates are calculated using the sine of the angle from the working line, just as in calculating an 8- or 16-ordinate miter. Only half of each piece is mitered in one direction, though on the header, the miters meet in the middle and mirror each other. The cutback for the header of a 4", 60-degree wye, then, is $\frac{5}{8}$". Since the goal does not involve figuring the radius of a turn, it is not necessary to figure the lengths of the miter pieces. The header is the easy piece with just two cuts that mirror each other. The branches have an opposite cut, because one side is matched to the header, while the other side matches up to the other branch. The second cut on the branch is at 60 degrees from the other branch, which turns 30 degrees from the line of the header, so the calculation for the cutback is 60 degrees.

Therefore, the second cut is calculated this way:

$$\frac{\text{Pipe OD} \times \tan 60}{2} = \frac{(4.5 \times 1.73205)}{2}$$

$$\text{Cutback} = 3.8971 = 3\,{}^{7}/_{8}\text{"}$$

when rounded to the nearest sixteenth

For this task, do the layout from the working line for both sides. That will establish the longest ordinate on each side. If the layout has 8 or 16 ordinates, the remainder of the ordinates are calculated just as they would be for a miter (with the understanding that miters do not form a single straight line).

Next, lay out the header. Then lay out the short cut on each branch. Finally, lay out the long cut on each branch. All of the layouts have to be done before doing any cutting, so that any mistakes in the cutting won't spoil the layout for the other cuts.

2.0.0 Section Review

1. The purpose of ordinate lines is to _____.
   a. regulate flow and velocity within the line
   b. indicate direction of the flow
   c. serve as guides for cutting miters and marking contours
   d. predetermine anchor points along the set of a piping run

2. The first step in laying out cutback lines is _____.
   a. laying out the correct number of ordinate lines
   b. drawing a working line around the pipe at the center of the turn
   c. measuring the cutback distance
   d. calculating cutback lengths

3. The number of segments in a 90-degree mitered turn (for pipe 6" and larger) will vary from _____.
   a. 2 to 7 segments
   b. 2 to 12 segments
   c. 4 to 8 segments
   d. 5 to 10 segments

4. When laying out and fabricating three-piece, 90-degree mitered turns, it's important to measure from the 0-0 line because _____.
   a. tack welding will become necessary
   b. the use of an oxyacetylene torch adds at least one inch to the cut
   c. the lengths will be incorrect if it's not cut square
   d. center punching will offset the sine of the radius

5. Each miter slants _____ the previous one.
   a. perpendicular to
   b. horizontally aligned to
   c. in the same direction as
   d. in the opposite direction from

6. When mitering a wye, the *first* calculation is the _____.
   a. miter angle
   b. miter radius
   c. pipe radius
   d. pipe angle

SECTION THREE

3.0.0 SADDLES AND SUPPORTS MADE OUT OF PIPE

Objective

Explain how to lay out and fabricate saddles and supports made out of pipe.

a. Explain how to lay out and fabricate a saddle.
b. Explain how to lay out and fabricate supports made out of pipe.

Performance Tasks

6. Using reference charts, lay out and fabricate a 45-degree lateral.
7. Lay out and fabricate a Type 1 pipe support.
8. Lay out a 45-degree lateral by performing a geometric layout.
9. Lay out and fabricate a saddle.

Trade Terms

Chord: A straight line crossing a circle that does not pass through the center of the circle.

Saddle: A fabricated 90-degree intersection of pipe; also known as fishmouth.

A **saddle** is a 90-degree intersection of two pipes involving the ID of the branch and the OD of the header.

3.1.0 Saddle

The information that is needed for a saddle is the same as that which is needed for a lateral. One way of visualizing the relationship of the two pieces is with two circles, one the size of the branch ID, and one the size of the header OD. An advanced pipefitter will transfer the branch ID to the header OD, while making the calculations.

After determining the radius of the branch, draw the line to the circle of the header. The point where the branch will contact the OD of the header forms a right triangle, with one side being the radius of the branch ID and the other (the hypotenuse) being the radius of the header OD (*Figure 28*). Using the Pythagorean theorem ($a^2 + b^2 = c^2$), the sum of the squares of two sides of a right triangle is equal to the square of the hypotenuse.

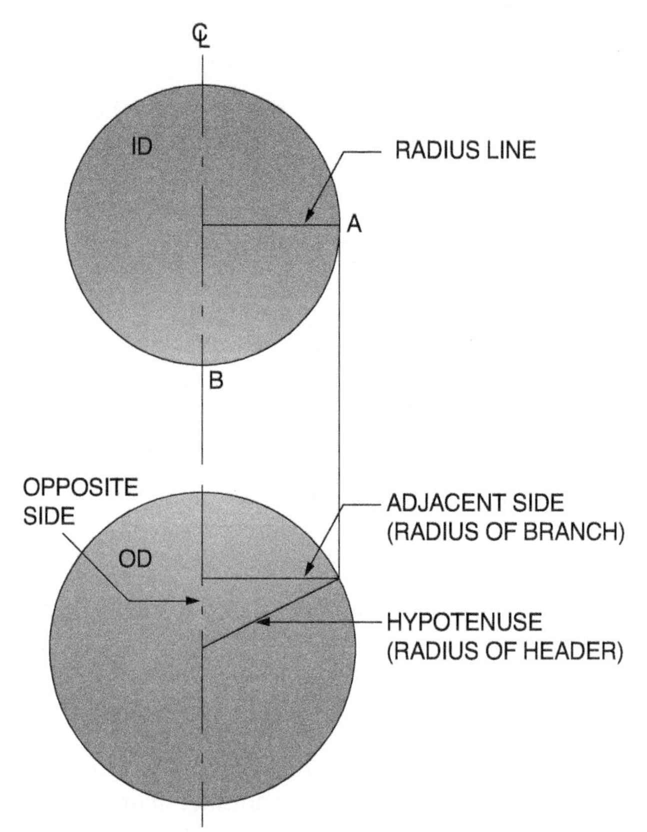

Figure 28 Calculating the first line.

To calculate the first line:

Step 1 Subtract the square of the radius of the ID of the branch from the square of the OD of the header.

- This will equal the square of the length from the center of the header to the line where the branch intersects with the header. The square root of that number is what is used to calculate the longest ordinate of the branch.

Step 2 Subtract this number from the radius of the OD of the header to know how long the long side of the branch will come from the working center line.

If only 4 ordinates are needed, all the calculations are complete. If 8 or 16 ordinates are needed, more calculations are in order to determine the lengths of the other ordinates. In a right triangle, the side that is opposite, or across from the 90-degree angle that is formed by the two sides, is called the hypotenuse. The right angle of the triangle is formed by a line drawn to the end of the radius from another radius at 45 degrees to the hypotenuse (*Figure 29*). That line will equal the radius (the hypotenuse) times the sine of 45 degrees. Now the same set of calculations that were used for the long ordinate can be done for the other ordinates, using the side that was calculated in conjunction with the Pythagorean theorem.

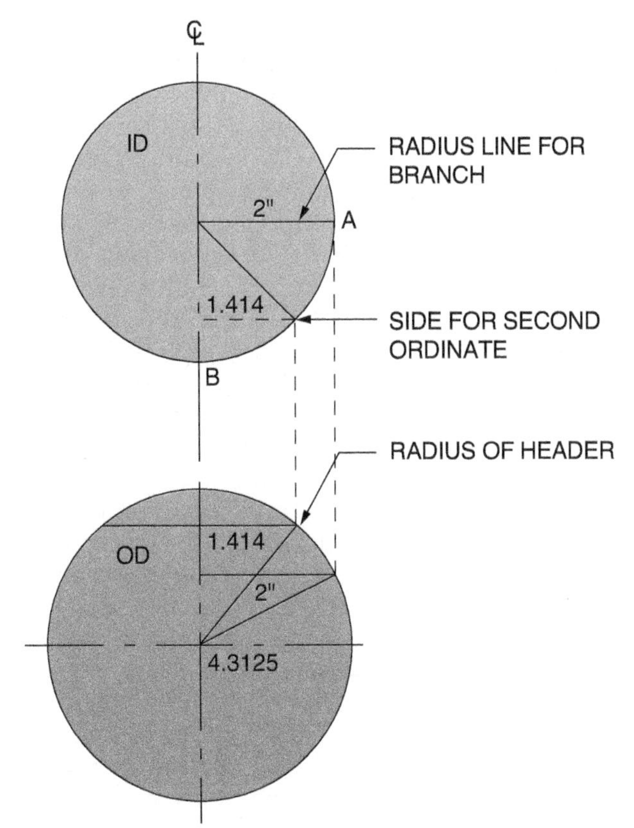

Figure 29 Calculating the second line.

To explore a typical, everyday form, assume the task of putting together a 4" nominal branch with an 8" standard weight nominal header. The first line is a 2" radius on the ID of the branch. The OD of the 8" header is 8.625", and the radius would be half that, or 4.3125". The first calculation, then, will be: side = $\sqrt{(4.3125^2 - 2^2)}$ = 3.8206… . The side is to be subtracted from the radius of the OD of the header, to provide the long ordinate, thus producing an ordinate which equals 4.3125 – 3.8206… = 0.4918. Rounding to nearest 16th increases this by ½". Next, draw a working line around the circumference of the branch ½" from the end of the branch to mark the end of the cut on the pipe.

> **NOTE**
>
> An ellipsis following a decimal number indicates that there was a continuation of the number that has been omitted to save space.

At this point, more ordinates are needed to produce a smooth, curved cut on the branch. Calculate one more point on the 4" branch to get a smoother line. A 45-degree angle is desired so use the branch radius times sine 45. Now the line = 2 × sine 45 = 1.4142.

Applying those results to the Pythagorean theorem, a side = $\sqrt{(4.3125^2 - 1.4142^2)}$ = 4.074… . Subtract that from the radius of the header to get 4.3125 – 4.074… = 0.2384…, which rounds out to ¼". Use that as the middle point of the layout on the branch and the header and the two will fit well.

The next group of steps calls for considering the fit when putting a 12" branch on a 16" header; additional ordinates are needed. The first calculation series is still square root of (header OD radius^2 – branch ID radius^2). That equates to $\sqrt{(8^2 - 6^2)}$ = 5.2915… . Subtract that answer from the radius of the header to get 2.7084…, or roughly 2$^{11}/_{16}$".

Next, get the ordinate lines from the shortest to the longest. Since the zero points on the working line are known, start with ordinate 1 = 6 × sine 22.5 degrees = 2.2961… . This means that $\sqrt{(8^2 - 2.2961^2)}$ = 7.6634…, so the next step is to subtract that from 8" to get approximately $^5/_{16}$".

Moving next to 45 degrees, start with the branch radius times sine 45 (6 × sine 45) = 4.2426… . Put this answer in the Pythagorean theorem, and see the result of $\sqrt{(8^2 - 4.24622^2…)}$ = 6.7823…, or 6$^{13}/_{16}$". Subtract that from 8" to get approximately 1.2176, or $^{13}/_{16}$".

The final step involves the last ordinate at 6 × sins 67.5 = 5.5432… . When used with the Pythagorean theorem, the result is $\sqrt{(8^2 - 5.432^2…)}$ = 5.7682, or 5¾". Subtract that from 8" to get approximately 2.2318, or 2¼". *Figure 30* shows the two circles with the triangles shown.

Now it is possible to lay out the set of marks and cope out the end of the riser, having confidence that it will fit against the surface of the header:

Step 1 Determine where the center of the riser will be on the header.

Step 2 Mark a center line on the top of the header and extend it on the top of the header past where the outside of the riser will be.

Step 3 Mark a center line on both sides of the riser, using a center head to transfer the line from one side to the other.

Step 4 Measure to the first ordinate that was calculated. This provides the highest point of the curve being cut. Put a working line around the riser at this point.

Step 5 Put the other calculated segment heights and chord lengths on the riser. This is done on all four quadrants, and the curves can be drawn with a spline or a wrap. The cuts should be made so that the curve is as smooth as possible, and they should be beveled so that the original cut line is where the edge of the hole is.

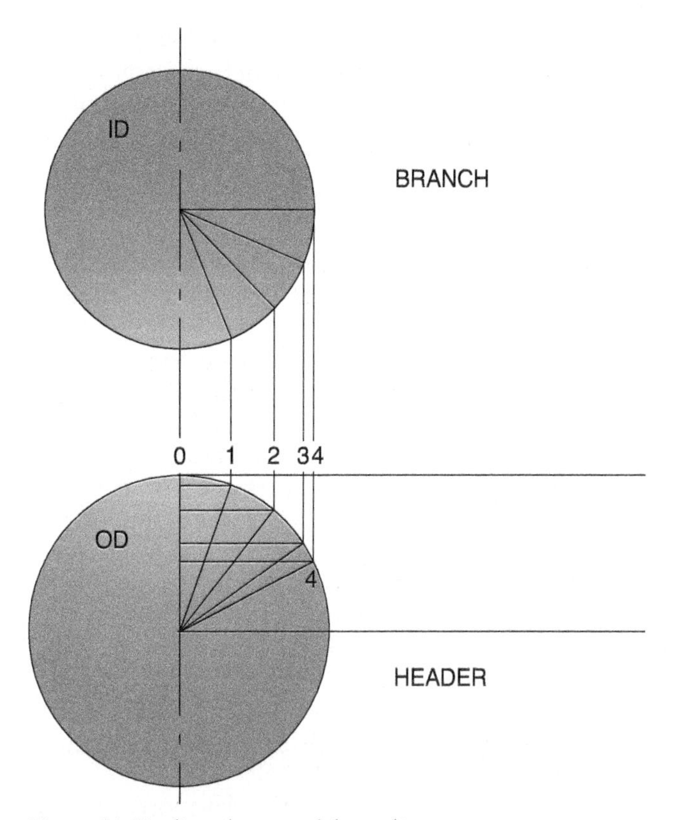

Figure 30 Finding the rest of the ordinates.

Step 6 Put a torpedo level on the side of the riser and use a framing square on the center line of the header to keep the riser at right angles to the header along that axis.

Step 7 Mark the point where the inside of the bevel touches the header with a soapstone.

Step 8 Cut the hole in the header and bevel it.

Step 9 Complete the fit-up and tack-weld it together.

3.1.1 Determining Lateral Dimensions

A lateral is the intersection of two pipes at less than 90 degrees. One-piece laterals are available in a variety of sizes from manufacturers, but pipefitters must often fabricate laterals in the field. The following sections explain how to lay out 45-degree laterals.

When fabricating angled laterals, ordinate reference charts are available. However, these charts may or may not always be available for a given angle, and it is best to be able to manually calculate the figures. The equations used for the calculations for risers on 90-degree intersections, supplemented with trigonometry, can provide the necessary measurements for any angle of laterals.

It is necessary to perform the calculations from OD to ID for the laterals, as the insides of the pipes still have to match smoothly and the OD will match relatively smoothly if the IDs are flush. If the lateral is the same ID on both pieces, the calculation is relatively simple. For a 45-degree lateral of 4" Schedule 40 pipe, the two cuts are at a 90-degree angle, meeting at the center of the pipe. One cut is at 22.5 degrees from the axis of the pipe and the other cut is 67.5 degrees from the axis of the pipe. The fact that the pipe is the same ID means that the pipe will meet in the middle of both pipes, so half of the diameter is the right angle measurement to the intersection of the two cuts. That allows for calculation of the length from the point of intersection in the middle to the point of intersection at the surface for both cuts.

The shorter cut will be calculated by the formula for the tangent of 22.5 degrees, because length to the surface is known to be half the OD. From the point of intersection to the surface is 2.25" for a 4" Schedule 40 pipe. Tan 22.5 degrees × 2.25 = 0.9319…, or $^{15}/_{16}$". The second, longer cut will be tan 67.5 degrees × 2.25 = 5.4319…, or $5^{7}/_{16}$". These are quick calculations with a scientific calculator.

For a reducing lateral, the calculation is a little more complex, but can still be done with a scientific calculator. For the smaller laterals, such as a 4" by 8" 45-degree lateral, it is possible to calculate two points on the branch, cut the branch, hold it on the header, and lay out the curve on the header. However, it is more accurate to cut the saddle curve on the branch, so that it fits neatly on the header, and then draw the cut on the header. There are a series of steps for each point. Follow these steps with reference to *Figure 31*.

Series 1 (Ordinate #1):

Step 1 Line A1 = the radius of the branch ID times the sine of 22.5 degrees = 2.013 × sine 22.5 = 0.7703… .

NOTE

Line B1 is the same size as Line A1.

Step 2 To calculate the length of Line C1, use the Pythagorean theorem, using the outside radius of the header as the length of the hypotenuse, so $C1^2 = (R^2 - B1^2)$. Therefore, $C1 = \sqrt{(4.3125^2 - 0.7703^2)} = 4.2431\ldots$.

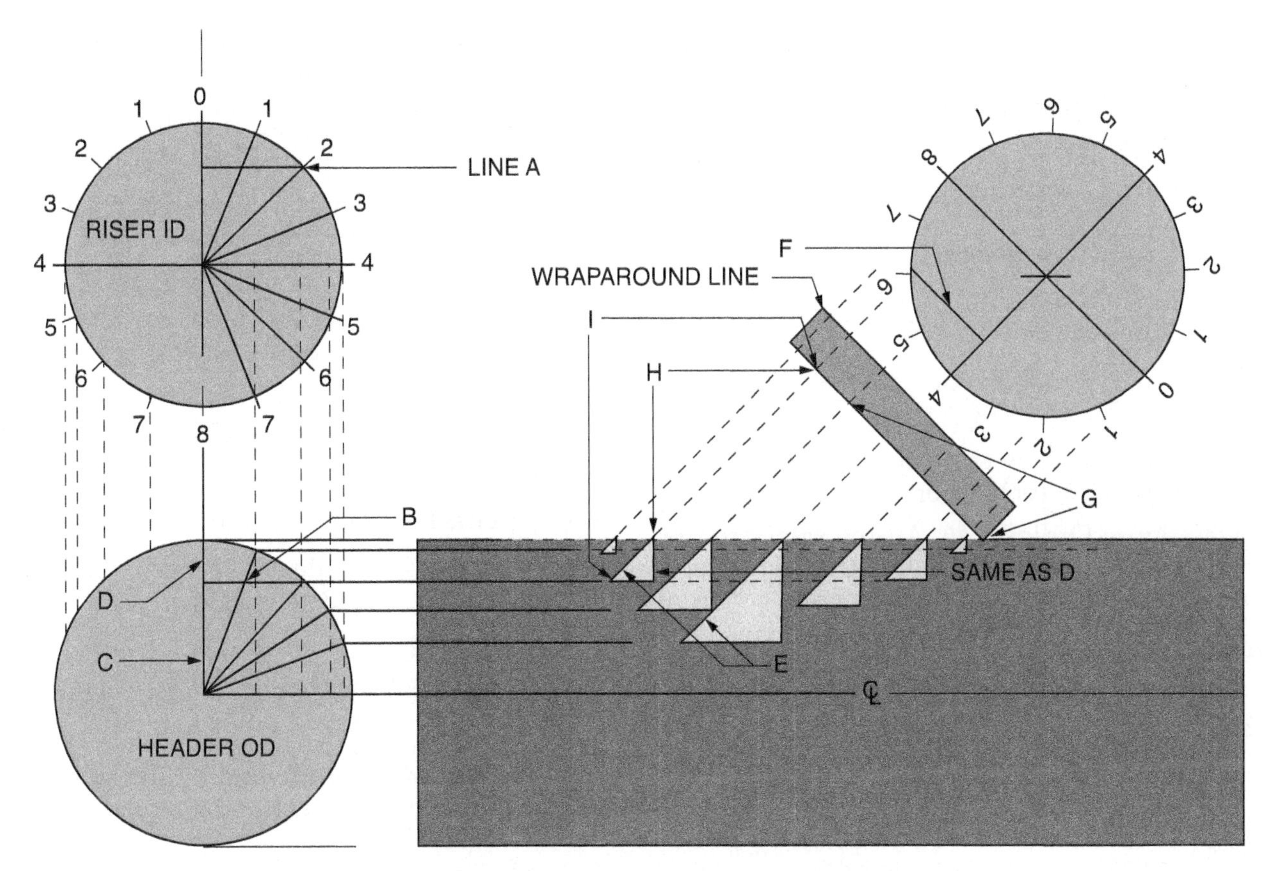

Figure 31 Layout for a lateral.

Step 3 Line D1, the height from Line B1 to the top of the circle at the longest point, is the radius of the header minus Line C1: D1 = R – C1 = 0.0693…

Step 4 This line, D1, is the same length as Line D1 on the branch ordinates. Divide D1 by the sine of the angle of the lateral to get the length of Line E1: 0.0693…÷ sine 45 = 0.0980… (Note E1 and E7 are the same length.)

Step 5 Find Line F1 by multiplying the radius of the branch ID by sine 67.5, F1 = sine 67.5 × 2.013 = 1.8597… (Note: F1 and F7 are the same length.)

Step 6 Subtract the inside radius of the branch from F1 to get G1: 2.013 – 1.8597 = 0.1532…

Step 7 Line H1 = Line G1 divided by the tangent of the angle of the lateral = 0.1532… ÷ tan 45 = 0.1532…, because the tangent of 45 degrees is 1.

Step 8 To find I1, add Line H1 to Line E1: 0.0981… + 0.1532… = 0.2513…, or $\frac{1}{4}$".

Now figure Ordinate #7. All the steps are the same until Step 6 is reached. Follow Steps 1 through 5 above (for Ordinate #1), then substitute the below Steps 6 through 8 to complete the calculation for Ordinate #7.

Step 6 Add the inside radius of the branch to F7 to get G7: 2.013 + 1.8597…= 3.8727…

Step 7 Line H7 = Line G7 divided by the tangent of the angle of the lateral = 3.8727… ÷ tan 45 = 3.8727… because the tangent of 45 degrees is 1.

Step 8 To find I7, add Line H7 to line E7: 0.0981… + 3.8727… = 3.9709… or 4".

Series 2 (Ordinate #2):

Step 1 Line A2 = radius of branch times sine 45 degrees = 2.013 × sine 45 = 1.4234…

Line B2 = Line A2

Step 2 Line $C2^2$ = radius of $header^2$ – Line $B2^2$. Therefore, C2 = $\sqrt{(4.3125^2 - 1.4234^2)}$ = 4.0708…

Step 3 Line D2 = radius of header – Line C2 = 4.3125 – 4.0708… = 0.2417…

Step 4 Line E2 = D2 divided by the sine of the angle of the lateral = 0.2417… ÷ sine 45 degrees = 0.3418… (Note E2 and E6 are the same length.)

Step 5 Line F2 = radius of branch × sine 45 = 2.013 × sine 45 = 1.423… (Note F2 and F6 are the same length.)

Step 6 Line G2 = radius of branch – Line F2 = 2.013 – 1.423… = 0.5895…

Step 7 Line H2 = Line G2 ÷ tan 45 degrees = 0.5895...

Step 8 Line I2 = Line H2 + Line E2 = 0.5895… + 0.3418… = 0.9313… or $^{15}/_{16}$".

Now figure Ordinate #6. All the steps are the same until you get to Step 6. Use the below steps to substitute for steps 6 through 8 above.

Step 6 Add the inside radius of the branch to F6 to get G6: 2.013 + 0.3418…= 2.3548…

Step 7 Line H6 = Line G6 divided by the tangent of the angle of the lateral = 2.3548…/ tan 45 = 2.3548…

Step 8 To find I7 add Line H7 to Line E7: 0.0981… + 2.3548… = 2.4529…. or $2^{7}/_{16}$".

Series 3 (Ordinate #3):

Step 1 Line A3 = radius of branch times sine 67.5 = 2.013 × sine 67.5 = 1.8597…

Line B3 = Line A3.

Step 2 Line $C3^2$ = radius of $header^2$ – Line $B3^2$ = $4.312^2 - 1.8597\ldots^2$ = 15.139…

Line C3 = $\sqrt{15.139\ldots}$ = 3.8909…

Step 3 Line D3 = radius of header – Line C3 = 4.3125 – 3.8909 = 0.42162…

Step 4 Line E3 = D3 divided by the sine of the angle of the lateral = 0.4216… ÷ sine 45 = 0.5962… (Note E3 and E5 are the same length.)

Step 5 Line F3 = radius of branch × sine 22.5 = 2.013 × sine 22.5 degrees = 0.7703… (Note F3 and F5 are the same length).

Step 6 Line G3 = radius of branch – Line F3 = 2.013 – 0.7703... = 1.2426…

Step 7 Line H3 = Line G3 ÷ tan 45 degrees = 1.2426…

Step 8 Line I3 = Line H3 + Line E3 = 1.2426…+ 0.5962… = 1.8389…= $1^{13}/_{16}$"

Now figure Ordinate #5. All the steps are the same until you get to Step 6:

Step 6 Add the inside radius of the branch to F5 to get G5: 2.013 + 0.7703…= 2.7833…

Step 7 Line H5 = Line G5 divided by the tangent of the angle of the lateral = 2.7833… ÷ tan 45 = 2.7833…

Step 8 To find I5 add Line H5 to Line E5: 0.5962… + 2.7833… = 3.3795… or $3^{3}/_{8}$".

Series 4 (Ordinate #4):

Step 1 Line A4 = radius of branch times sine 90 degrees = 2.013 × sine 90 = 2.013

Line B4 = Line A4

Step 2 Line $C4^2$ = radius of $header^2$ – Line $B4^2$ = $4.3125^2 - 2.013^2$ = 14.5454…

C4 = $\sqrt{14.5454\ldots}$ = 3.8138…

Step 3 Line D4 = radius of header – Line C4 = 4.3125 – 3.8138… = 0.4986…

Step 4 Line E4 = D4 divided by the sine of the angle of the lateral = 0.4986… ÷ sine 45 = 0.7051…

Step 5 Line F4 = radius of the branch × sine 90 degrees = 2.013 × sine 90 degrees = 2.013

Step 6 Line G4 = radius of branch = 2.013 (Note that nothing has been added to the branch radius. Point 4 is on the radius; it won't be added here.)

Step 7 Line H4 = Line G4 ÷ tan 45 degrees = 2.013

Step 8 Line I4 = Line H4 + Line E4 = 2.013 + 0.7051… = 2.7181…or $2^{11}/_{16}$"

Series 5 (Ordinate #8):

Step 1 Line A8 is zero.

Step 2 Since A8 is equal to B8, it is also zero.

Step 3 Since B8 is zero, C8 is equal to the radius of the header OD and there is no D8.

Step 4 Since there is no D8, there is no E8 as well, since it equals zero.

Step 5 Line F8 is equal to the radius of the branch; therefore, F8 equals 2.013".

Step 6 Line G8 is equal to F8 + the radius of the branch. G8 = 2.013 + 2.013 = 4.026"

Step 7 Line H8 = G8 divided by the tangent of the angle of the lateral. H8 = 4.026 ÷ tan 45 degrees = 4.026".

Step 8 Line I8 = H8 + E8 = 4.026 + 0 = 4.026 or 4".

When references are available, they make the job of laying out laterals easier and quicker than laying them out on a flat surface and then transferring them to the pipe. The following exercise calls for laying out a 45-degree lateral using 4" carbon steel pipe.

To lay out a 45-degree lateral, using references:

Step 1 Obtain a 45-degree lateral reference chart, such as the one shown in *Figure 32*.

Step 2 Measure 5" or 6" from the end of the pipe and mark a straight line around the pipe using a wraparound. This is the base line.

Step 3 Divide the outside circumference of the pipe by 8 to obtain 8 ordinates.

Step 4 Mark the 8 ordinates on the pipe, 0 to 4 on each side, as shown in *Figure 33*.

Step 5 Obtain the ordinate 1 cutback measurement from *Figure 32*.

Step 6 Measure 1 5/16" from the base line along ordinate 1 and mark the pipe.

Step 7 Obtain the ordinate 2 cutback measurement from *Figure 32*.

Step 8 Measure 3 3/4" from the base line along ordinate 2 and mark the pipe.

Step 9 Obtain the ordinate 3 cutback measurement from *Figure 32*.

Step 10 Measure 4 1/8" from the base line along ordinate 3 and mark the pipe.

Step 11 Obtain ordinate 4 cutback measurement from *Figure 32*.

Step 12 Measure 4" from the base line along ordinate 4 and mark the pipe.

Step 13 Place a wraparound on the lateral and mark a line connecting ordinates 1, 0, and 1, using a soapstone.

Step 14 Place a wraparound on the lateral and mark a line connecting ordinates 1, 2, and 3, using a soapstone.

Step 15 Place a wraparound on the lateral and mark a line connecting ordinates 3, 4, and 3, using a soapstone. *Figure 34* shows the lateral ready to cut.

Step 16 Cut the lateral along the cut line, using a cutting torch, keeping the torch at a 90-degree angle to the pipe.

45° Laterals
Standard Weight Pipe Risers
Eight Ordinates

SIZE OF RISER	SIZE OF HEADER												ORDINATE NO.
	3"	4"	6"	8"	10"	12"	14"	16"	18"	20"	22"	24"	
3"	1	13/16	11/16	5/8	5/8	9/16	9/16	9/16	9/16	9/16	1/2	1/2	1
	2 13/16	2 3/8	2 1/16	1 15/16	1 7/8	1 13/16	1 3/4	1 3/4	1 3/4	1 11/16	1 11/16	1 11/16	2
	3 1/8	3	2 7/8	2 13/16	2 3/4	2 3/4	2 3/4	2 3/4	2 11/16	2 11/16	2 11/16	2 11/16	3
	3 1/16	3 1/16	3	3 1/16	3 1/16	3 1/16	3 1/16	3 1/16	3 1/16	3 1/16	3 1/16	3 1/16	4
4"	—	**1 5/16**	1 1/16	15/16	7/8	13/16	13/16	3/4	3/4	3/4	3/4	11/16	1
	—	**3 3/4**	3	2 11/16	2 9/10	2 1/2	2 7/16	2 3/8	2 15/16	2 15/16	2 1/4	2 1/4	2
	—	**4 1/8**	3 7/8	3 3/4	3 11/16	3 11/16	3 5/8	3 5/8	3 5/8	3 9/16	3 9/16	3 9/16	3
	—	**4**	4	4	4	4	4	4	4	4	4	4	4
6"	—	—	2	1 11/16	1 1/2	1 7/16	1 3/8	1 5/16	1 1/4	1 3/16	1 3/16	1 3/16	1
	—	—	5 13/16	4 13/16	4 3/8	4 1/8	4	3 7/8	3 3/4	3 11/16	3 5/8	3 9/16	2
	—	—	6 5/16	6	5 5/16	5 11/16	5 5/8	5 9/16	5 9/16	5 1/2	5 1/2	5 5/16	3
	—	—	6 1/16	6 1/16	6 1/16	6 1/16	6 1/16	6 1/16	6 1/16	6 1/16	6 1/16	6 1/16	4
8"	—	—	—	2 5/8	2 5/16	2 1/8	2	1 7/8	1 13/16	1 3/4	1 11/16	1 5/8	1
	—	—	—	7 3/4	6 1/2	6	5 3/4	5 1/2	5 5/16	5 3/16	5 1/16	4 15/16	2
	—	—	—	8 5/16	7 15/16	7 3/4	7 5/8	7 9/16	7 7/16	7 3/8	7 5/16	7 5/16	3
	—	—	—	8	8	8	8	8	8	8	8	8	4

Figure 32 45-degree lateral reference chart.

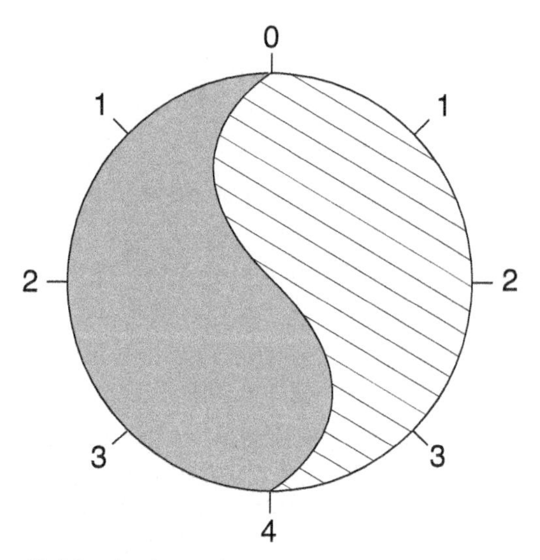

Figure 33 Numbering ordinates.

Step 17 Clean the slag off the lateral using a grinder.

Step 18 Place the lateral on the header and use a torpedo level to establish a 45-degree angle.

Step 19 Mark around the outside diameter of the lateral on the header using a soapstone.

Step 20 Cut along the inside of the mark on the header using a cutting torch. Be sure to make the cut at a 90-degree angle with the header.

Step 21 Grind off any slag to clean the header.

Step 22 Place the lateral on the header and check the fit.

Step 23 Make any necessary adjustments and bevel the lateral and header.

Step 24 Fit up the lateral to the header using spacer wires to obtain the correct welding gap; verify that the header is level and use a torpedo level to establish a 45-degree angle.

Step 25 Tack-weld the lateral to the header.

Step 26 Verify that the lateral is still at a 45-degree angle to the header.

Step 27 Complete the weld.

3.2.0 Fabricating Supports Made out of Pipe

Pipefitters are sometimes required to fabricate project-made supports, also known as supports made out of pipe, to help hold a piping system in place. These are usually welded directly to the pipe on a 90-degree, long-radius elbow and then welded to an anchor plate on the floor, wall, or ceiling. Supports made out of pipe provide horizontal and vertical support. They come in three types and the one that is used depends on the area of the elbow to which the support is attached. *Figure 35* shows three types of supports.

The layout for each type of support is the same, although each type requires different dimensions. The dimensions for each type of support can be found in *The Pipe Fitters Blue Book*. The procedure for laying out the supports is very similar to the procedure for laying out 45-degree laterals, using reference charts. *Figure 36* shows the layout for a support on the back of an elbow.

This procedure calls for fabricating a Type 1 support using 3", carbon steel, 40-weight pipe that will support a 6", long-radius, 90-degree elbow. The center line of the support is in line with the center line of the elbow. The support must be cut

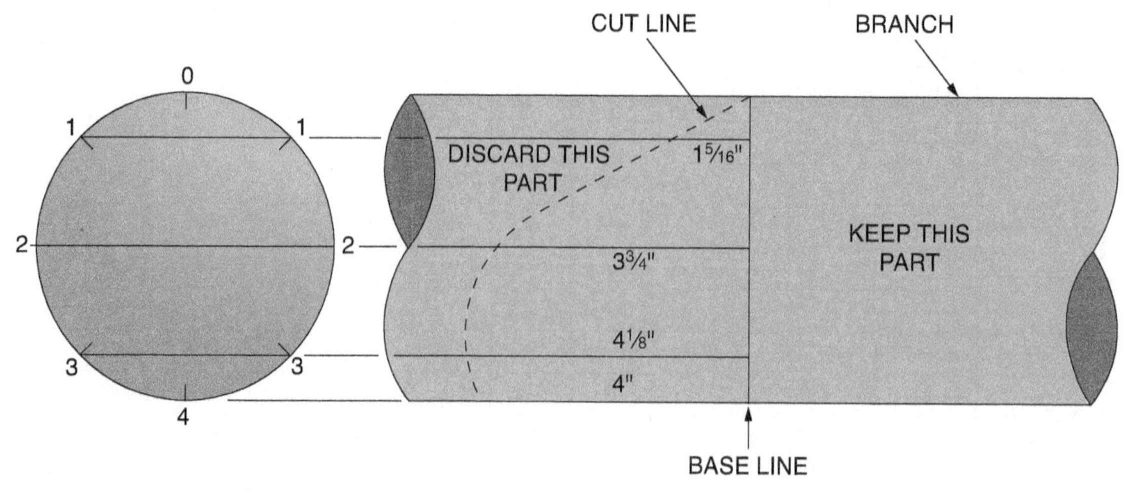

Figure 34 Lateral ready to cut.

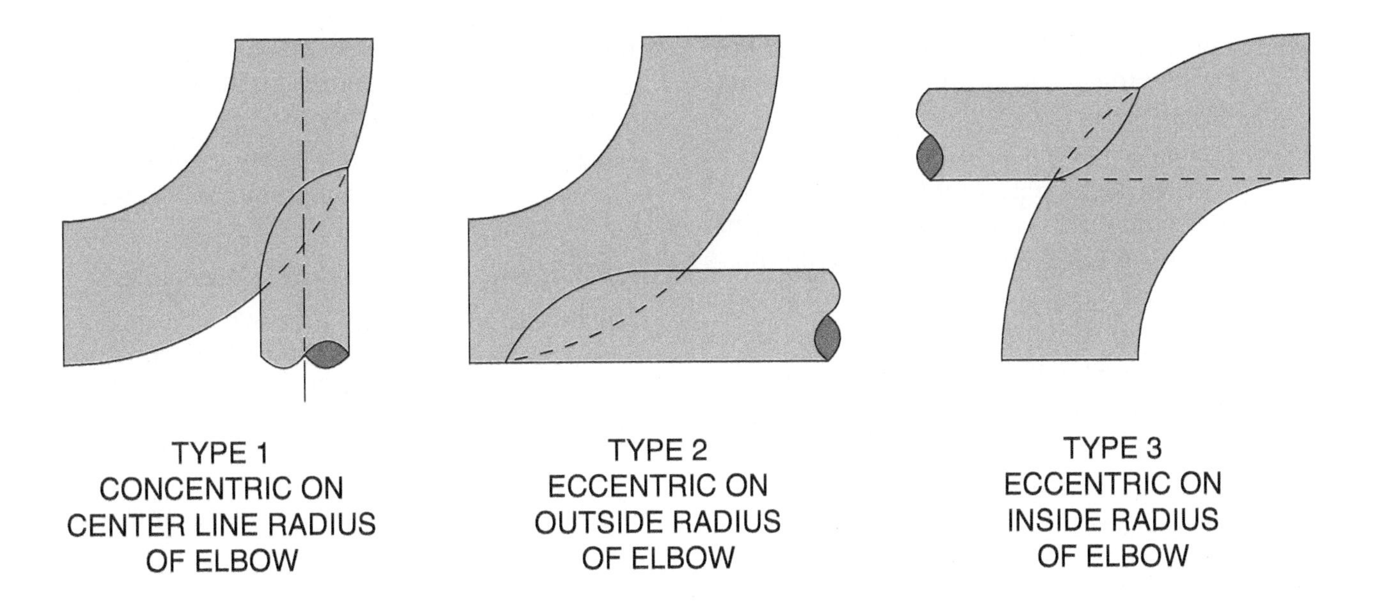

Figure 35 Three types of supports.

so that the center line of the pipe is 2'6" above grade elevation. A 6" base and ½" baseplate will be used with the support. The 90-degree elbow is part of a spool consisting of the elbow and a straight length of pipe. *Figure 37* shows the support.

Follow these steps to fabricate the support:

Step 1 Obtain the support detail sheet and specifications.

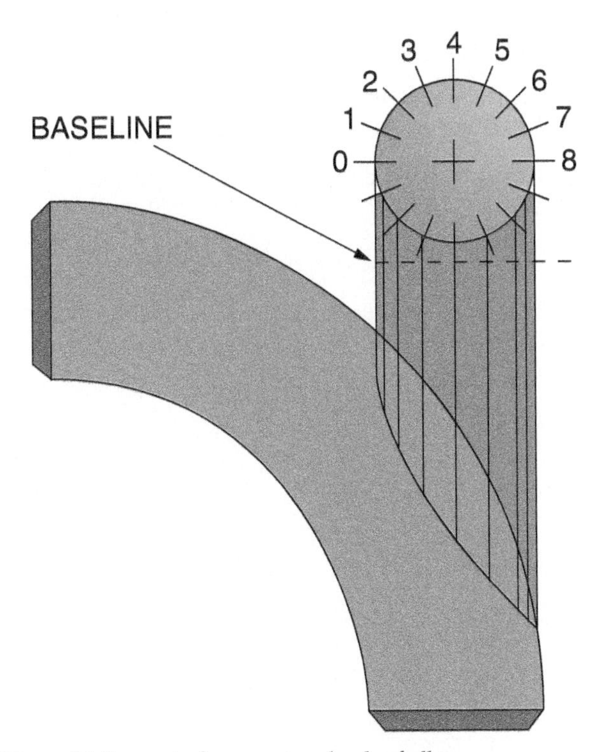

Figure 36 Layout of support on back of elbow.

Step 2 Review the support detail sheet to determine the size of pipe to be used to fabricate the support.

Step 3 Determine what type of support the specifications require.

Step 4 Obtain the pipe to be used for the support.

Step 5 Secure the spool with the elbow in a pipe vise and jack stands.

Step 6 Obtain a reference chart for the type and size support needed. *Figure 38* shows a reference chart for a Type 1 concentric support on the back of a 90-degree, long-radius elbow.

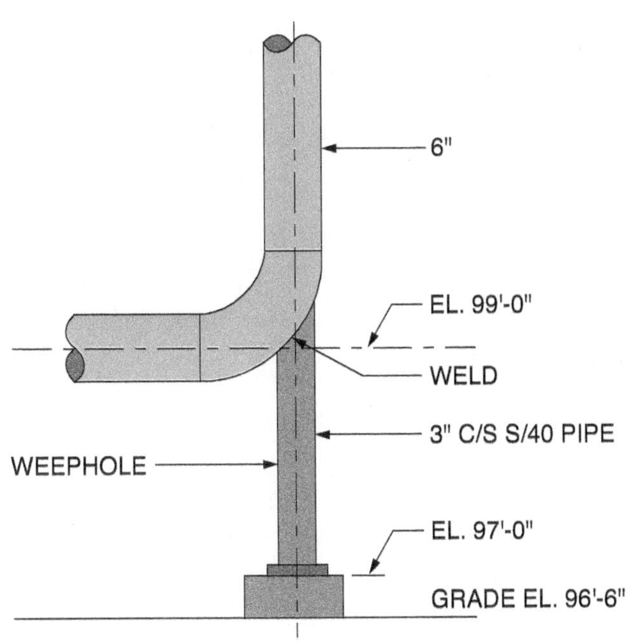

Figure 37 Support.

Step 7 Locate the ordinate measurements for the pipe being used on the reference chart and find the longest measurement.

Step 8 Measure 12" from the end of the 3" pipe and mark the pipe at this point. The dimension of 12" is used as a reference point past the longest ordinate measurement.

Step 9 Place a wraparound on the pipe at this mark and scribe a line around the pipe. This line is the base line.

Step 10 Divide the outside circumference of the pipe by 16 to obtain 16 ordinates.

Step 11 Mark the 16 ordinate lines on the pipe 0 to 8 on each side using an angle iron as a straight edge.

Step 12 Obtain the ordinate 1 cutback measurement from the reference chart.

Step 13 Measure $8^3/_8$" from the base line along ordinate 1 and mark the pipe.

Step 14 Obtain the ordinate 2 cutback measurement from the reference chart.

Step 15 Measure $8^3/_{16}$" from the base line along ordinate 2 and mark the pipe.

Step 16 Continue to measure and mark the ordinates on the pipe according to the reference chart.

Step 17 Connect all the ordinate marks using the wraparound.

Step 18 Cut the support along the cut line, using a cutting torch, keeping the torch at a 90-degree angle to the pipe.

Step 19 Clean the support to remove any slag, using a grinder.

NOTE

It is frequently the case that a support leg is required to have a weep hole. Check local standards to be sure, but if it is not forbidden, drill a small hole before welding the support leg to the elbow.

Step 20 Align the support with the elbow on the pipe, using a framing square and levels, as shown in *Figure 39.*

Step 21 Fit up the branch to the header using spacer wires to obtain the correct welding gap.

Step 22 Tack-weld the branch to the header.

Step 23 Verify that the support is square and properly aligned with the center line of the elbow.

Step 24 Have the welder weld the support to the elbow.

Step 25 Transfer a mark from the center line of the pipe to the support.

Step 26 Determine the length that the support needs to be cut. Subtract the height of the base and the thickness of the baseplate from the measurement between the center line of the pipe and the grade elevation to determine the length. For this example, the length between the baseplate and the center line of the pipe is $23^1/_2$".

Step 27 Measure the support from the center line mark and mark the support at this point.

Step 28 Scribe a line around the support at this mark using a wraparound.

Step 29 Cut the support at this line using a cutting torch.

Step 30 Clean the slag from the cut pipe end.

Step 31 Square the baseplate to the end of the support.

Step 32 Weld the baseplate to the pipe end.

	Size Of Elbow							
	2"	3"	4"	6"	8"	10"	12"	NO.
2" PIPE	2 5/16	3 13/16	5 1/16	8 5/8	11 15/16	15 3/16	18 1/2	0
	2 7/16	3 15/16	5 1/2	8 11/16	12	15 1/4	18 9/16	1
	2 13/16	4 3/16	5 13/16	9	12 3/16	15 1/2	18 7/8	2
	3 3/8	4 11/16	6 1/4	9 3/8	12 11/16	15 15/16	19 1/4	3
	4 1/16	5 3/16	6 11/16	9 13/16	13 1/8	16 3/8	19 11/16	4
	4 1/2	5 9/16	7 1/8	10 1/4	13 9/16	16 13/16	20 7/8	5
	4 3/4	5 7/8	7 7/16	10 9/16	13 7/8	17 1/8	20 7/16	6
	4 7/8	6 1/16	7 5/8	10 13/16	14 1/8	17 3/8	20 11/16	7
	4 7/8	6 1/8	7 11/16	10 7/8	14 3/16	17 7/16	20 3/4	8
3" PIPE	—	3 1/2	5 1/16	8 13/16	11 1/2	14 11/16	18	0
	—	3 11/16	5 1/4	8 3/8	11 5/8	14 7/8	18 3/16	1
	—	4 1/4	5 3/4	8 13/16	12 1/16	15 1/4	18 9/16	2
	—	5 1/8	6 7/16	9 7/16	12 11/16	15 7/8	19 3/16	3
	—	6 1/8	7 1/4	10 3/16	13 3/8	16 9/16	19 13/16	4
	—	6 13/16	7 15/16	10 13/16	14	17 3/16	20 1/2	5
	—	7 3/16	8 3/4	11 1/4	14 1/2	17 11/16	21	6
	—	7 15/16	8 9/16	11 9/16	14 13/16	18	21 3/16	7
	—	7 3/8	8 5/16	11 5/8	14 7/8	18 1/16	21 7/16	8
4" PIPE	—	—	4 3/4	7 7/8	11 1/8	14 5/16	17 3/8	0
	—	—	5	8 1/16	11 5/16	14 1/2	17 13/16	1
	—	—	5 3/4	8 11/16	11 7/8	15 1/16	18 3/8	2
	—	—	6 15/16	9 5/8	12 3/4	15 7/8	19 1/8	3
	—	—	8 3/8	10 5/8	13 11/16	16 13/16	20 1/16	4
	—	—	9 3/8	11 1/2	14 9/16	17 5/8	20 7/8	5
	—	—	9 13/16	12 1/16	15 3/16	18 1/4	21 9/16	6
	—	—	10	12 3/8	15 1/2	18 11/16	21 15/16	7
	—	—	10 1/16	12 1/2	15 5/8	18 13/16	22 1/16	8
6" PIPE	—	—	—	7 1/4	10 3/8	13 1/2	16 3/4	0
	—	—	—	7 5/8	10 3/4	13 13/16	17 5/16	1
	—	—	—	8 3/4	11 3/4	14 3/4	18	2
	—	—	—	10 5/8	13 1/4	16 1/8	19 5/16	3
	—	—	—	12 15/16	14 7/8	17 5/8	20 3/4	4
	—	—	—	14 7/16	16 1/4	18 15/16	22	5
	—	—	—	15 1/16	17 1/16	19 13/16	22 15/16	6
	—	—	—	15 3/16	17 9/16	20 3/8	23 1/2	7
	—	—	—	15 3/8	17 11/16	20 1/2	23 11/16	8

Figure 38 Type 1 support reference chart.

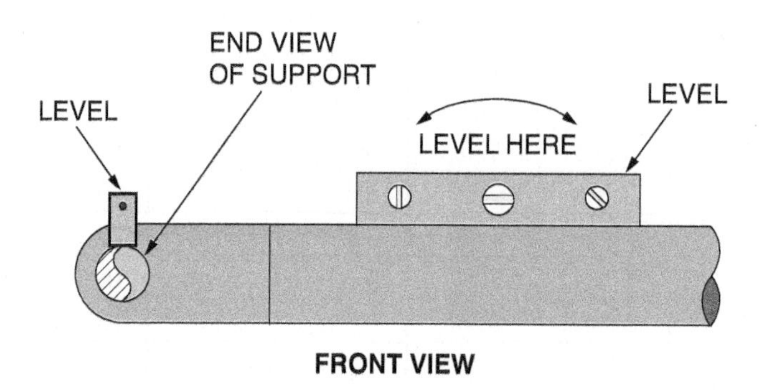

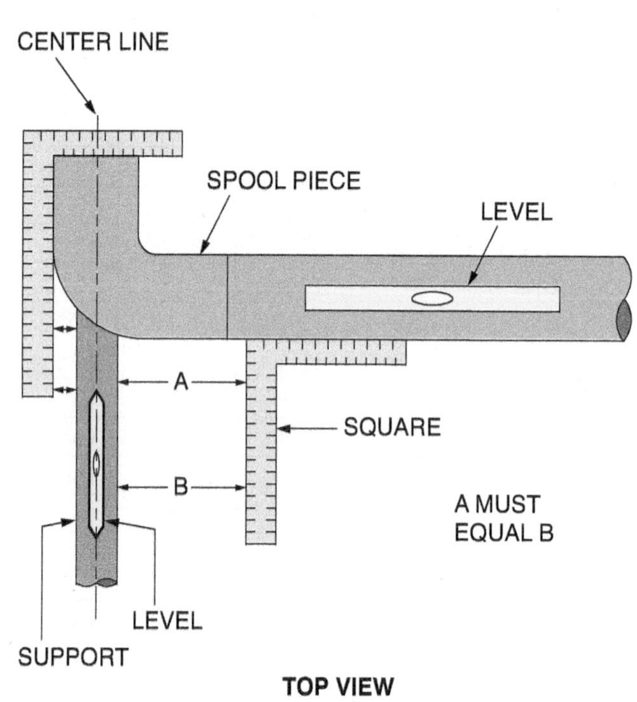

Figure 39 Aligning support to elbow.

3.0.0 Section Review

1. A fabricated, 90-degree intersection of pipe is called a _____.
   a. lateral
   b. saddle
   c. chord
   d. wye

2. When using supports made of pipe, the type that's used depends on _____.
   a. the diameter of the pipe on which welds will be made
   b. the type and size of the pipe being used
   c. the area of the elbow to which the support is attached
   d. the diameter of the cross-section hinged to the braces

SECTION FOUR

4.0.0 LAYING OUT LATERALS AND SUPPORTS WITHOUT USING REFERENCES

Objective

Explain how to lay out laterals without using references.

Laying out laterals can be performed even when references are not available. The process involves a series of steps designed to carefully ascertain the information needed. Drafting tools are needed to carry out the procedure.

When references are not available, laying out laterals can be accomplished by laying out the lateral on a flat surface and then transferring this to the pipe. A compass, 45-degree right triangle, T-square, dividers, and ruler are needed for the following exercise.

To layout a 45-degree lateral without using reference charts:

Step 1 Measure the exact inside and outside diameter of the pipe.

Step 2 Draw a vertical straight line that is at least $2^1/_2$ times the outside diameter of the pipe on a sheet of drafting paper. This line will become the vertical center line of two circles representing the inside diameter branch and outside diameter header of the pipe.

Step 3 Use a compass to draw a circle that is the exact size of the ID of the pipe, using a point on the straight line as the center.

Step 4 Use a compass to draw a circle that is the exact size of the OD of the pipe, using a point on the straight line below the ID circle as the center (*Figure 40*).

Step 5 Use the right triangle or T-square to draw a 90-degree line from the center of the ID circle to the outside edge of the circle (*Figure 41*).

Step 6 Using dividers, find the center between points A and B on the circumference of the circle and draw a line from the center of the circle to this point on the edge of the circle.

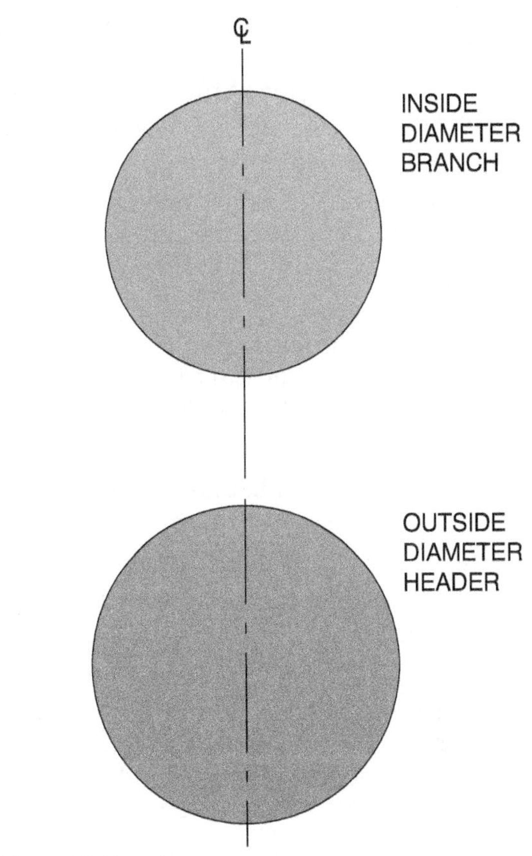

Figure 40 Step 4.

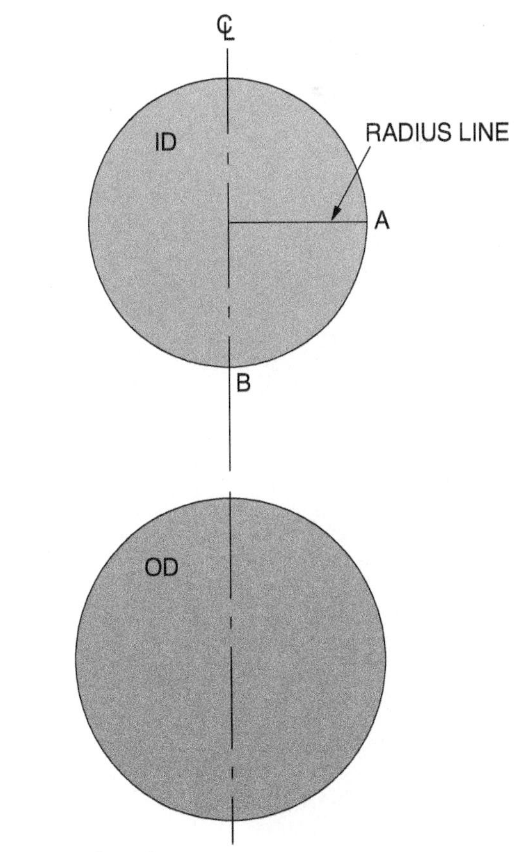

Figure 41 Step 5.

Step 7 Using dividers, find the center between points A and C and B and C and draw a line from the center of the circle to these points on the edge of the circle (*Figure 42*).

Step 8 Transfer points A, B, C, D, and E to the circumference of the pipe OD circle. Make sure that these lines are parallel to the straight line drawn in Step 2.

Step 9 Label the points where these lines intersect the circumference of the OD circle starting with the vertical center line being 0 and numbering 1, 2, 3, and 4 around the circle (*Figure 43*).

Step 10 Draw lines perpendicular to the vertical line at point 0 and at the center point of the pipe OD circle (*Figure 44*).

Step 11 Draw a line 45 degrees off of the two horizontal lines.

Step 12 Draw another ID circle half toward the end of the 45-degree line and mark the center line of this circle half with a line that is perpendicular to the 45-degree line (*Figure 45*).

Step 13 Using the dividers, divide this half circle into four equal parts.

Step 14 Using the dividers, divide each part in half so that the half circle is divided into eight equal parts.

Step 15 Label these sections with ordinate numbers 0 through 8, as shown on *Figure 46*.

Step 16 Draw lines perpendicular to the vertical line off the OD circle points 1, 2, 3, and 4 (*Figure 47*).

Step 17 Transfer the points from the ID circle half sections to the horizontal lines coming off the OD circle points 0, 1, 2, 3, and 4 (*Figure 48*).

Step 18 Draw a wraparound line perpendicular to these lines. This creates a reference point for measuring the cutback distances.

Step 19 Measure from the wraparound line down to the points where the transfer lines meet the ordinate lines (*Figure 49*). These measurements provide the ordinate cutback measurements for laying out the lateral.

Step 20 Lay out 16 ordinates, labeled 0 through 8, as well as a working line on the lateral pipe.

Step 21 Use the measurements found above to lay out the cut line on the lateral pipe.

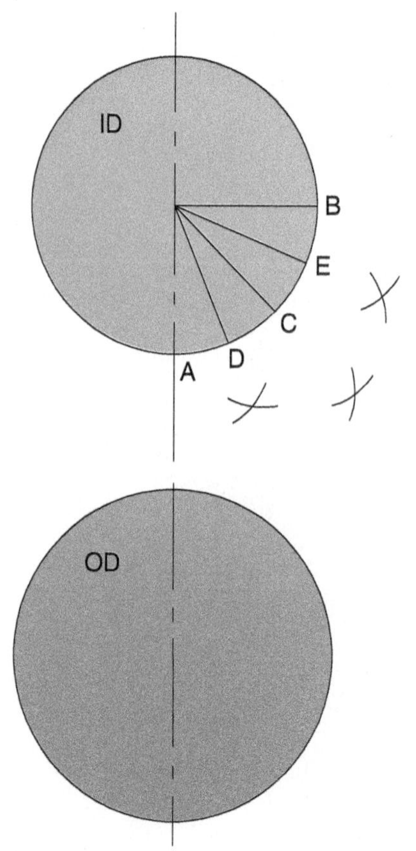

Figure 42 Step 7.

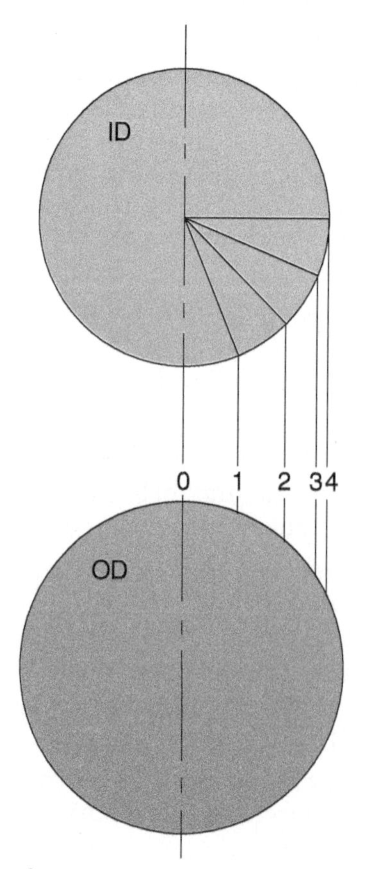

Figure 43 Step 9.

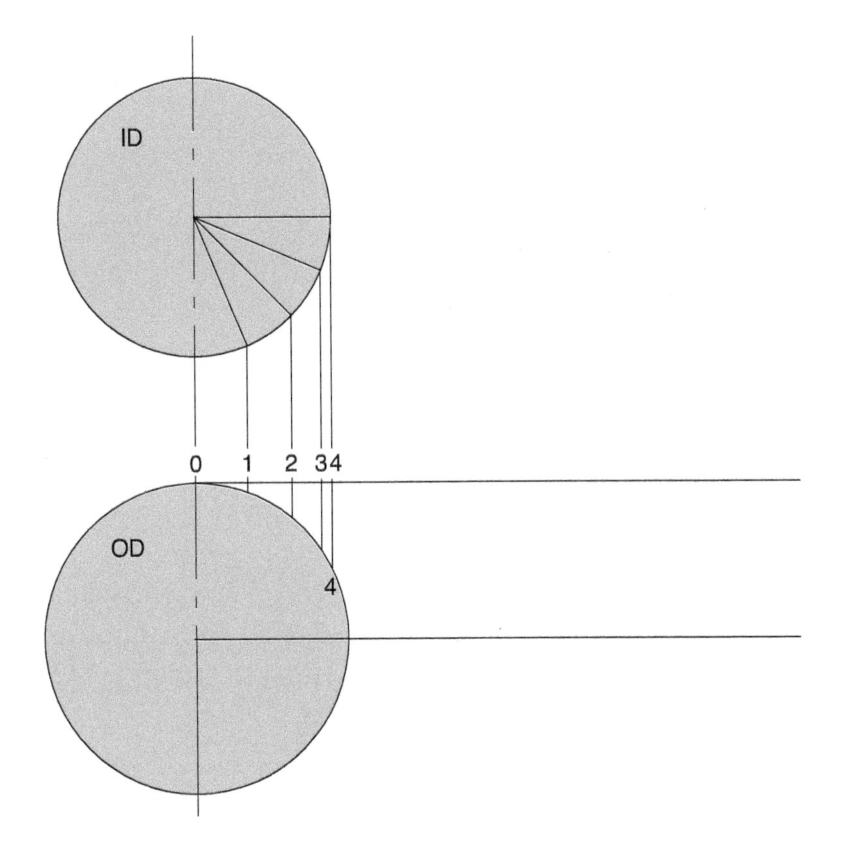

Figure 44 Step 10.

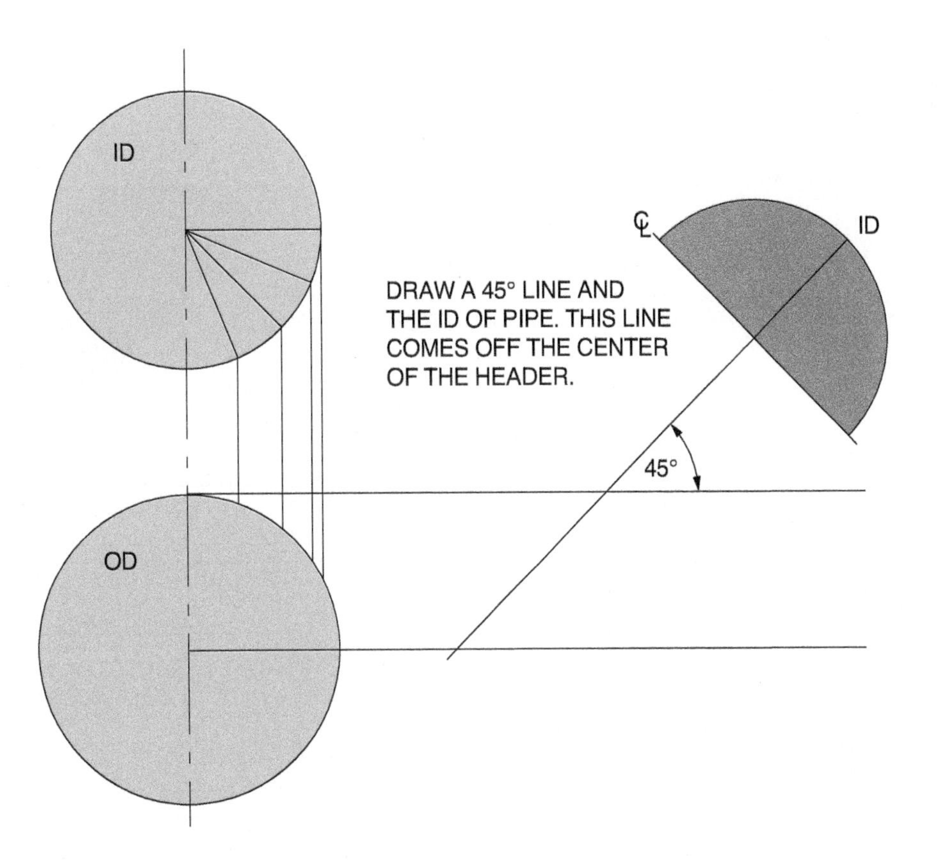

Figure 45 Step 12.

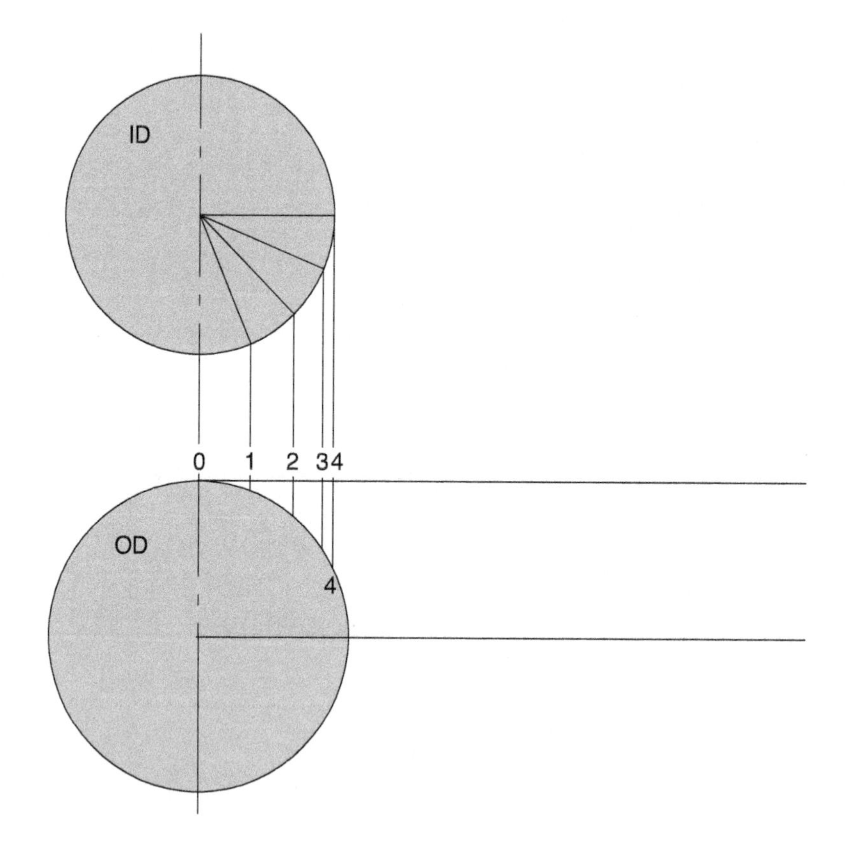

Figure 46 Step 15.

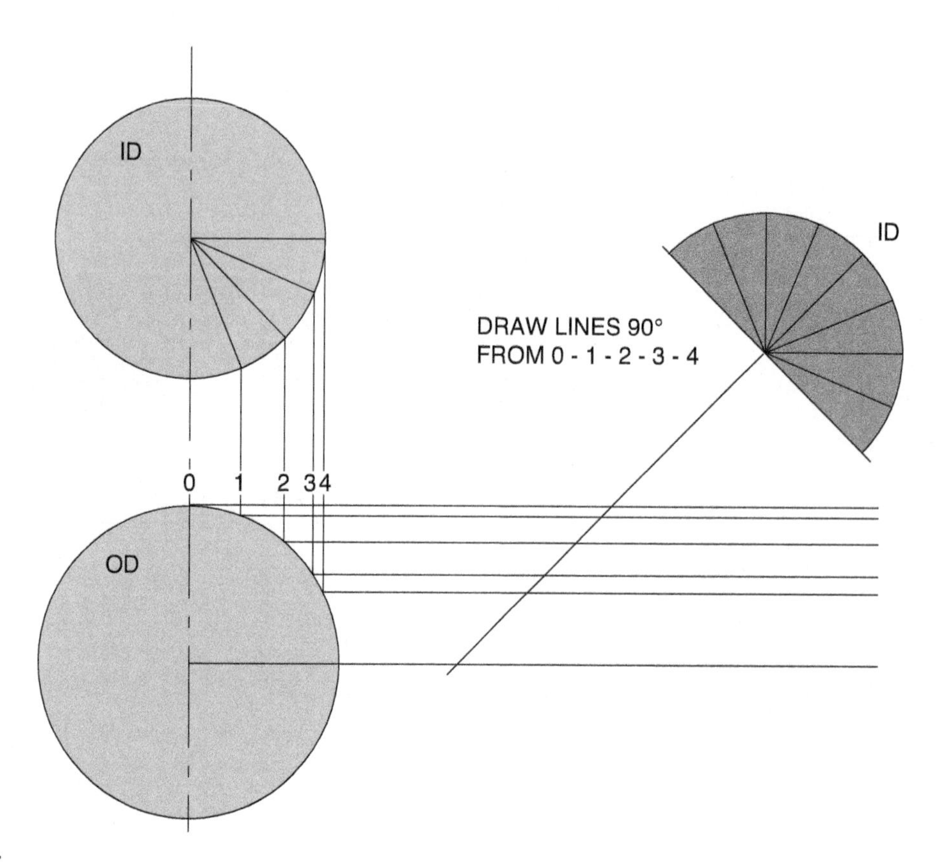

Figure 47 Step 16.

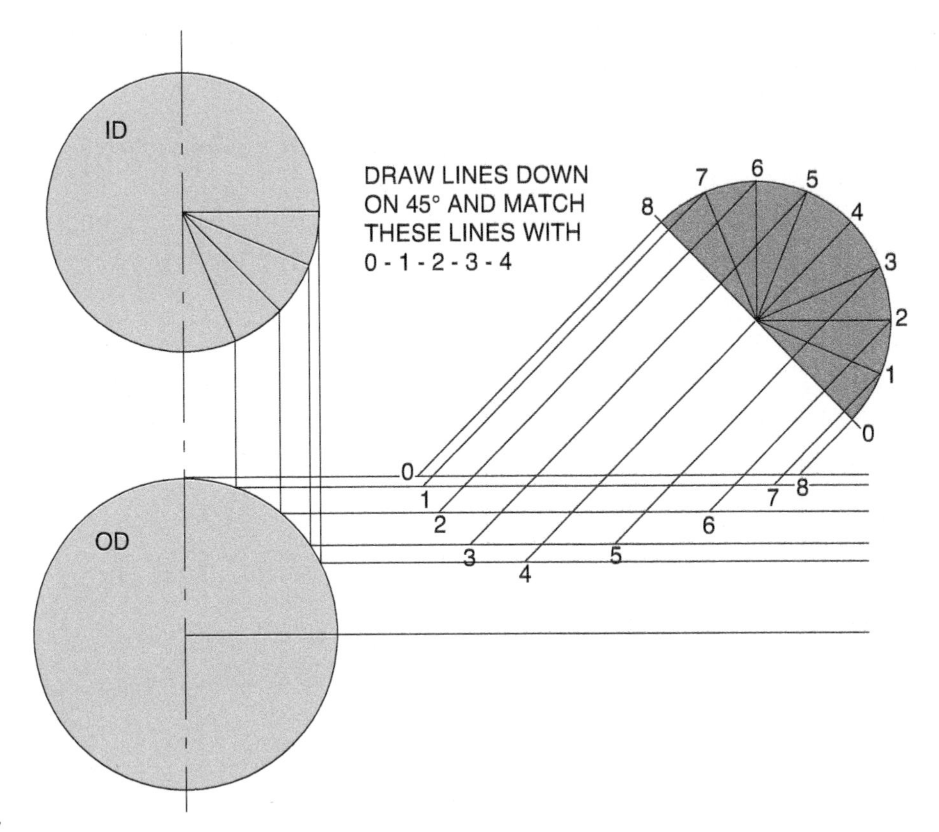

Figure 48 Step 17.

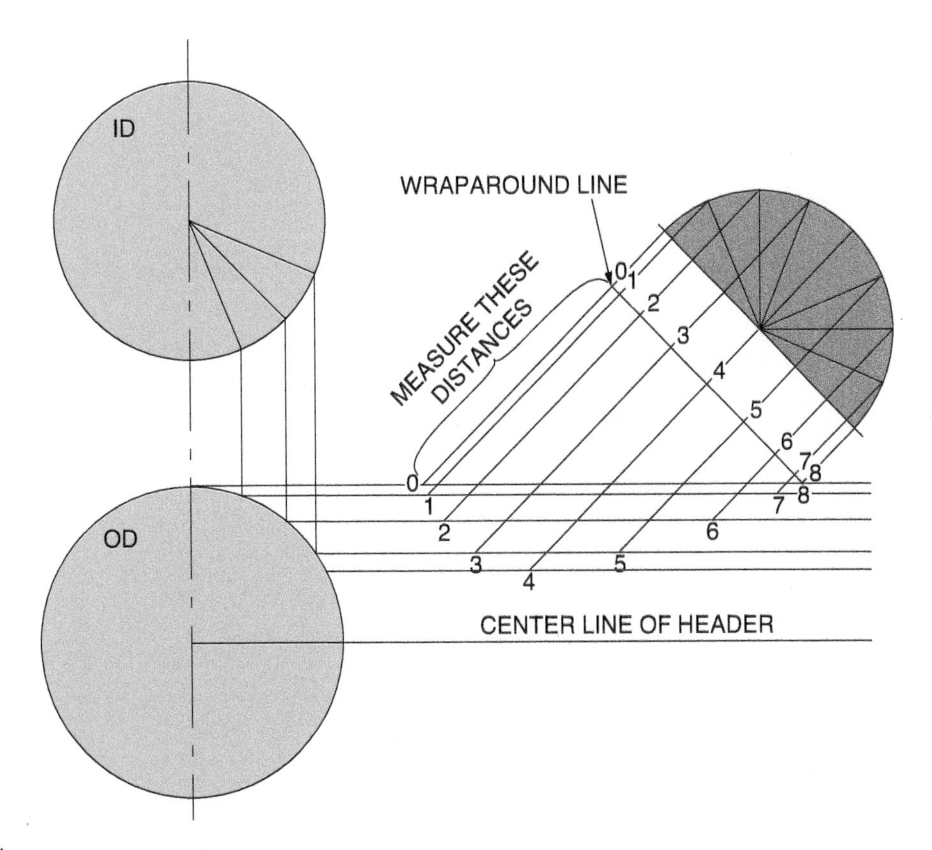

Figure 49 Step 19.

4.0.0 Section Review

1. When references are not available, pipefitters can lay out laterals if ____ are available.
   a. precast materials
   b. site plans
   c. drafting tools
   d. engineering records

Review Questions

1. The sum of the angles of a triangle is _____ degrees.
   a. 90
   b. 180
   c. 270
   d. 360

2. In order to solve any right triangle, you *must* know the length of two sides or the _____.
   a. degrees of one angle
   b. degrees of two angles and the length of one side
   c. length of all three sides
   d. length of one side

3. To find the length of travel for a simple 45-degree offset when the set is known, multiply the set by _____.
   a. 0.7071
   b. 1
   c. 1.414
   d. 2

4. The distance from the center of a vessel to the starting point of an offset is equal to the distance from the center of the vessel to the center line of the pipe times the _____.
   a. sine of the offset angle
   b. tangent of the offset angle
   c. tangent of half the offset angle
   d. cosine of the offset angle

5. The set and travel of all three lines in an equal-spread offset is the same, but the _____ are different.
   a. runs
   b. angles of offset
   c. starting points
   d. tangents of the angles

6. For 45-degree equal-spread offsets, the difference in starting points can be obtained by multiplying the distance between adjacent lines by _____.
   a. 0.4142
   b. 0.7071
   c. 1.414
   d. 2.414

7. For three-line unequal-spread offsets, you need to know the center-to-center distance between the lines before the offset _____.
   a. and not after the offset
   b. and after the offset
   c. except for line 2
   d. except for line 1

8. For a three-line unequal-spread 45-degree offset, lines 1 and 2 are 16" apart, and lines 2 and 3 are 20" apart. At the other end, lines 1 and 2 are 16" apart, and lines 2 and 3 are 32" apart. The starting point for line 3 is _____ past the start for line 1.
   a. 6.34"
   b. 14.91"
   c. 20"
   d. 31⅞"

9. The center-to-center lengths for pipes in a coil can be obtained by multiplying 2 times the radius of the coil by the _____.
   a. cosine of the angle
   b. sine of the angle
   c. tangent of ½ the angle
   d. number of pipes in the coil

10. Pipe laterals that are 4" through 10" are laid out with _____.
    a. 4 ordinates
    b. 8 ordinates
    c. 16 ordinates
    d. 32 ordinates

11. The outside diameter of the pipe is multiplied by the _____ to obtain the cutback for a miter.
    a. sine of the angle
    b. tangent of the angle, then divide by two
    c. sine of the angle, then divide by two
    d. cosine of the angle

12. The middle ordinate for an eight-point miter is calculated by multiplying the long cutback by _____.
    a. sine $22\frac{1}{2}$ degrees
    b. tangent 45 degrees
    c. sine 45 degrees
    d. cosine $22\frac{1}{2}$ degrees

13. The rule of thumb for building a miter is that the radius should be _____.
    a. half the OD
    b. the same as a long radius elbow of the same OD
    c. the length of the pipe
    d. the same as the radius of a coil

14. A miter in 16" pipe would have _____.
    a. 2 to 4 segments
    b. 3 to 5 segments
    c. 4 to 7 segments
    d. 10 to 12 segments

15. The number of degrees in a miter cut equals _____.
    a. the number of segments
    b. half the degrees of the full turn
    c. twice the number of segments
    d. the degrees in the turn divided by twice the number of welds

16. The center-to-center length for the segments of a mitered turn is found by _____.
    a. two times sine of miter
    b. radius times sine of miter
    c. tangent of miter
    d. two times radius times tangent of miter

17. The length of the end piece of a mitered turn is _____.
    a. two times the sine of the miter
    b. half the sine of the miter
    c. half the center-to-center length
    d. two times the OD

18. A saddle is calculated based on the geometry of _____.
    a. spherical bodies
    b. segments of a circle
    c. conical sections
    d. pyramids

19. A lateral is the intersection of two pipes at _____.
    a. 45 degrees
    b. 90 degrees or more
    c. more than 45 degrees
    d. less than 90 degrees

20. A 6" × 8" 45-degree lateral is to be calculated. What is the number 6 ordinate out of 8?
    a. 0.9936"
    b. 3"
    c. 6"
    d. 12"

Trade Terms Introduced in This Module

Base line: A straight line drawn around a pipe to be used as a measuring point.

Branch: A line that intersects with another line.

Chord: A straight line crossing a circle that does not pass through the center of the circle.

Cutback: The point at which a miter fitting is to be cut.

Miter: A specified angle to which a piece of pipe is cut.

Ordinate lines: Straight lines drawn along the length of the pipe connecting the ordinate marks.

Ordinates: Divisions of segments obtained by dividing the circumference of a pipe into equal parts.

Saddle: A fabricated 90-degree intersection of pipe.

Segments: Parts of a circle that are defined by a chord and the curve of the circumference.

Additional Resources

This module presents thorough resources for task training. The following reference material is recommended for further study.

Trigonometry. 2018. S.O.S. Mathematics Trigonometry. Available at: **www.sosmath.com/trig/trig.html**.

Section Review Answer Key

Section 1.0.0

Answer	Section Reference	Objective
1. c	1.1.0	1a
2. b	1.2.0	1b
3. a	1.3.0	1c
4. d	1.4.0	1d

Section 2.0.0

Answer	Section Reference	Objective
1. c	2.1.0	2a
2. a	2.2.0	2b
3. a	2.3.0	2c
4. c	2.4.0	2d
5. d	2.5.0	2e
6. a	2.6.0	2f

Section 3.0.0

Answer	Section Reference	Objective
1. b	3.1.0	3a
2. c	3.2.0	3b

Section 4.0.0

Answer	Section Reference	Objective
1. c	4.0.0	4

This page is intentionally left blank.

NCCER CURRICULA — USER UPDATE

NCCER makes every effort to keep its textbooks up-to-date and free of technical errors. We appreciate your help in this process. If you find an error, a typographical mistake, or an inaccuracy in NCCER's curricula, please fill out this form (or a photocopy), or complete the online form at **www.nccer.org/olf**. Be sure to include the exact module ID number, page number, a detailed description, and your recommended correction. Your input will be brought to the attention of the Authoring Team. Thank you for your assistance.

Instructors – If you have an idea for improving this textbook, or have found that additional materials were necessary to teach this module effectively, please let us know so that we may present your suggestions to the Authoring Team.

NCCER Product Development and Revision
13614 Progress Blvd., Alachua, FL 32615
Email: curriculum@nccer.org
Online: www.nccer.org/olf

❑ Trainee Guide ❑ Lesson Plans ❑ Exam ❑ PowerPoints Other ____________

Craft / Level: Copyright Date:

Module ID Number / Title:

Section Number(s):

Description:

Recommended Correction:

Your Name:

Address:

Email: Phone:

This page is intentionally left blank.

Stress Relieving and Aligning

Overview

Stress relieving is the process of preheating and post weld heat treatment to keep welds from distorting a pipe assembly. Alignment is the reason for stress relieving, because if the pipe will not fit up accurately to machinery, dynamically balanced pumps will be unbalanced by the distortions of the piping attachments. A skilled pipefitter takes charge of these situations, working with millwrights and others to prevent misalignments and problems which stem from them.

Module 08403

Trainees with successful module completions may be eligible for credentialing through the NCCER Registry. To learn more, go to **www.nccer.org** or contact us at 1.888.622.3720. Our website, **www.nccer.org**, has information on the latest product releases and training.

Your feedback is welcome. You may email your comments to **curriculum@nccer.org**, send general comments and inquiries to **info@nccer.org**, or fill in the User Update form at the back of this module.

This information is general in nature and intended for training purposes only. Actual performance of activities described in this manual requires compliance with all applicable operating, service, maintenance, and safety procedures under the direction of qualified personnel. References in this manual to patented or proprietary devices do not constitute a recommendation of their use.

08403
Stress Relieving and Aligning

Objectives

Successful completion of this module prepares you to do the following:

1. Describe thermal expansion in piping systems and approaches to accommodating the imposed stress.
   a. Explain the role of flexibility in accommodating expansion.
   b. Explain the installation of expansion loops.
   c. Describe cold springing and how it accommodates expansion.
2. Identify and describe methods used to minimize stress in pipe welds.
   a. Describe the relationship between temperature and the structure of metal.
   b. Identify metals that require preheating prior to welding.
   c. Identify and describe metal-preheating methods.
   d. Identify and describe the results of various post weld heat treatment processes.
   e. Explain how pipe temperatures are best measured.
3. Describe how piping stress is avoided when a connection to rotating equipment is required.
   a. Explain the purpose of grouting and its roles in supporting machinery.
   b. Describe how preliminary alignments are completed on rotating equipment.

Performance Tasks

Under supervision, you should be able to do the following:

1. Identify three methods used to stress-relieve welds.
2. Indicate the area of a pipe that needs to be stress-relieved.

Trade Terms

Alloy
Carbon
Constituents
Critical temperature
Hardenability
Heat-affected zone (HAZ)
Induction heating
Interpass heat
Post-weld heat treatment (PWHT)
Preheat
Preheat weld treatment (PHWT)
Quenching
Stress
Stress relieving
Tempering

Industry Recognized Credentials

If you are training through an NCCER-accredited sponsor, you may be eligible for credentials from NCCER's Registry. The ID number for this module is 08403. Note that this module may have been used in other NCCER curricula and may apply to other level completions. Contact NCCER's Registry at 1.888.622.3720 or go to **www.nccer.org** for more information.

Contents

Figures and Tables

This page is intentionally left blank.

SECTION ONE

1.0.0 THERMAL EXPANSION IN PIPING

Objective

Describe thermal expansion in piping systems and approaches to accommodating the imposed stress.

a. Explain the role of flexibility in accommodating expansion.
b. Explain the installation of expansion loops.
c. Describe cold springing and how it accommodates expansion.

Trade Terms

Stress: The load imposed on an object.

Stress relieving: The even heating of a structure to a temperature below the critical temperature followed by a slow, even cooling.

All materials change in size, to some extent, when the temperature changes. Because they expand and contract at different rates, this can cause problems when various types of materials are used in the same piping system. Pipefitters must therefore be able to account for thermal expansion and contraction while working out the details of the fabrication of a pipeline.

Under normal conditions, metallic piping will change less than an inch in a 100' length with a temperature change of 300°F. If only one point in a pipeline were kept fixed during a change in length, movement would take place in perfect freedom, and no stress would be imposed on the pipe or the connection. However, piping systems have more than one fixed point.

Most process piping systems are routed between storage tanks and pumps. They are nearly always restrained at terminal points by anchors, guides, stops, rigid hangers, or sway braces. These restraining points resist expansion and put the line under stress, which causes it to deform when the temperature changes.

Pipefitters must be able to properly align piping to pumps and other rotating equipment to avoid excessive stress on the pipe, joints, and rotating equipment. When pipe and rotating equipment are not properly aligned, the stress that exists between these components causes leaks in the piping system, damage to the equipment, and excessive pressure and stress to the pipe hangers and supports. All of these conditions decrease the life of the piping system and equipment and also present serious safety hazards to those who work in the vicinity of the system.

Stress imposed by expansion must be kept within the strength capability of the pipe and the points to which it is anchored. Equipment, such as a pump or turbine, that is fastened solidly will be put under excessive stress if the pipe fastened to it expands toward it. This must be prevented by allowing expansion away from the equipment. Failure to do so can result in ruptured pipe or damage to the structure or the equipment to which the pipe is anchored. Before stress relieving any pipe, verify by checking policies and procedures.

To find out how much a length of pipe would expand for a given change in temperature, multiply the length of the pipe (in inches) by the temperature change (in degrees Fahrenheit). Take that number and multiply it by the coefficient of expansion for that material from *Table 1*, which shows the expansion of different types of pipe.

For example, 100' of carbon steel pipe (1,200") that is heated from 100°F to 500°F would expand by 3.024" ([length] 1,200 × [temperature change] 400 × [coefficient of expansion] 0.0000063 = 3.024).

Charts may not always be available to show the exact linear expansion for each degree of heat of a particular material. However, a close estimate that can be used for all temperatures is to allow ¾" per 100' per 100°F.

Table 1 Coefficient of Expansion

Material	Inches
Carbon steel	0.0000063
Aluminum	0.0000124
Cast iron	0.0000059
Nickel steel	0.0000073
Stainless steel	0.0000095
Concrete	0.0000070

1.1.0 Flexibility in Layout

Flexibility can often be built into the layout of a piping system. If a pipeline is built with small-diameter pipe and it is not expected to have great temperature changes, the system can be designed to allow for expansion. Flexibility in layout consists of making many changes in direction between anchor points. These changes in direction allow for some thermal expansion of the line without damaging the joints and fittings. *Figure 1* shows an example of flexibility in layout of a pipeline constructed of 4", Schedule 40, steel pipe carrying steam indoors at 400°F.

1.2.0 Installing Expansion Loops

A second method used to account for expansion is expansion loops that are installed in the pipe run to provide extra flexibility between anchors. Expansion loops accommodate thermal expansion and contraction as well as sideways bowing. These loops absorb the movement of the line caused by expansion and contraction. The pipe is anchored along its length and sometimes in the loop itself. Guides are placed at designated intervals along the pipe run to ensure that the pipe moves toward the expansion loop as it expands.

Expansion loops are made with pipe bends or welded pipe and fittings. They should not be fabricated with flanges or threaded fittings because these types of joints cannot withstand the stress and they will leak. Fabricated expansion U-loops made of pipe welded to long-radius elbows are usually very simple in shape and can be fabricated in the shop according to specifications on the layout drawing. The number and type of expansion loops that are used will differ with each installation. *Figure 2* shows an expansion loop.

1.3.0 Cold Springing

A third way to deal with expansion is to use cold springing along with one of the other two methods. Cold springing involves shortening the pipe run when it is installed so that when the pipe heats up, it expands to its intended length. This also relieves the stress that thermal expansion exerts on the pipe.

The cold-spring gap refers to the difference between the length of the installed pipe and the length that the pipe should be. The location and the size of the cold-spring gap must be calculated accurately, and the piping must be installed according to those calculations. Cold springing pipe at the wrong location or by the wrong amount can damage the pipe joints, anchors, valves, or attached equipment, so it is not advisable to guess. *Figure 3* shows examples of cold springing.

The most common method of cold springing a pipe assembly is to use a jack to push the pipe out to the specified point to compensate for expansion, and then weld the pipe there. Expansion will move the pipe the desired amount.

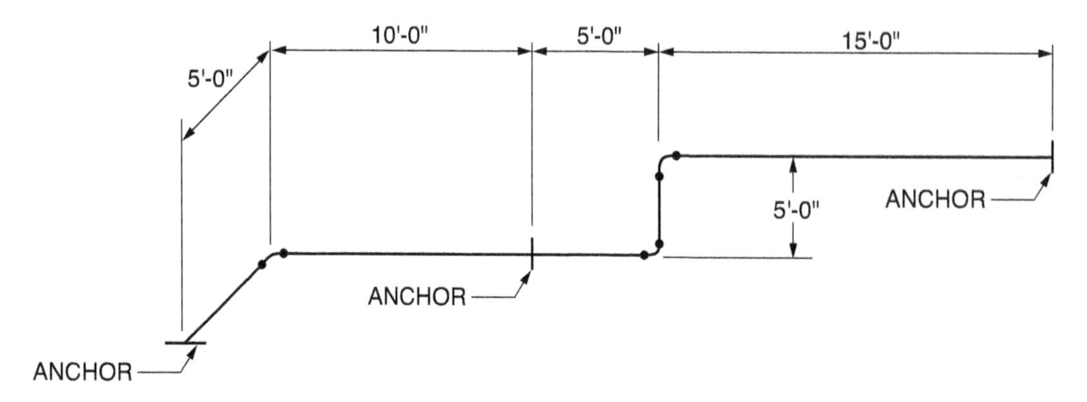

Figure 1 Flexibility in layout.

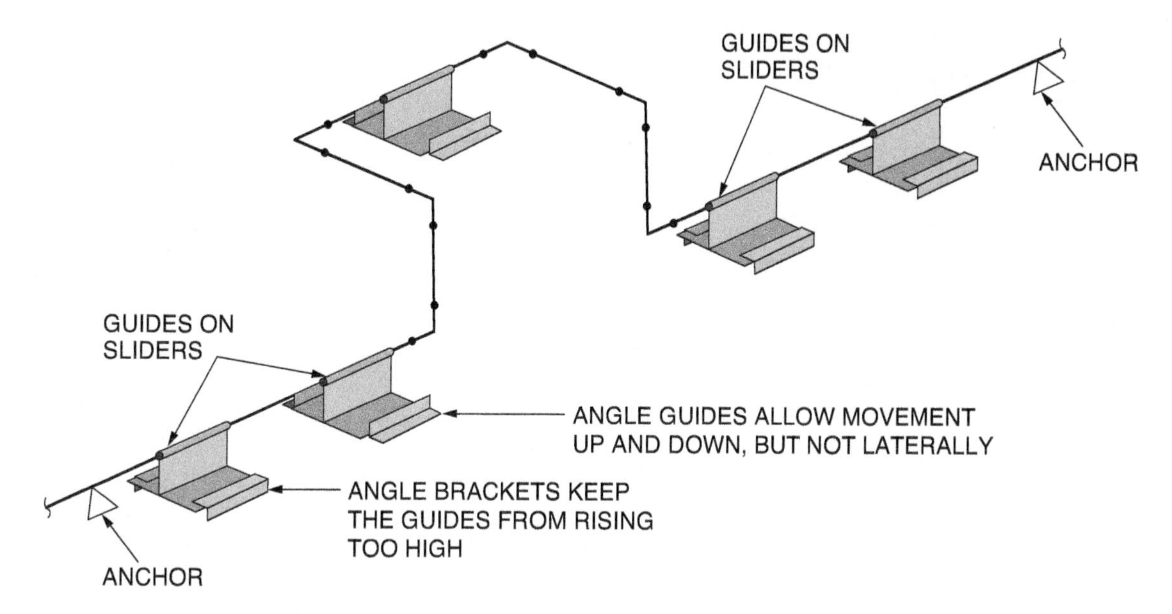

Figure 2 Expansion loop.

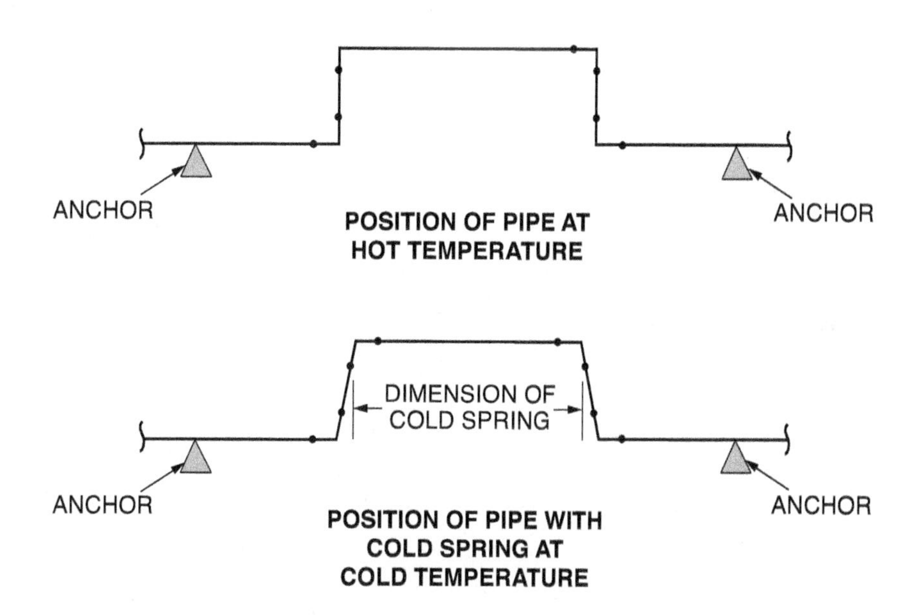

Figure 3 Cold springing.

1.0.0 Section Review

1. Building flexibility into a layout is important for protecting the ______.
   a. hangers and rods
   b. joints and fittings
   c. anvils and flanges
   d. flanges and valves

2. Expansion loops are made with pipe bends or welded pipe and _____.
   a. flanges
   b. valves
   c. fittings
   d. dapplers

3. With cold springing, approximations are allowed if they fall within two standard deviations of the norm, when trigonometry is applied.
   a. True
   b. False

SECTION TWO

2.0.0 PERFORMING STRESS RELIEF

Objective

Identify and describe methods used to minimize stress in pipe welds.

a. Describe the relationship between temperature and the structure of metal.
b. Identify metals that require preheating prior to welding.
c. Identify and describe metal-preheating methods.
d. Identify and describe the results of various post-weld heat treating processes.
e. Explain how pipe temperatures are best measured.

Performance Tasks

1. Identify three methods used to stress-relieve welds.
2. Indicate the area of a pipe that needs to be stress-relieved.

Trade Terms

Alloy: A metal that has had other elements added to it, which substantially changes its mechanical properties.

Carbon: An element which, when combined with iron, forms various kinds of steel. In steel, it is the carbon content that affects the physical properties of the steel.

Constituents: The elements and compounds, such as metal oxides, that make up a mixture or alloy.

Critical temperature: The temperature at which iron crystals in a ferrous-based metal transform from being face-centered to body-centered. This dramatically changes the strength, hardness, and ductility of the metal.

Hardenability: A characteristic of a metal that makes it able to become hard, usually through heat treatment.

Heat-affected zone (HAZ): The area of the base metal that is not melted but whose mechanical properties have been altered by the heat of the welding.

Induction heating: The heating of a conducting material by means of circulating electrical currents induced by an externally applied alternating magnetic field.

Interpass heat: The temperature to which a metal is heated while an operation is performed on the metal.

Post-weld heat treatment (PWHT): The temperature to which a metal is heated after an operation is performed on the metal.

Preheat: The temperature to which a metal is heated before an operation is performed on the metal.

Preheat weld treatment (PHWT): The controlled heating of the base metal immediately before welding begins.

Quenching: Rapid cooling of a hot metal using air, water, or oil.

Tempering: Increases the toughness of quenched steel and helps avoid breakage and failure of heat-treated steel; also called drawing.

Pipe stresses are locked into fabricated pipe assemblies by the rapid heating and cooling of pipe that occurs during the welding process. Heat treatment and stress relieving of piping are necessary to relieve stresses created during welding and to prevent stress cracks from forming along the weld. Stresses are caused by temperature differentials between the hot welding zone and the cooler zones. Heat treatment and stress relieving maintain or restore the original crystalline structure of the base metal in the pipe and maintain certain desired properties, such as corrosion resistance.

To stress relieve steel pipe, heat about 6" of the pipe, on either side of the weld, to the temperature that specifications allow. Maintain this temperature for a period of time according to the size and thickness of the pipe. Cool the pipe slowly and uniformly to room temperature. Uniform cooling prevents development of new stresses, which can equal those created during the welding process.

Time and temperature affect stress relief. Maintaining a lower pipe temperature for a longer period of time achieves the same result as maintaining a higher temperature for a shorter period of time. The heat treatments used in welding piping are **preheat**, **interpass heat**, and **post-weld heat treatment (PWHT)**.

Preheat weld treatment (PHWT) is the controlled heating of the base metal immediately before welding begins. Its main purpose is to keep the weld and the base metal from cooling too fast within the heat-affected zone (HAZ) along the weld line. In some cases, the whole structure is preheated. Sometimes, it may not be possible or practical to heat the whole structure. When this is the case, only the section near the weld is preheated.

Interpass temperature control is used to maintain the temperature of the weld zone during welding. It is used when the cooling rate is too high or too low to maintain the correct temperature of the weldment between weld passes. The minimum interpass temperature is usually the same as the preheat temperature. If welding is ever interrupted, the temperature of the weldment must be brought back to the interpass temperature before continuing. Usually, there is also a maximum interpass temperature, which is especially important in preventing overheating from the heat buildup in smaller weldments.

2.1.0 Temperature and Metal Structure

At room temperature, most metals are solid with a crystalline structure. They become liquified if heated to a high enough temperature. Temperature is a measure of molecular activity, so when the temperature is high enough, the forces that hold the molecules in their crystalline structures break down and the metal becomes a liquid. At even higher temperatures, metals become gaseous.

The temperature at which a metal becomes liquid is its melting point, and this point is different for every metal. In the case of an alloy metal (metal mixtures), the various constituents melt at different temperatures. Some, like carbon, may not melt at all.

Below the melting point, there is a temperature zone at which crystalline changes take place. Depending on the metal or alloy, these can seriously affect the metal's physical structure and mechanical characteristics. Quick cooling (quenching) of a metal while it is in this temperature zone can freeze some of the modified crystalline structure. The result is that the mechanical properties and physical traits of the cooled metal will be very different than if the same metal were cooled more slowly. Quenching does not have much of an effect on low-carbon and mild steels because they do not contain much carbon. However, quenching medium- and high-carbon steels produces a hard and brittle crystalline structure, which reduces the metal's toughness. *Table 2* shows the minimum preheat temperatures for various base materials. When medium-carbon and high-carbon steel are welded, the temperature in the weld zone always reaches the melting point (above the critical temperature), and the crystalline structure in the weld zone changes. If the metal is not preheated enough, the large mass of the cool base metal will quench the weld zone and cause localized hardness and brittleness. If the metal is adequately preheated, the weld zone cools much more slowly, allowing the crystalline structure to transform back to the crystalline form of the surrounding base metal.

Preheating may be required to:

- Reduce shrinkage in the weld and adjacent base metal
- Prevent excessive hardening and reduced ductility of weld filler and base metals in the HAZ
- Reduce hydrogen gas in the weld zone
- Ensure compliance with required welding procedures

Table 2 Minimum Preheat Temperatures for Various Base Materials

Base Materials	Welding Processes	Material Thickness	Minimum Preheat Temperature (°F)
ASTM A36 ASTM A53 GR. B ASTM A106 GR. B API 5L GR. B API 5L GR. X42	Shielded metal arc welding with other than low hydrogen electrodes	Less than or equal to 3/4"	None (see note)
		Over 3/4" through 1 1/2"	150
		Over 1 1/2" through 2 1/2"	225
		Over 2 1/2"	300
ASTM A36 ASTM A53 GR. B ASTM A106 GR. B ASTM A572 GR. 42 ASTM A572 GR. 50 ASTM A586 API 5L GR. B API 5L GR. X42	Shielded metal arc welding with low hydrogen electrodes, submerged arc welding, gas metal arc welding, and flux cored arc welding	Less than or equal to 3/4"	None (see note)
		Over 3/4" through 1 1/2"	50
		Over 1 1/2" through 2 1/2"	150
		Over 2 1/2"	225
ASTM A572 GR. 60 ASTM A572 GR. 65 ASTM 5L GR. X52	Shielded metal arc welding with low hydrogen electrodes, submerged arc welding, gas metal arc welding, and flux cored arc welding	Less than or equal to 3/4"	50
		Over 3/4" through 1 1/2"	150
		Over 1 1/2" through 2 1/2"	225
		Over 2 1/2"	300
ASTM A514 ASTM A517	Shielded metal arc welding with low hydrogen electrodes, submerged arc welding, gas metal arc welding, and flux cored arc welding	Less than or equal to 3/4"	50
		Over 3/4" through 1 1/2"	125
		Over 1 1/2" through 2 1/2"	175
		Over 2 1/2"	275
NOTE: If below 32°F, preheat to 70°F			

2.1.1 Excessive Hardening from Quenching

All welding processes use or generate temperatures higher than the melting point of the base metal. If the welded metal is a high-carbon alloy steel or cast iron that is quenched, a hard, brittle metal called martensite will form in the weld zone. This causes the entire HAZ parallel to the weld fusion line to become hard and brittle (*Figure 4*). The large temperature difference also causes differential thermal expansion and creates stresses in the weld region. Heavy plate, pipe, and castings have a high heat-absorption capacity, which will cause the weld metal and adjacent base metal to quench unless they are adequately preheated.

2.1.2 Reducing Hydrogen Gas in the Weld Zone

The welding arc can split moisture in the weld zone into hydrogen and oxygen. The hydrogen dissolves in the molten base metal, and the oxygen combines with other elements to form slag. As the metal begins to cool, the hydrogen forms gas bubbles in the base metal along the weld boundary. These bubbles create pressure that strains the metal and causes underbead cracking along the weld boundary and at the toe. Preheating helps remove moisture from the weld zone and results in slower cooling, which allows the hydrogen more time to diffuse up through the filler

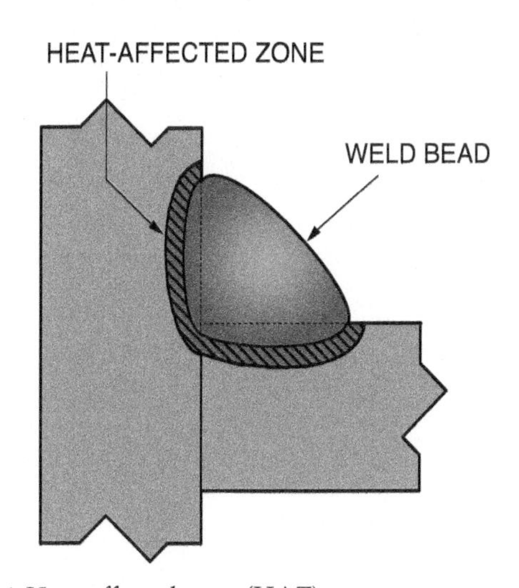

Figure 4 Heat-affected zone (HAZ).

metal and down into the adjacent base metal. The more time the hydrogen has to diffuse, the less pressure it causes and, therefore, the less chance that underbead cracking will occur. *Figure 5* is a simplified diagram showing underbead cracking caused by hydrogen bubbles.

2.1.3 Complying with Required Welding Procedures

If a site has welding procedure specifications (WPS) or quality specifications for particular welding procedures, it is usually mandatory that they be followed exactly. Check with the site supervisor if there are any questions about the specifications.

2.2.0 Metals That Require Preheating

It can be difficult to determine whether metal assemblies and conditions require preheating. Preheating and interpass temperature control depend on the base metal composition, thickness, and degree of restraint. Some base metals only require preheating when they exceed a certain thickness or are too cold. For example, metals exposed to freezing winter temperatures usually require some preheating. Complex-shaped welding assemblies and castings usually require preheating to avoid severe stresses or warping. Some alloys, such as high-carbon or alloy steels, require preheating to avoid brittleness and cracking. Not all metals require preheating. Low-carbon steels rarely need preheating. *Table 3* shows the metals that usually require preheating.

> **NOTE**
>
> Refer to WPS or client specifications for preheat requirements.

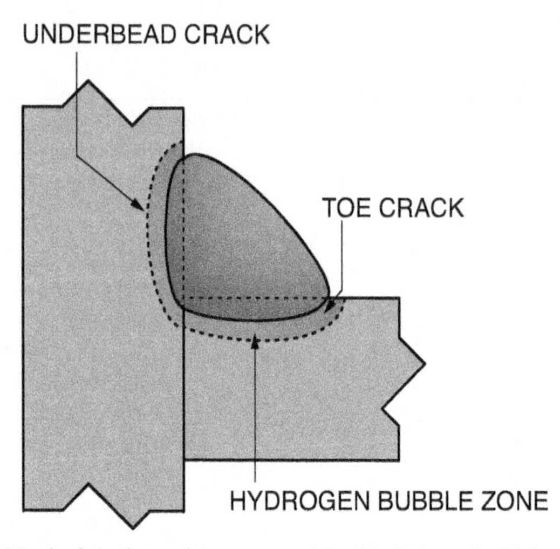

Figure 5 Underbead cracking caused by hydrogen bubbles.

2.2.1 Preheating Temperatures

Preheating requirements are affected by the alloy composition and thickness of the base metal, the welding process to be used, and the ambient temperature. The **hardenability** of a steel is directly related to its carbon content and alloying elements. Because different types of steel have different amounts of carbon and alloying elements, they have different preheat requirements. As a general rule, the higher the carbon content, the higher the preheating temperature required. *Table 4* lists preheat requirements for steels.

Table 3 Metals That Usually Require Preheating

Metal or Alloy	Conditions or Forms
Aluminum	Large or thick section castings
Copper	All (prevents too-rapid heat loss)
Bronze, copper-based	All (prevents too-rapid heat loss)
Mild carbon steels	Restricted joints, complex shapes, freezing temperatures, and carbon content over 0.30%
Cast irons	All types and all shapes
Cast steels	Higher carbon content or complex shapes
Low-alloy steels	Thicker sections or restricted joints
Low-alloy steels	Heating temperature dependent on carbon or alloy content
Manganese steels	Heating temperature dependent on carbon or alloy content
Martensitic stainless steels	Carbon content over 0.10%
Ferritic stainless steels	Thick plates

Table 4 Preheat Requirements for Steels

Steel Type	Shapes or Conditions	Preheat Temperature
Low-carbon steel (0.10%–0.30% C)	Freezing temperatures	Above 70°F
Low-carbon steel (0.10%–0.30% C)	Complex large shapes	100°F–300°F
Medium-carbon steel (0.30%–0.45% C)	Complex large shapes	300°F–500°F
High-carbon steel (0.45%–0.80% C)	All sizes and shapes	500°F–800°F
Low-alloy steel	Lower C or alloy %	100°F–500°F
	Higher C or alloy %	500°F–800°F
Manganese steel	Lower C or alloy %	100°F–200°F
	Higher C or alloy %	200°F–500°F
Cast iron	All shapes	200°F–400°F
Martensitic stainless steel	0.10%–0.20% C	500°F
	0.20%–0.50% C	500°F and postheat
Ferritic stainless steel	¼" and thicker	200°F–400°F and postheat

Temperature-indicating crayons, or temp sticks (*Figure 6*), are often used to measure the preheat temperature of the base metal. They are used by marking the workpiece before heating or by stroking the workpiece with the crayon during heating. When the rated temperature has been reached, a distinct melt or smear will become evident.

2.2.2 Interpass Temperature

When preheating is important, interpass temperature for multiple-pass welds is usually also important. Interpass temperature control is often simply a continuation of preheating. Heating equipment and techniques are the same.

The interpass temperature is usually the same as the preheat temperature. However, some specifications list both a maximum and a minimum interpass temperature. Sometimes, the interpass temperature may be specified as a maximum temperature. This might be the case with small weldments and frequent weld passes, where too much heat could build up.

With large, high-mass weldments and infrequent weld passes, the base metal can cool below the required preheat temperature if additional heat is not supplied. If the base metal cools below the specified minimum preheat temperature, it must be reheated to the preheat temperature before welding continues.

2.3.0 Preheating Methods

Preheating is most effective when the entire welding assembly is preheated (general heating) in an oven or preheating furnace, because the heating is even. However, because of site conditions or the size or shape of the welding assembly, preheating in this manner is not always possible. Localized heating of specific regions can be done with gas torches, blowtorches, electric resistance heaters, radio frequency induction heaters, or partial ovens that are built only around the region to be heated. Parts of the oven or enclosure can be removed so that the welder can have access and still maintain the preheating temperature.

Most welding shops contain one or more preheating devices, which may include:

- Oxyfuel torches
- Portable preheating torches
- Open-top preheaters
- Electric resistance heaters
- Induction heaters

Figure 6 Temperature-indicating crayon.

NOTE

The choice of preheating device is usually specified in the QC requirements for the job. Propane torches are sometimes used when oxyfuel torches are too hot. Refer to the WPS for details.

2.3.1 Oxyfuel Torches

If properly monitored, oxyfuel torches equipped with specialized heating tips can be used for localized preheating.

NOTE

Because an oxyfuel torch has a localized flame, it is not recommended for preheating large objects.

The specialized heating tips are designed to produce blowtorch flame patterns. Heating tips, like the rosebud-style, multi-flame heating heads, (*Figure 7*) are designed to mount directly to torch handles like welding tips. Heating tips designed to replace the cutting tips on straight cutting torches or combination cutting torch cutting attachments are also available.

WARNING!

Always make sure that the correct type of tip is employed for the fuel gas being used. Using a tip with the wrong gas can cause the tip to explode. Observe the pressure and flow rates recommended by the manufacturer.

2.3.2 Portable Preheating Torches

Portable preheating torches are designed to burn either oil or gas combined with compressed air. Some use just gas and have an air blower to increase the intensity of the flame. *Figure 8* shows a gas and compressed air torch suitable for preheating large weldments.

Figure 7 Rosebud-style heating tip.

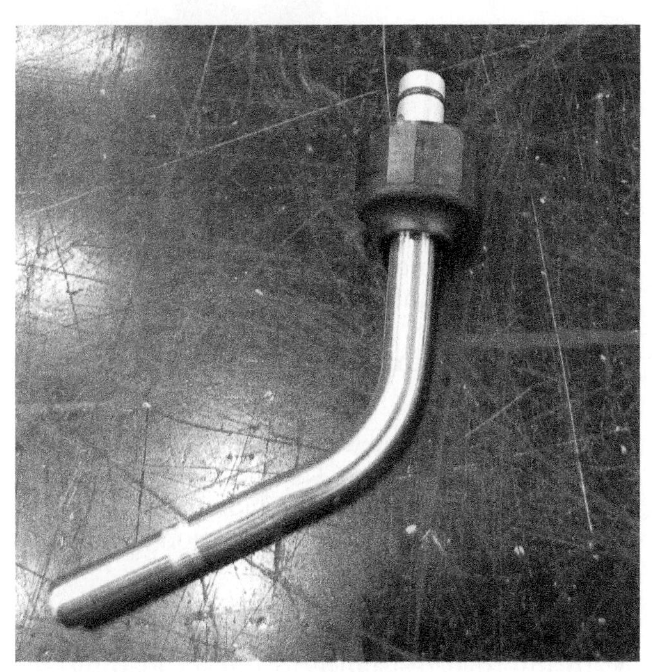

Figure 8 Gas preheating torch.

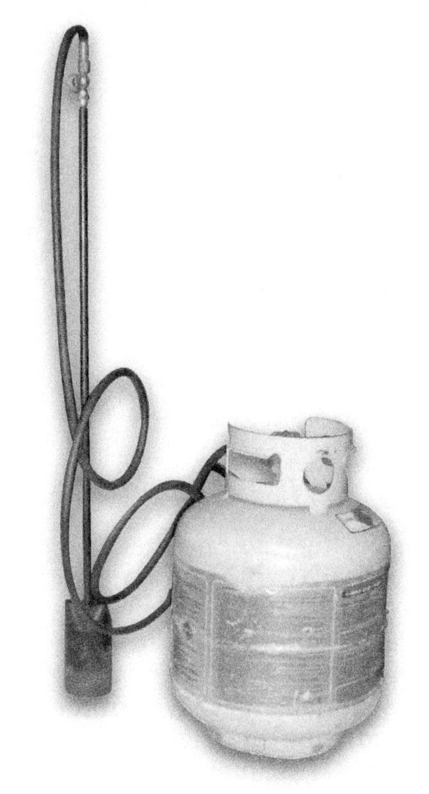

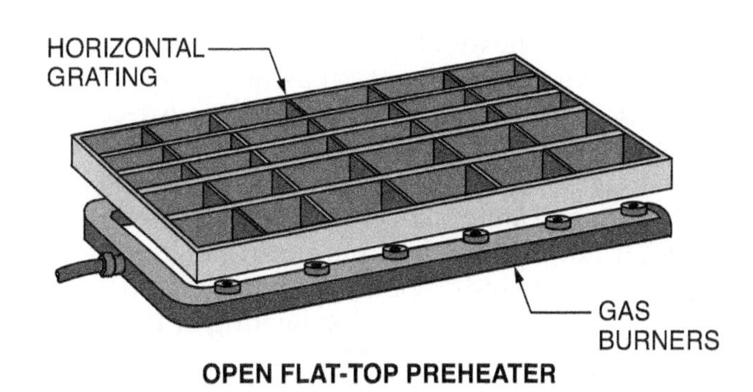

Figure 9 Open flat-top preheater.

2.3.3 Open-Top Preheaters

Open-top preheaters come in several basic designs. The simplest is a large, horizontal gas burner with an adjacent or overhead support structure. The object to be heated is mounted on the support structure directly over the gas burners. A variation is the open, flat-top preheater that consist of a horizontal grate or series of bars with gas burners located underneath (*Figure 9*). The item to be preheated is placed on the grating. This type of preheater works similarly to a kitchen gas cooking range.

2.3.4 Electric Resistance Heaters

Electric resistance heaters consist of tubular resistance elements that are formed in several shapes or styles to fit various applications (*Figure 10*). They are temporarily fastened against the surface of the metal to be heated. Electricity passing through the elements will heat the weldment.

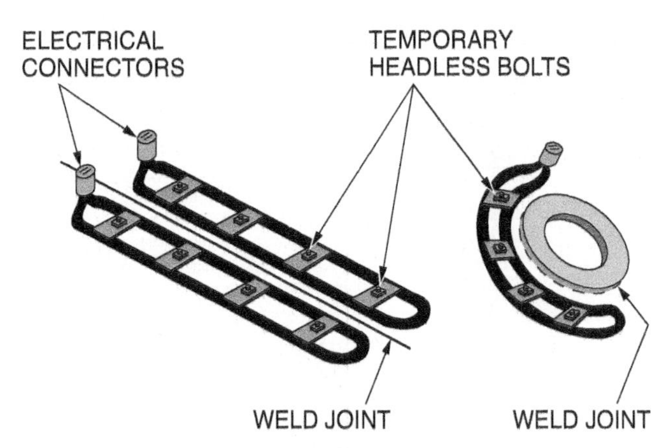

Figure 10 Resistance heating elements.

2.3.5 Induction Heaters

In induction heating, an alternating magnetic field is generated within the metal body to be heated. A magnetic coil is placed around the metal object and a strong alternating current is passed through the coil. The resulting alternating magnetic field generated within the base metal creates circulating electrical currents known as eddy currents. The interaction of these currents with the electrical resistance of the base metal produces an electrical heating effect, raising the temperature of the base metal.

Figure 11 shows an air-cooled induction heating system consisting of a power supply, induction blanket, and associated cables. Air-cooled systems are typically used for preheat applications up to 400°F and may use a controller to monitor and automatically control the temperature. If not equipped with a controller, the use of a temperature indicator is required. The existence of ceramic blanket induction heating systems has made this a desirable option.

2.4.0 Post-Weld Heat Treating

Post-weld heat treatment (PWHT) is the heating of a weldment or assembly after welding is completed. It can be performed on most types of base metals. PWHT equipment and techniques are the same as those used for preheating.

PWHT treatments include the following:

- Stress relieving
- Annealing
- Normalizing
- Tempering

Figure 11 Induction heating system.

2.4.1 Stress Relieving

Stress relieving is the most commonly used PWHT. It is done below the critical (recrystallization) temperature, usually in the 1,050°F to 1,200°F range.

Stress relief treatment is done for the following reasons:

- To reduce residual (shrinkage) stresses in weldments and castings. This is especially important with highly restrained joints.
- To improve resistance to corrosion and caustic embrittlement.
- To improve dimensional stability during machining operations.
- To improve resistance to impact loading and low-temperature failure.

Some codes cover stress relieving to specify the heating rate, holding time, and cooling rate. The heating rate is usually 300°F to 350°F per hour. The holding time is usually one hour for each inch of thickness. The cooling rate is also usually 300°F to 350°F per hour. Stress relieving requires the use of temperature indicators and temperature-control equipment.

As discussed in this module, the most commonly used method for stress relieving weldments is by PWHT. However, stress relief can also be accomplished mechanically. One method involves attaching a mechanical vibrating device to the weldment during or immediately after welding. When activated, the device vibrates the weldment at a specific resonant frequency. This acts to even the stress distribution within the weldment by means of plastic deformation of the metal's grains. The weight of the weldment generally determines the length of time the weldment is subjected to the vibrations.

2.4.2 Annealing

Annealing relieves stresses much more than a stress relieving PWHT can, but it results in a steel of lower strength and higher ductility. Annealing is done at temperatures approximately 100°F above the critical temperature, with a prolonged holding period. The heating is followed by slow cooling in the furnace or by covering, wrapping, or burying the item, although austenitic stainless steel requires rapid cooling. Annealing is used to relieve the residual stresses associated with welds in carbon-molybdenum pipe and also to relieve stresses in welded castings that contain casting strains. Generally, annealing is considered to leave the metal in its softest condition, with good ductility.

2.4.3 Normalizing

Normalizing is the process in which heat is applied to remove strains and reduce grain size. Normalizing is done at temperatures and holding periods comparable to those used in annealing, but the cooling is usually done in still air outside the furnace, and the cooling rate is slightly faster. Normalizing usually results in higher strength and less ductility than annealing. It can be used on mild steel weldments to form a uniform austenite solid solution, to soften the steel, and to make it more ductile. Normalized metal is not as soft and free of stresses as fully annealed metal.

2.4.4 Tempering

Tempering, also called drawing, increases the toughness of quenched steel and helps avoid breakage and failure of heat-treated steel. It reduces both the hardness and the brittleness of hardened steel. Tempering is done at temperatures below the critical temperature, much lower than those used for annealing, normalizing, or stress relieving. Tempering is commonly done to tool steel.

2.4.5 Time-at-Temperature Considerations

In most heat treatment procedures, temperature and time-at-temperature are both specified. Time is very important because metallurgical changes are sluggish at temperatures below the critical temperatures. The maximum temperature is determined by the metal alloy. The holding time at the maximum temperature is based on the metal's thickness, usually one hour for each inch of thickness. The cooling rate is determined by the specific treatment or any applicable code. Temperature control must also be used for PWHT. PWHT temperatures are determined in the same manner as preheating temperatures.

2.5.0 Measuring Temperatures

Controlling preheating, interpass, and PWHT temperatures depends on accurately determining the temperature of the weld zone or assembly. Both underheating and overheating can negatively affect a metal's physical and mechanical characteristics. There are a number of ways to determine the surface temperature of an item being heated. Some involve complex manufactured products, while others use less sophisticated materials.

Devices and materials that can be used to determine surface temperatures include the following:

- Pyrometers
- Thermocouple devices
- Temperature-sensitive indicators

2.5.1 Infrared Pyrometers

An infrared pyrometer is a handheld device that can measure temperature from a distance, based on the detection of infrared energy emitted from a surface. They are common tools found in manufacturing plants and industrial facilities, as well as in the tool bags of craftworkers, since the ability to measure the temperature of a distant surface is a distinct advantage.

Laser sighting typically projects a red dot for targeting the surface to be measured. The temperature is displayed instantly. Note that they are less effective on reflective surfaces. Some infrared pyrometers have an adjustable setting to accommodate reflectivity, improving the accuracy of the measurement. However, the temperature of a shiny surface can still be measured by applying a matte paint or non-reflective tape to the surface where the measurement will be taken.

One important consideration is the size of the target area that an infrared pyrometer "sees" in its path. This refers to the distance-to-spot ratio (D:S) of infrared thermometers. The D:S is the ratio of the distance to the object being measured compared to the diameter of the temperature-measurement area (the spot). For example, if the D:S ratio is 12:1, the measurement of an object 12" (30.5 cm) away will average the temperature over a 1" (25 mm) diameter spot on the target.

The greater the distance, the larger the averaged spot becomes. When using an infrared unit with a 12:1 D:S from 12' (3.7 m) away, the spot in view becomes 1' (0.3 m) in diameter. As a result, the measurement may be inaccurate, as the pyrometer will average the temperature of two or more unrelated surfaces. This is not a problem, of course, if the surface being measured is larger than the target area.

When using an infrared pyrometer, it is important to be aware of its specifications and limitations. Devices that provide a 50:1 D:S ratio are more expensive, but they offer much better performance. At a range of 12', a 50:1 D:S ratio measures a target about 2.9" in diameter—far more practical for the industrial environment.

2.5.2 Thermocouple Devices

A thermocouple device is a bimetallic device (usually twisted wires of two dissimilar metals) placed in physical contact with the material whose temperature is to be measured. Thermocouples work on the principle that when two dissimilar metals are joined, a predictable voltage will be generated that relates to the difference in temperature between the measuring junction and the reference junction. Leads from the thermocouple are connected to a device that can measure extremely small electric voltages and currents. The thermocouple generates a weak electric voltage and current that is in direct proportion to the thermocouple's temperature. The measured voltage or current is fed to a calibrated meter and read directly in degrees of temperature (*Figure 12*).

2.5.3 Temperature-Sensitive Indicators

Temperature-sensitive indicators are devices that change form or color at specific temperatures or temperature ranges. These include both commercially made materials and devices such as crayon sticks (*Figure 13* [A]), liquids (*Figure 13* [B]), chalks, powders, pellets, and labels (*Figure 13* [C]) designed to change color or form at precise temperatures, often within a one percent tolerance. An example of a commonly used indicator is the temperature stick, mentioned earlier in this module. Temperature sticks are a series of graduated temperature-indicating crayons that cover the temperature range from 100°F to 1,150°F. These sticks are used to make a mark on the base metal. At a specific temperature, the mark melts or changes color. *Table 5A* shows a preheating chart and the temperature-indicating crayons to use for each type of metal.

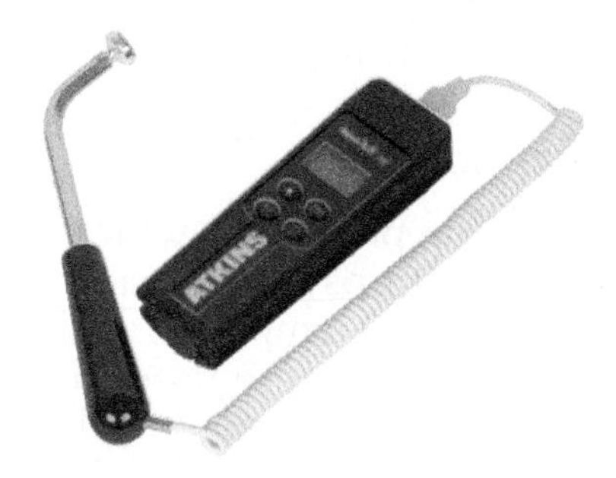

Figure 12 Thermocouple pyrometer.

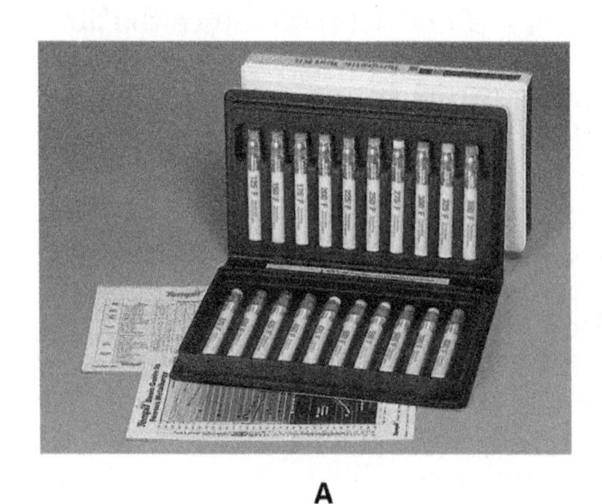

A

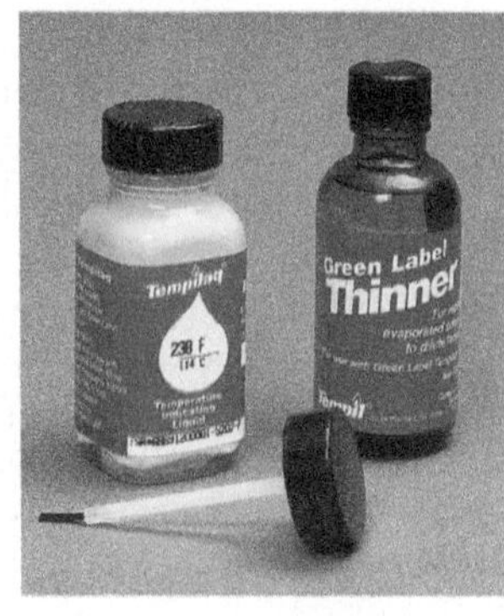

B

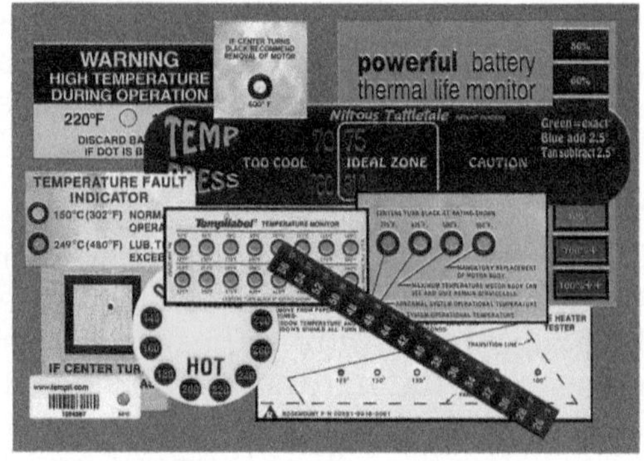

C

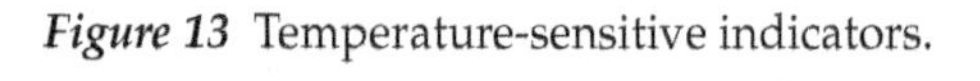

Figure 13 Temperature-sensitive indicators.

Table 5A Preheating Chart and Temperature-Indicating Crayons (1 of 2)

METAL GROUP	METAL DESIGNATION	APPROXIMATE COMPOSITION (PERCENTAGE)								RECOMMENDED PREHEAT TEMPERATURE	TEMPERATURE-INDICATING CRAYONS (°F)
		C	MN	BI	CR	NI	MO	CU	OTHER		
PLAIN CARBON STEELS	PLAIN CARBON STEEL	<0.20								UPTO 150°F	150
	PLAIN CARBON STEEL	0.20–0.30								150°F–300°F	150-200-250-300
	PLAIN CARBON STEEL	0.30–0.45								250°F–450°F	250-300-350-400-450
	PLAIN CARBON STEEL	0.45–0.80								450°F–750°F	450-500-600-700-750
PLAIN MOLY STEELS	CARBON MOLY STEEL	0.10–0.20					0.50			150°F–250°F	150-200-250
	CARBON MOLY STEEL	0.20–0.30					0.50			200°F–400°F	200-250-300-350-400
MANGANESE STEELS	SILICON STRUCTURAL STEEL	0.35	0.80	0.25						300°F–500°F	300-400-500
	MEDIUM MANGANESE STEEL	0.20–0.25	1.0-1.75							300°F–500°F	300-400-500
	SAE 1330 STEEL	0.30	1.75							400°F–800°F	400-500-600
	SAE 1340 STEEL	0.40	1.75							500°F–800°F	500-600-700-800
	SAE 1045 STEEL	0.50	1.75							800°F–900°F	600-700-800-900
	12% MANGANESE STEEL	1.25	12.0							USUALLY NOT REQUIRED	

Table 5A Preheating Chart and Temperature-Indicating Crayons (2 of 2)

METAL GROUP	METAL DESIGNATION	APPROXIMATE COMPOSITION (PERCENTAGE)								RECOMMENDED PREHEAT TEMPERATURE	TEMPERATURE-INDICATING CRAYONS (°F)
		C	MN	BI	CR	NI	MO	CU	OTHER		
HIGH TENSILE STEELS	MANGANESE MOLY STEEL	0.20	1.65	0.20			0.35			300°F–500°F	300-400-500
	JALTEN STEEL	0.35 MAX	1.50	0.30				0.40		400°F–800°F	400-500-600
	MANTEN STEEL	0.30 MAX	1.35	0.30				0.20		400°F–800°F	400-500-800
	ARMCO HIGH TENSILE STEEL	0.12 MAX				0.50 MIN	0.05 MIN	0.35 MIN		UP TO 200°F	200
	DOUBLE STRENGTH #1A STEEL	0.12 MAX	0.75			0.50–1.25	0.10 MIN	0.50–1.50		300°F–800°F	300-400-500-600
	DOUBLE STRENGTH #1A STEEL	0.30 MAX	0.75			0.50–1.25	0.10 MIN	0.50–1.50		400°F–700°F	400-500-600-700
	MAYARIR STEEL	0.12 MAX	0.75	0.35	0.2-1.0	0.25–0.75		0.80		UP TO 300°F	200-300
	OTISCOLOY STEEL	0.12 MAX	1.25	0.10 MAX	0.10 MAX			0.50 MAX		200°F–400°F	200-300-400
	NAX HIGH TENSILE STEEL	0.15–0.25		0.75	0.60	0.17	0.15 MAX	0.25 MAX	ZR.12	UP TO 300°F	200-300
	CROMANSIL STEEL	0.14 MAX	1.25	0.75	0.50					300°F–400°F	300-400
	A.W. DYN-EL STEEL	0.11–0.14						0.40		UP TO 300°F	200-300
	CORTEN STEEL	0.12 MAX		0.25–1.0	0.5–1.5	0.55 MAX		0.40		200°F–400°F	200-300-400

Table 5B Preheating Chart and Temperature-Indicating Crayons (1 of 2)

METAL GROUP	METAL DESIGNATION	APPROXIMATE COMPOSITION (PERCENTAGE)								RECOMMENDED PREHEAT TEMPERATURE	TEMPERATURE-INDICATING CRAYONS (°F)
		C	MN	BI	CR	NI	MO	CU	OTHER		
HIGH TENSILE STEELS (CONT.)	CHROME COPPER NICKEL STEEL	0.12 MAX	0.75		0.75	0.75		0.55		200°F–400°F	200-300-400
	CHROME MANGANESE STEEL	0.40	0.90		0.40					400°F–600°F	400-500-600
	YOLOY STEEL	0.05–0.35	0.3–1.0			1.75		1.0		200°F–600°F	200-300-400-500-600
	HI-STEEL	0.12 MAX	0.6	0.3 MAX		0.55		0.9–1.25		200°F–500°F	200-300-400-500
NICKEL STEELS	2½% NICKEL STEEL	0.25				2.25				200°F–400°F	200-300-400
	3½% NICKEL STEEL	0.23				3.50				200°F–400°F	200-250-300-350-400
MOLY BEARING CHROMIUM AND NICKEL STEELS	SAE 4140 STEEL	0.40			0.90		0.20			600°F–800°F	600-700-800
	SAE 4340 STEEL	0.40			0.80	1.85	0.25			700°F–900°F	700-800-900
	SAE 4615 STEEL	0.15				1.80	0.25			400°F–800°F	400-500-600
	SAE 4820 STEEL	0.20				3.50	0.25			600°F–800°F	600-700-800
LOW CHROME MOLY STEELS	1¼% CR–½% MO STEEL	0.17 MAX			1.25		0.50			250°F–400°F	250-300-350-400
	2¼% CR–1% MO STEEL	0.15 MAX			2.25		1.0			300°F–500°F	300-400-500
MEDIUM CHROME MOLY STEELS	5% CR–½% MO STEEL	0.15 MAX			5.0		0.5			400°F–800°F	400-500-600
	7% CR–½% MO STEEL	0.15 MAX			7.0		0.5			400°F–800°F	400-500-600
	9% CR–1% MO STEEL	0.15 MAX			9.0		1.0			400°F–800°F	400-500-600

Table 5B Preheating Chart and Temperature-Indicating Crayons (2 of 2)

METAL GROUP	METAL DESIGNATION	APPROXIMATE COMPOSITION (PERCENTAGE)								RECOMMENDED PREHEAT TEMPERATURE	TEMPERATURE-INDICATING CRAYONS (°F)
		C	MN	BI	CR	NI	MO	CU	OTHER		
PLAIN HIGH CHROMIUM STEELS	11½–13% CR TYPE 410	0.15 MAX			12.0					400°F–800°F	400-500-800
	16–18% CR TYPE 430	0.12 MAX								300°F–500°F	300-400-500
	23–27% CR TYPE 448	0.20 MAX								300°F–500°F	300-400-500
CHROME NICKEL STAINLESS STEELS	18% CR 8% NI TYPE 304	0.07				8.0				USUALLY DO NOT REQUIRE PREHEAT BUT IT MAY BE DESIRABLE TO REMOVE CHILL	200
	25-12 TYPE 309	0.07				12.0					
	25-20 TYPE 310	0.10				20.0					
	18-8 CB TYPE 347	0.07				8.0			CB 10xC		
	18-8 MO TYPE 316	0.07				8.0	2.5				
	18-8 MO TYPE 317	0.07				8.0	3.5				
IRONS	CAST IRON									700°F–900°F	700-800-900
	NI RESIST									500°F–1,000°F	500-700-900-1000
NON FERROUS	NICKEL, MONEL, INCONEL									PREHEAT NOT USUALLY REQUIRED FOR THIN SECTIONS. 300°F–400°F PREHEAT MAY BE DESIRABLE FOR THICK SECTIONS.	
	ALUMINUM-COPPER										

2.0.0 Section Review

1. How are pipe stresses locked into fabricated pipe assemblies?
   a. By double-tacked welds
   b. With bolts that correspond to the type of metal being used
   c. With anti-corrosive metal glue
   d. By rapid heating and cooling of pipe during welding

2. The quick cooling of metal that can freeze some of the modified crystalline structure is called _____.
   a. dousing
   b. cryogenesis
   c. quenching
   d. rapid refrigeration

3. The alloy composition, thickness of the base metal, welding process, and ambient temperature all affect a metal's _____.
   a. hardenability
   b. preheating requirements
   c. postheating requirements
   d. interpass temperature

4. Why is it preferable to preheat in an oven or furnace?
   a. It's more economical.
   b. It's faster and more reliable.
   c. The item gets heated evenly.
   d. Safety is heightened.

5. Pyrometers, thermocouple devices, and temperature-sensitive indicators are used to determine _____.
   a. interior temperatures
   b. ambient temperatures
   c. surface temperatures
   d. functionality of an indicating gauge

SECTION THREE

3.0.0 STRESS RELIEF FOR ALIGNING PIPE TO ROTATING EQUIPMENT

Objective

Describe how piping stress is avoided when a connection to rotating equipment is required.

a. Explain the purpose of grouting and its roles in supporting machinery.
b. Describe how preliminary alignments are completed on rotating equipment.

Pipefitters must be able to make the connection between the pipe flange and equipment nozzle without forcing the pieces together. If the pipe flange is not properly aligned to equipment flanges, excessive stresses are imposed on the equipment and the pipeline. If it is necessary to use force to align the flange and equipment nozzle, the pipe will pull the pump and motor out of alignment and decrease the lives of both.

Pipe alignment to rotating equipment takes coordination of work activities between the pipefitter installing the pipe and the millwright installing the pump and motor. Before final pipe alignment can be made, the equipment must be set and grouted in place by the millwright. The millwright will normally perform a preliminary alignment between the pump and motor before the final alignment.

3.1.0 Grouting

The three primary reasons for grouting under the bases of machinery are:

- To spread the impact of equipment operation over a larger area
- To provide higher compressive strength than reinforced concrete
- To give a more uniform support by filling in all the voids along the base of the equipment

Grout is also used for the installation of baseplates, anchor bolts, bearing plates, and other types of fasteners. It must be vertically confined until the drying process is complete. Equipment that is resting on isolators should not be grouted because the grouting defeats the purpose of the isolators. A concrete mixture should never be used in place of grouting.

3.2.0 Performing Preliminary Alignment

Rotating shafts are usually aligned with the coupling halves in place. Measurements are taken from the coupling halves and adjustments are made accordingly. When aligning two coupling halves so that the shafts will be aligned, there are two ways that they must be lined up. First, the outer diameters, or rims, of the couplings must be lined up all the way around; then the faces of the couplings must be lined up. For this reason, conventional alignment is also called rim-and-face alignment. Although there are only two ways that a coupling can be misaligned, there are four ways of looking at the misalignment.

The four basic types of misalignment are as follows:

- Vertical offset
- Horizontal offset
- Vertical angularity
- Horizontal angularity

3.2.1 Vertical Offset

Vertical offset is also called parallel misalignment. Vertical offset occurs when one of the coupling halves is not in line with the other when viewed from the side. In vertical offset, one coupling is higher than the other. In this type of misalignment, one of the units must be raised or lowered to align the couplings. *Figure 14* shows vertical offset.

3.2.2 Horizontal Offset

Horizontal offset occurs when the outer diameters of the couplings are misaligned from side to side as viewed from the top. In this type of misalignment, one of the units must be moved to one side or the other to align the couplings. *Figure 15* shows horizontal offset.

3.2.3 Vertical Angularity

Vertical angularity is also called angular misalignment. Vertical angularity means that the faces of the couplings are not square with one another when viewed from the side. In this type of misalignment, one of the units must be tilted on its base to align the couplings. *Figure 16* shows vertical angularity.

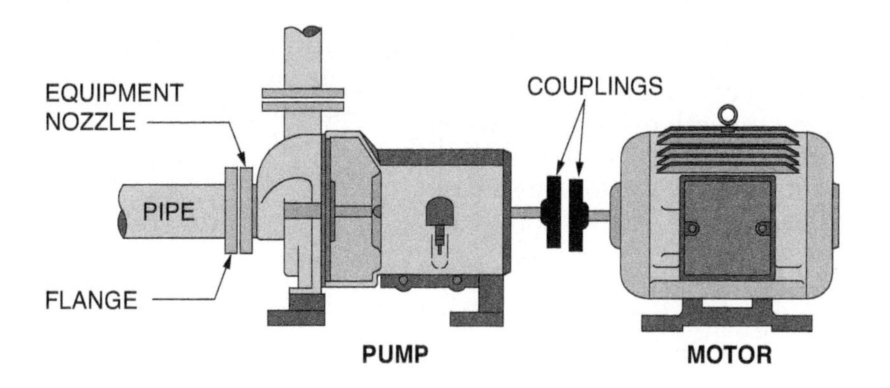

Figure 14 Vertical offset.

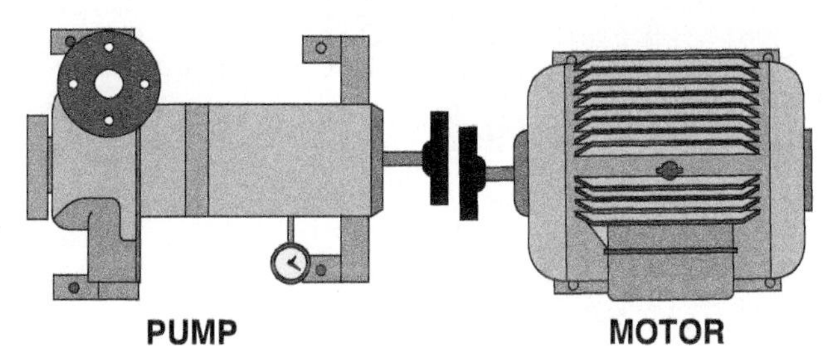

Figure 15 Horizontal offset (top view).

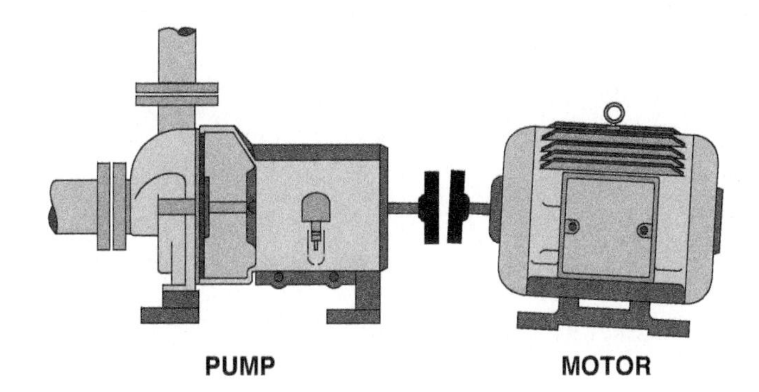

Figure 16 Vertical angularity.

3.2.4 Horizontal Angularity

Horizontal angularity means that the faces of the couplings are not square with one another when viewed from the top. In this type of misalignment, one of the units must be rotated on its base to align the couplings. *Figure 17* shows horizontal angularity.

Although these are the four basic types of misalignment, any variation of the four types may occur. It is possible to have horizontal offset and vertical angularity, or a misalignment problem that includes all four types.

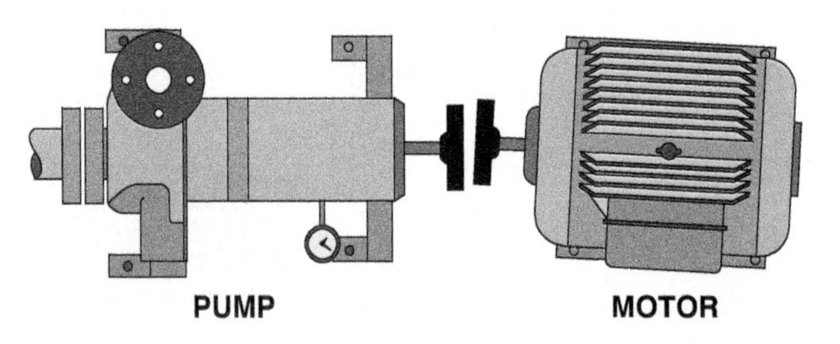

Figure 17 Horizontal angularity (top view).

3.0.0 Section Review

1. "Rim-and-face" alignment is another term for ______.
   a. angular alignment
   b. offset alignment
   c. conventional alignment
   d. non-conventional alignment

Review Questions

1. When temperature changes, all materials change in _____.
   a. color
   b. light absorption
   c. size
   d. texture

2. To determine the expansion of a given piece of material, multiply the length of the pipe by the temperature change in degrees Fahrenheit by the _____ for that material.
   a. coefficient of expansion
   b. modulus of elasticity
   c. crystallization
   d. diameter of the pipe

3. Cold springing, installation of expansion loops, and changes in direction between anchor points are all methods of dealing with _____.
   a. poor weldments
   b. expansion
   c. compression
   d. thermal isolation

4. Flexibility in layout consists of making many _____ between anchor points.
   a. secondary anchors
   b. hangers
   c. changes in direction
   d. pipe breaks

5. Expansion loops should *not* be fabricated with _____.
   a. socket weld fittings
   b. flanged or threaded joints
   c. butt weld joints
   d. mitered joints

6. Pipe stresses are locked into fabricated pipe assemblies by the _____ that occurs in the welding process.
   a. alignment change
   b. welding gap
   c. rapid heating and cooling
   d. uniform cooling

7. The main purpose of PHWT is to keep the weld and the base metal from _____.
   a. heating too fast
   b. cooling too fast
   c. melting
   d. getting too hot

8. Quenching medium- and high-carbon steels reduces the _____.
   a. diameter of the pipe
   b. toughness of the metal
   c. weldability of the metal
   d. length of cooling required

9. When medium- and high-carbon steels are not preheated enough, the large mass of base metal will _____ the weld zone.
   a. bend
   b. quench
   c. melt with
   d. crack

10. In induction heating, _____ is generated within the metal.
    a. DC current
    b. AC current
    c. an alternating magnetic field
    d. liquid coolant

11. Stress relief can be accomplished mechanically, but the *most* common method is _____.
    a. air compression
    b. post-weld heat treatment (PWHT)
    c. quenching
    d. cooling by induction of ambient air

12. To increase the toughness of quenched steel and prevent failure of heat-treated steel, _____ or drawing is used.
    a. tempering
    b. tampering
    c. annealing
    d. normalizing

13. If grouting is not an option for providing higher compressive strength under the bases of machinery, the use of a concrete mixture is an acceptable alternative.
    a. True
    b. False

14. Offset misalignment is also called _____ misalignment.
    a. shaft
    b. parallel
    c. face
    d. angular

15. If the faces of couplings are not square with one another when viewed from the side, it is referred to as _____.
    a. vertical offset
    b. horizontal offset
    c. vertical angularity
    d. horizontal angularity

Trade Terms Introduced in This Module

Alloy: A metal that has had other elements added to it, which substantially changes its mechanical properties.

Carbon: An element which, when combined with iron, forms various kinds of steel. In steel, it is the carbon content that affects the physical properties of the steel.

Constituents: The elements and compounds, such as metal oxides, that make up a mixture or alloy.

Critical temperature: The temperature at which iron crystals in a ferrous-based metal transform from being face-centered to body-centered. This dramatically changes the strength, hardness, and ductility of the metal.

Hardenability: A characteristic of a metal that makes it able to become hard, usually through heat treatment.

Heat-affected zone (HAZ): The area of the base metal that is not melted but whose mechanical properties have been altered by the heat of the welding.

Induction heating: The heating of a conducting material by means of circulating electrical currents induced by an externally applied alternating magnetic field.

Interpass heat: The temperature to which a metal is heated while an operation is performed on the metal.

Post-weld heat treatment (PWHT): The temperature to which a metal is heated after an operation is performed on the metal.

Preheat: The temperature to which a metal is heated before an operation is performed on the metal.

Preheat weld treatment (PHWT): The controlled heating of the base metal immediately before welding begins.

Quenching: Rapid cooling of a hot metal using air, water, or oil.

Stress: The load imposed on an object.

Stress relieving: The even heating of a structure to a temperature below the critical temperature followed by a slow, even cooling.

Tempering: Increases the toughness of quenched steel and helps avoid breakage and failure of heat-treated steel; also called drawing.

Additional Resources

This module presents thorough resources for task training. The following reference material is recommended for further study.

Cold Springing of Pipes. 2018. Piping Engineering. Available at: **https://www.pipingengineer.org/cold-springing-of-pipes/**.

Welding Trainee Guide, Contren Series, Prentice Hall, 2003.

Figure Credits

Miller Electric Mfg. Co., Figure 11

Pyrometer Instrument Company, Inc., Figure 12

Tempil, Inc., Figure 13

Section Review Answer Key

SECTION 1.0.0

Answer	Section Reference	Objective
1. b	1.1.0	1a
2. c	1.2.0	1b
3. b	1.3.0	1c

SECTION 2.0.0

Answer	Section Reference	Objective
1. d	2.0.0	2
2. c	2.1.0	2a
3. b	2.2.2	2b
4. c	2.3.0	2c
5. c	2.5.0	2e

SECTION 3.0.0

Answer	Section Reference	Objective
1. c	3.2.0	3b

This page is intentionally left blank.

NCCER CURRICULA — USER UPDATE

NCCER makes every effort to keep its textbooks up-to-date and free of technical errors. We appreciate your help in this process. If you find an error, a typographical mistake, or an inaccuracy in NCCER's curricula, please fill out this form (or a photocopy), or complete the online form at **www.nccer.org/olf**. Be sure to include the exact module ID number, page number, a detailed description, and your recommended correction. Your input will be brought to the attention of the Authoring Team. Thank you for your assistance.

Instructors – If you have an idea for improving this textbook, or have found that additional materials were necessary to teach this module effectively, please let us know so that we may present your suggestions to the Authoring Team.

NCCER Product Development and Revision

13614 Progress Blvd., Alachua, FL 32615

Email: curriculum@nccer.org

Online: www.nccer.org/olf

❑ Trainee Guide ❑ Lesson Plans ❑ Exam ❑ PowerPoints Other ______________

Craft / Level: Copyright Date:

Module ID Number / Title:

Section Number(s):

Description:

Recommended Correction:

Your Name:

Address:

Email: Phone:

This page is intentionally left blank.

In-Line Specialties

Overview

Special fittings and instruments used in process piping equipment are known as *in-line specialties* and described in system documentation as "specials." This category includes steam traps, desuperheaters, bursting discs, strainers, and related equipment. Steam traps protect the lines against water hammer while desuperheaters reduce the temperature of steam. Strainers are used with many types of fluids to keep solids from clogging pipes. Bursting discs provide emergency pressure relief to prevent very high-pressure surges from damaging equipment. Each in-line specialty item serves a specific purpose and must be installed, monitored, and removed by knowledgeable pipefitters who may be working in tandem with other professionals for coordinated activities across the pipe run.

Module 08405

Trainees with successful module completions may be eligible for credentialing through the NCCER Registry. To learn more, go to **www.nccer.org** or contact us at 1.888.622.3720. Our website, **www.nccer.org**, has information on the latest product releases and training.

Your feedback is welcome. You may email your comments to **curriculum@nccer.org**, send general comments and inquiries to **info@nccer.org**, or fill in the User Update form at the back of this module.

This information is general in nature and intended for training purposes only. Actual performance of activities described in this manual requires compliance with all applicable operating, service, maintenance, and safety procedures under the direction of qualified personnel. References in this manual to patented or proprietary devices do not constitute a recommendation of their use.

08405
In-Line Specialties

Objectives

Successful completion of this module prepares you to do the following:

1. Identify and describe the operation of various inline specialties.
   a. Identify and describe the operation of snubbers.
   b. Identify and describe the operation of ball joints.
   c. Identify and describe the operation of bleed rings.
   d. Identify and describe the operation of drip legs.
   e. Identify and describe the operation of steam traps.
   f. Identify and describe the operation of expansion joints.
   g. Identify and describe the operation of filters and strainers.
   h. Identify and describe the operation of flowmeters.
   i. Identify and describe the operation of level measurement devices.
   j. Identify and describe the operation of flow pressure switches.
   k. Identify and describe the operation of rupture discs.
   l. Identify and describe the operation of thermowells.
   m. Identify and describe the operation of desuperheaters.
   n. State common safety and storage practices related to inline specialties.
2. Explain how to troubleshoot and maintain steam traps.
   a. Explain and discuss diagnostic methods and maintenance procedures related to steam traps.

Performance Tasks

Under supervision, you should be able to do the following:

1. Identify a number of specialties at the discretion of the instructor.
2. Identify specific problems and corrective actions required for faulty steam traps.
3. Install steam traps.

Trade Terms

Condensate
Conductivity
Differential pressure
Drip legs
Flash steam
Saturated steam
Superheated steam
Tracer

Industry Recognized Credentials

If you are training through an NCCER-accredited sponsor, you may be eligible for credentials from NCCER's Registry. The ID number for this module is 08405. Note that this module may have been used in other NCCER curricula and may apply to other level completions. Contact NCCER's Registry at 1.888.622.3720 or go to **www.nccer.org** for more information.

Contents

Figures

SECTION ONE

1.0.0 TYPES OF IN-LINE SPECIALTIES

Objective

Identify and describe the operation of various inline specialties.

a. Identify and describe the operation of snubbers.
b. Identify and describe the operation of ball joints.
c. Identify and describe the operation of bleed rings.
d. Identify and describe the operation of drip legs.
e. Identify and describe the operation of steam traps.
f. Identify and describe the operation of expansion joints.
g. Identify and describe the operation of filters and strainers.
h. Identify and describe the operation of flowmeters.
i. Identify and describe the operation of level measurement devices.
j. Identify and describe the operation of flow pressure switches.
k. Identify and describe the operation of rupture discs.
l. Identify and describe the operation of thermowells.
m. Identify and describe the operation of desuperheaters.
n. State common safety and storage practices related to inline specialties.

Performance Tasks

1. Identify a number of specialties at the discretion of the instructor.
3. Install steam traps.

Trade Terms

Condensate: The liquid byproduct of cooling steam.

Conductivity: A measure of the ability of a material to transmit electron flow.

Differential pressure: The measurement of one pressure with respect to another pressure, or the difference between two pressures.

Drip legs: Drains for condensate in steam lines placed at low points in the line and used with steam traps.

Flash steam: Steam formed when hot condensate is released to lower pressure and re-evaporated.

Saturated steam: Pure steam without water droplets that is at the boiling temperature of water for the given pressure.

Superheated steam: Saturated steam to which heat has been added to raise the working temperature.

Tracer: A steam line piped beside product piping to keep the product warm or prevent it from freezing.

In-line specialty devices are installed in a piping system for metering, measuring, relieving, controlling, filtering, and monitoring products. In-line specialty equipment is delicate in nature, and care must be taken when it is handled to prevent damage.

Common types include:

- Snubbers
- Ball joints
- Bleed rings
- Drip legs
- Steam traps
- Expansion joints
- Filters and strainers
- Flowmeters
- Level measurement devices
- Flow pressure switches
- Rupture discs
- Thermowells
- Desuperheaters

1.1.0 Snubbers

Snubbers (*Figure 1*) are special fittings that are installed between piping or vessels and pressure gauges. The purpose of snubbers is to protect gauges from pressure surges and shocks in the system. Snubbers smooth and dampen the pressure impulses, thereby prolonging the life and accuracy of the pressure gauges. Pressure dampening or snubbing is accomplished through the use of a porous, stainless steel element assembled into the body of the snubber. This element absorbs and dampens the pressure surges moving through the snubber. Snubbers are widely used with pressure gauges during hydrostatic testing. Without a snubber, the positive displacement action of the test pump causes the gauge needle to move erratically and the gauge to give false readings.

1.2.0 Ball Joints

Ball joints (*Figure 2*) are flanged or welded pipe connections that provide pivot points to allow for expansion and contraction of a piping system. They provide movement in all planes from a single joint and solve many problems with alignment, expansion, contraction, vibration, and shock. Ball joints are designed to swivel or rotate 360 degrees with 30 to 40 degrees of flexing or angular movement.

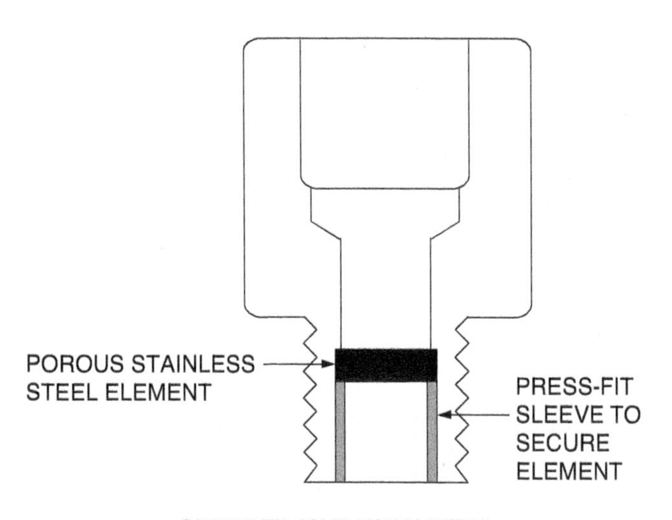

Figure 1 Snubber.

Figure 2 Ball joint.

1.3.0 Bleed Rings

Bleed rings provide a leak-proof means of stopping flow in pipelines. They are used when there is a change in the process material flowing through the line and cross-contamination must be avoided. Bleed rings are installed between block valves and have a valved bleed connection attached. Before the process flow is changed, the valved bleed connection drains the area between the block valves. *Figure 3* shows a typical pipe arrangement for a bleed ring.

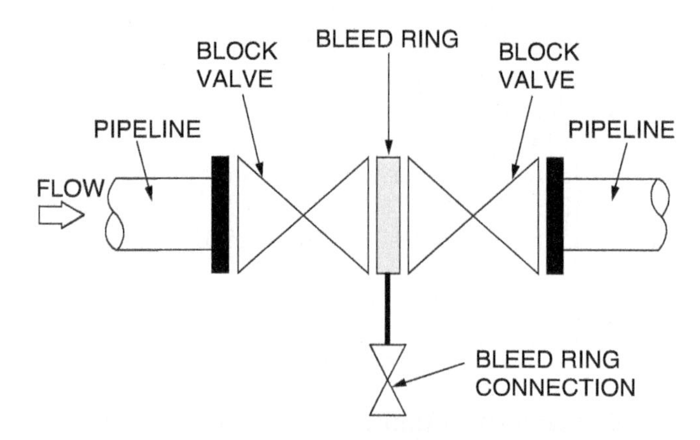

Figure 3 Typical pipe arrangement for bleed ring.

1.4.0 Drip Legs

Drip legs collect condensate and sediments from steam systems. They are made from pipe, fittings, and valves and are installed between the steam header and the equipment at a low point in the line. They can be connected directly to the header or in the line and should be installed at every elevation change. At the base of the drip leg is a drain valve that discharges collected sediment. A pipeline and an isolation valve are normally attached to the side of a drip leg that is routed to a steam trap. Steam traps that serve a drip leg are installed at a lower level than the drip leg. *Figure 4* shows typical drip leg arrangements.

When installing a drip leg:

- Install the drip leg to the steam header line.
- Install a valve at the base of the drip leg to allow condensate to be drained.
- Install a shutoff valve upstream from the drip leg to cut off discharge of condensate from equipment and to allow for maintenance work.
- Install an isolation valve close to the drip leg exit flow to allow maintenance work to be performed without shutting down the supply line.

Steam traps and strainers are located downstream from the drip leg.

1.5.0 Steam Traps

Steam is used in process plants because it efficiently supplies large quantities of heat. For steam to remain efficient, it must be dry (containing no suspended water particles) and of the proper temperature and pressure; this is where steam traps become important.

There are three general types of steam traps: mechanical, thermostatic, and thermodynamic (*Figure 5*). They act as self-actuating drain valves and can be made of cast iron, stainless steel, forged steel, or cast steel. It is important to have the correct trap for the application. A steam trap of the right size and type will operate reliably, prevent steam loss, and maximize product output. The wrong trap will wear out prematurely and fail to drain the condensate.

The most common use of traps in a steam system is to clear the distribution headers of condensate. Traps maintain high energy-transfer rates by constantly purging condensate and gases; this is because air and carbon dioxide can cause problems in a steam line. Air is an excellent insulator and can reduce the efficiency of the steam at its working point on the heat transfer surfaces. Carbon dioxide can dissolve in the condensate and form carbonic acid, which is extremely corrosive. Scale and debris are also transferred with the condensate to areas that are difficult to repair when clogged. A well-placed strainer and steam trap can prevent these problems.

Steam traps are fitted to the pipe with screwed, flanged, or socket weld fittings. Headers are sometimes sloped to a drip leg that is drained by the trap. A low-capacity trap that operates near the temperature of saturated steam is adequate and operates under varying steam pressures. *Figure 6* shows a steam header draining system.

All types of steam traps automatically open an orifice, drain the condensate, then close before steam is lost. With the exception of the inverted bucket-type, no steam trap releases any steam when it's new or in good condition.

1.5.1 Mechanical Steam Traps

Mechanical steam traps respond to the difference in density between steam and condensate. They open to condensate and close to steam. The first mechanical trap was the orifice trap, which contained a small orifice to allow condensate to escape. The most typical modern mechanical trap is called an inverted bucket trap (*Figure 7*), and it has only two moving parts: the valve lever assembly and the bucket. The trap is normally installed between the source of steam and the condensate return header.

The cycle starts with the bucket down and the valve all the way open. Gases and air flow through first, then steam raises the bucket and closes the valve. The gas slowly bleeds out through a small vent in the top of the bucket. In operation the condensate fills the body and the bucket, causing the bucket to sink and open the discharge valve. The valve remains open until enough steam has collected to float the bucket again.

Since the inverted bucket trap operates on pressure differential, it must be primed before being put into service. The inverted trap can handle high-pressure steam; it resists damage from pressure surges and water hammer; and it tolerates freezing if the body is made of a ductile material. Its disadvantages are that it has limited air discharge capacity and a tendency to be noisy.

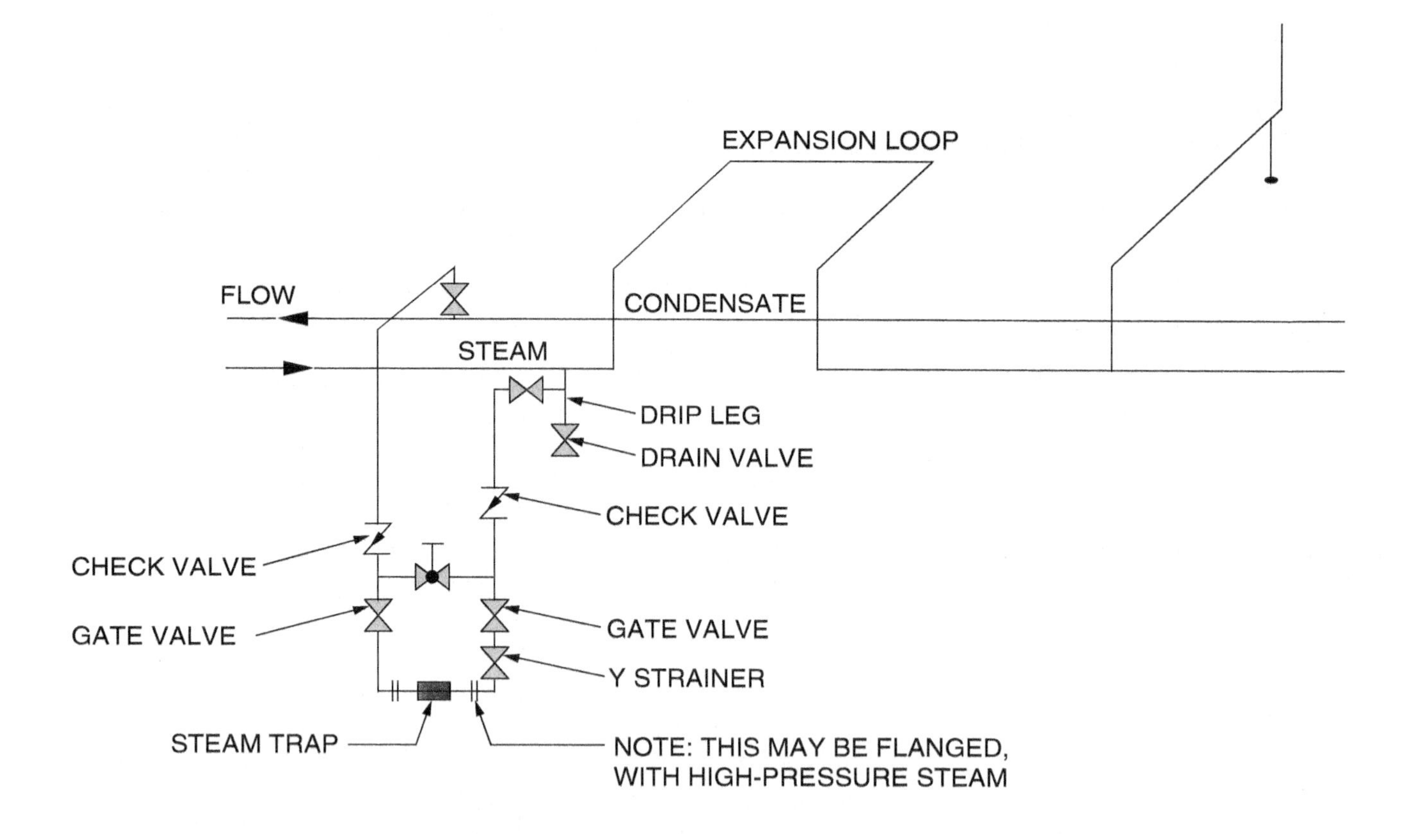
EXPANSION LOOP
FLOW
CONDENSATE
STEAM
DRIP LEG
DRAIN VALVE
CHECK VALVE
CHECK VALVE
GATE VALVE
GATE VALVE
Y STRAINER
STEAM TRAP
NOTE: THIS MAY BE FLANGED, WITH HIGH-PRESSURE STEAM

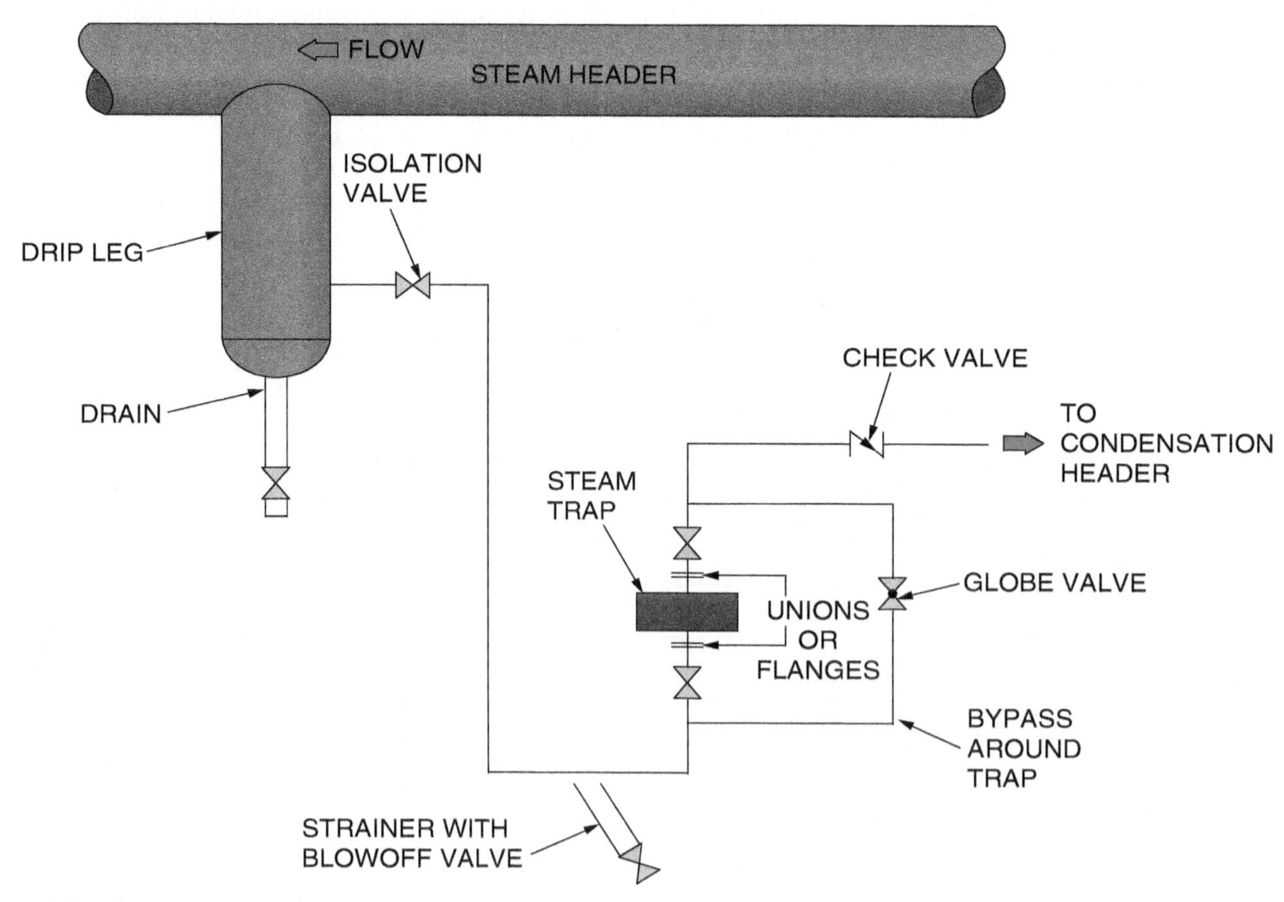
FLOW
STEAM HEADER
ISOLATION VALVE
DRIP LEG
DRAIN
CHECK VALVE
TO CONDENSATION HEADER
STEAM TRAP
GLOBE VALVE
UNIONS OR FLANGES
BYPASS AROUND TRAP
STRAINER WITH BLOWOFF VALVE

Figure 4 Drip leg arrangements.

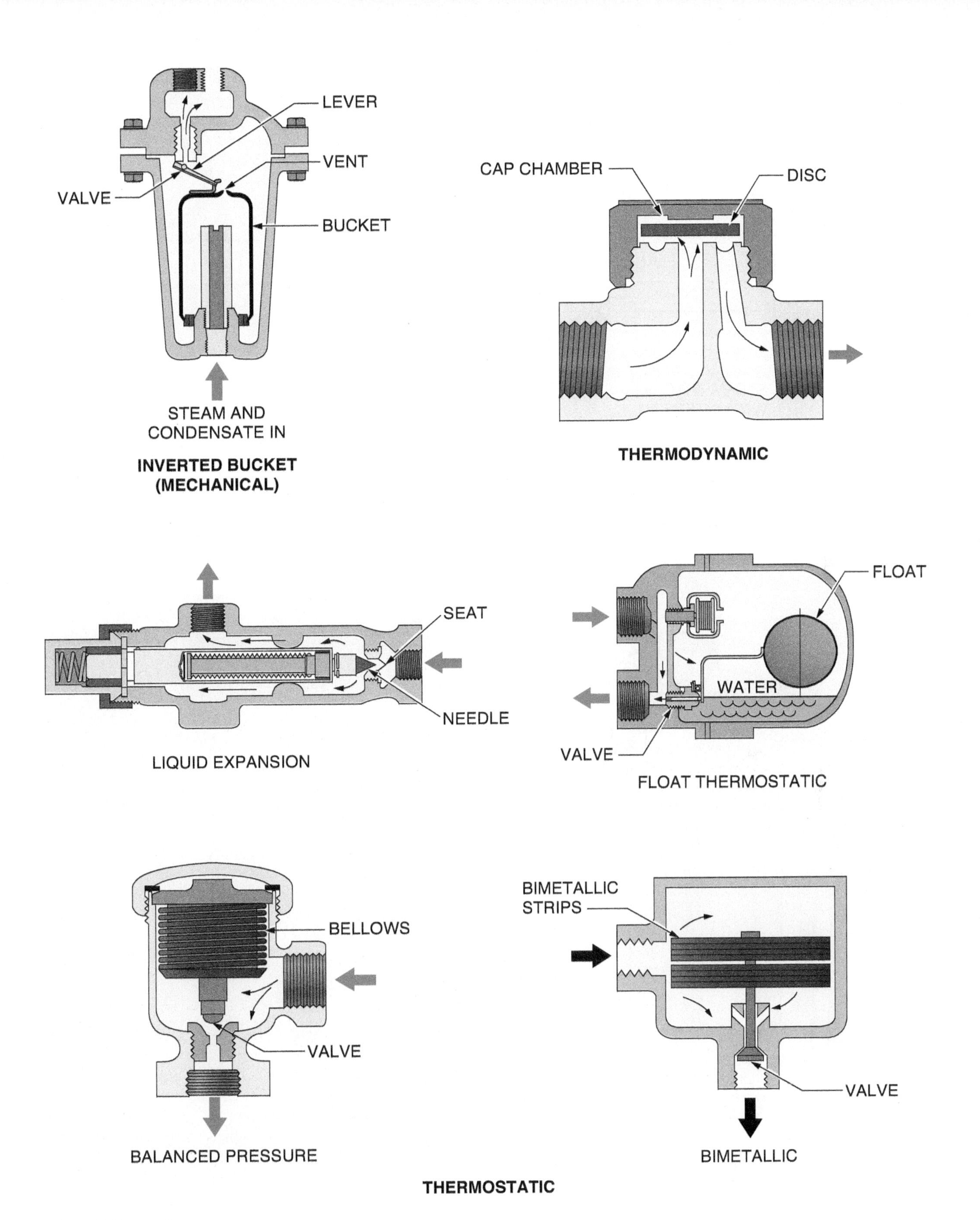
LEVER
VENT
VALVE
BUCKET
STEAM AND CONDENSATE IN
INVERTED BUCKET (MECHANICAL)
CAP CHAMBER
DISC
THERMODYNAMIC
SEAT
NEEDLE
LIQUID EXPANSION
FLOAT
WATER
VALVE
FLOAT THERMOSTATIC
BELLOWS
VALVE
BALANCED PRESSURE
BIMETALLIC STRIPS
VALVE
BIMETALLIC
THERMOSTATIC

Figure 5 Types of steam traps.

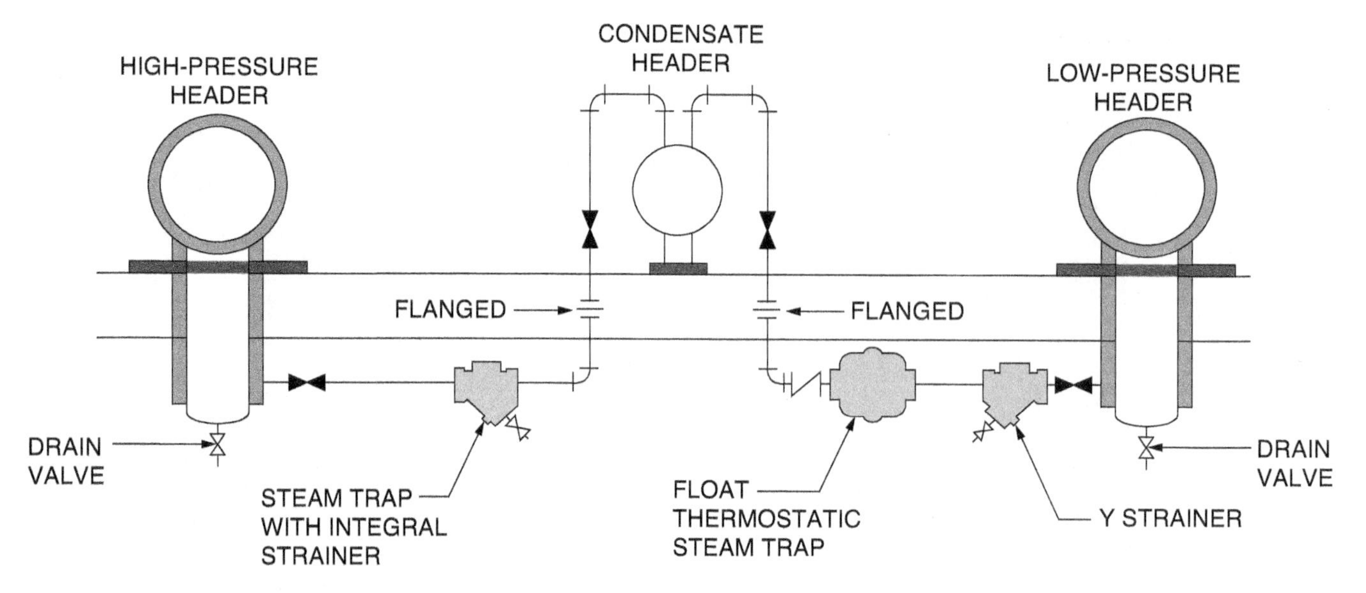

NOTE: Because of the pressure involved, the connection of high and low pressure headers to the condensate header would have a flanged joint.

Figure 6 Steam header draining system.

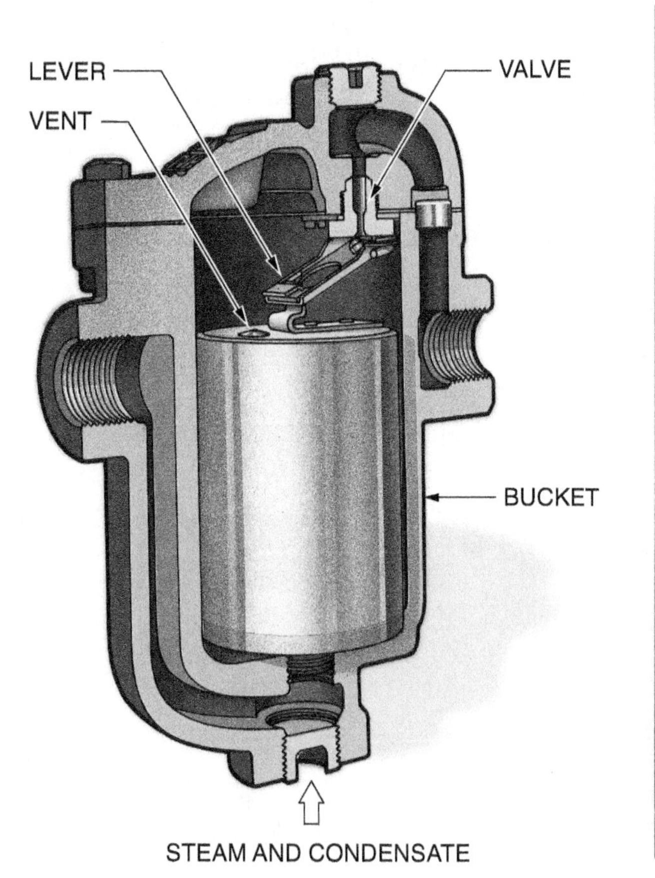

Figure 7 Inverted bucket trap.

1.5.2 Thermostatic Steam Traps

Thermostatic steam traps respond to temperature changes in the steam line. They open when cooler condensate is present and close to higher steam temperatures. The four types of thermostatic steam traps are:

- Liquid expansion
- Balanced pressure
- Bimetallic
- Float

Figure 8 shows the types of thermostatic steam traps.

The liquid expansion thermostatic trap remains closed until the condensate cools below 212°F, opening the trap. This discharges only cool condensate at a constant temperature, which improves heat recovery from submerged tank coils and **tracer** lines.

The balanced pressure trap opens when the condensate cools slightly below the steam saturation temperature at any pressure within the trap's range. Controlled by a liquid-filled bellows, the discharge valve closes when hot condensate vaporizes the liquid in the bellows and opens when the condensate cools enough to lower the pressure inside the element. Balanced pressure traps are smaller than mechanical traps. They open wide when cold, readily purge gases, and are unlikely to freeze.

The bimetallic thermostatic trap has bimetallic strips that respond to cool condensate by bending to open the condensate valve. When the steam hits the strips, they expand and close the valve. They require considerable cooling to open back up, and pressure-compensating characteristics vary among the models. The closing force of some designs varies with the steam pressure, approximating the response of a balanced pressure trap. This trap has a large capacity but responds slowly to process load changes. It vents gases well; is not easily damaged by freezing, water hammer, or corrosion; and handles the high temperatures of **superheated steam**.

Process heating applications call for a steam trap system that vents air well, withstands outside temperature extremes, continuously discharges hot condensate, responds quickly to load changes, and is unaffected by changing inlet pressure. The float thermostatic steam trap, in series with a liquid expansion steam trap, works well in these conditions. *Figure 9* shows an outdoor process trap system.

The float thermostatic trap has a float that rises when condensate enters and opens the main valve that lies below the water level. At the same

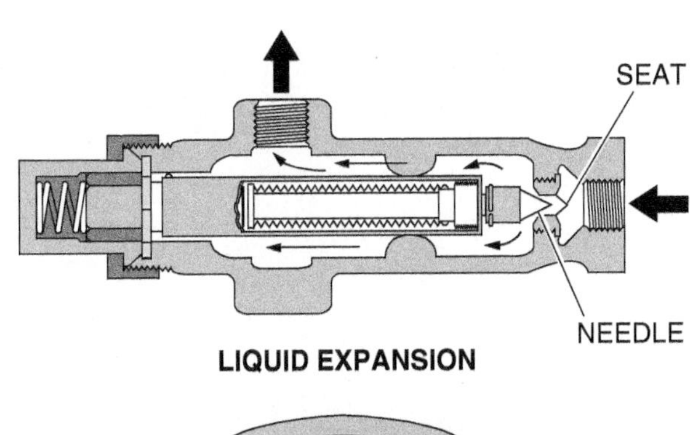

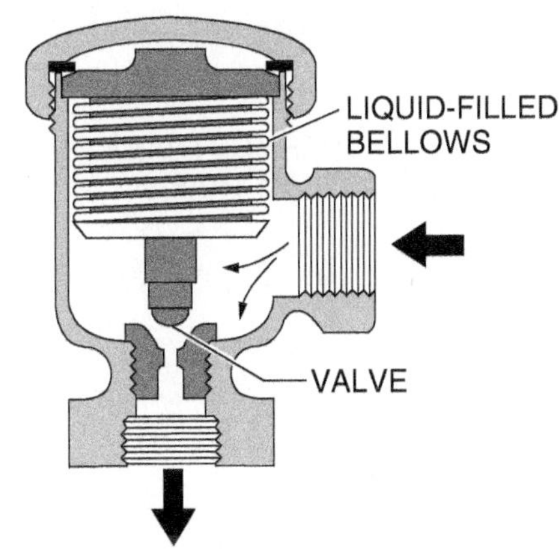

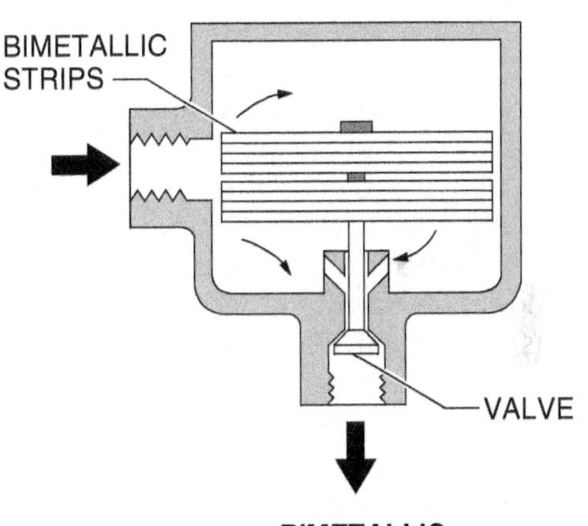

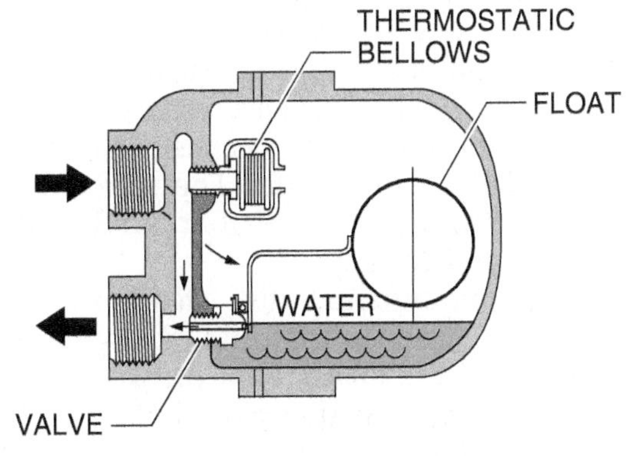

Figure 8 Thermostatic steam traps.

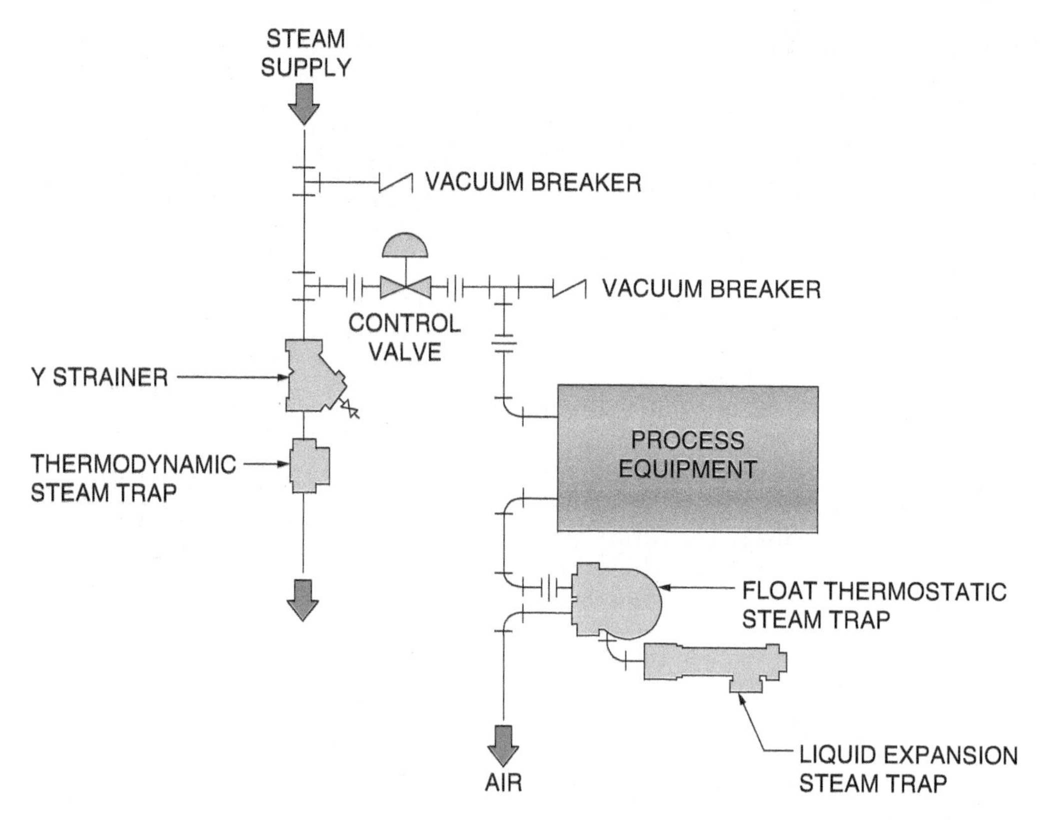

Figure 9 Outdoor process trap system.

time, another discharge valve at the top of the trap, operated by a thermostatic bellows, opens to release cooler gases. This trap vents gases well and responds to wide and sudden pressure changes, discharging condensate continuously. In the failure mode, the main valve is closed but the air vent is usually open. Float thermostatic traps should always be protected from freezing since they usually contain water.

1.5.3 Thermodynamic Steam Traps

Thermodynamic steam traps (*Figure 10*) use the heat energy in hot condensate and steam to control the opening and closing of the trap. They operate well at higher pressures and can be installed in any position because their function does not depend on gravity. They are most often used to drain tracer lines that are in the pressure range of 15 to 200 psig.

The thermodynamic disc steam trap is the most widely used. In this unit, the only moving part is the disc. **Flash steam** pressure from hot condensate keeps the disc closed. When the flash steam above the disc is condensed by cooler condensate, the disc is pushed up and remains open until hot condensate flashes again to build up pressure in the cap chamber. At the same time, flashing condensate on the underside of the disc is discharged at high velocity, lowering the pressure in this area and slamming the disc shut before steam can escape.

The thermodynamic steam trap can withstand freezing, high pressure, superheating, and water hammer without damage. The audible clicking of the disc indicates when the discharge cycle occurs. These traps last longer if they are not oversized for the load.

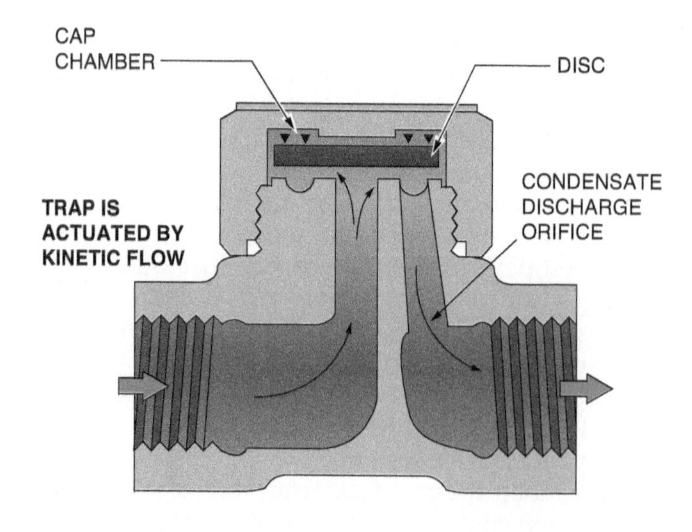

Figure 10 Thermodynamic steam trap.

1.5.4 Steam Trap Installation

To mount a steam trap anywhere other than the correct position would negate its use. Two general rules for steam trap mountings are that they should be lower than any line in the system and that a strainer has to be upstream of the trap. Isolation drains, couplings, check valves, and drip leg placement are also important to each application, and their use depends on the needs and contents of the system. Each steam trap manufacturer has resource books that detail how to install steam trap systems. Some of these are referenced at the back of this module.

For proper steam trap installation:

- Provide a separate trap for each piece of equipment or apparatus. Short-circuiting may occur if more than one piece of equipment is connected to a single trap.
- Tap the steam supply off the top of the steam header to obtain dry steam and avoid steam line condensate.
- Install a supply valve close to the steam header to allow maintenance or revisions to be performed without shutting down the steam header.
- Install a steam supply valve close to the equipment entrance to allow equipment maintenance work to be performed without shutting down the supply line.
- Connect the condensate discharge line to the lowest point in the equipment to avoid water pockets and water hammer.
- Install a shutoff valve upstream of the condensate discharge piping to cut off discharge of condensate from equipment and allow service work to be performed.
- Install a strainer and strainer flush valve ahead of the trap to keep rust, dirt, and scale out of working parts and to allow blow-down removal of foreign material from the strainer basket.
- Provide unions on both sides of the trap for its removal or replacement.
- Install a test valve downstream from the trap to allow observation of discharge when testing.
- Install a check valve downstream from the trap to prevent condensate flow-back during shutdown or in the event of unusual conditions.
- Install a downstream shutoff valve to cut off equipment condensate piping from the main condensate system for maintenance or service work.
- If a bypass around the trap is installed, trap replacement can occur without loss of flow.

A steam trap should have a provision for inspection without interrupting the process flow. The gate valves, as arranged in the piping schematic, show a generic steam trap and strainer layout. *Figure 11* shows basic trap installation.

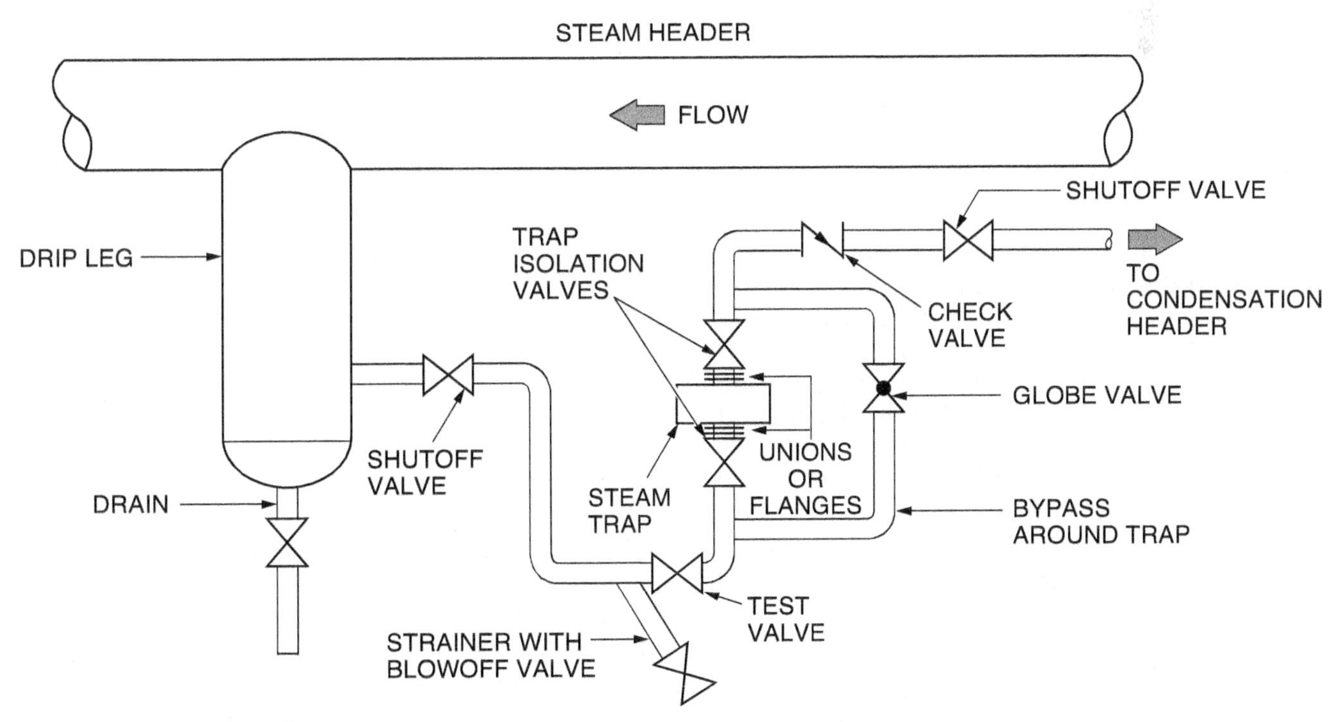

Figure 11 Basic trap installation.

Most steam traps fail in the open position, which is difficult to pinpoint because it does not affect equipment operation. When the trap fails open, both steam and condensate flow freely from the steam header to the condensate header, keeping the heat transfer high. The traps that fail in the closed position are easy to identify because the backup of condensate cools the system.

An effective preventive maintenance (PM) program includes scheduled checks of the entire system. Clogged strainers, leaking joints, leaking valve packing, or missing insulation are some of the PM items to check. When the internal parts of a steam trap wear out, the water seal deteriorates, steam flows through the valve assembly, and this erodes the seal into a worsening condition.

Common causes of steam trap failure are:

- Scale, rust, or corrosive buildup preventing the valve from closing
- Valve assembly wear
- Defective or damaged valve seat
- Physical damage from severe water hammer
- Foreign material lodged between seat and valve
- Blocked, clogged, or damaged strainers
- Increased back pressure
- Other failures are specific to the type of steam trap. The two indications that there is a failure in a steam trap are that the trap blows steam and the trap will not pass condensate.

> **NOTE**
> Remember that many process systems require piping to slope toward one part of the system or another, so that drainage in the absence of pressure will occur in the correct direction.

1.6.0 Expansion Joints

Expansion joints (*Figure 12*) control linear expansion and contraction of a pipe. Linear expansion occurs because of the difference in temperature between the fluid being carried through the pipe and the surrounding air; expansion joints compensate for that. Pipefitting charts showing the exact linear expansion for each degree of heat, but when they are not available the craftworker may estimate the expansion for any length of steel pipe by using ¾" per 100' per 100°.

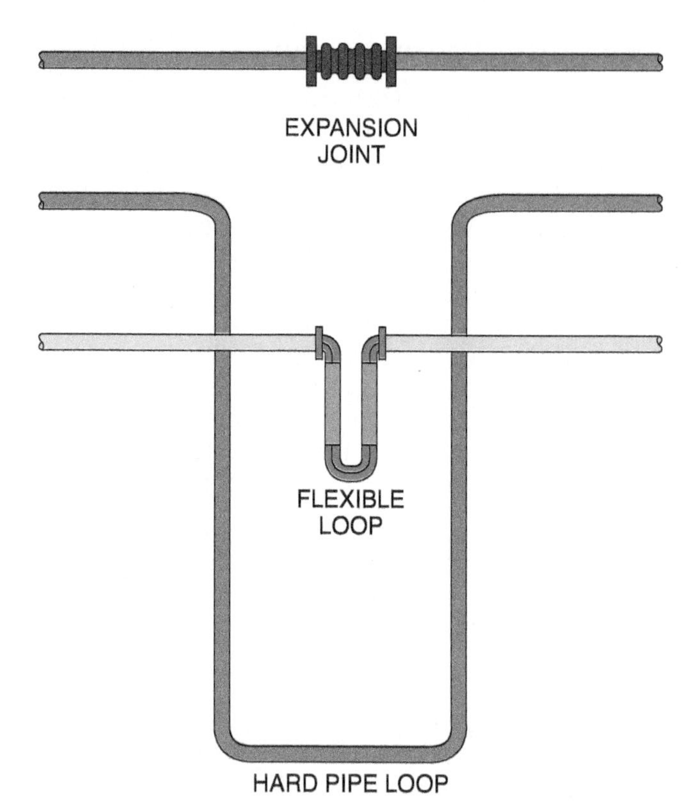

Figure 12 Expansion joints.

1.7.0 Filters and Strainers

Filters are located upstream from various devices to collect suspended solids that need to be removed. Pressure elements are located both upstream and downstream from the filters. The elements take fluid pressure readings at the entrance of the filter and at their points of discharge; these readings are then compared to produce a **differential pressure** reading.

As filters age and collect more particles, differential pressure increases. When this happens, it may be necessary to replace the filter. On critical filter systems, a range of differential pressure is established for the selected system and the range is programmed into the differential pressure monitor. The monitor indicates the low end of the range when the filter is clean and the high end of the range when it becomes saturated or dirty. When installing filters in a filter housing, be sure to install the correct filter with the proper temperature rating. *Figure 13* shows a filter diagram.

1.7.1 Steam Trap Strainers

In all systems, a strainer should be installed upstream of the steam trap. The scale and corrosion in any steam system has to be stopped before it enters the trap, or it will become clogged and damaged. Steam trap strainers need to be cleaned and inspected on a regular basis. Strainers are available in many styles and are sometimes incorporated into the steam trap. *Figure 14* shows three types of strainers.

Temporary or start-up strainers, also known as witches' hats and top hats (*Figure 15*), are installed during system startup between the first

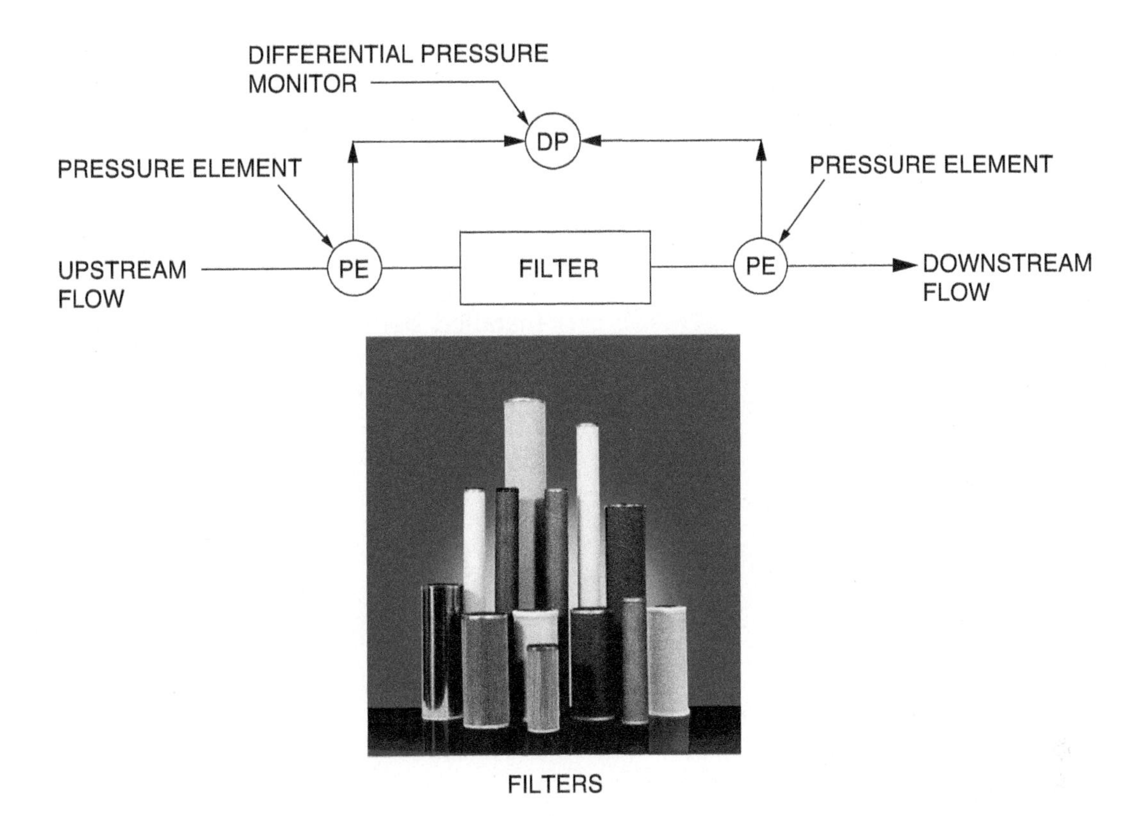

Figure 13 Filter diagram.

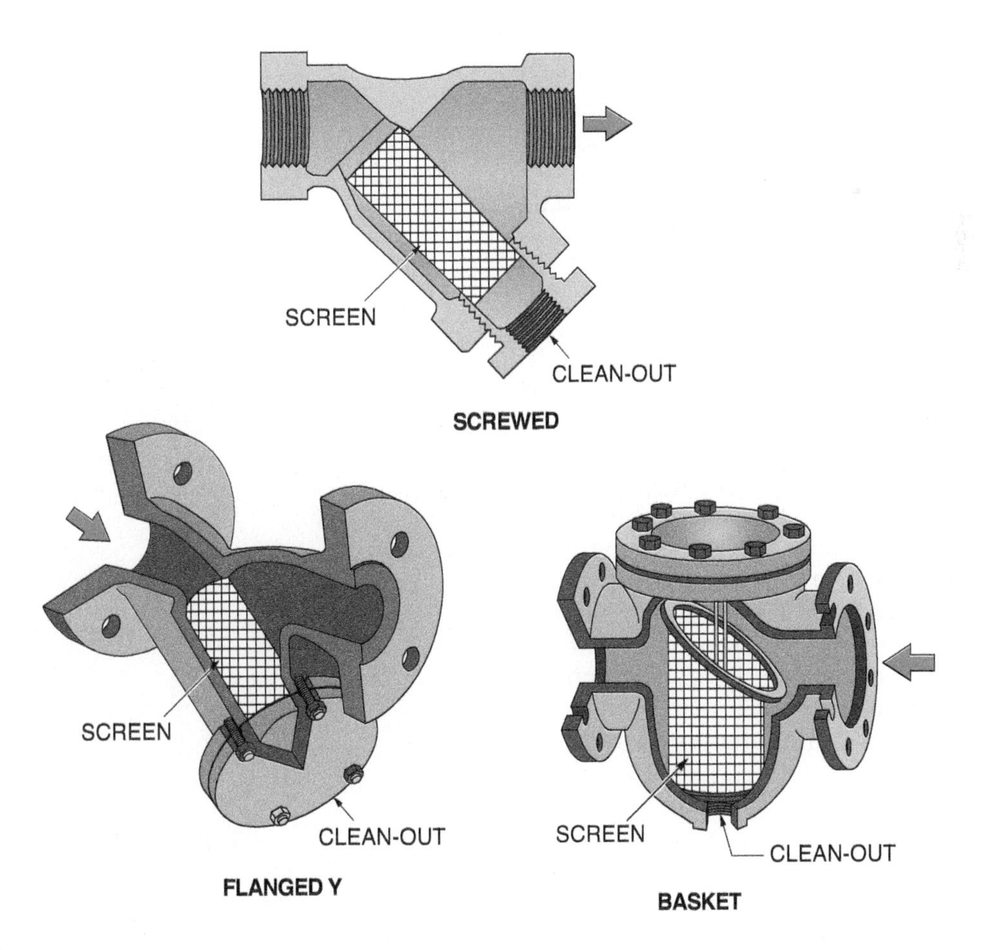

Figure 14 Strainers.

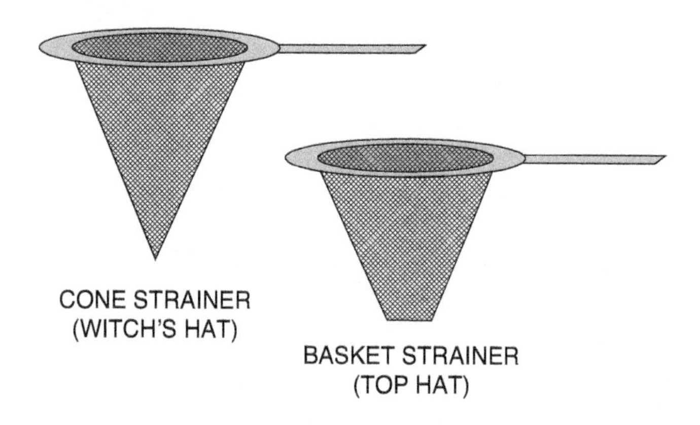

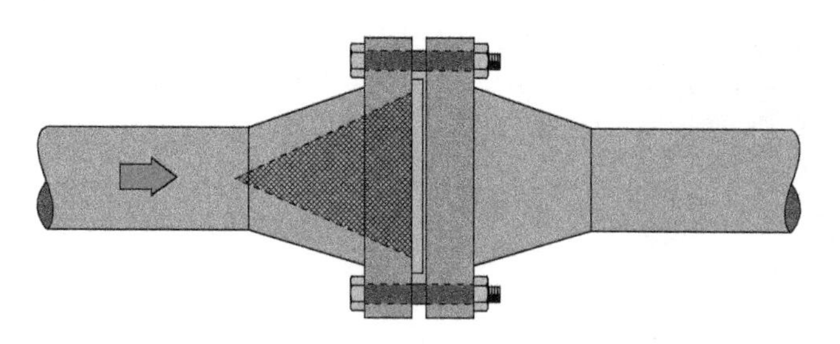

Figure 15 Temporary strainers.

set of flanges upstream from the pump suction to catch any type of construction debris in the system. These strainers can be removed after startup or left in the system permanently, depending on the plant conditions. In the case of conical strainers, whether cone or basket, the strainer should be installed with the smaller end pointed into flow, so that the cone will not fill up too quickly.

1.8.0 Flowmeters

Flowmeters measure the amount of fluid being transferred in a piping system. Methods for measuring flow fall into two categories: invasive and noninvasive. Invasive measurement devices project into a pipe that carries material. Noninvasive measurement devices do not enter the pipe. Because they are outside, noninvasive measurement devices must measure by inference, detecting outside the pipe some property of the material inside or producing a signal to which the inside material can react.

The four basic types of flowmeters include:

- Orifice plates
- Rotameters
- Venturi tubes
- Pitot tubes

1.8.1 Orifice Plates

The orifice plate (*Figure 16*) is the simplest and most commonly-used type of invasive device for flow measurement. It is used to determine flow rates by taking pressure readings above and below the plate. The pressure will vary dependent on the rate of flow through the orifice.

Orifice plates, which are made of metal, get installed between two flanges that have orifice taps. Each plate has a hole smaller than the inside diameter of the pipes in which it is installed.

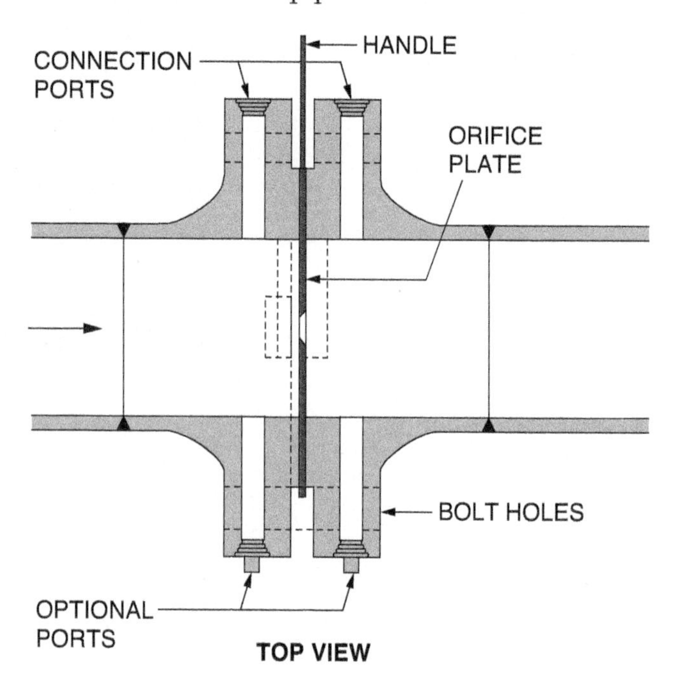

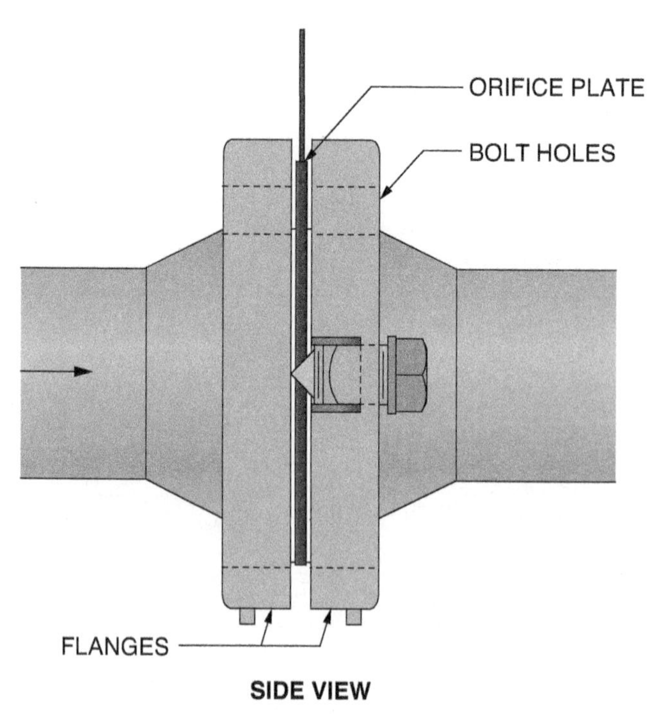

Figure 16 Orifice plate.

The tapped holes, located in the flange rims, are used as connection ports for tubing and pressure gauges.

All orifice plates have a specific orientation to direction of flow. Because of the pressure differential and possible flow characteristic effects, they are to be installed, unless otherwise specified, with a straight horizontal run of pipe that is 10′ long upstream of the plate and 5′ long below the plate.

1.8.2 Rotameters

A rotameter (*Figure 17*) is a type of flowmeter that works by measuring and reporting the turning speed of a propeller or turbine placed in the flow of fluid. One type of rotameter consists of a transparent tube with a tapered and calibrated bore, arranged vertically, wide end up, supported in a casing or framework with end connections. The instrument is connected so the flow enters at the lower end and leaves at the top. A ball or spinner rides on the moving gas inside the tapered tube. As the flow rate increases, the ball or spinner is lifted higher to indicate the flow rate through the rotameter. Isolation and bypass valves are used in conjunction with the device. Another type of rotameter is inserted directly into the pipe from the side with a propeller or turbine protruding into the flow.

1.8.3 Venturi Tubes

A venturi tube is a tube whose diameter is narrower in the center to restrict the flow through the pipe (*Figure 18*). It recovers pressure drop downstream from the restriction and is therefore considered highly efficient in terms of conserving energy. At the front end of the tube is a suction tap. The sloped area leading up to the restriction in a venturi tube is called the approach area; the restriction is called the throat; and the tube following the throat is called the recovery area. The throat tap is located at the throat.

The throat usually contains one of the pressure-measuring pickup points, and because of the long slope behind the restriction, most of the pressure drop is recovered. The pressure drop reading can be viewed at the recovery tap.

Venturi tubes are used mainly on special applications because they are expensive.

1.8.4 Pitot Tubes

The pitot tube (*Figure 19*) is a fluid velocity measuring tool. The flow-measurement device compares the dynamic pressure taken in by the tube to the static pressure in a static port orthogonal (at a right angle) to the direction of flow. It is used to monitor and meter fluid and steam flow in relatively large pipes. It can be used in a variety of difficult piping scenarios, including those involving high temperatures, high pressures, and the conveyance of corrosive materials.

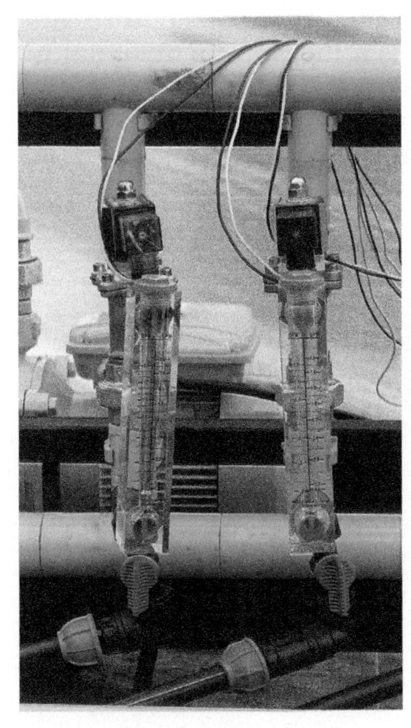

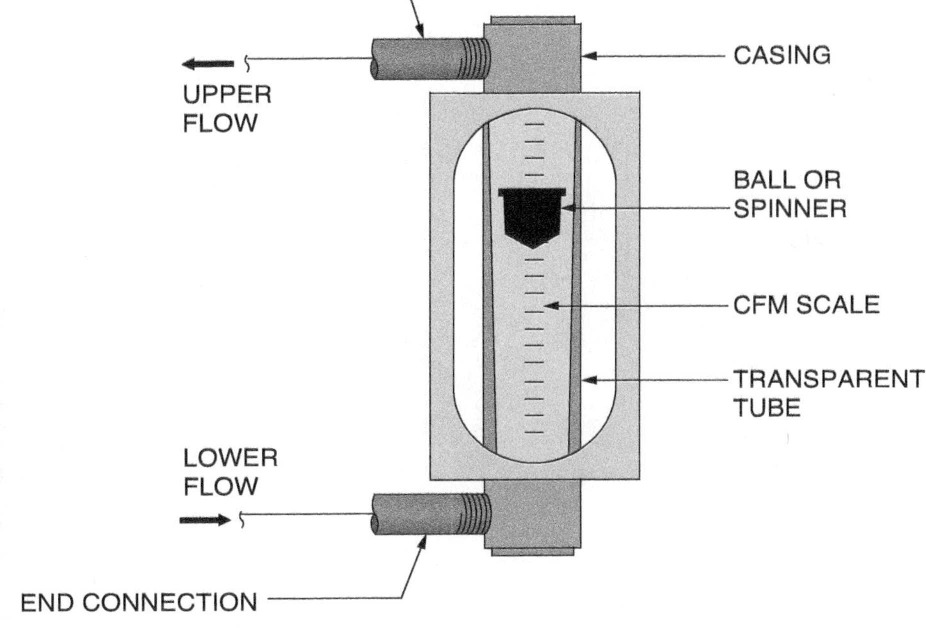

Figure 17 Rotameters.

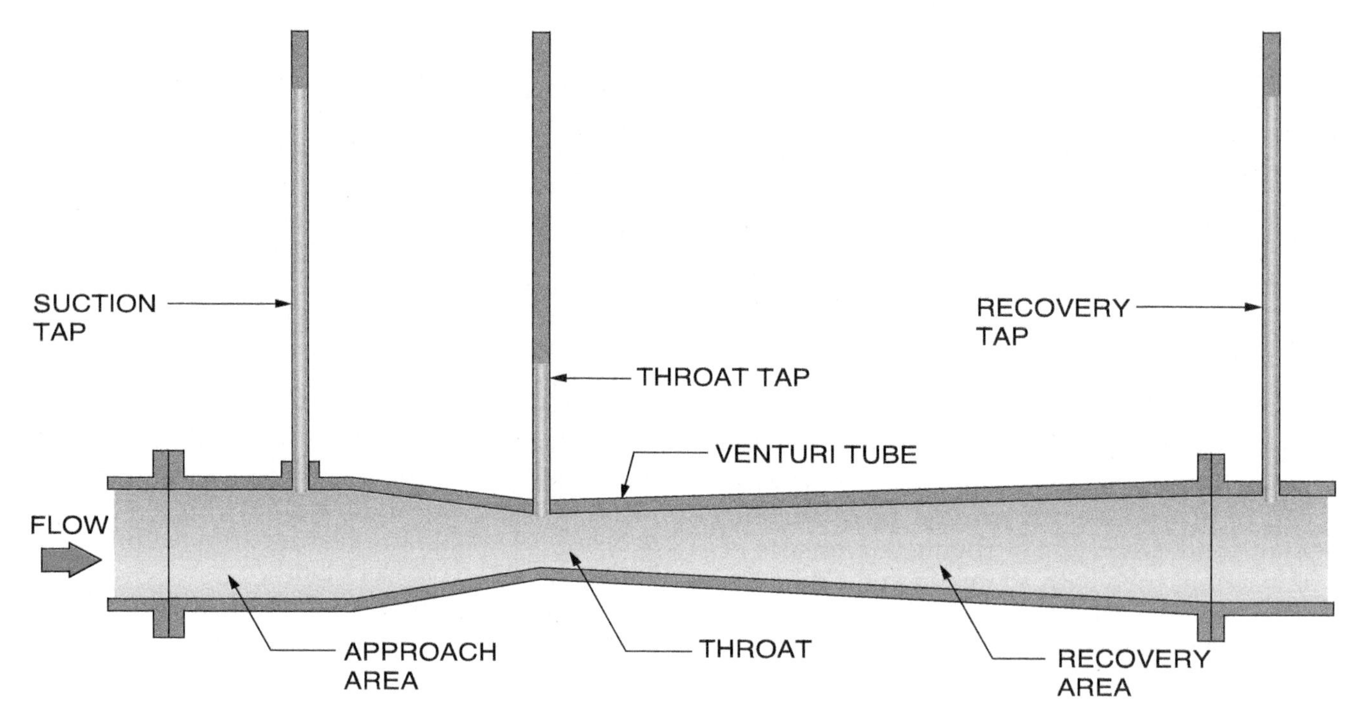

Figure 18 Venturi tube flow.

Two types of pitot tubes are the screw type and the insertion type. The insertion type is generally mounted into the pipeline through a valve and then pressure-fitted to the pipeline. This allows the pitot tube to be removed, inspected, and repaired without interrupting the process flow.

1.9.0 Level Measurement Devices

Level measurement devices are sensing instruments used in determining the level of material in a vessel. The fluid level of the contents in vessels is critical in many processes. Two types of level measurement devices are **conductivity probes** and sight glasses.

1.9.1 Conductivity Probes

Conductivity probes (*Figure 20*) measure the conductive characteristics of various processes throughout a job site. Typical applications include rinse tanks, water and waste treatment, condensate monitoring, and desalination units. The conductivity probe is inserted into the pipeline and provides a measurement signal to a conductivity transmitter. The electrode of the conductivity probe must remain wet at all times by the service within the pipeline to provide reliable and accurate measurement signals.

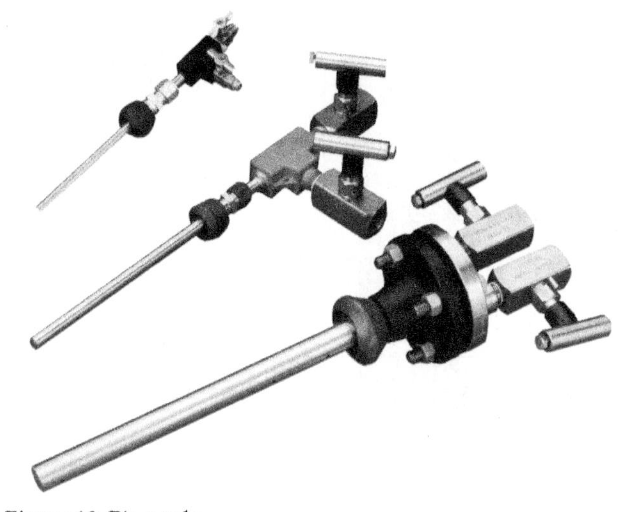

Figure 19 Pitot tubes.

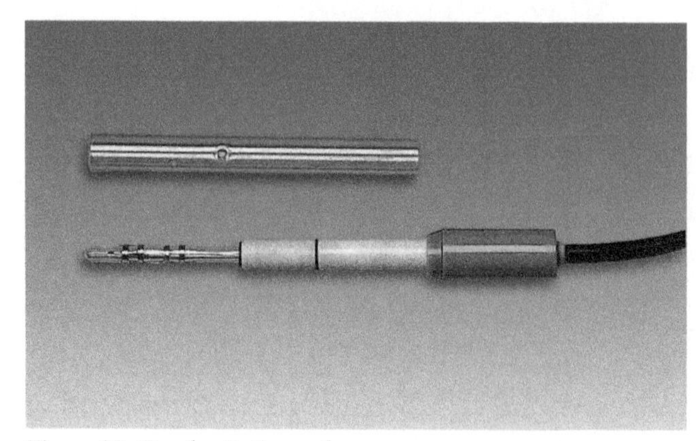

Figure 20 Conductivity probe.

1.9.2 Sight Glasses

Sight glasses provide a visual indication of the liquid level in a vessel or tank for both routine observations and level controller adjustments. Sight glasses are usually attached to a pipe that is connected to the vessel. They come in a variety of lengths and can be installed in staggered positions, so that when the liquid level exceeds the range of the glass, it is immediately indicated in the next glass. The combination of glasses must cover the complete float range required by the process. *Figure 21* shows typical sight glass arrangements.

1.10.0 Flow Pressure Switches

Flow pressure switches control liquid transfer from one storage area to another. The transfer of liquid often takes a long time; therefore, it is advantageous to provide automatic control of the operation. Once the transfer process has begun, the flow pressure switch takes over and automatically controls the operation until it's finished. Flow pressure switches are installed in the pipeline between the supply tank and the transfer pump. After the transfer pump has been started, the flow pressure switch keeps the pump operating until the transfer of liquid is complete.

Pressure differential sensors (*Figure 22*) measure flow or level pressures. The measured span of pressure may vary from 0 to 1,000 in Hg (inches of mercury), depending on the nature of the application. The physical dimensions of the sensing element vary in proportion to the pressure range of the transmitter.

The sensing element measures the pressure of the flow or level and provides an input signal to the transmitter. The transmitter contains an electronic assembly that receives the input signal and converts it to a proportional output signal.

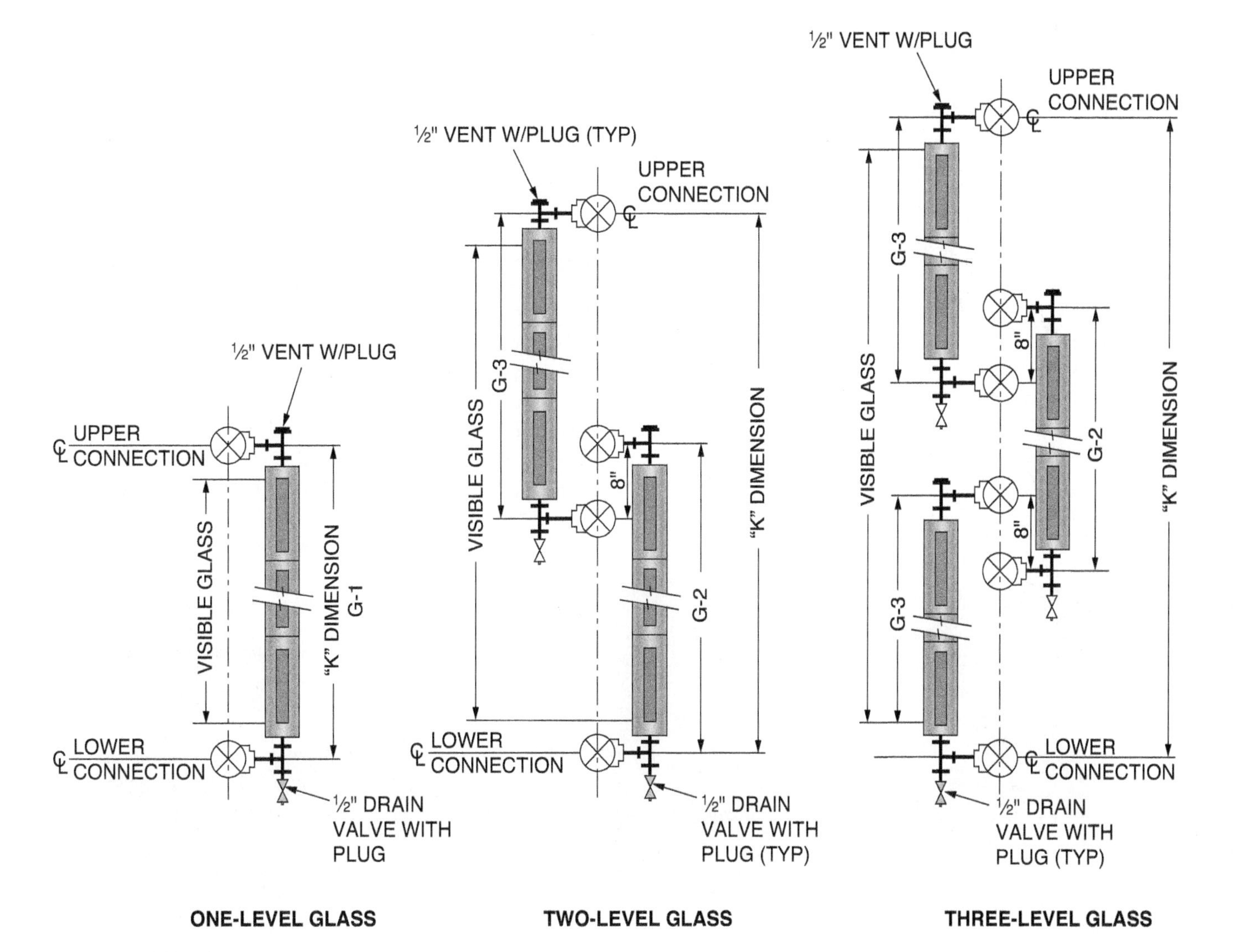

Figure 21 Sight glass arrangements.

1.11.0 Rupture Discs

A rupture disc is a safety device that is designed to burst at a certain operating pressure to release gas or liquid from a system. The rupture disc is usually a replaceable metal disc that is held between two flanges. It isolates and protects valves from corrosive process fluids and serves as a supplemental pressure-relief valve.

> **WARNING!**
> Before installing a rupture disc, make sure you understand the direction of flow. Rupture discs only break open in one direction. If a rupture disc is installed against the flow, it will not operate correctly and it will cause a massive, explosive failure in the system.

Normally, the installation of a rupture disc is a hold point, which means that its installation must be witnessed by a quality control inspector or plant representative before continuing a maintenance or installation process. The installation happens after all tests of the system have been completed. *Figure 23* shows a typical rupture disc setup.

There are several types of rupture discs, including the following:

- Conventional
- Scored tension-loaded
- Composite
- Reverse-acting
- Graphite

1.11.1 Conventional Rupture Discs

The conventional rupture disc is a pre-bulged, solid metal disc designed to burst when it is over-pressurized on the concave side. The conventional disc (with a flat seat or angular design) is used when operating conditions are 70 percent or less of the rated burst pressure of the disc and when limited pressure cycling and temperature variations are present. For vacuum or back pressure conditions, the conventional disc should be furnished with a support to prevent flexing or implosion.

1.11.2 Scored Tension-Loaded Rupture Discs

The scored tension-loaded rupture disc allows a close ratio of system-operating pressure to burst pressure. It is thicker than the conventional disc and opens along scored lines. The scored tension-loaded disc does not fragment and withstands full vacuum without the addition of a separate vacuum support.

1.11.3 Composite Rupture Discs

The composite disc is usually metal and designed to fail at a certain pressure, dependent on the construction of the disc; the model number of the disc indicates the pressure at which it will fail. Composite rupture discs have a leak-preventing seal on the pressure side. It bursts when over-pressurized on the concave side. There are curved composite discs and flat composite discs. The flat disc is designed to burst when it is over-pressurized on the side designated by the manufacturer. Composite rupture discs are available in burst pressures lower than the conventional rupture disc and may offer a longer service life as a result of the corrosion-resistant properties of the seal material selected.

Figure 22 Pressure differential transmitter.

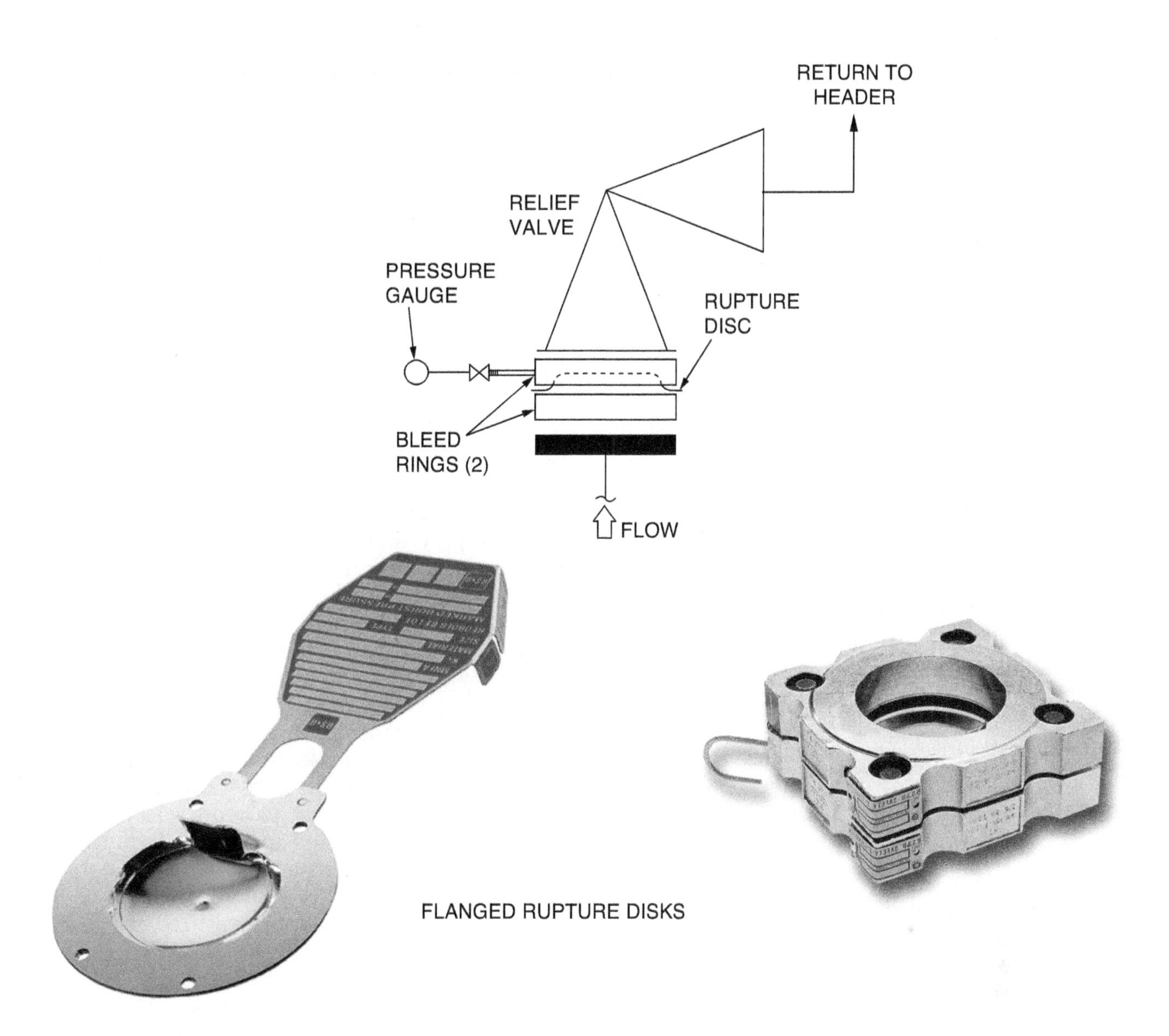

Figure 23 Rupture disc setup.

1.11.4 Reverse-Acting Rupture Discs

A reverse-acting rupture disc is a domed, solid metal device designed to burst when over-pressurized on the convex side. It allows up to 90 percent of the rated burst pressure. Reverse-acting rupture discs do not fragment and do not need vacuum support. They provide good service under vacuum pressure cycling conditions and temperature fluctuations.

1.11.5 Graphite Rupture Discs

A graphite rupture disc is made from graphite-impregnated material with a binder and is designed to burst by bending or shearing. It is resistant to most acids, alkaline items, and organic solvents and operates to 70 percent of the rated burst pressure. Graphite rupture discs are helpful because metallic discs will corrode in these atmospheres, even within design pressures. A support is required for a disc that is rated 15 psig or less and when used in systems with high back pressures. When a disc ruptures, a metal screen or similar catchment for capturing fragmentation must be provided.

1.12.0 Thermowells

Thermowells (*Figure 24*) are metal housings that hold temperature-measuring devices and protect them from the process environment. They are placed in contact with the fluid. The main causes of thermowell failure are mechanical breakdowns, corrosion, erosion, and vibration fatigue.

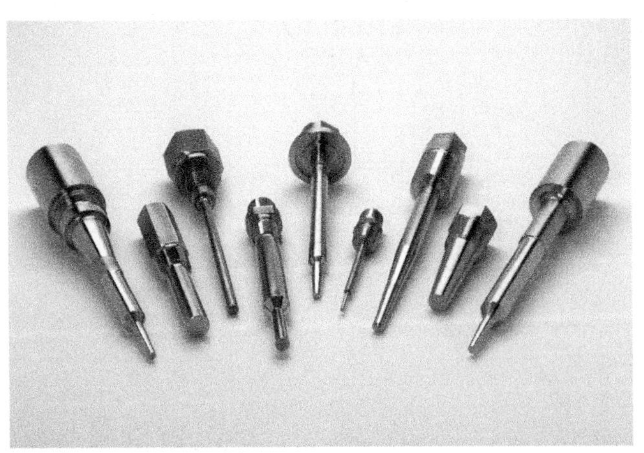

Figure 24 Thermowells.

Mechanical breakdowns are the result of breaking or bending due to applied force that exceeds the thermowell's yield strength. Corrosion is caused by chemicals and elevated temperatures. Erosion is caused by particles in the system that strike the thermowell at high speeds. Vibration fatigue is caused by the turbulent wake produced when the fluid flows past the thermowell; when this wake frequency equals or exceeds its natural frequency, the thermowell vibrates to its breaking point.

1.13.0 Desuperheaters

Steam is a powerful and predictable form of stored energy. Superheated steam refers to steam that has absorbed heat beyond that necessary for its creation. Changes in pressure result in a change in the boiling point of water. For example, at a pressure of 20 psi, water must be heated to a temperature of 239°F before it begins to boil and become steam. The change of state from water to steam, or from steam back to water, is not instantaneous though. To create steam, heat must be continually added after the temperature has risen to 239°F. Note that, due to the laws of thermodynamics, the temperature of the steam/water mixture will remain at 239°F until all the water nearby has completed the change of state. Assuming there is no change in pressure from 10 psi, only at 239°F can water and steam coexist.

After all the water in a space has been converted to steam, the temperature of the steam can be increased above 239°F. Steam is said to be superheated when the temperature rises above its saturation point, which in this example is 239°F. Steam can be superheated to very high temperatures if desired. For example, steam at 10 psi can easily be heated to 400°F, 161°F above its saturation temperature. However, even a rise above the saturation temperature of one degree Fahrenheit is technically considered superheat.

Superheating steam is essentially a method of storing additional heat (energy), allowing it to do more work downstream. It also ensures the steam supply is very "dry" and does not carry large water droplets with it as it travels through the piping system to its destination. Superheated steam is particularly well suited for driving turbines.

Unfortunately, steam carrying a great deal of superheat is not ideal for all applications. Although highly superheated steam is valued for turning power-generation turbines, it creates problems in other, more common process applications. For example, the elevated temperature may damage some components and require others to carry a higher temperature rating, driving up the cost. As a result, when superheated steam is what an energy facility predominantly produces, the steam supply often needs to be cooled before it can be put to work in additional processes.

Desuperheaters, as the name implies, are used to reduce the temperature of superheated steam to a usable level for other processes. There are several types of desuperheaters available to system engineers. In most cases, water is used in some way to transfer the heat out of the superheated steam. Of course, the water becomes significantly warmer in the process, and the heated water can often be put to work elsewhere in the process or even used for domestic purposes. Using water heated in a desuperheater for another purpose is just one small example of the many energy conservation measures at work in an industrial facility.

1.13.1 Large Tank Desuperheaters

The large tank desuperheater is also referred to as a direct-contact desuperheater, since the cooling water and steam supply mix directly. This type injects the steam into the bottom of a tank of water. A steam outlet is placed well above the water level at the top of the tank. Some of the water flashes to steam as the steam bubbles up through the water, removing energy from the steam supply. Since some water flashes to steam and leaves through the outlet, make-up water must be added to keep up with the loss. This approach offers a great deal of capacity, and it is relatively simple to operate and maintain.

1.13.2 Heat Exchanger Desuperheaters

The heat-exchanger style of desuperheater does not mix the incoming steam with the cooling water, as does the tank desuperheater. Shell-and-tube heat exchangers consist of a large tank with an isolated tube bundle passing through the water in the tank. The steam flows through the tubes, and water that flashes to steam escapes through an outlet in the shell. These systems can also handle a large volume of steam, as well as very high temperatures and pressures.

1.13.3 Water Spray Desuperheaters

Another approach is the water spray desuperheater, which breaks the cooling water supply into a spray comprised of very tiny droplets. The spray is then injected into the superheated steam. The water absorbs heat from the steam and flashes to vapor quickly, cooling the steam flowing by in the process. Two examples of cooling water spray nozzles are shown in *Figure 25*. The example variable-nozzle style is positioned by a pneumatic actuator mounted above the control valve.

Another style of cooling-water nozzle is inserted directly through the walls of the steam line at a 45-degree angle (*Figure 26*). The cooling water supply is directed through the small nozzle at high pressure, atomizing the water into a foggy vapor, which is then injected into the steam supply stream. The water flashes to steam

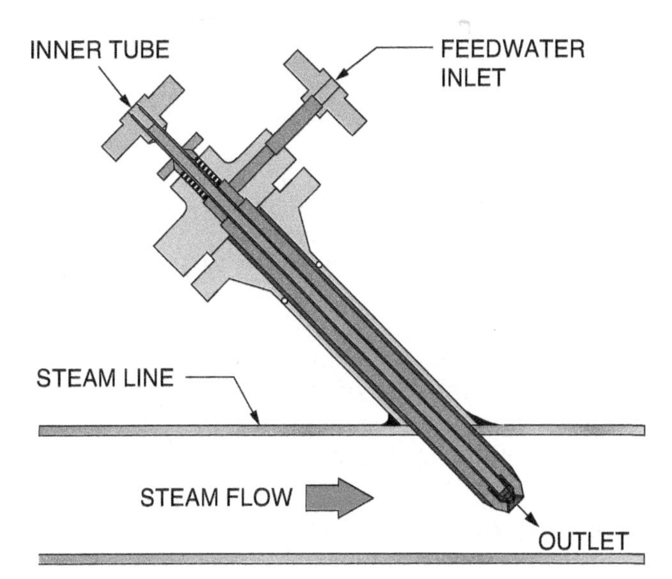

Figure 26 Injection desuperheater.

almost instantly, absorbing heat and cooling the steam supply. A similar version routes the steam through the nozzle as well, where it mixes with the water stream, allowing some of the heat transfer to occur even before the water has left the nozzle.

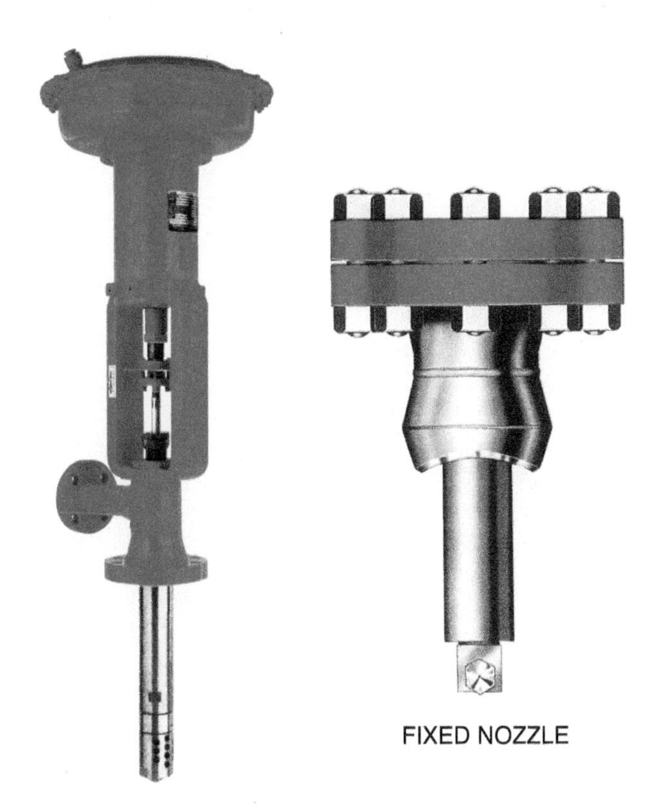

Figure 25 Desuperheating spray nozzles.

1.14.0 Safety and Handling Practices

The installation and removal of in-line specialty equipment presents critical safety hazards because the system itself may be potentially hazardous. Examples include the possibility of being splashed with or exposed to corrosive, toxic, or hot substances, possibly at high pressures. Extreme care must be taken when cutting into or welding on gas lines that contain highly flammable and explosive substances.

WARNING!

Only qualified welders who have been properly trained to perform welding operations in explosive environments should attempt to weld on active gas line or any type of line that contains flammable and/or explosive substances.

CAUTION

Be sure to follow the company's line break procedures.

Always check permit requirements and company regulations to see which of the following pieces of personal protective equipment must be worn when performing line entries:

- Full-face shield
- Safety glasses with side shields or monogoggles
- Self-contained breathing apparatus or in-line-supplied respirator
- A complete chemical-resistant suit
- Chemical gloves with cuffs long enough to fit up under the sleeves of the suit
- Rubber boots high enough for the tops to be covered by the legs of the suit pants
- Hard hat

Follow these safety precautions when performing any type of line entry:

- Lock out and tag all electrical switches and disconnects on the equipment as required.
- Before the line is opened, clear and barricade an area large enough to ensure the safety of nearby personnel.
- Provide a container to catch any drained material from the pipeline.
- When a line containing flammable liquids is to be opened, ensure that a dry chemical fire extinguisher and a connected water hose are brought to the area and made ready for use.
- If the equipment is activated by mechanical or control devices, these devices are to be manually operated to be sure that the equipment is indeed taken out of service.
- If the line to be opened contains flammable or hazardous vapors, the line must be purged with nitrogen, water, steam, or air for a long enough period of time (as specified by plant procedures) to ensure that all toxic and flammable vapors have been removed.

1.14.1 Storing and Handling In-Line Specialties

The proper functioning of in-line specialty equipment depends not only on the inherent quality of each instrument but also on how well the equipment is stored and handled.

When storing and handling in-line specialty equipment:

- Store pitot tubes in reinforced boxes.
- Keep pitot tubes clean.
- Keep ball joints crated until they are installed.
- Handle conductivity probes with extreme care.
- Do not flex, puncture, or contaminate filters.
- Dispose of contaminated filters according to EPA regulations.
- Store flow pressure switches in a dry area and in their original packing until they are ready to be installed.
- Handle flow pressure switches with extreme care.
- Do not dent, scratch, or alter orifice plates.
- Do not strain, flex, or bend rupture discs.
- Store rupture discs in their packing until they are ready to be installed.
- Do not remove any identification tags or plates from any in-line equipment.

1.0.0 Section Review

1. The *most* common type of modern mechanical trap is called _____.
   a. an orifice trap
   b. an inverted bucket trap
   c. a hatch trap
   d. a thermostatic trap

2. Which type of trap opens when cold, readily purges gases, and is unlikely to freeze?
   a. Liquid expansion
   b. Balanced pressure
   c. Bimetallic
   d. Float

3. What is the purpose of a strainer, in relation to a steam trap?
   a. To remove scale and corrosion before it enters the trap
   b. To separate hydrogen from oxygen in the production of steam
   c. To help moderate temperatures by slowing the flow
   d. To establish boundaries between two portions of the pipeline

4. Snubbers are important because they _____.
   a. prevent chemicals from mixing within cross-functional systems
   b. prolong the life and accuracy of pressure gauges
   c. block particulates from entering valves
   d. send an electronic signal when problems erupt in the piping system

5. What is the range of angular movement for a ball joint?
   a. 10 to 20 degrees
   b. 20 to 30 degrees
   c. 30 to 40 degrees
   d. 50 to 60 degrees

6. Where do bleed rings get installed?
   a. Between flanges
   b. Between entry points of the initial pipe run
   c. Between block valves
   d. Above the compressor

7. Drip legs should be installed _____.
   a. every 4 to 6 months, depending on the material in the system
   b. everywhere that leakage is expected
   c. at every elevation change
   d. beneath the pressure gauge

8. What is the role of an expansion joint in a piping system?
   a. To control linear expansion and contraction
   b. To clear a filter of catchment debris
   c. To obfuscate the difference among fluids in a system
   d. To stop the flow of accumulated liquids

9. What is the role of a pressure element?
   a. To provide data for producing a differential pressure reading
   b. To increase pressure in systems where fluid is moving too slowly
   c. To lower the pressure in high-pressure systems
   d. To provide pressure relief at specified points along the run

10. Venturi tubes, pitot tubes, and orifice plates are all types of _____.
    a. thermowells
    b. desuperheaters
    c. flowmeters
    d. drip legs

11. To provide accurate signals, a conductivity probe *must* remain _______.
    a. stationary
    b. idle
    c. wet
    d. dry

12. In a flow pressure system, what is the optimal type of control?
    a. Manual
    b. Automatic
    c. Semi-automatic
    d. Digital

13. An in-line specialty used to protect valves from corrosive fluids and supplement pressure relief is called _____.
    a. a flow pressure switch
    b. an expansion valve
    c. a rupture disc
    d. a drip leg

14. Mechanical breakdowns, corrosion, erosion, and vibration fatigue are the main causes of _____.
    a. thermowell failure
    b. desuperheater failure
    c. steam trap failure
    d. rupture disc failure

15. A large tank desuperheater may also be called _____.
    a. a heat exchanger desuperheater
    b. a water spray desuperheater
    c. an inert spray desuperheater
    d. a direct contact desuperheater

SECTION TWO

2.0.0 TROUBLESHOOTING AND MAINTAINING STEAM TRAPS

Objective

Explain how to troubleshoot and maintain steam traps.

a. Explain and discuss diagnostic methods and maintenance procedures related to steam traps.

Performance Task

2. Identify specific problems and corrective actions required for faulty steam traps.

To diagnose and solve problems with steam traps requires listening to the noises they make and measuring temperature and pressure. The temperature of steam can be measured with handheld or remote-sensing pyrometers that read in degrees centigrade or Fahrenheit. Steam is produced at 212°F when at atmospheric pressure. When steam systems are functioning, the temperature will be substantially higher since it is confined inside the system; likewise, the pressure will be higher as well.

Pressure readings are in two scales: absolute (psi) and gauge (psig). Absolute pressure is read from 0 up, which is to say that normal atmospheric pressure is 14.7 psi at sea level. Gauge pressure is relative to atmospheric pressure at sea level; therefore, 0 psig is equal to 14.7 psi absolute. Pressures below zero gauge are expressed in inches of mercury.

2.1.0 Diagnostic Methods

Three basic diagnostic methods of reading a steam system are sight, sound, and temperature. The criteria observed include pressure. The other process, less used to date, is a conductance probe, which acts by comparing the difference in electrical conductivity between condensate and live steam.

Each of the three main methods has advantages and limitations. Sight is fairly immediate; it is easy to tell whether there is condensate coming from a trap. If the condensate line is open, or if a test valve has been added on the condensate line, the condensate can be vented to atmosphere, and it will be obvious that the trap is removing condensate from the line. If live steam is going into the condensate line, the trap has failed open, and it is necessary to resolve the problem. If no condensate is coming from the line, the trap may have failed closed. The limitation of sight diagnosis is that the worker must be able to tell the difference between flash steam and live steam, and it is necessary for the worker to check by activating a test valve. It is also possible for the trap to be overloaded or loaded from the return system, producing an apparent open failure.

The two most commonly used diagnostic tools are temperature sensing and sound, either at human hearing ranges or ultrasound. Note that temperature and sound together are a much more sensitive test than one or the other alone.

From a diagnostic point of view, there are two general types of traps. Traps either flow or dribble, continuously or intermittently (on and off). Float thermostat (F and T) traps discharge continuously, if working properly. Inverted bucket, bellows thermostatic, thermodynamic, and most thermostatic traps flow intermittently. This allows the person checking the trap to visually determine if the trap is failing, in the more extreme cases. However, visual inspection is not an absolute solution because a partial failure or overloading may cause misinterpretation.

When checking traps, be aware:

- A thermal analysis will show the inlet as a high temperature area for any of the traps. In most functioning traps, the condensate outlet will be cooler than the inlet. Remember that the condensate will most frequently be at steam temperature, but it should be lower than the inlet. The live steam outlet should be hotter than the condensate outlet. If this is not the case, the trap is most likely malfunctioning.
- Normal failure for an inverted bucket trap is open failure. The trap loses its prime and steam and condensate steadily blow through. If this happens, the continuous rushing sound of the blow-through may be accompanied by the bucket banging against the inside of the trap. The temperature of the trap and of the condensate line will rise, since it is no longer holding condensate. The sound of the bucket linkage rattling inside the trap may indicate that the trap is beginning to loosen up and needs checking. With ultrasound equipment the inverted bucket or thermodynamic traps will show a cyclic curve on the screen in normal function.
- Normal failure for a thermodynamic trap is open; that is, passing steam. Usually, the disc clicks shut audibly once for each cycle. When the disc is cycling normally, it will shut about

4 to 10 times a minute. When the trap fails, the disc no longer clicks as it rises and falls, and steam blows through. If the disc produces a continuous, rapid, rattling sound, the disc is worn and additional problems are likely to develop. If a continuous hissing sound is produced, the trap is not cycling. Again, the condensate line will heat up from escaping steam and the absence of cooler condensate.

- Bellows or bimetallic thermostatic traps operate on a difference in temperature between condensate and steam. Both types are intermittent in operation. When there is little condensate, they stay closed most of the time. When there is a lot of condensate, for example at startup, they may run continuously for a long time. In the case of bimetallic traps, misalignment may allow steam leaks. With sound testing equipment, a constant rushing sound in the closed part of the cycle would indicate a leak.
- Float thermostatic traps tend to fail closed. A leaky float will not float on the condensate, or the float may have been crushed by water hammer. In either case, the F and T trap will remain closed and not cycle. The trap will remain cool and a temperature sensor will confirm the finding. With listening equipment, a failed float will be silent, in the absence of discharging condensate.
- The thermostatic element in an F and T trap is normally quiet; a rushing sound indicates the element has failed open. A rattling or metallic clanging noise might mean the mechanical linkage has sustained some damage.

The sight method uses a test valve that vents the process steam to the atmosphere for visual inspection. This test is subjective and depends on skill and experience. Since most traps are dealing with condensate that is at steam temperature (higher than the boiling point of water), when condensate is released a certain amount will immediately turn into flash steam. The difference is fairly visible to pipefitters who have observed this a few times; the live steam comes out at first as a hard, blue-white straight flow, often with a clear, bluish area at the outlet. Flash steam usually billows and spreads more quickly, as much of the pressure is dissipated in the initial release.

Diagnostic sound probes are either audible-range or ultrasonic-range, picking up noise from the system and sending an audible signal to headphones. Technicians must be able to interpret the noises as leaks, discharges, or other problems. *Figure 27* shows a portable ultrasonic tester.

Figure 27 Ultrasonic tester.

Temperature-sensing tests may involve handheld pyrometers (*Figure 28*) for external readings or mounted thermowells for reading actual steam temperatures. A thermowell is a permanently installed well or cavity in a process pipe or tank into which a glass thermometer or thermocouple can be inserted. The input from these and other test methods must be applied to problems and causes of problems in the steam trap system.

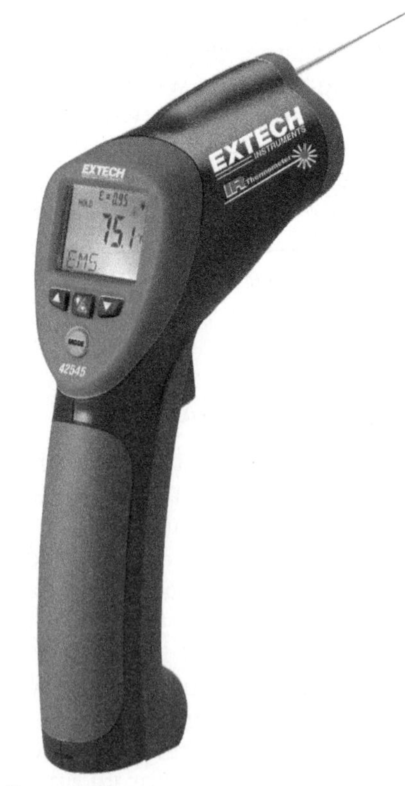

Figure 28 Pyrometer.

2.1.1 Maintaining Steam Traps

Most steam traps fail in the open position, which is difficult to pinpoint because it does not affect equipment operation. When the trap fails open, both steam and condensate flow freely from the steam header to the condensate header, keeping the heat transfer high. The traps that fail in the closed position are easy to identify because the backup of condensate cools the system.

An effective preventive maintenance program includes scheduled checks of the entire system. Pipefitters should be on the lookout for clogged strainers, leaking joints, leaking valve packing, and missing insulation. When the internal parts of a steam trap wear out, the water seal deteriorates, steam flows through the valve assembly, and the seal becomes eroded.

Common causes of steam trap failure are:

- Scale, rust, or corrosive buildup preventing the valve from closing
- Valve assembly wear
- A defective or damaged valve seat
- Physical damage from severe water hammer
- Foreign material lodged between the seat and valve
- Blocked, clogged, or damaged strainers
- Increased back pressure

Other failures are specific to the type of steam trap; these are indicated when the trap blows steam and/or when it will not pass condensate. Several options are available for addressing these issues; however, when a steam trap fails, it is best to just replace it.

2.0.0 Section Review

1. Steam systems are generally diagnosed using sight, sound, and _____.
   a. smell
   b. touch
   c. temperature
   d. induction

2. If a steam trap fails, the *best* option is to _____.
   a. consult with a steam trap repair specialist for advice on saving it
   b. contact the manufacturer for a repair and maintenance kit
   c. replace it
   d. wait until it starts causing problems and then repair it

Review Questions

1. Snubbers smooth and dampen _____.
   a. heat waves
   b. steam surges
   c. pressure impulses
   d. process flow

2. Ball joints are installed in piping systems to allow for _______.
   a. expansion and contraction
   b. chemical absorption
   c. pressure relief
   d. opening and closing

3. Drip legs collect condensate and sediments from _____.
   a. hot-water systems
   b. process water systems
   c. gas systems
   d. steam systems

4. The *most* commonly used invasive device for measuring flow is a(n) _____.
   a. rotameter
   b. venturi tube
   c. pitot tube
   d. orifice plate

5. Which type of flowmeter uses propellers and turbines inserted in the flow of fluid?
   a. Rotameter
   b. Venturi tube
   c. Pitot tube
   d. Orifice plate

6. Which of the following is a level measurement device?
   a. Pitot tube
   b. Sight glass
   c. Thermowell
   d. Flowmeter

7. The conventional rupture disc is designed to burst when over-pressurized on the _____.
   a. convex side
   b. concave side
   c. top face
   d. bottom face

8. A desuperheater is designed to _____.
   a. increase the pressure of steam
   b. decrease the acidity of steam
   c. increase the temperature of steam
   d. decrease the temperature of steam

9. Until they are ready to be installed, flow pressure switches should be stored in a(n) _____.
   a. saline solution
   b. dry area
   c. damp area
   d. open area

10. Pitot tubes should be stored in _____.
    a. reinforced boxes
    b. dry areas
    c. crates
    d. packing from other specialties

11. When the potential for cross-contamination is introduced due to a change in the material flowing through a line, what are used to provide a leak-proof means of stopping the flow?
    a. Ball joints
    b. Snubbers
    c. Steam traps
    d. Bleed rings

12. Process plants rely on steam because _____.
    a. it is less expensive than forced mechanical power
    b. moisture is essential to maintaining the pipefitting environment
    c. it efficiently supplies large quantities of heat
    d. it works well with all types of components, including asbestos

13. Traps in a steam system are usually present to _____.
    a. clear the distribution headers of condensate
    b. reduce or reroute the flow of inorganic materials
    c. stabilize the internal conductive environment
    d. assist the valves with the management of intermittent fluctuations in temperature

14. All steam traps work to release any steam when it's new or in good condition, *except* for the _____.
    a. mechanical type
    b. thermostatic type
    c. thermodynamic type
    d. inverted bucket type

15. A mechanical steam trap is usually placed _____.
    a. at the start of the pipe run
    b. between flanges within parallel sections of pipe
    c. downstream of the next-to-last valve
    d. between the source of steam and the condensate return header

16. Disadvantages of using inverted bucket traps are that they tend to be noisy and they _____.
    a. have limitless air discharge capacity
    b. have limited air discharge capacity
    c. cannot be remotely controlled
    d. are difficult to repair once installed

17. The purpose of expansion joints is to _____.
    a. compensate for the difference in temperature between conveyed fluids and surrounding air
    b. allow for easy entry into piping systems that convey hazardous materials
    c. reduce bicarbonate expulsion of petrochemicals and nuclear waste
    d. conduct and redirect kinetic energy as solids and condensates make their way to the steam traps

18. Strainers should be installed _____.
    a. upstream of the steam trap
    b. downstream of the steam trap
    c. alongside the steam trap
    d. parallel to the steam trap

19. Automatic control of flow pressure switch operations is recommended because _____.
    a. the training required for manual control is cost-prohibitive
    b. interference with electromagnetic processes may be avoided
    c. manual interventions introduce special hazards
    d. the transfer of liquid is usually a lengthy process

20. What is 0 psig equal to in absolute?
    a. 7.1 psi
    b. 14.7 psi
    c. 21.3 psi
    d. 28.4 psi

Trade Terms Introduced in This Module

Condensate: The liquid byproduct of cooling steam.

Conductivity: A measure of the ability of a material to transmit electron flow.

Differential pressure: The measurement of one pressure with respect to another pressure, or the difference between two pressures.

Drip legs: Drains for condensate in steam lines placed at low points in the line and used with steam traps.

Flash steam: Steam formed when hot condensate is released to lower pressure and re-evaporated.

Saturated steam: Pure steam without water droplets that is at the boiling temperature of water for the given pressure.

Superheated steam: Saturated steam to which heat has been added to raise the working temperature.

Tracer: A steam line piped beside product piping to keep the product warm or prevent it from freezing.

Additional Resources

This module presents thorough resources for task training. The following reference material is recommended for further study.

Armstrong Steam Conservation Guidelines for Condensate Drainage, Armstrong Steam Specialty Products, Three Rivers, MI 49093, (616) 273-1415.

Design of Fluid Systems, Steam Utilization, Spirax Sarco Inc., P.O. Box 119, Allentown, PA 18105, (610) 797-5830.

Preventing Steam/Condensate System Accidents. 1995. The National Board of Boiler and Pressure Vessel Inspectors. Available at:**http://www.nationalboard.org/Index.aspx?pageID=164&ID=208.**

Safety Lifecycle Workbook. 2003. Geneva, Switzerland: International Electrotechnical Commission. Available at: **http://www3.emersonprocess.com/deltav/safetyworkbook/**.

Steam Engineering Tutorials. 2018. Spirax Sarco. **www.spiraxsarco.com/resources/steam-engineering-tutorials.asp**.

Figure Credits

Applied Technical Services, Module opener, Figure 27

Courtesy of EBAA Iron, Inc., Figure 2

The Mack Iron Works Company, Sandusky Ohio, Figure 3

Filter-Fab Corporation, Figure 13

iStock@psisa, Figure 17

Meriam Process Technologies, Figure 19

HANNA® Instruments USA, Figure 20

Photo courtesy of Emerson Process Management and North American Manufacturing, Figure 22

BS&B Safety Systems, Figure 23

Weed Instrument Co., Figure 24

Courtesy of SPX Cooling Technologies, Inc., Figure 25

Applied Technical Services, Figure 27

Courtesy of Extech Instruments, a FLIR Company, Figure 28

Section Review Answer Key

Section 1.0.0

Answer	Section Reference	Objective
1. b	1.5.1	1e
2. b	1.5.2	1e
3. a	1.7.1	1g
4. b	1.1.0	1a
5. c	1.2.0	1b
6. c	1.3.0	1c
7. c	1.4.0	1d
8. a	1.6.0	1f
9. a	1.7.0	1g
10. c	1.8.0	1h
11. c	1.9.0	1i
12. b	1.10.0	1j
13. c	1.11.0	1k
14. a	1.12.0	1l
15. d	1.13.1	1m

Section 2.0.0

Answer	Section Reference	Objective
1. c	2.1.0	2a
2. c	2.1.1	2a

NCCER CURRICULA — USER UPDATE

NCCER makes every effort to keep its textbooks up-to-date and free of technical errors. We appreciate your help in this process. If you find an error, a typographical mistake, or an inaccuracy in NCCER's curricula, please fill out this form (or a photocopy), or complete the online form at **www.nccer.org/olf**. Be sure to include the exact module ID number, page number, a detailed description, and your recommended correction. Your input will be brought to the attention of the Authoring Team. Thank you for your assistance.

Instructors – If you have an idea for improving this textbook, or have found that additional materials were necessary to teach this module effectively, please let us know so that we may present your suggestions to the Authoring Team.

NCCER Product Development and Revision

13614 Progress Blvd., Alachua, FL 32615

Email: curriculum@nccer.org
Online: www.nccer.org/olf

❏ Trainee Guide ❏ Lesson Plans ❏ Exam ❏ PowerPoints Other ____________

Craft / Level: Copyright Date:

Module ID Number / Title:

Section Number(s):

Description:

Recommended Correction:

Your Name:

Address:

Email: Phone:

This page is intentionally left blank.

Special Piping

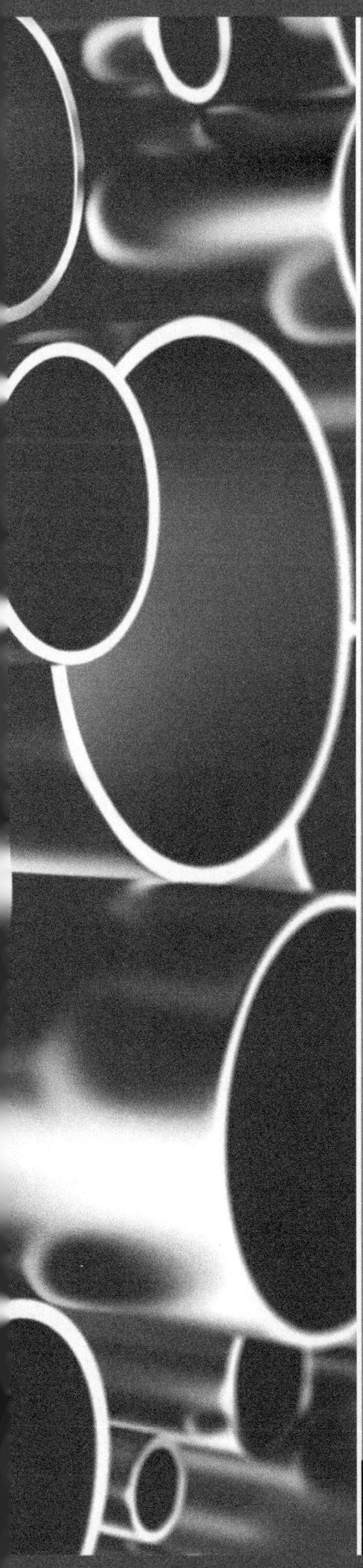

Overview

Pipefitters must be prepared to assemble small piping and tubing. While these jobs are infrequent, they still require a clear understanding of the skills needed to correctly and safely connect and route pipe on a smaller scale. Some of these skills include brazing, soldering, pipe bending, and installing various fittings made of copper, stainless steel, aluminum, and brass.

Module 08406

08406
Special Piping

Objectives

Successful completion of this module prepares you to do the following:

1. Explain how to assemble flared and compression joints using copper tubing.
   a. Describe flared and compression methods of joining tubing.
   b. Identify and describe flared and compression fittings.
   c. Explain how to assemble a flared joint.
   d. Explain how to assemble a compression joint.
2. Explain how to solder and braze copper tubing and fittings.
   a. Explain how to solder copper tubing and fittings.
   b. Explain how to braze copper tubing and fittings.
3. Explain how to lay out and create bends in pipe and tubing.
   a. Explain how to calculate pipe bends.
   b. Explain how to lay out pipe bends.
   c. Describe various methods of bending pipe and tubing.
4. Explain how to remove, install, and maintain glass-lined piping systems.
   a. Explain how to remove and install sections of glass-lined piping systems.
   b. Explain how to maintain glass-lined piping systems.
5. Identify and describe how to assemble Lokring® and other crimped joints on tubing.
   a. Identify and explain how to create Lokring® joints.
   b. Identify and explain how to create crimped joints on copper tubing.

Performance Tasks

Under supervision, you should be able to do the following:

1. Install flared fittings using copper tubing.
2. Install compression fittings using copper tubing.
3. Solder copper tubing joints.
4. Braze copper tubing joints.
5. Bend pipe or tubing to a specified radius.
6. Fabricate a Lokring® or a crimped copper-tubing joint.

Trade Terms

Acetylene
Capillary action
Compression coupling
Cup depth engagement
Ferrules
Flaring
Inside diameter (ID)
Nonferrous metal
Outside diameter (OD)
Oxidation
Reamer
Solder
Soldering flux
Wetting

Industry Recognized Credentials

If you are training through an NCCER-accredited sponsor, you may be eligible for credentials from NCCER's Registry. The ID number for this module is 08406. Note that this module may have been used in other NCCER curricula and may apply to other level completions. Contact NCCER's Registry at 1.888.622.3720 or go to **www.nccer.org** for more information.

Contents

Figures and Tables

Figures and Tables (Continued)

This page is intentionally left blank.

SECTION ONE

1.0.0 COPPER-TUBING FLARED AND COMPRESSION JOINTS

Objective

Explain how to assemble flared and compression joints using copper tubing.

a. Describe the flared and compression methods of joining tubing.
b. Identify and describe flared and compression fittings.
c. Explain how to assemble a flared joint.
d. Explain how to assemble a compression joint.

Performance Tasks

1. Install flared fittings using copper tubing.
2. Install compression fittings using copper tubing.

Trade Terms

Compression coupling: A mechanical fitting that is compressed onto a pipe, tube, or hose.

Ferrules: Bushings placed over a tube to tighten it.

Flaring: Increasing the diameter at the end of a pipe or tube.

Inside diameter (ID): Measurement of the inside diameter of the pipe.

Outside diameter (OD): Measurement of the outside diameter of the pipe.

Solder: An alloy, such as zinc and copper, or tin and lead, that is melted to join metallic surfaces.

The term *special piping* refers to a complete network of special pipes, fittings, valves, and other components that are designed to work together to convey a specific material. Special piping systems must be designed and fabricated according to local building codes, plans, specifications, and component installation instructions. Where conflicts between these sources are identified, use the one that is most conservative and provides the greatest margin of safety. When working with any type of special piping, it is important to get the necessary training from the piping manufacturer prior to installing or repairing the pipe.

Copper tube fittings are common, but compression fittings are used with plastic, steel, and stainless steel tubing, as well, and are applied across many industries. Some utilize multiple **ferrules**, which are bushings forced by the compression nut to bite into the surface of the tubing with a wedge-like action. Equipment manufactured outside the US is likely to be built with hardware under the metric standard, including the use of metric compression fitting systems.

> **CAUTION**
>
> Although compression fittings often appear to be the same, it is important that compression hardware from different manufacturers is not used interchangeably. Doing so can result in joint failure and possible injury due to the differing standards among manufacturers and nearly imperceptible differences in construction.

Many compression fittings have very specific assembly instructions regarding torque and how they are to be tightened to prevent failure. Always ensure that compression fitting components are properly matched, and that there is a thorough understanding of the assembly instructions associated with the specific product being installed or manipulated. Some compression fittings can be removed and retightened many times, while others cannot.

Copper tubing can be connected using flared or compression joints. Flared joints typically are used when the connection needs to be disassembled and reassembled often or when the tubing content cannot be removed completely for repairs. They are also used when **solder** cannot be used to join the tubing because of the presence of hazardous liquid or fuel within the tube or when the tubing is run underground. Flared joints are used underground because solder is destroyed by chemicals in the ground.

Compression fittings are used for quick-finish connections that require limited equipment and finish work. Two types of quick-finish connections are made on ice makers and humidifiers that need to be connected to the structure's water supply.

While both soft copper tubing and hard-drawn tubing are available, flared and compression joints are generally used with soft copper tubing. Hard-drawn tubing is made by drawing the copper through a series of dies to reduce its diameter; soft tubing is heated and cooled slowly, leaving it less brittle, more pliable, and less likely to tear.

These characteristics make soft copper tubing the best choice for flared or compression joint applications. Another key advantage of soft tubing is that it is produced in long lengths, which reduces the number of connections, and it can be bent to fit almost all conditions. Heavier tubing has a thicker wall and can withstand greater pressures.

After the tubing is produced, the manufacturer identifies it by printing the company name or trademark and the tubing type on the entire length of coil. Soft copper tubing is available in three types that differ in wall thickness; each is color-coded for convenience and instant identification. The three different types are as follows:

- Type K has the thickest wall and is color-coded green. This type of tubing is usually used for underground services or high-pressure applications.
- Type L has a thinner wall and is color-coded blue. This type of tubing is typically used for aboveground services but can be used underground.
- Type ACR also has a thinner wall and is also color-coded blue. It is typically used for aboveground services but can be used underground. This type, which is used in refrigerator and air conditioning applications, is purged, pre-charged with nitrogen, and sealed at the point of manufacture to prevent copper oxide from forming. When used in refrigeration circuits, ensure that the inside of the copper tube is clean.

Many sizes of soft copper are available. Types K and L are manufactured in coils of 40', 60', and 100'. Coils of 100' and 60' are manufactured in all thicknesses up to and including 1". The 60' coils are also available in the 2" size.

Tubing is sized by its **outside diameter (OD)** so that all soldered fittings can be used on any type of tubing. *Table 1* lists copper pipe and tubing specifications.

1.1.0 Flared and Compression Methods

Flaring involves expanding the end of copper tubing. Flared joining involves placing a flare nut on the end of copper tubing and expanding the end of the tubing using a flaring tool. The flared tubing is then placed against the tapered surface of the fitting, and the flare nut is tightened onto the fitting. Long or short flare nuts are available. Short nuts are used on many types of tubing, while long nuts are used more often on refrigeration and fuel oil piping where vibration could affect the performance. The flare nut forces the flared tubing and the fitting into a leakproof seal. An advantage of this assembly is that it can be easily disassembled to make repairs. *Figure 1* illustrates a typical flare joint using a long flare nut, while *Figure 2* shows a typical short and long flare nut for comparison.

In some applications, the angle of the flare formed on the tubing may be different. Most situations require a flare angle of 45 degrees, but 37-degree flare angles exist, as well. The Society of Automotive Engineers primarily governs the standards of flare fittings and their construction. In some cases, double-flare fittings are used. These fittings are made by turning the end of the tubing inward first, then forming it outward, creating a flared opening of double thickness.

Compression fittings are comprised of a threaded body, a threaded nut, and a brass or plastic ring or sleeve called a ferrule. The compression ring, or ferrule, is compressed between the nut and the fitting to produce a leakproof seal.

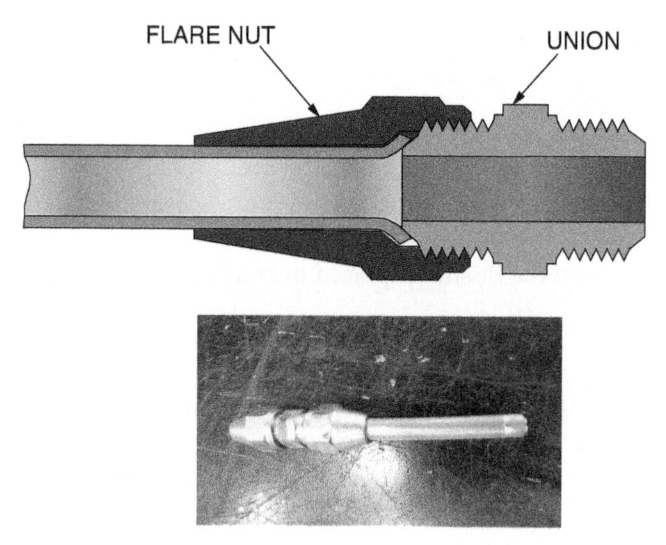

Figure 1 Flared joint.

Figure 2 Typical flare nuts.

Table 1 Copper Pipe and Tubing Specifications

Type K		
Nominal Size (inches)	**Actual OD**	**Wall Thickness**
1/4	0.375	0.035
3/8	0.5	0.049
1/2	0.625	0.049
5/8	0.75	0.049
3/4	0.875	0.065
1	1.125	0.065
1 1/4	1.375	0.065
1 1/2	1.625	0.072
2	2.125	0.083
2 1/2	2.625	0.095
3	3.125	0.109
3 1/2	3.625	0.12
4	4.125	0.134
5	5.125	0.16
6	6.125	0.192
Type L		
Nominal Size (inches)	**Actual OD**	**Wall Thickness**
1/4	0.375	0.3
3/8	0.5	0.035
1/2	0.625	0.04
5/8	0.75	0.042
3/4	0.875	0.045
1	1.125	0.05
1 1/4	1.375	0.055
1 1/2	1.625	0.06
2	2.125	0.07
2 1/2	2.625	0.08
3	3.125	0.09
3 1/2	3.625	0.1
4	4.125	0.114
5	5.125	0.125
6	6.125	0.14

This compression method is quick and easy and requires few tools or equipment. Compression joining makes it easy to attach humidifiers, ice makers, and other fixtures to a water system. *Figure 3* shows a compression joint.

1.2.0 Fittings

The same types of fittings are available for flared and compression methods of joining. To select and install fittings, it is important to understand the sizing, labeling, and types of fittings.

1.2.1 Sizing of Fittings

The size of fittings should correspond to the tube size. Tube sizes, set by ASTM, are nominal. A nominal size is one that is larger or smaller than the actual measurement. People generally refer to nominal sizes even though actual size is different.

In refrigeration, tube size is determined by measuring the outside diameter (OD) of the pipe. The measured size of the tube is always 1⁄8" larger than the **inside diameter (ID)** of the tube. The ID can vary because of differences in wall thickness. For example, Type K, which is thick-walled tubing, has a smaller ID than Type L tubing given equal sizes in OD.

Figure 3 Compression joint.

1.2.2 Labeling of Fittings

Industrial tubing and fittings are labeled according to a manufacturer scheme. While labels usually identify the size, type, and alloy, these systems have not been nationally systematized or standardized.

1.2.3 Types of Fittings

Many types of fittings are manufactured for flared or compression joints (*Figure 4*). The most common fittings are the following:

- Tee
- 90-degree elbow (ell)
- 45-degree elbow
- Union
- Cross

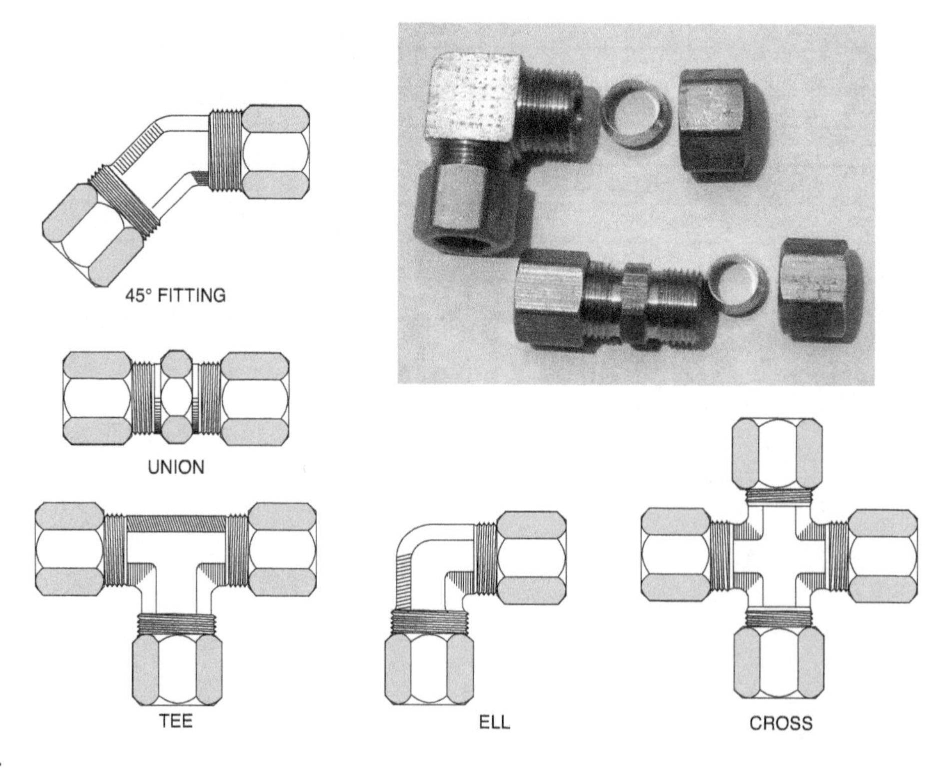

Figure 4 Fittings.

In addition to the fittings above, special fittings are available to connect flared and compression tubing to brazed, soldered, plastic, or threaded pipe. The only type of material that cannot be used with compression fittings is glass-lined pipe.

Flared and compression fittings should be ordered by their OD dimensions, not their ID dimensions. For example, a coupling with a $^3/_8$" ID tubing takes a $^1/_2$" **compression coupling**, which is a fitting that is compressed onto the pipe.

1.3.0 Assembling Flared Fittings

Remember that flared fittings are for low-pressure systems. Flaring tools (*Figure 5*) are used to spread the ends of tubing joined by flare nuts and fittings. The flared ends of the tubing, along with the flare nuts, provide a method of coupling tubing without welds, solder, or leaks. The most common type of flaring tool is the screw-in type.

Before forming the tubing for a flare joint, the nut of the fitting must be placed on the tube. The tube is then flared using a flaring tool.

Follow these steps to install flared fittings with a screw-in type flaring tool:

Step 1 Determine the size of the tubing to be flared.

Step 2 Select the flaring tool based on the size of the tubing.

Step 3 Inspect the selected flaring tool for obvious damage or excessive wear on the screw tip or the anvil clamping bars.

Step 4 Inspect and clean the end of the tubing to be flared.

CAUTION

The end of the tubing must be square and free of burrs—inside and outside—or the flared tubing may cause leaks when used.

Step 5 Place one of the flare nuts over the end of the tubing and move it out of the way with the threaded end of the nut facing in the same direction as the tubing end.

Step 6 Loosen the locking nut on the anvil clamping bars.

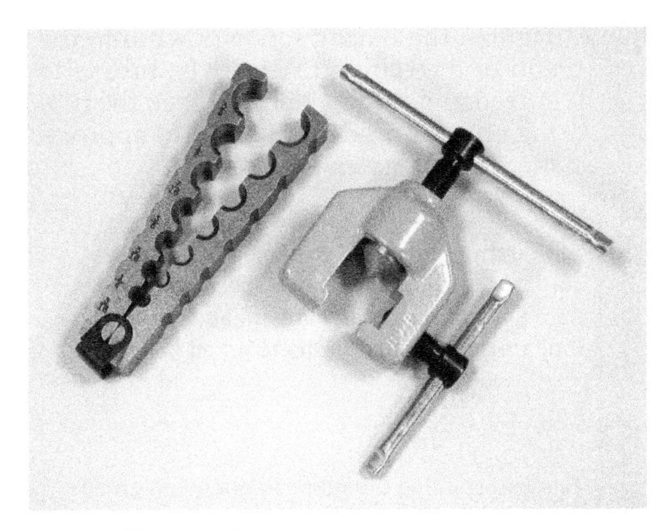

Figure 5 Flaring tools.

Step 7 Open the bars wide enough to allow the tubing end to be placed inside one of the flared holes.

Step 8 Move the end of the selected tubing between the open bars and into the flared hole closest to the tubing diameter.

Step 9 Close the bars loosely on the tubing.

Step 10 Position the tubing so that the end extends toward the flaring screw tip above the face of the bars by a distance equal to about half the tubing diameter. Allow for less than half a diameter extension for tubing $^1/_2$" diameter or more.

Step 11 Tighten the locking nut on the clamping bars while holding the tubing in position.

CAUTION

It is important to tighten the bars thoroughly so that the tubing cannot twist or slip while it is being flared.

Step 12 Position the flaring pin screw tip over the end of the tubing. The screw may have to be raised before it is positioned over the tube end. A drop of light oil may be used on the tip to reduce the danger of the tube splitting.

Step 13 Tighten the flaring screw down into the end of the tubing to cause the tubing to spread and stop tightening when the tubing end is flared to an angle of approximately 45 degrees.

> **CAUTION**
> Although 45-degree flare angles are very common, some applications specify a different flare angle. Ensure the flare nut and the flaring tool are appropriate for the required flare angle.

> **CAUTION**
> To avoid splitting the tube, do not over-tighten the flare.

Step 14 Unscrew and remove the flaring screw tip.

Step 15 Loosen the locking nut and open the clamping bars.

Step 16 Remove the flared tubing from the tool.

Step 17 Inspect the tubing flare.

> **CAUTION**
> The flared tubing must be evenly spread without any tears or splits. Flared tubing with flaws will result in a leaky joint.

Step 18 Slide the flare nut over the flared end until the flared end is seated inside the connector. The flared end should appear well-matched to the inside of the connector.

Step 19 Screw the fitting into the flare nut.

Step 20 Hold the fitting using a combination wrench and use another combination wrench to tighten the flare nut onto the fitting.

1.4.0 Assembling Compression Fittings

Follow these steps to install a compression fitting:

> **CAUTION**
> Never mix compression fitting components of different manufacturers. Ensure that all hardware components, including fittings, nuts, and ferrules, are compatible and adhere to manufacturer-specific assembly instructions for the type of fittings being used.

Step 1 Determine the size of the tubing to be fitted.

Step 2 Clean and ream the end of the tubing.

Step 3 Place the fitting over the end of the tubing and slide it out of the way with the threaded end facing in the same direction as the tubing end.

Step 4 Place a ferrule over the end of the tubing ensuring that it is properly oriented for the style of fitting being installed.

Step 5 Slide the tubing end into the appropriate fitting. It must remain firmly seated against the bottom of the fitting throughout the assembly process.

> **NOTE**
> Ensure the tubing end is squarely bottomed out inside the fitting and remains in that position during the remainder of the assembly process.

Step 6 Slide the ferrule into position against the fitting.

Step 7 Screw the threaded compression nut onto the fitting and firmly hand-tighten.

Step 8 Hold the fitting using one combination wrench while turning the compression nut with a second combination wrench. Tighten the nut $1\frac{1}{4}$ turns beyond firmly hand-tight. Do not overtighten the nut, as that will produce a leak. There are go-no-go gauges for checking fittings.

1.0.0 Section Review

1. A flaring tool is used to _____.
   a. form the end of the tubing
   b. tighten the flare nut
   c. ream the end of the tubing
   d. cut and form the tubing in a single action

2. The size of fittings should correspond to the ________.
   a. tubing size
   b. job size
   c. ferrule shape
   d. metal gauge

3. Although 45-degree flare angles are very common, _____.
   a. some applications specify a different flare angle
   b. most are made at a 90-degree angle
   c. they cannot be used on copper
   d. they cannot be formed with a flaring tool

4. When assembling a compression fitting, it is appropriate to _____.
   a. flare the tubing slightly first
   b. use components from different manufacturers
   c. tighten the nut $1\frac{1}{4}$ turns beyond firmly hand-tight
   d. lubricate the tubing first

SECTION TWO

2.0.0 SOLDERING AND BRAZING COPPER TUBING AND FITTINGS

Objective

Explain how to solder and braze copper tubing and fittings.

a. Explain how to solder copper tubing and fittings.
b. Explain how to braze copper tubing and fittings.

Performance Tasks

3. Solder copper tubing joints.
4. Braze copper tubing joints.

Trade Terms

Acetylene: A gas composed of two parts carbon and two parts hydrogen, commonly used in combination with oxygen to cut, weld, and braze steel.

Capillary action: The flow of liquid (solder) into a small space between two surfaces.

Cup depth engagement: The distance the tubing penetrates the fitting.

Nonferrous metal: A metal that contains no iron and is therefore nonmagnetic.

Oxidation: A chemical reaction that increases the oxygen content of a compound.

Reamer: A tool used to enlarge, shape, smooth, or otherwise finish a hole.

Soldering flux: A chemical substance that aids the flow of solder. Flux removes and prevents the formation of oxides on the pieces to be joined by soldering.

Wetting: Spreading liquid filler metal or flux on a solid base metal.

Soldering and brazing are two other methods used for joining copper tubing and fittings. Soldered joints are used for domestic water lines, sanitary drain lines, hot water heating systems, and other applications where the temperature does not exceed 250°F. Brazed joints are used where greater strength is required or where temperatures exceed 250°F. Brazing is used in some steam and hot water heating lines, refrigeration lines, compressed air lines, vacuum lines, oil lines, and some chemical lines that need extra corrosion-resistance in the piping joints. The main difference in the soldering and brazing procedures is the temperature at which each is performed. Soldering is most often performed at temperatures below 800°F; brazing is performed at temperatures greater than 800°F.

When assembling long runs of tubing, the tubing must be properly supported to eliminate any danger of poor alignment. Smaller diameter tubing requires more frequent support than tubing of a larger diameter due to the structural integrity of the tubing. Proper support of tubing and piping near any change in direction is also critical.

Both vertical and horizontal support requirements are governed by code. For example, for vertical support of copper tubing systems, both the *Uniform Plumbing Code* and the *BOCA National Plumbing Code* require support at maximum intervals of 10'. The *Standard Plumbing Code* requires support at each story for tubing equal to or greater than 1½" in diameter. For tubing equal to or less than 1¼" in diameter, support must be provided at maximum intervals of 4'.

For horizontal support, both the *Standard Plumbing Code* and the *Uniform Plumbing Code* require support at 6' intervals for tubing equal to or less than 1½" in diameter. For tubing greater than 1½" in diameter, horizontal support must be provided at 10' intervals. The *BOCA National Plumbing Code* requires horizontal support at 6' intervals for tubing equal to or less than 1¼" in diameter. For diameters greater than 1½", horizontal support must be provided at 10' intervals.

CAUTION

These are minimum standards. Always consult local codes and job specifications to ensure compliance. In many situations, more support may be necessary.

It is also important to check for proper alignment. Poor alignment affects joint integrity by changing the space and gap between the tubing and fitting walls. It also leaves a poor final appearance. Wires or other temporary hangers can be used to achieve proper alignment during installation.

2.1.0 Soldering Copper Tubing and Fittings

Soldering is the most common method of joining copper tubing and fittings. Soldering involves joining two metal surfaces by using heat and a nonferrous metal. A nonferrous metal does not contain iron and is, therefore, non-magnetic. The melting point of the filler metal must be lower than that of the two metals being joined.

Soldered joints depend on capillary action to pull the melted solder into the small gap between the fitting and the tubing. Capillary action is the flow of liquid, in this case solder, into a small space between two surfaces. Capillary action is most effective when the space between the tubing and the fitting is between 0.002" and 0.005". *Figure 6* shows capillary action.

To properly solder copper tubing and fittings, pipefitters need to understand the following:

- Solders and soldering fluxes
- Preparing tubing and fittings for soldering
- Soldering joints

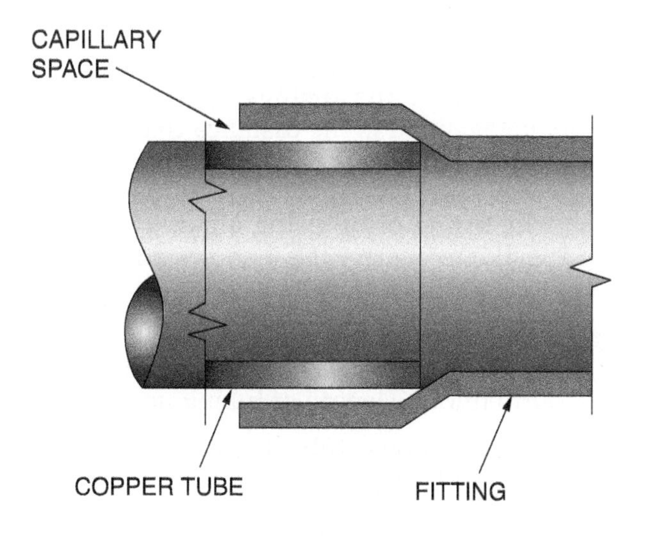

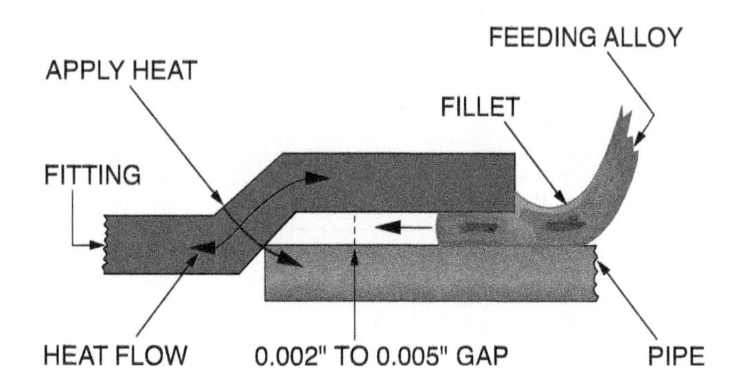

Figure 6 Capillary action.

2.1.1 Solders and Soldering Fluxes

Solder is a nonferrous metal or metal alloy with a melting point below 800°F. An alloy is any substance made up of two or more metals. The most common solder used on copper tubing is an alloy made of 95 percent tin and 5 percent antimony. This type of solder is used for potable water connections because it contains no lead. It is usually recommended for applications requiring greater joint strength. The tin-antimony alloy solder melts at 464°F and solidifies rapidly. Ultimately, proper use and selection of low-temperature solders can result in a superior joint when compared to high-temperature brazing. Excessive heat applied during brazing can further anneal copper, weakening the metal.

The *Federal Safe Drinking Water Act of 1986* mandated the use of lead-free (less than 0.2 percent) solders for drinking water supply piping. Solders containing alloys of antimony and cadmium must also be avoided for drinking water applications. Two types of solder are commonly applied in both drinking water and refrigerant piping applications. One contains 95.5 percent tin, 4 percent copper and 0.5 percent silver; and the other has 96 percent tin and 4 percent silver. The addition of silver to the solder increases overall strength significantly and, due to its elongation properties, works well where temperature and pressure causes expansion of the tubing. Another solder option has nickel added for extra strength.

Other solders are available for various applications, including 95 percent tin/5 percent antimony, which is commonly used on low-pressure refrigeration piping and non-drinking water applications. The combination of 50 percent tin/50 percent lead is still readily available for non-drinking water applications, but 95/5 (95 percent tin, 5 percent lead) can withstand roughly 2.5 to 3 times higher working pressures at temperatures up to 250°F. It should be noted that applications over 250°F generally require brazing instead of soldering to ensure joint integrity and strength.

Before selecting solder for a given application, it is essential to understand the proper use and application of the product and ensure that its composition meets the required standards. Technical specification sheets and MSDS reports are available for all such products and should be studied carefully when using a new product.

Choosing the proper soldering flux is important because the wrong flux can ruin a soldered joint. Flux performs the following functions:

- Chemically cleans and protects the surfaces of the tubing and its fitting from oxidation. Oxidation occurs when oxygen in the air combines with the recently cleaned metal. The result is tarnish or rust on metal, which prevents solder from adhering.
- Allows the soldering alloy or filler metal to flow easily into the joint.
- Floats out remaining oxides ahead of the molten filler metal.
- Promotes wetting of the metal. Wetting is the process that decreases the surface tension so that the molten solder flows evenly throughout the joint.

Fluxes can be classified as highly corrosive, less corrosive, and non-corrosive. All fluxes need to be somewhat corrosive; however, the fluxing process must render the flux inert. If it is not inert, the flux gradually destroys the soldered joint. As is the case with solder, many different types of flux are available. Solder-flux pastes are also available. These pastes contain finely ground particles of a solder mixed with a paste flux. The addition of filler metal during the soldering process is still generally required, but it is dependent upon the application and product.

A flux should be chosen from the same manufacturer as the solder. Although many solder and flux products are interchangeable, using a flux provided by the solder manufacturer ensures that the two products have been carefully tested together and will likely offer the best result.

The best fluxes for joining copper tubing and copper fittings are non-corrosive fluxes. The most common noncorrosive fluxes have petroleum bases and are compounds of mild concentrations of zinc and ammonium chloride.

An oxide film begins forming on copper immediately after it has been mechanically cleaned. Therefore, apply flux immediately to all recently cleaned copper fittings and tubing. Flux should be applied to clean metal with a brush or swab but never with fingers. Doing so may result in a cut, and there is also a chance that flux could be carried to the eyes or mouth. Body contact with the cleaned fittings and tubing adds to unwanted contamination of the metal.

If a can of flux is not closed immediately after use, or if a can is not used for a considerable length of time, the chlorides separate from the petroleum base. The flux must be stirred before each use.

CAUTION

Brazing flux and soldering flux are not the same. Do not allow these fluxes to become mixed or interchanged. Carelessness can ruin quality work.

2.1.2 Preparing Tubing and Fittings for Soldering

To prepare tubing and fittings for soldering, the tubing must be measured, cut, and reamed; the tubing and fittings must also be cleaned. Proper cleaning techniques are essential for producing a leakproof joint. Follow these steps to prepare the tubing and fittings for soldering:

Step 1 Measure the distance between the faces of the two fittings using the face-to-face method (*Figure 7*).

Step 2 Determine the cup depth engagement of each of the fittings. The cup depth engagement is the distance that the tubing penetrates the fitting. This measurement can be determined by measuring the fitting or by using a manufacturer's makeup chart. *Table 2* shows a manufacturer's makeup chart.

Step 3 Add the cup depth engagement of both fittings to the measurement found in *Step 1* to find the length of tubing needed.

Step 4 Cut the copper tube to the correct length using a tubing cutter.

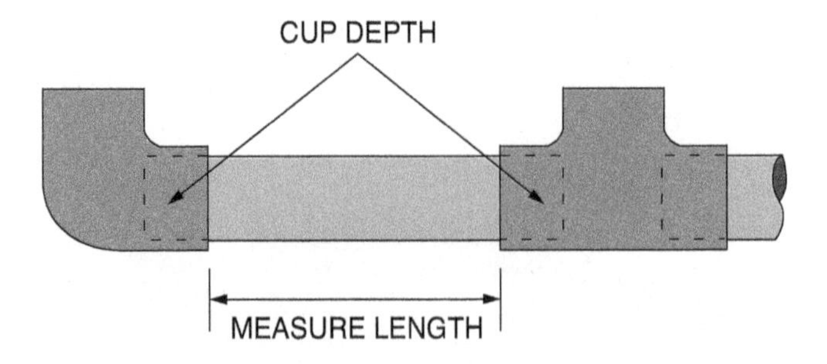

Figure 7 Face-to-face method.

Table 2 Manufacturer's Makeup Chart

Pipe Size	Depth of Cup	Pipe Size	Depth of Cup
1/4	5/16	2	1 11/32
3/8	3/8	2 1/2	1 15/32
1/2	1/2	3	1 21/32
5/8	5/8	3 1/2	1 29/32
3/4	3/4	4	2 5/32
1	29/32	5	2 21/32
1 1/4	31/32	6	3 3/32
1 1/2	1 3/32	–	–

Step 5 Ream the inside and outside of both ends of the copper tube using a reamer, which is used to finish the hole.

Step 6 Clean the tubing and the fitting using No. 00 steel wool, emery cloth, or a special copper-cleaning tool.

CAUTION

Be careful when cleaning copper tubing and fittings to remove abrasions on the copper without removing a large amount of metal. Abrasions can weaken or ruin a copper joint. Do not touch or brush away filings from the tube or fitting with your fingers because it may contaminate the freshly cleaned metal.

Step 7 Apply flux to the copper tubing and to the inside of the copper fitting socket immediately after cleaning them.

Step 8 Insert the tube into the fitting socket, then push and turn the tube into the socket until the tube touches the inside shoulder of the fitting.

Step 9 Wipe away any excess flux from the joint.

Step 10 Check the tube and fitting for proper alignment before soldering.

2.1.3 Soldering Joints

Because soldering requires relatively low heat, heating equipment that mixes acetylene, butane, or propane directly with air is sufficient. This means only one tank of gas and a torch are needed. Follow these steps to solder a joint:

WARNING!

Always solder in a well-ventilated area because fumes from the flux can irritate your eyes, nose, throat, and lungs. Point the torch away from your body when lighting it and use only a spark lighter to light the torch. Do not use a match, cigarette lighter, or cigarette to light the torch.

Step 1 Obtain either an acetylene tank and related equipment or a propane bottle and torch.

Step 2 Set up the equipment according to the manufacturer's instructions.

Step 3 Light the heating equipment according to the type of equipment being used and the manufacturer's instructions.

Step 4 Heat the tubing first, then move the flame onto the fitting. *Figure 8* depicts heating the tubing and fitting.

CAUTION

The inner cone of the flame should barely touch the metal being heated. Do not direct the flame into the socket because this will burn the flux. Be sure to keep the flame moving on the metals instead of holding the flame in one place.

Step 5 Move the flame away from the joint.

Step 6 Touch the end of the solder to the area between the fitting and the tube. The solder will be drawn into the joint by the capillary action. The solder can be fed upward or downward into the joint. If the solder does not melt on contact with the joint, remove the solder and heat the joint again. Do not melt the solder with the flame.

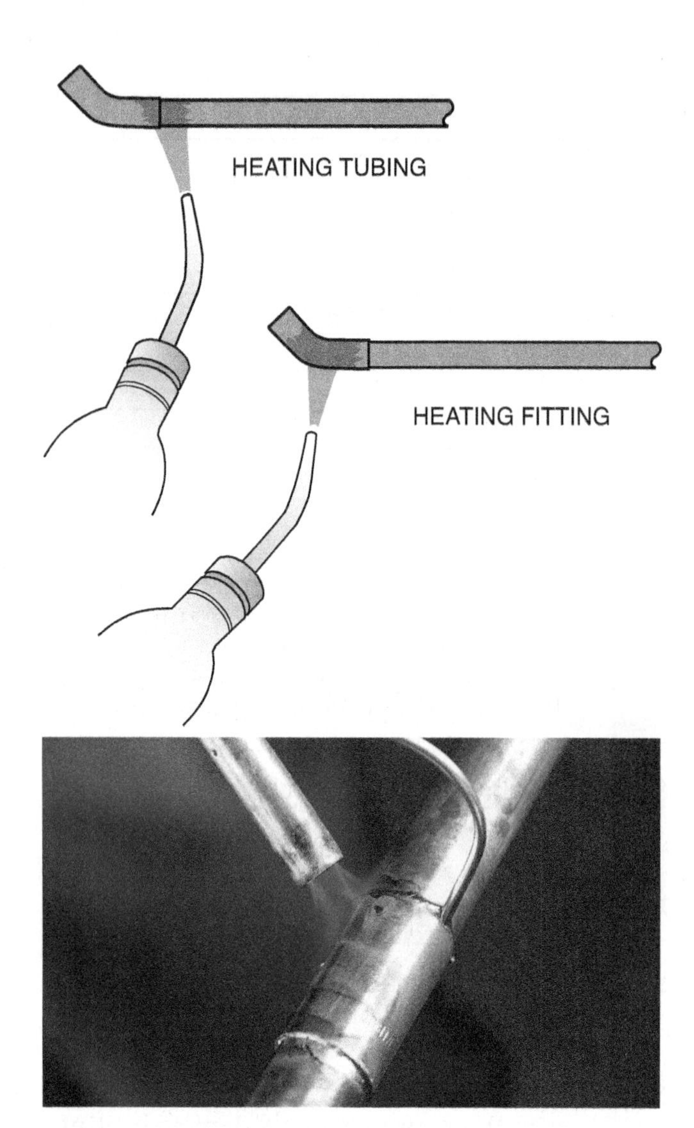

Figure 8 Heating tubing and fitting.

Step 7 Continue to feed the solder into the joint until a ring of solder appears around the joint, which indicates that the joint is filled. On 3⁄4" diameter tubing and smaller, the solder can be fed into the joint from one point. On larger tubing, the solder should be applied from the 6-o'clock position to the 12-o'clock position on the tubing. Generally, the amount of solder used is equal to the diameter of the tubing. For example, with 3⁄4" tubing, 3⁄4" of solder will fill the joint.

Step 8 Allow the joint to cool. Water can be applied to the joint to speed the cooling process.

Step 9 Wipe the joint clean with a cloth after the joint has cooled.

2.2.0 Brazing Copper Tubing and Fittings

Brazing, like soldering, uses nonferrous filler metals to join base metals that have a melting point above that of the filler metals. Brazing is performed above 800°F. Pipefitters braze tubing and fittings used in saltwater pipelines, low-pressure steam lines, refrigeration lines, medical gas lines, compressed air lines, vacuum lines, fuel lines, or other chemical lines that need extra corrosion resistance in the piping joints. Cast-bronze fittings are sometimes used to connect tubing made of copper, brass, copper-nickel, and steel.

Brazing produces mechanically strong, pressure-resistant joints. The strength of a brazed joint is partially dependent upon the ability of the filler metal to penetrate the base metal. Penetration can only occur if the base metals are properly cleaned, the proper flux and filler metal are selected, and the clearance gap between the outside of the tubing and the inside of the fitting is 0.003" to 0.004".

To properly braze copper tubing and fittings, pipefitters need to understand the following:

- Filler metals and fluxes
- Preparing tubing and fittings for brazing
- Brazing joints

2.2.1 Filler Metals and Fluxes

There are two groups of filler metals used to join copper tubing: alloys that contain 30 to 60 percent silver (the BAg series) and copper alloys that contain phosphorous (the BCuP series). *Table 3* lists brazing filler materials.

WARNING!

BAg-1 and BAg-2 contain cadmium. Heating when brazing can produce highly toxic fumes. Use adequate ventilation and avoid breathing the fumes.

Both groups of filler metal have different melting, fluxing, and flowing characteristics. Consider these characteristics when selecting a filler metal. When joining copper tubing, any of these filler metals can be used; however, the most often used filler metals for close tolerances are BCuP-3 and BCuP-4. BCuP-5 is used where close tolerances cannot be held, and BAg-1 is used as a general-purpose filler metal.

Brazing fluxes are applied using the same methods and rules as soldering fluxes. Brazing fluxes are more corrosive than soldering fluxes, so be careful not to mix a soldering flux with a brazing flux. For best results, use the flux

Table 3 Brazing Filler Materials

Percent of Principal Element						
AWS Classification	**Silver**	**Phosphorus**	**Zinc**	**Cadmium**	**Tin**	**Copper**
BCuP-2	–	7–7.50	–	–	–	Balance
BCuP-3	4.75–5.25	5.75–6.25	–	–	–	Balance
BCuP-4	5.75–6.25	7–7.50	–	–	–	Balance
BCuP-5	14.5–15.50	4.75–5.25	–	–	–	Balance
BAg-1	44–46	–	14–18	23–25	–	14–16
BAg-2	34–36	–	19–23	17–19	–	25–27
BAg-5	44–46	–	23–27	–	–	29–31
BAg-7	55–57	–	15–19	–	4.5–5.5	21–23

recommended by the manufacturer of the brazing filler metals. Fluxes best suited for copper tube joining should meet *American Welding Society (AWS) Standard A5.31, Type FB3-A* or *FB3-C*.

When copper tubing is joined to wrought copper fittings with copper-phosphorus alloys (BCuP series), flux can be omitted because the copper-phosphorus alloys are self-fluxing on copper. Fluxes are required for joining all cast fittings.

2.2.2 Preparing Tubing and Fittings for Brazing

To prepare tubing and fittings for brazing, follow the same procedures used to prepare tubing and fittings for soldering. Proper cleaning techniques must be used to produce a solid, leakproof joint. Follow these steps to prepare the tubing and fittings for brazing:

Step 1 Measure the distance between the faces of the two fittings.

Step 2 Determine the cup depth engagement of each of the fittings. Cup depths for fittings used with brazing are shorter than the cup depths of fittings used with soldering since less penetration is needed for brazing than with soldering. This measurement can be found by measuring the fitting or by using a manufacturer's makeup chart. *Figure 9* shows a manufacturer's brazing fitting makeup chart.

Step 3 Add the cup depth engagement of both fittings to the measurement found in *Step 1* to find the length of tubing needed.

Step 4 Cut the copper tubing to the correct length using a tubing cutter.

Step 5 Ream the inside and outside of both ends of the copper tubing using a reamer.

Step 6 Clean the tubing and the fitting using No. 00 steel wool, emery cloth, or a special copper-cleaning tool.

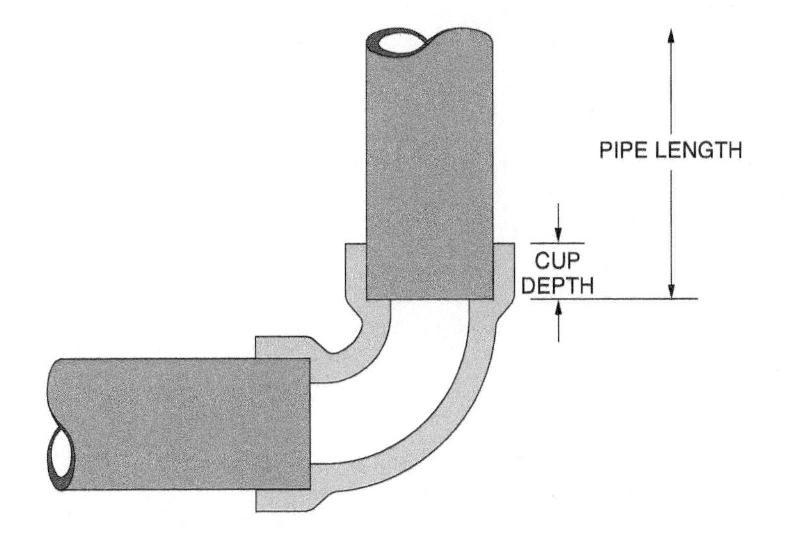

PIPE SIZE (in)	CUP DEPTH (in)
1/4	17/64
3/8	5/16
1/2	3/8
3/4	13/32
1	7/16
1 1/4	1/2
1 1/2	5/8
2	21/32
2 1/2	25/32
3	53/64
3 1/2	7/8
4	29/32
5	1
6	1 7/64
7	1 7/32
8	1 5/16

Figure 9 Manufacturer's brazing fitting makeup chart.

> **CAUTION**
>
> When cleaning the tubing and fittings, remove all abrasions on the copper without removing a large amount of metal. The abrasions can weaken or ruin a copper joint. Do not touch or brush away filings from the tube or fitting with your fingers because your fingers will also contaminate the freshly cleaned metal.

Step 7 Apply flux to the copper tubing and to the inside of the copper fitting socket immediately after cleaning them.

Step 8 Insert the tube into the fitting socket, then push and turn the tube into the socket until the tube touches the inside shoulder of the fitting.

Step 9 Wipe away any excess flux from the joint.

Step 10 Check the tube and fitting for proper alignment before brazing.

2.2.3 Heating Equipment

The brazing heating procedure differs from soldering in that different equipment is required to raise the temperature of the metals above 800°F. Oxyacetylene brazing equipment is used for heating brazed joints. The flame is produced by burning acetylene mixed with pure oxygen. *Figure 10* shows oxyacetylene brazing equipment.

MAP-Pro gas is mostly propylene with a small amount of propane mixed in. Due to its combustion temperature of 5,201°F in oxygen, it is considered a much better choice for brazing than propane or other propane-based bottled gases. However, it has a lower heat value than oxyacetylene and offers less flexibility in flame control. Generally used for smaller tubing only, it is readily available in small containers (1#) for hobbyists and other small projects. It is safer to store and transport than acetylene and significantly safer at higher operating pressures such as those required for underwater welding applications.

When setting up the torch, set the oxygen at about 40 pounds per square inch (psi) and do not use the trigger on the torch if it has one.

> **CAUTION**
>
> Do not run the acetylene regulator above 15 psi. At this pressure, acetylene can become unstable when mixed with air.

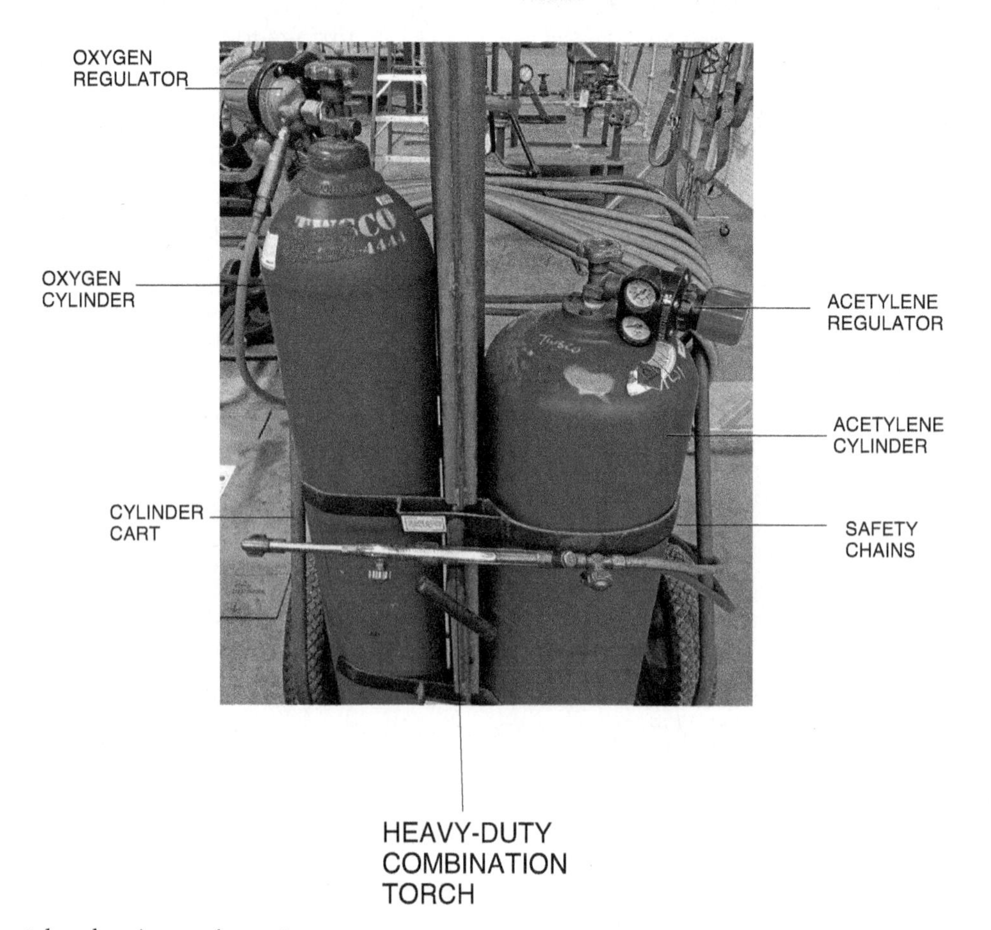

Figure 10 Oxyacetylene brazing equipment.

2.2.4 Brazing Joints

Follow these steps to braze a joint:

Step 1 Set up the oxyacetylene brazing equipment.

Step 2 Ensure that the tubing to be brazed is properly supported.

Step 3 Ensure that the tubing and fitting are properly aligned.

Step 4 Apply heat to the tubing about 1" from the joint and observe the flux. The flux will bubble, turn white, then melt into a clear liquid.

Step 5 Move the heat to the fitting and hold it there until the flux on the fitting turns clear.

Step 6 Continue to move the heat back and forth over the tubing and the fitting. Allow the fitting to receive more heat than the tubing by pausing at the fitting while the flame is moved back and forth. When heating tubing that is $1^1/_2$" in diameter or larger, move the heat around the entire circumference of the joint so that the entire joint is heated to brazing temperature.

Step 7 Touch the filler metal rod to the joint when a puddle of molten copper forms at the joint.

NOTE

If the filler metal melts upon contact with the joint, the brazing temperature has been met. Proceed to Step 8. If the filler metal does not melt upon contact, continue to heat and test the joint until the filler metal melts.

Step 8 Hold the filler metal rod to the joint and allow the filler metal to enter into the joint while holding the torch slightly ahead of the filler metal and directing most of the heat to the shoulder of the fitting.

Step 9 Continue to fill the joint with the filler metal until the filler metal inside the joint fills to a depth of at least three times the thickness of the tubing. This will not fill the joint, but a joint with this amount of penetration will be stronger than the tubing. If the tubing is 2" or more in diameter, two torches can be used to evenly distribute the heat. For larger joints, small sections of the joint can be heated and brazed. Be sure to overlap the previously brazed section while continuing to move around the fitting. *Figure 11* shows working in overlapping sectors.

Step 10 Wash the joint with warm water to clean excess, dried, or hardened brazing flux from the joint. If the flux is too hard to be removed with water, chip the excess flux off using a small chisel and light ball-peen hammer, then wash the joint with warm water. Joints clean best when they are still warm.

Step 11 Allow the joints to cool naturally.

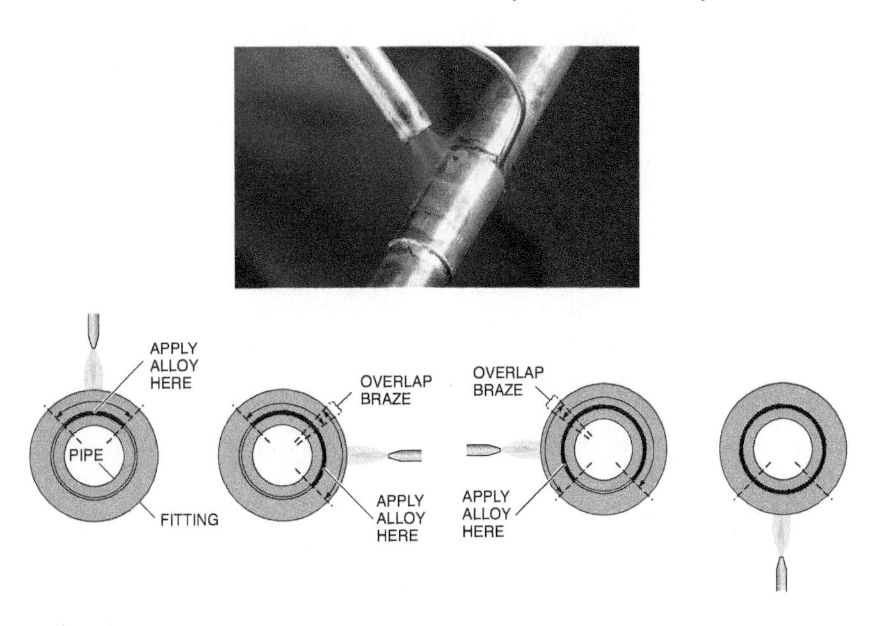

Figure 11 Working in overlapping sectors.

2.0.0 Section Review

1. Soldering involves joining two metal surfaces by using heat and _____.
   a. flux
   b. an iron alloy
   c. a nonferrous metal or alloy
   d. a torch

2. The strength of a brazed joint is partially dependent upon the ability of what material to penetrate the base metal?
   a. More base metal
   b. Brazing gases
   c. Filler metal
   d. Flux

SECTION THREE

3.0.0 BENDING PIPE AND TUBING

Objective

Explain how to lay out and create bends in pipe and tubing.

a. Explain how to calculate pipe bends.
b. Explain how to lay out pipe bends.
c. Describe various methods of bending pipe and tubing.

Performance Task

5. Bend pipe or tubing to a specified radius.

When designing a piping system, an engineer has two choices to make when determining how to change directions of the piping system. One is to buy fittings (butt weld, socket weld, or threaded) to change the directions of the piping system, and the other is to use pipe benders to bend the pipe. Bending the pipe allows the process to flow more freely through the system without restrictions caused by fittings. *Figure 12* shows a 90-degree screw fitting and a 90-degree bend.

Pipe bending can be accomplished by either cold or hot bending methods. Some of the factors that influence pipe bending process are the size of the pipe, the material composition of the pipe, and the number of identical bends required for the piping system.

Both cold and hot bending methods can cause distortion to occur. This happens because the outside of the bend stretches until the pipe flattens or collapses, causing the inside of the pipe to buckle or wrinkle. To prevent distortion, dry sand or a mandrel is often placed inside the pipe during the bending process. *Figure 13* shows different types of bends.

3.1.0 Calculating Pipe Bends

There are tables that give the dimensions for commonly used bends, but these tables do not provide information for all possible bends. If the information for the fabrication or replacement of bends is not available in tables or on the fabrication drawings, the pipefitter must know how to calculate the pipe bend. Formulas require the angle of the bend shown in degrees and represented by the letter D in the formula. To calculate pipe bends, it is important to recognize a few pipe-bending terms (*Figure 14*).

Pipefitters must know how to calculate the following:

- Bend allowance
- Angle of bend
- Radius of bend

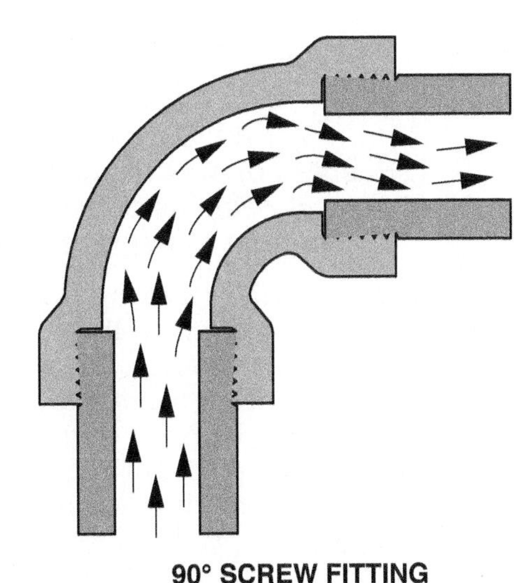

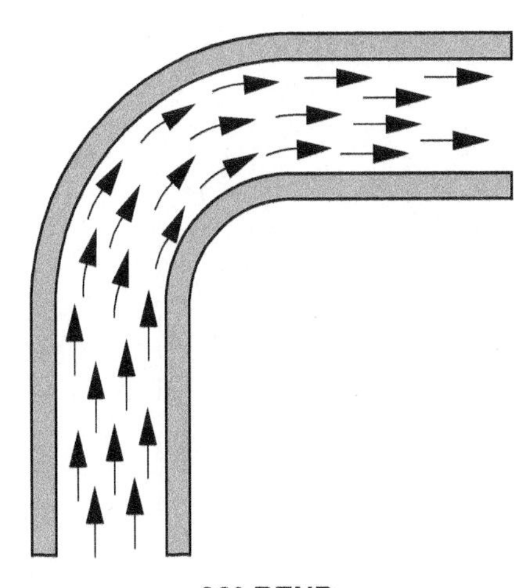

Figure 12 90-degree screw fitting and 90-degree bend.

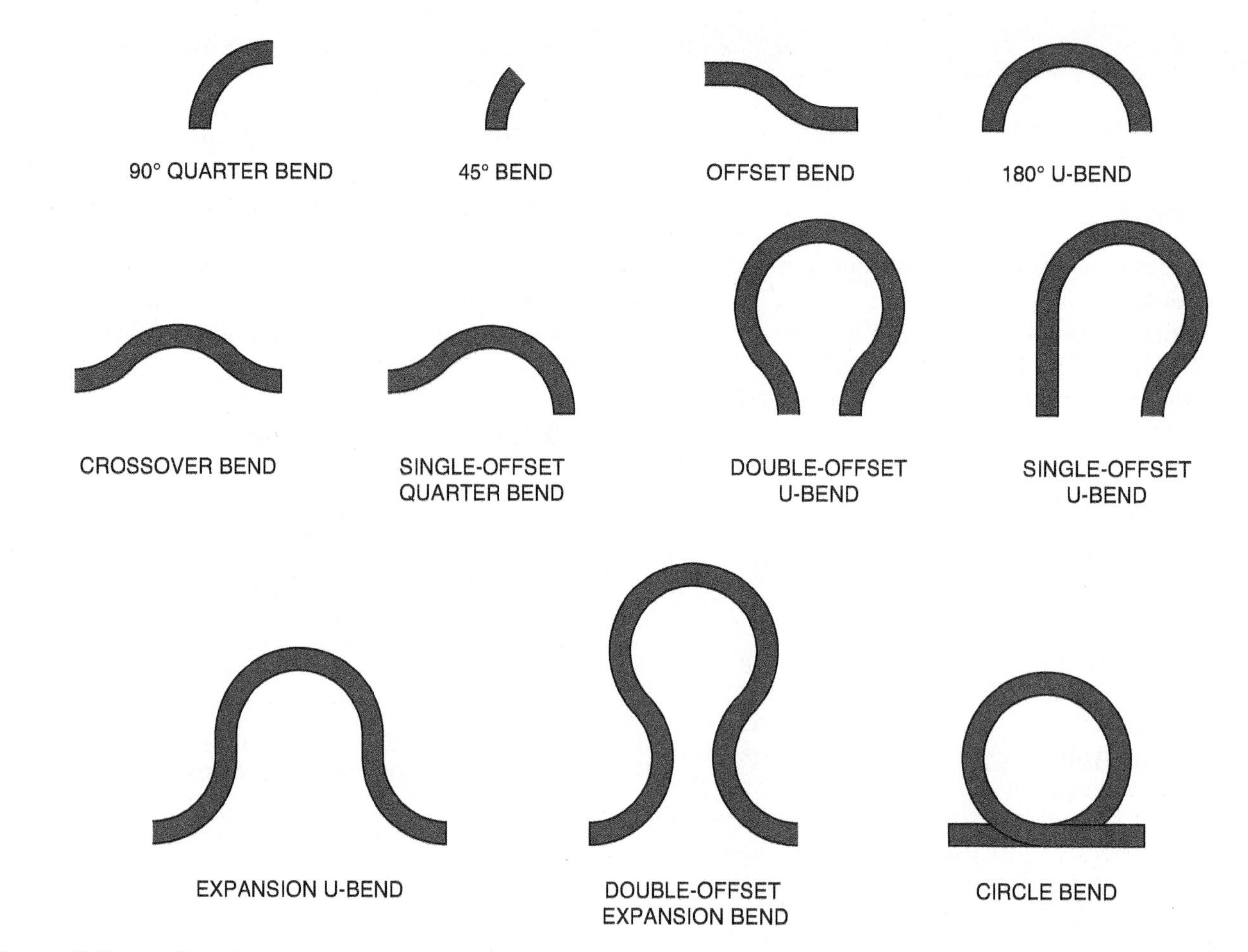

Figure 13 Types of-bending terms.

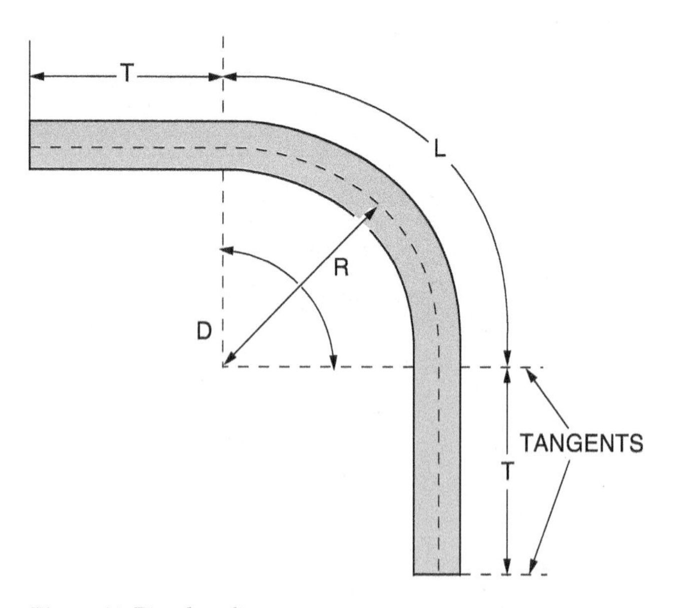

Figure 14 Pipe-bending terms.

3.1.1 Calculating Bend Allowance

The length of straight pipe needed to form a bend is called the *bend allowance.* The pipefitter can determine how much straight pipe is needed by using the following formula:

$$L = R \times D \times 0.01745$$

L = Length of pipe needed for bend
R = Radius of bend
D = Number of degrees in bend

For example, when making a 90-degree bend, D is 90, so L = R × 90 × 0.01745.

Large pipe must be bent with a long radius. To prevent the pipe from overstretching or collapsing, the radius of the bend should be five times the diameter of the pipe. The radius may be larger but should not be smaller. According to this rule, when bending 2" pipe, the bend radius should be 5" × 2", or 10". If the bend radius is 10", R is 10, and L = 10 × 90 × 0.01745.

Once the radius of the bend (R) and degrees in bend (D) are known, the length of pipe needed for bend (L) can be determined by multiplying as follows:

$$L = 10 \times 90 \times 0.01745\text{, where}$$
$$10 \times 90 = 900$$
$$L = 900 \times 0.01745$$
$$900 \times 0.01745 = 15.705$$
$$L = 15.705$$

According to the formula, the length of pipe needed for the bend, or the bend allowance, is 15.705".

To find the total length of a pipeline containing a bend, add the bend allowance to the length of straight pipe on each side of the bend. These straight lengths are called *tangents*. In *Figure 15*, the total length of the line is 27.705" since 7" + 5" + 15.705" = 27.705".

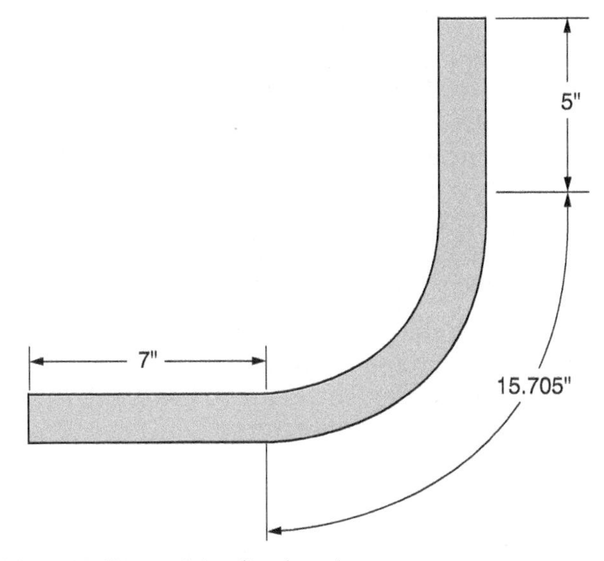

Figure 15 Determining line length.

Follow these steps to calculate the bend allowance and total length of line needed for a 70-degree bend in 1" pipe (*Figure 16*):

Step 1 Write down the bend allowance formula.

$$L = R \times D \times 0.01745$$

Step 2 Find D.

$$L = R \times 70 \times 0.01745$$

Step 3 Calculate R.

$$1\text{" pipe} \times 5 = 5\text{"}$$
$$L = 5 \times 70 \times 0.01745$$

Step 4 Multiply R × D.

$$5 \times 70 = 350$$

Step 5 Multiply the total by 0.

$$01745.350 \times 0.01745 = 6.1075$$
The bend allowance (L) = 6.1075

Step 6 Add both tangents.

$$15\text{"} + 10\text{"} = 25\text{"}$$

Step 7 Add the length of both tangents to the bend allowance.

$$6.1075\text{"} + 25\text{"} = 31.1075\text{"}$$

The total length of line needed for the run is 31.1075".

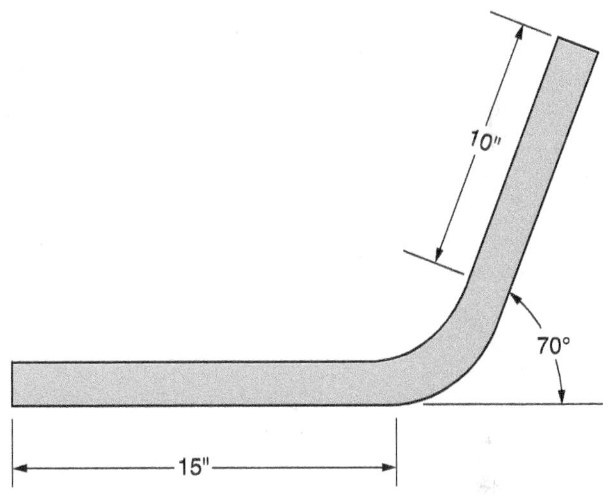

Figure 16 Bent pipe run.

3.1.2 Calculating Angle of Bend

A pipefitter must sometimes replace a bend in a pipe run. The new bend is often the same size and angle as the old bend. If drawings showing the angle and radius of the bend are not available, the pipefitter can find the angle of the bend by using two straightedges and a protractor.

Follow these steps to determine the bend angle:

Step 1 Place two straightedges on the tangents of the bend, which are the straight pieces of the pipe run. *Figure 17* shows how to set up the straightedges and protractor.

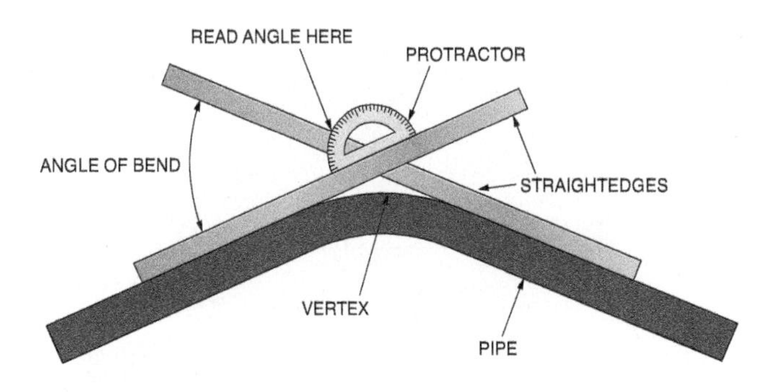

Figure 17 Finding bend angle.

Step 2 Place the center point of the protractor at the point where the two straightedges cross. This point, called the vertex, is the point where two straight lines meet to form an angle. When the protractor is lined up on one straightedge and the center point is placed at the vertex, the angle is easy to read. *Figure 18* shows two examples of a vertex.

Step 3 Mark the point where the second straightedge crosses the dial of the protractor. This mark is the angle of the bend. *Figure 19* shows a protractor.

3.1.3 Calculating Radius of Bend

Follow these steps to find the radius of a pipe bend with a square and rule:

Step 1 Place the end of the square on one tangent at the point where the bend begins.

Step 2 Draw a line from point A to point B as shown in *Figure 20.*

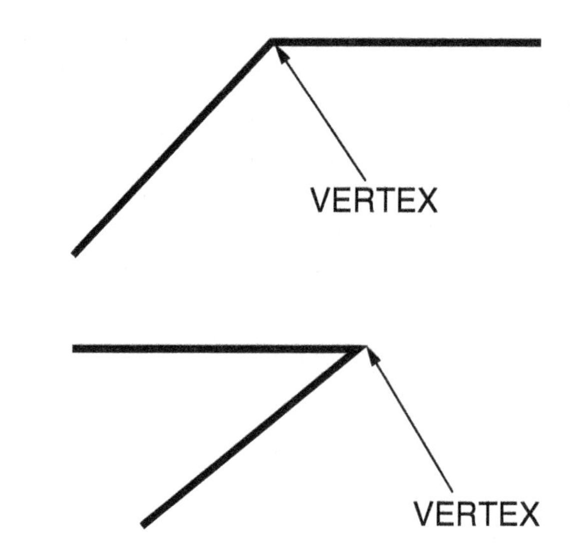

Figure 18 Vertex.

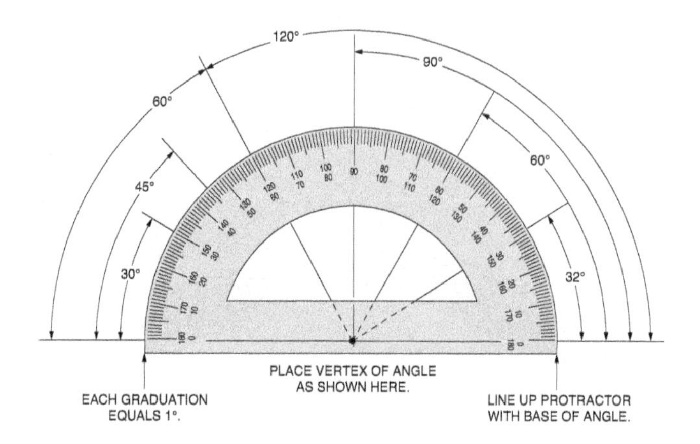

Figure 19 Protractor.

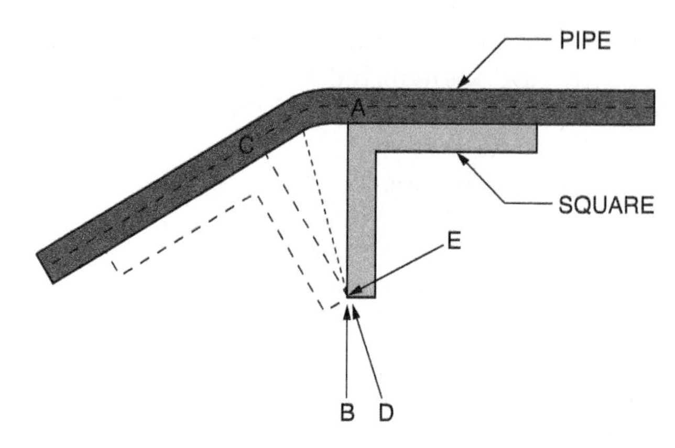

Figure 20 Finding the radius.

Step 3 Repeat *Step 1* and *2* for the other tangent and draw a line from point C to point D.

Step 4 Measure the distance from point E to where the two lines cross to the center of the pipe bend. This distance is the radius of the bend. *Figure 20* shows finding the radius.

3.2.0 Laying Out Bends

Dimensions of the pipeline are usually included on pipe drawings. The dimensions are given from the center lines of the pipe. The radius of a bend is also given from the center line of the pipe. Once the radius and the angle of the bend are known, the needed layout and the length of pipe needed can be determined. To lay out a pipe bend, use a protractor and the blade from a combination set. Remember to measure the inside radius of the bend and that the calculations need to account for the radius.

Follow these steps to lay out a 45-degree bend on a 4" pipe with a 36" radius:

Step 1 Subtract one-half of the OD of the 4" pipe from the 36" radius center line to obtain the inside radius of the bend. Note that the OD of a 4" pipe is $4\frac{1}{2}$".

$$4\frac{1}{2}'' \div 2 = 2\frac{1}{4}''$$

$$36'' \times 2\frac{1}{4}'' = 33\frac{3}{4}''$$

Step 2 Draw an arc with a radius of $33\frac{3}{4}$" (*Figure 21*).

Step 3 Draw a straight line to meet the arc and label the line AB (*Figure 22*). Line AB is the tangent of the arc, which means it touches the arc in one place.

Step 4 Lay out a 45-degree angle from line AB using the combination set protractor and blade.

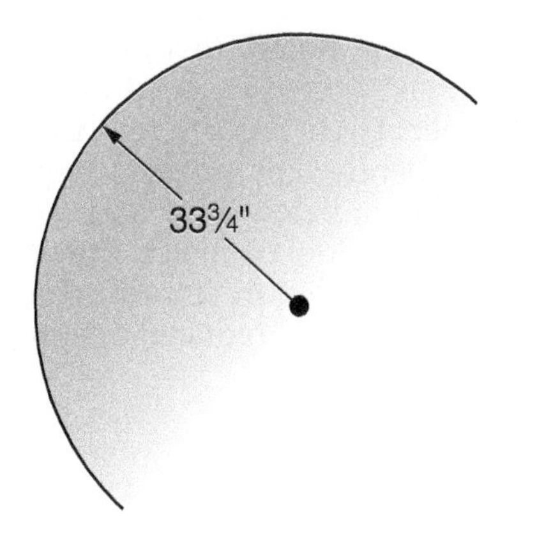

Figure 21 Arc with radius of $33\frac{3}{4}$".

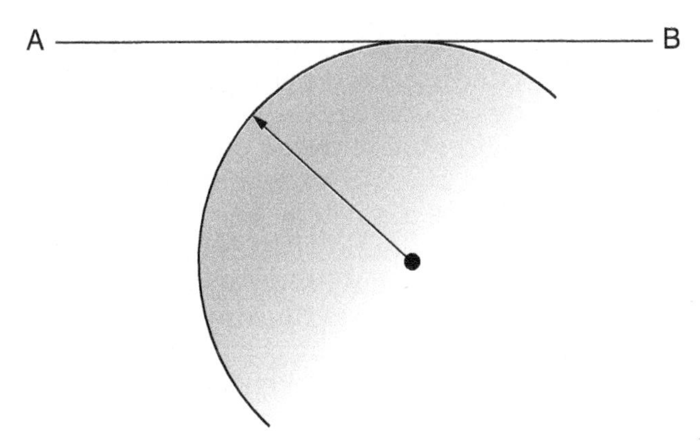

Figure 22 Line AB.

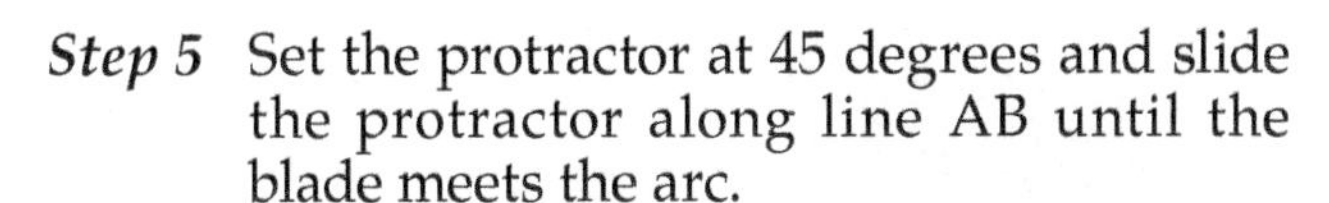

Step 5 Set the protractor at 45 degrees and slide the protractor along line AB until the blade meets the arc.

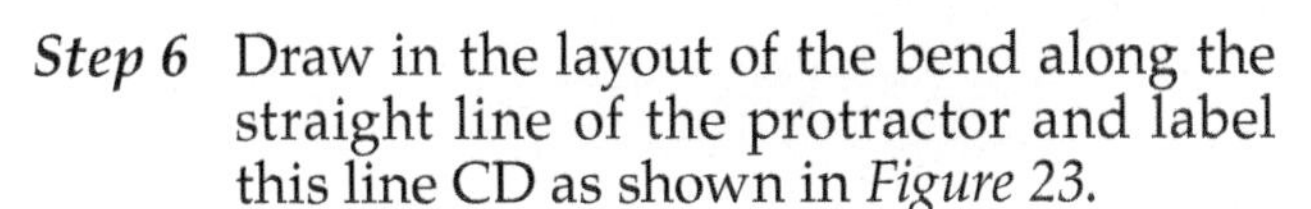

Step 6 Draw in the layout of the bend along the straight line of the protractor and label this line CD as shown in *Figure 23*.

Step 7 Measure $2\frac{1}{4}$" ($\frac{1}{2}$ the OD) at right angles to lines AB and CD to draw the center lines (*Figure 24*).

Step 8 Draw extension lines at right angles from the center of the bend on each center line and another extension line at each end of the bend. *Figure 25* shows the length of the bend (LB).

3.3.0 Methods of Pipe Bending

There are different kinds of benders used to bend either cold or hot pipe. Bending cold pipe has several advantages. It requires less time because no preheating or special handling is needed. Also, cold bending does not anneal the pipe and soften it. It produces a stiffer, more rigid pipe. Pipes can be cold bent using either hand-bending or machine-bending methods.

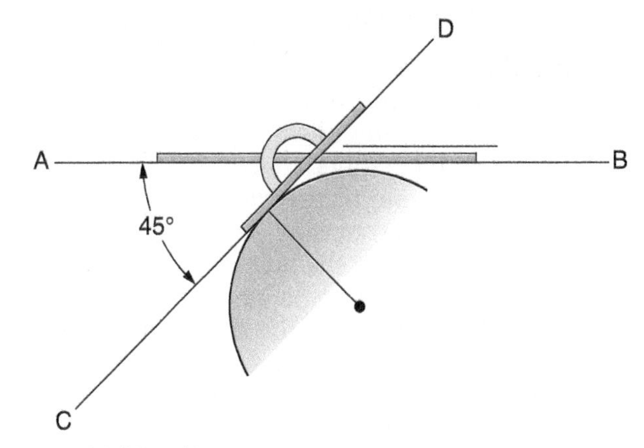

Figure 23 Line CD.

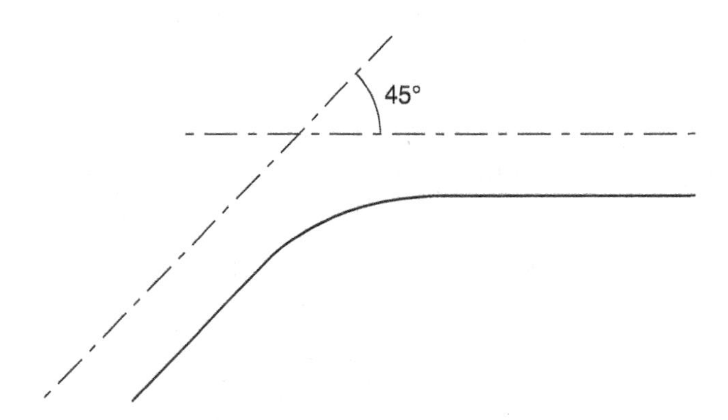

Figure 24 Center lines.

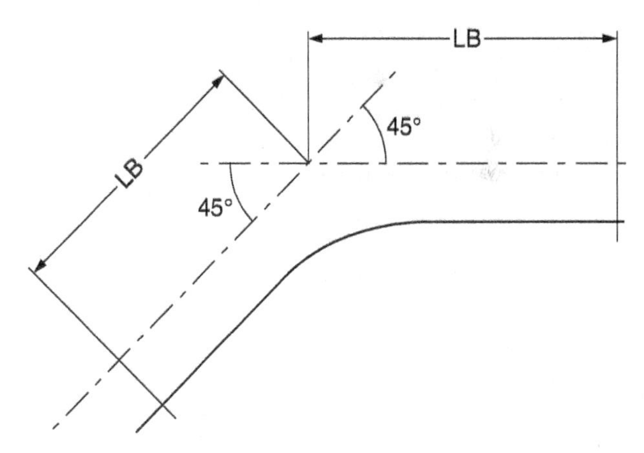

Figure 25 Layout of LB.

It is sometimes necessary to heat thick wall piping and tubing before bending. Hot-bending requires less force than cold-bending, but heating does soften the metal. When hot-bending, extra steps must be taken to prevent the pipe walls from collapsing.

Pipefitters use two methods to bend pipe: hydraulic and manual bending.

3.3.1 Hydraulic Bending

The hydraulic bender is a power-driven tool that uses a ram that fits into the bending frame. The hydraulic bender is one of the easiest benders to use. Most hydraulic benders have a bending shoe that fits over the head of the ram. Different size pipe requires a different size bending shoe and swivel shoes that fit into the holes for the nominal pipe size to be bent. An angle gauge is connected to the swivel shoe to indicate the angle of the bend.

To use a hydraulic pipe bender, the pipe must be inserted at the proper position where the bend is needed. Make sure your hands are away from the hydraulic ram then activate the ram until the stroke completes the bend. Release the hydraulic ram and remove the pipe. *Figure 26* shows a typical hydraulic bender. Follow manufacturer's instructions on these and any other machines.

Some jobs require a minimal amount of pipe bending. In these cases, manual or light-duty hydraulic benders are the first choice. Most light-duty, highly portable hydraulic benders utilize a compact hand pump to provide the hydraulic pressure, although electric hydraulic pumps are available that can increase speed and productivity. Other projects may require hundreds, or even thousands, of bends. For projects requiring this level of production, heavy-duty hydraulic or electric benders are the tool of choice; they offer much higher levels of production, improved consistency for repetitive bends, and reduced operator fatigue.

MODEL 3 TUBING BENDER

MODEL 3 TUBING BENDER (SIDE VIEW)

Figure 26 Hydraulic and electric benders.

3.3.2 Manual Bending

Manual benders are normally used on electromagnetic tubing (EMT), electrical conduit, and 1" or smaller rigid pipe due to the manual effort required to work with larger sizes. Typical benders feature a foot pedal to provide the necessary leverage for bending; no other equipment is required. Hickey benders are designed exclusively for bending rigid pipe with the pipe placed in a sturdy vise. They are also made in different sizes to fit standard pipe diameters. The head of each type has an internal radius that is shaped to fit the contour of the pipe with an arm or hook to pull the pipe around the radius during bending. Manual benders are generally constructed of aluminum or steel, while Hickey benders are usually constructed of steel or iron. *Figure 27* illustrates both types of benders.

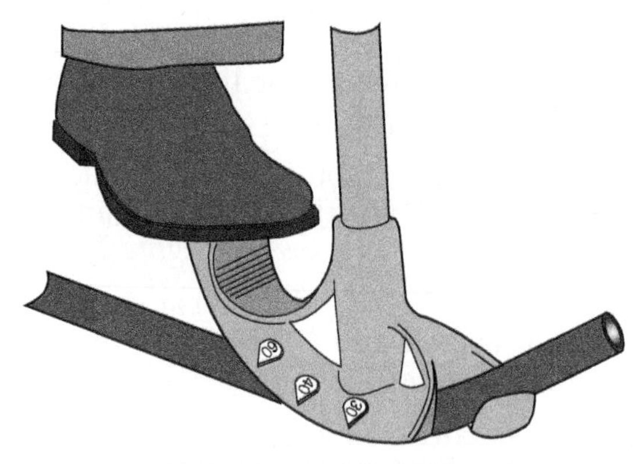

THIN-WALL CONDUIT BENDER

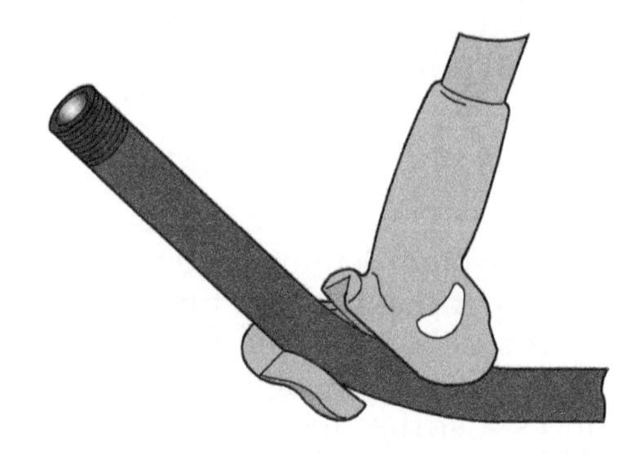

PIPE AND HEAVY-WALL CONDUIT BENDER

Figure 27 Hand benders.

The following procedure explains how to use a Hickey bender to bend a 1" pipe that is 12" long to a 90-degree bend, starting 8" from the end-to-center after the bend is made:

Step 1 Measure from the end of the pipe 8" and mark the pipe at this point. This is measurement A (*Figure 28*).

Step 2 Calculate the radius of the pipe.

$$R = 1" \times 5 = 5"$$

Step 3 Measure back $5\frac{3}{4}$" along the pipe from measurement A and label the radius on the pipe as B (*Figure 29*). This is the point where the bend or arc begins.

Step 4 Calculate the length of the bend (LB), where LB = 9". This is the distance from B past A to C. *Figure 30* shows measurement BC.

Step 5 Place the pipe in a vise at point B.

Step 6 Place the Hickey bender at point C.

Step 7 Bend the pipe 90 degrees.

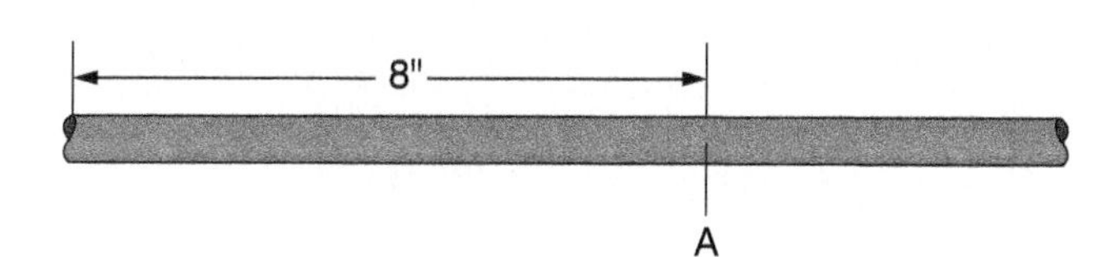

Figure 28 Measurement A.

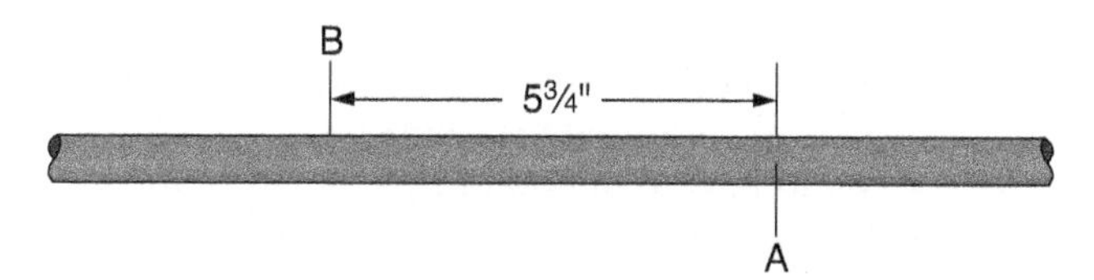

Figure 29 Marking radius on pipe.

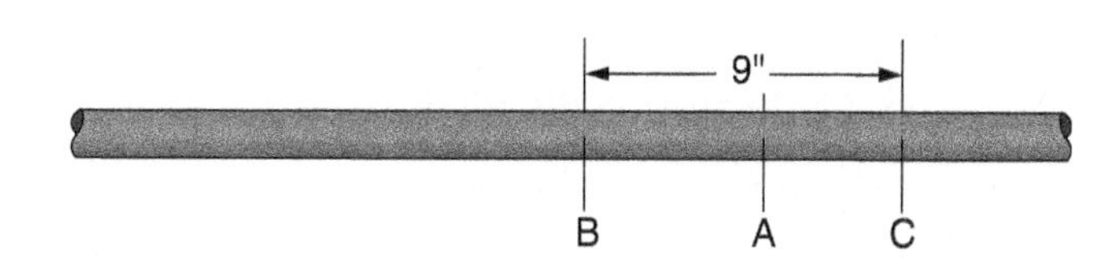

Figure 30 Measurement BC.

3.0.0 Section Review

1. To prevent the pipe from overstretching or collapsing, the radius of the bend should be equal to the diameter multiplied by _____.

   a. two
   b. five
   c. seven
   d. ten

2. Once the radius and the angle of the bend are known, the needed layout can be determined, as well as the _____.

   a. cost of fabrication
   b. tangent of the arc
   c. length of pipe needed
   d. bending force required

3. To use a Hickey bender, you will need _____.

   a. thin tubing less than $\frac{1}{2}$" diameter
   b. a torch to heat the pipe
   c. a sledgehammer
   d. a sturdy vise

SECTION FOUR

4.0.0 GLASS-LINED PIPING

Objective

Explain how to remove, install, and maintain glass-lined piping systems.

a. Explain how to remove and install sections of glass-lined piping systems.
b. Explain how to maintain glass-lined piping systems.

Glass-lined pipe refers to cast iron, carbon steel, stainless steel, and alloy pipe that is lined with a specified grade and thickness of glass. Carbon steel is the most widely used for glass lining because of its relatively low cost, its availability, and its ease of fabricating and glassing. Glass can only be applied to smooth, accessible surfaces. Machined surfaces cannot be glass lined. Glass-lined piping is widely used in many chemical and food-processing industries because of the following advantages:

- *Cleanliness* — Glass does not contaminate products, and it protects the color and purity of products.
- *Smoothness* — Glass lining minimizes friction, thus reducing wear on agitating equipment. The smoothness of the glass also makes it resistant to viscous or sticky product buildup.
- *Superior vacuum service* — Although glass linings are comparable in pressure and temperature ratings to other linings, the tight bond between the glass and steel provides superior vacuum service.
- *Corrosion resistance* — With the exception of hydrofluoric acid, glass resists corrosion in most acid solutions. It also resists corrosion in alkali solutions at moderate temperatures.

The disadvantages of glass piping are that it is costly and very fragile, plus it can be very troublesome to work with.

Three of the major suppliers of glass-lined pipe in North America include Pfaudler, De Dietrich, and The Ceramic Coating Company. Engineers at these companies can recommend and specify glass formulations and thicknesses based on the way it will be used. These companies also offer specific training on the installation and repair of glass-lined piping and equipment.

4.1.0 Removing and Installing Glass-Lined Piping

Follow these steps to remove and install a section of glass-lined piping:

Step 1 Drain the pipe section to be removed.

Step 2 Valve out, tag, and lock the pipe section to be removed.

Step 3 Support the pipe section.

Step 4 Loosen and remove the nuts and bolts from the first flange joint.

Step 5 Remove the old gasket from the first flange joint and record the size and type of gasket before discarding it.

Step 6 Loosen and remove the nuts and bolts from the second flange joint.

Step 7 Remove the old gasket from the second flange joint and record the size and type of the gasket before discarding it.

Step 8 Disconnect the pipe section from all pipe supports present.

Step 9 Remove the old pipe section from the pipeline.

CAUTION

Never strike glass-lined pipe and fittings with tools or other hard objects because damage can result. Mechanical stresses caused by unequal or excessive bolt torque, springing or pulling the pipe into position, or clamping the pipe too tightly (which prevents uniform thermal expansion) can damage the glass.

Step 10 Remove the old pipe section from the restraining device.

Step 11 Obtain the new pipe section.

Step 12 Install the new pipe section into the restraining device.

Step 13 Obtain the new gaskets.

NOTE

Self-centering Teflon® gaskets are often recommended. However, check specifications for the product in use to ensure the appropriate gasket is used.

Step 14 Depending on application and/or specification, apply anti-seize compound to the hardware.

Step 15 Ensure that the gaskets match the size and type of the original gasket exactly, as recorded in *Steps 5* and *7*. Do not use gaskets that are not of the correct size and type. Doing so can result in leaks.

Step 16 Ensure that any foreign material on the gasket, such as grease, product, or adhesive tape, is removed before installing the gasket.

Step 17 Install the pipe into position, but do not make the flange joint connections.

CAUTION

Never pull or spring a pipe section into place. This could cause the pipe to warp, resulting in an improper fit. Likewise, never weld to the exterior of a glass-lined pipe, because the heat from welding may separate the glass lining from the pipe.

Step 18 Ensure that all joints are on the same center line, with the flanges nearly touching (normally within $\frac{1}{4}$" and never more than $\frac{1}{2}$" apart).

Step 19 Install the pipe section into the pipe hangers.

Step 20 Ensure that the clamping ring used in the pipe hanger is loose enough around the outside surface of the pipe to allow for the thermal stress.

NOTE

On vertical runs, use riser clamps above or below floors. Where possible, install the clamp below a joint. Do not support the piping too rigidly. Allow for some freedom of movement to relieve thermal stress.

Step 21 Ensure that the pipe supports are positioned at 10' intervals along the length of the pipe to keep the joints from sagging.

Step 22 Measure the distance between the flanges at several points around the circumference and adjust the piping or shim the gaskets to correct misalignment between the two flanges.

Step 23 Check the alignment of the flanges.

Step 24 Install the gaskets into position in the flange joints while ensuring the gaskets are kept dry during installation. Depending on the material, wet gaskets may slip under pressure.

Step 25 Ensure that the position of the gasket allows the entire flange seating area to be covered by the gasket.

Step 26 Ensure that all joints, including the mating flanges, are parallel and have enough freedom of movement to permit uniform tightening, which helps prevent stress when the flange joints are connected.

Step 27 Install the nuts and bolts into the first flange joint, tightening the bolts by hand at first.

Step 28 Tighten two bolts that are 180 degrees apart, then advance in the bolt-tightening sequence shown in *Figure 31*. Ensure that the bolt is held in place when the nut is tightened. To prevent twisting the gasket out of alignment, be sure not to rotate the bolts. Always use a torque wrench to tighten the bolts.

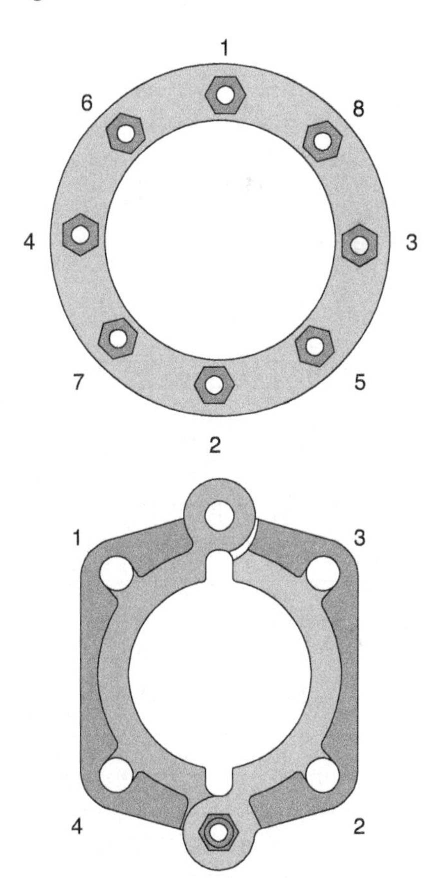

Figure 31 Bolt-tightening sequence.

Step 29 Tighten the bolts to 25 percent of the recommended torque value, using a torque wrench. *Table 4* lists some recommended torque values for several types of steel hardware. *Table 5* provides common torque values for graded steel bolts.

NOTE

Ensure that nuts, bolts, and lubrication comply with the job specifications.

Step 30 Continue to tighten the bolts in increments of 25 percent of the recommended torque value, using the crossover tightening method, until the recommended torque value is reached.

Step 31 Install the nuts and bolts onto the second flange joint, tightening the bolts by hand at first.

Step 32 Repeat *Steps 28* through *30* to tighten the bolts to the correct torque.

Step 33 Retighten all bolts and clamps after one operating cycle or after at least 24 hours at rest.

Step 34 Remove all locks and tags.

Step 35 Check the flanges for leaks once the pipeline is back online and full of product.

Table 4 Torque Specifications

TORQUE IN FOOT POUNDS				
Fastener Size	**Threads Per Inch**	**Mild Steel**	**Stainless Steel 18-8**	**Alloy Steel**
1/4	20	4	6	8
5/16	18	8	11	16
3/8	16	12	18	24
7/16	14	20	32	40
1/2	13	30	43	60
5/8	11	60	92	120
3/4	10	100	128	200
7/8	9	160	180	320
1	8	245	285	490

Table 5 Torque Specifications for Graded Steel Hardware

SUGGESTED TORQUE VALUES FOR GRADED STEEL BOLTS					
Grade		**SAE 1 or 2**	**SAE 5**	**SAE 6**	**SAE 8**
Tensile Strength		**64,000 PSI**	**105,000 PSI**	**130,000 PSI**	**150,000 PSI**
Grade Mark					
Bolt Diameter	**Threads Per Inch**	**Foot Pounds Torque**			
1/4	20	5	7	10	10
5/16	18	9	14	19	22
3/8	16	15	25	34	37
7/16	14	24	40	55	60
1/2	13	37	60	85	92
9/16	12	53	88	120	132
5/8	11	74	120	169	180
3/4	10	120	200	280	296
7/8	9	190	302	440	473
1	8	282	466	660	714

4.2.0 Performing Preventive Maintenance on Glass-Lined Piping

Perform the following preventive maintenance procedures during routine inspections of the pipeline:

- Regularly check bolt torque values at all joints.
- Check all joints for leaks. If a joint leaks, do the following:
  - Confirm the bolts are tightened evenly.
  - Replace the gasket if necessary.
  - Correct any possible cause of strain at the joint.
- Watch for acid spills on the outside of the pipe or fitting. Hose off any spills immediately. Acid spills may react with the steel, causing chipping or fish-scaling of the glass on the inside.
- Check periodically for any external damage that may have broken the glass on the inside. Be sure to replace any damaged part.
- Inspect all gasketed joints at regular intervals for weeping, leaks, or deterioration. Because a gasket is subjected to temperature and pressure cycling, material relaxation, and corrosive environments, an initial three-month inspection is recommended. Bolt torque should also be checked at this time.

4.0.0 Section Review

1. Although you should always check the specifications to determine what type of gasket to use, the type often recommended for glass-lined piping is a _____.

   a. self-centering Teflon® gasket
   b. natural rubber gasket
   c. cork gasket
   d. duck and rubber gasket

2. After installation, gasket joints should be inspected within _____.

   a. two weeks
   b. one month
   c. three months
   d. six months

SECTION FIVE

5.0.0 CRIMPED PIPE AND TUBING JOINTS

Objective

Identify and describe how to assemble Lokring® and other crimped joints on tubing.

a. Identify and explain how to create Lokring® joints.
b. Identify and explain how to create crimped joints on copper tubing.

Performance Task

6. Fabricate a Lokring® or a crimped copper-tubing joint.

Crimped joints offer significant advantages in many applications. They can be used on small or large projects, where flammable or explosive conditions exist that prohibit the use of an open flame, and even under water. Several manufacturers have developed patented systems for joining tubing using this family of fittings.

It is common for crimped fittings to be referred to as *compression fittings*. The name is not inaccurate, since these fittings do rely on compression to assemble a joint. However, they should not be confused with standard compression joints presented earlier in this module. Standard compression joints rely on a compression nut to compress a ferrule against the tubing. Crimped joints require a tool to apply significant pressure around the circumference of the fitting to seal the joint. Crimped fittings may also be referred to as *press-to-connect fittings* or simply *press fittings*.

> **NOTE**
> Press-to-connect fittings should not be confused with push-to-connect fittings, which require no tools for assembly.

One popular and unique product line was developed by Lokring®. This versatile system can be used on a variety of tubing materials with fittings available in brass, stainless steel, carbon steel, and a copper-nickel alloy. Many other manufacturers have developed a line of different fittings and tools for copper tubing that are used on domestic water lines, refrigerant piping, and similar applications where copper is commonly used.

5.1.0 Lokring® Connection System

The Lokring® connection system relies on a precision-machined metallic fitting body and two swaging rings that are factory-assembled onto either end of the fitting body. The patented design has no O-ring seals, threads, or ferrules, and requires no heating or threading. *Figure 32* shows the Lokring® fitting components. Although this system is used extensively for smaller sizes of tubing, fittings and crimping tools that can accommodate tubing and pipe sizes up to 4" are available.

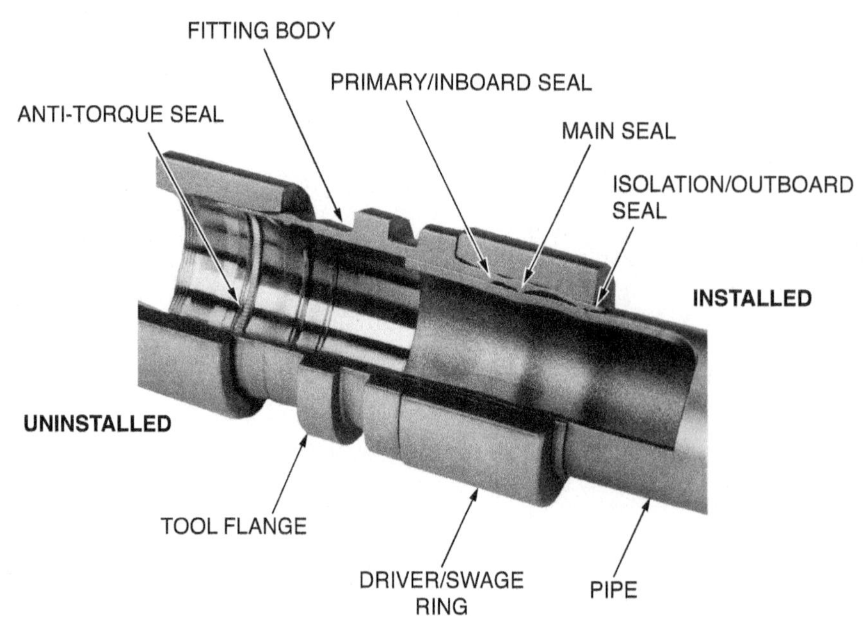

Figure 32 Lokring® fitting components.

Tubing is inserted into the ends of the fitting body, and the hydraulic crimping tool is engaged onto the fitting. As pressure is applied around the circumference of the fitting, the tool advances each swage ring axially over the fitting body until the leading edge of the swage ring contacts the fitting-body tool flange. *Figure 33* shows the movement of the Lokring® fitting.

As the swaging rings pass over the tapered section of the fitting body, they move over a slightly increased diameter, compressing the fitting body down onto the tubing surface. The compression causes the sealing lands of the fitting body to grip and seal on the tubing surface. This results in a permanent, gas-tight, metal-to-metal seal between the pipe surface and the fitting body.

1. GAUGE AND MARK PIPE ENDS.
2. POSITION FITTINGS OVER MARKS.

3. ENGAGE LOKTOOL® HEAD.
4. MAKE UP SWAGE RING.

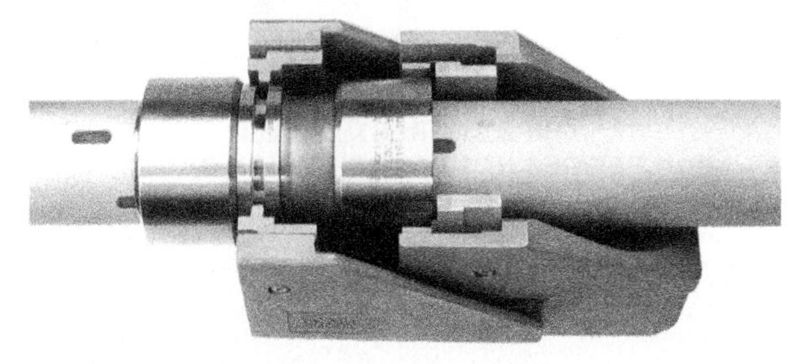

5. REMOVE LOKTOOL® HEAD. INSPECT COMPLETED JOB.

Figure 33 Movement of Lokring® fitting.

Lokring® fittings require a Loktool® installation tool. The Loktool® system is designed around a hydraulic pump operating at a maximum pressure of 10,000 psi. Various attachments and tools are available for use with the many fitting sizes and construction materials. The following general safety instructions apply to the use of high-pressure hydraulic equipment:

- Verify that all hydraulic connection hoses and fittings are rated for the 10,000-psi maximum pressure developed by Lokring® hydraulic pumps.
- Do not overtighten the threaded hydraulic hose connections.
- Purge the air from the hydraulic hoses before use.
- Use only hydraulic pumps, hoses, and fittings supplied or recommended by the equipment manufacturer.
- Ensure that the hydraulic hose is fully engaged on the fitting before actuating the hydraulic pressure.
- Release the hydraulic power when the fitting swaging rings seat against the fitting center stop.
- Do not drop heavy objects on the hydraulic hoses.
- Do not use hydraulic hoses that are kinked or sharply bent.
- Keep hydraulic equipment and hoses away from flames and heat.
- Always provide adequate clearance between hoses and couplers to avoid moving objects that may cause abrasions or cuts.

Remember that there is no better source of information than the manufacturer's documentation. Always follow the operating and installation instructions available from the manufacturer of the product in use.

5.1.1 Installing Lokring® Fittings

Refer to the Lokring® appropriate installation document. Installation guidance, such as *LP-101* or *LP-105 Installation Procedure for Lokring® Fittings* must be followed when installing Lokring® fittings.

Follow these steps to install Lokring® fittings using the dry method:

Step 1 Collect the following tools and materials:
- 120-grit aluminum oxide sand cloth as well as a 60-grit cloth in case there are any deep pits or scratches on the tubing
- Loctite Compound PST 567, an optional pipe sealant containing Teflon® (This compound may be applied as an aid to seal where the pipe surface is somewhat poor, and its use is suggested for all applications using Schedule 10 pipe or thinner.)
- A squaring and facing tool, also optional, similar to the one shown in *Figure 34* (Although this tool is not always required, a square and smooth face on the tubing or pipe is required. On some projects, the specifications may require the use of this tool.)
- Tubing of the desired size and material
- Lokring® fittings compatible with the tubing material

Step 2 Cut the tubing to the desired length using a tubing cutter that will provide a square cut and will not flatten or deform the pipe end. A square face with a maximum angle of 5 degrees from square is required.

Step 3 Remove the inside and outside burrs. A pipe squaring and facing tool may be used to ensure a proper, burr-free face (*Figure 34*).

Step 4 Smooth the face and ends of the tubing using the 120-grit aluminum oxide cloth. Clean the end a minimum of 1½ diameters back from the end of the tubing. For example, when working with ½" tubing, clean an area ¾" long.

Figure 34 Facing and squaring tool for tubing.

CAUTION

Always sand around the circumference of the pipe. Sanding parallel to the axis of the pipe can result in undesirable flat spots on the pipe end surface and scratches that can cause leaks. Extra care must be taken to avoid gouging or flattening the pipe surface during the preparation process. If deep pits or scratches persist, use 60-grit aluminum oxide cloth to remove them, followed by the 120-grit cloth for final finish.

Step 5 Inspect the cleaned pipe end in a well-lit area and wipe the pipe end clean to remove abrasive residue.

Step 6. Obtain a Lokring® NO GO gauge.

Step 7 Attempt to pass the sanded ends of the tubing through the end of the NO GO gauge at two points, 90 degrees apart. Try it once, then rotate the tubing 90 degrees and try again. If the tubing passes through at either point, it is likely deformed and should not be used. *Figure 35* shows the use of a NO GO gauge.

NOTE

You can use the NO GO gauge in various places along a length of tubing to see if the entire piece or a section of it has been damaged beyond use for Lokring® applications.

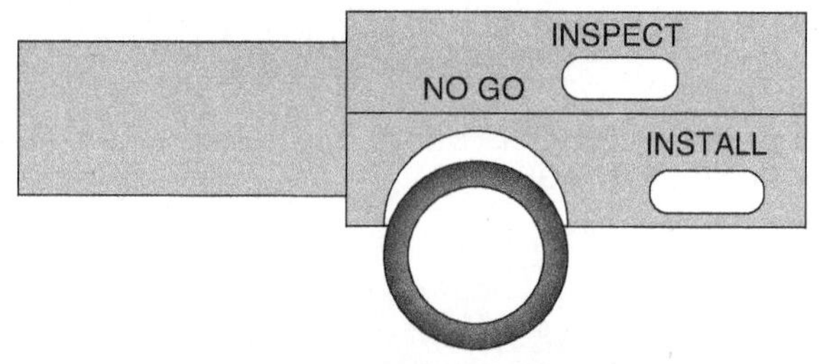

PIPE CORRECTLY SIZED
(PIPE DOES NOT PASS THROUGH GAUGE)

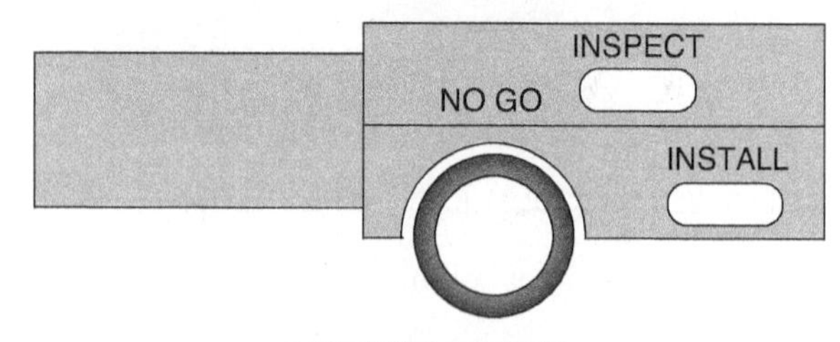

PIPE UNDERSIZED
(PIPE PASSES THROUGH GAUGE)

USED WITH PERMISSION OF LOKRING CORPORATION.

Figure 35 Testing tubing ends using a NO GO gauge.

Step 8 Slide the larger (hex) end of the NO GO gauge over the end of the tubing until the gauge bottoms out. If the tube does not fit into the NO GO gauge, it is oversized or deformed and should not be used.

Step 9 Mark the tubing through the slots on the gauge using a permanent marker as shown in *Figure 36*. Both ends of the tubing being joined must be marked so that the quality of the joint can be verified upon completion.

Step 10 Apply Loctite PST 567 pipe sealant (*Figure 37*). This is an optional step unless specified for the task or the tubing surface has some damage that cannot be removed by sanding. Apply it to the end of the tubing, just back from the end and up to the inspection mark. Care must be taken to prevent excess sealant material from entering the tubing and potentially contaminating the system; apply the sealant only within the shaded regions as shown in *Figure 37*.

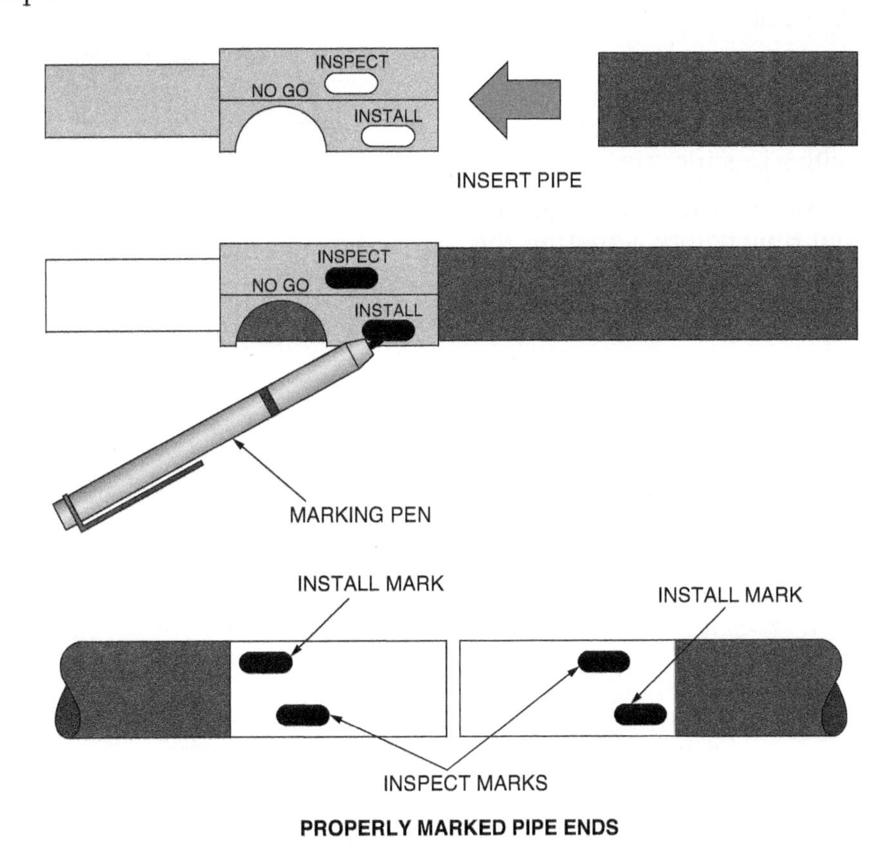

Figure 36 Marking tubing.

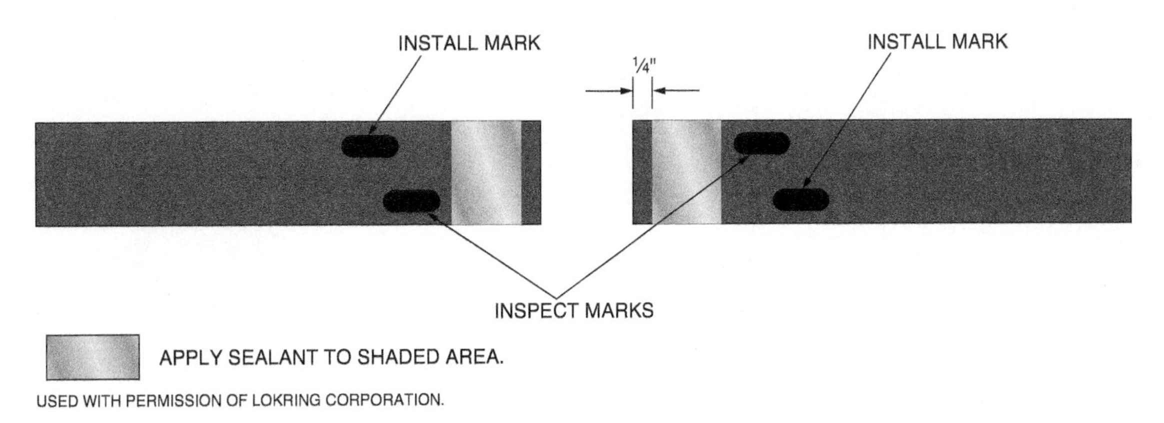

Figure 37 Application of anaerobic sealant.

Step 11 Place the Lokring® fitting onto the end of one section of tubing.

Step 12 Align the second piece of tubing with the first and butt the ends inside the fitting.

Step 13 Position the Lokring® fitting centrally over the tubing ends so that the inspection marks on both are covered while part of the installation marks remain visible. *Figure 38* shows a proper alignment and fit-up of the tubing prior to assembly.

Step 14 Set up the Loktool® system components (*Figure 39*).

Step 15 Install the specified Loktool® head on the Lokring® fitting (*Figure 40*). Each fitting is associated with a specific crimping tool.

Step 16 Check the position of the fitting on the tubing a final time before actuating the crimping tool.

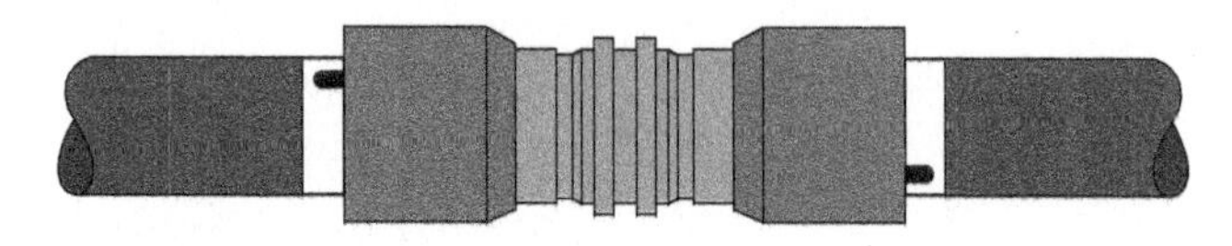

USED WITH PERMISSION OF LOKRING CORPORATION.

Figure 38 Proper alignment and fit-up.

Step 17 Actuate the Loktool® until the moving jaw forces the fitting against the external center stop. *Figure 41* shows the first end being made up. The fitting makeup is complete when the drive ring contacts the tool flange.

Step 18 Remove the Loktool® from the installed fitting end.

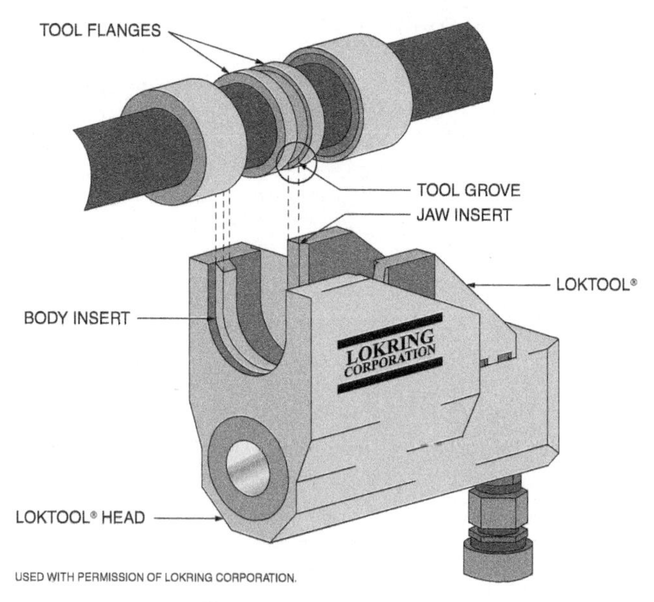

USED WITH PERMISSION OF LOKRING CORPORATION.

Figure 40 Loktool® positioning.

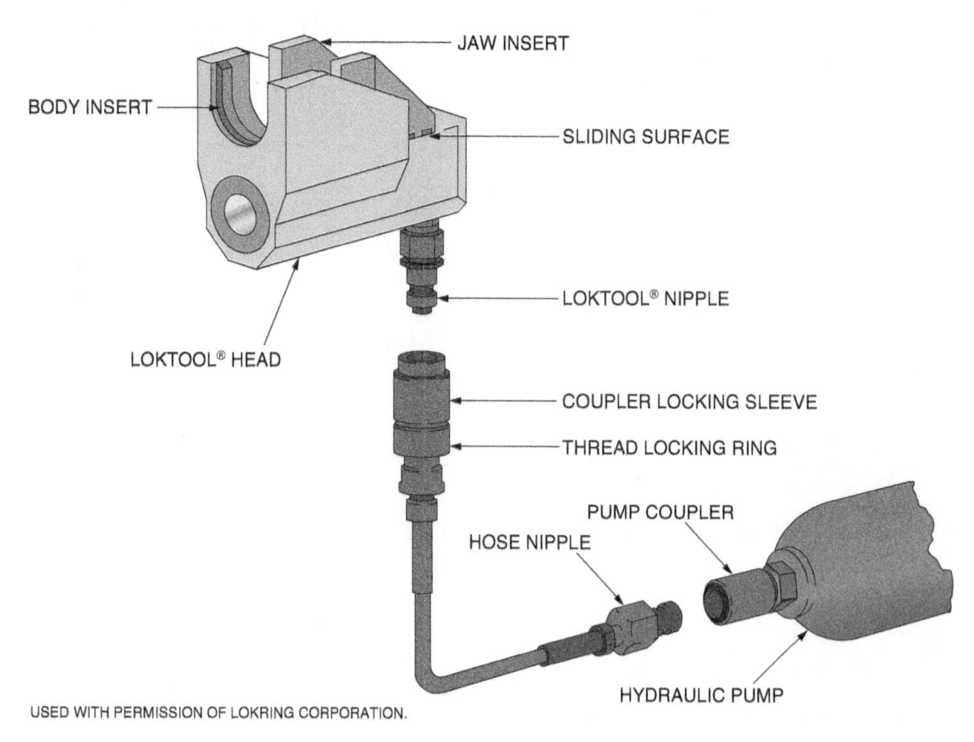

USED WITH PERMISSION OF LOKRING CORPORATION.

Figure 39 Loktool® system components.

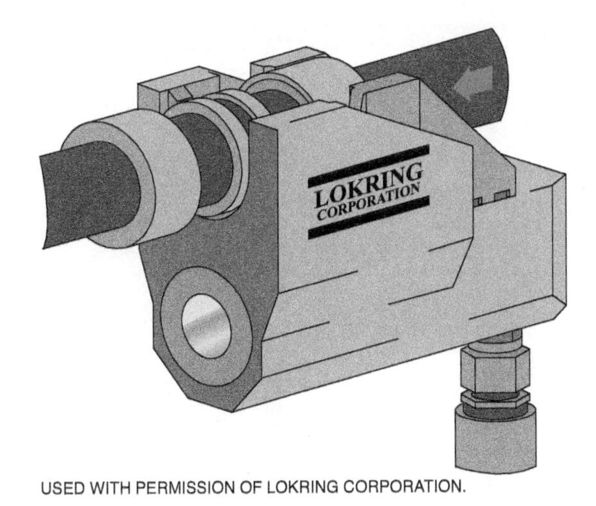

USED WITH PERMISSION OF LOKRING CORPORATION.

Figure 41 First-end makeup.

Step 19 Inspect the fitting. The outboard *Inspect* mark should be completely uncovered, but the inboard *Install* mark should be partially covered by the trailing edge of the fitting body. Check the drive ring—it should contact the tool plane. Finally, ensure the end of the fitting body protrudes beyond the drive ring. *Figure 42* shows an acceptable first-end installation.

Step 20 Before crimping the opposite end, ensure that the tubes remain properly aligned.

Step 21 Repeat *Step 15* through *Step 19* to crimp the Lokring® fitting on the adjoining tubing.

Step 22 Examine the fitting and marks to ensure that the fitting has been properly installed.

5.2.0 Crimped Copper Fittings

Crimped copper fittings rely on a special tool to crimp a seal-equipped fitting into place around the end of the tubing. This is one feature that separates them from the Lokring® fittings presented previously—a seal is used to prevent leakage. The seal is constructed of materials like EPDM (ethylene propylene diene monomer), which is also used for O-rings and seals in many other applications.

Figure 43 shows some common crimped fittings. The required seal can be seen inside the opening of several fittings. Standard fittings like those shown here are designed for operating pressures up to 200 psi and 250°F. Other fittings can be chosen for applications operating at a higher pressure, such as refrigerant piping. Fittings chosen for that purpose can typically handle pressures up to 700 psi.

Always follow the manufacturer's instructions for the fittings in use. Although there are similarities among the products, the installation requirements may differ and are always subject to change. The following is a general procedure for installing crimped fittings:

Step 1 Examine the copper tubing and fittings for cracks, distortion, or similar defects. Do not use tubing or fittings with defects. It is also best to avoid placing the sealing area of the connector across text that is etched or stamped into the surface of the tubing; printed text is not a concern.

Step 2 Cut the tubing using a wheeled tubing cutter or other approved cutting tool. Cut the tubing square to permit proper joining inside the fittings.

Step 3 Ream the tubing using an appropriate tool and ensure the face is square and free of burrs. Clean the ends of the tubing to remove any grease or other substances but do *not* sand the tube surface.

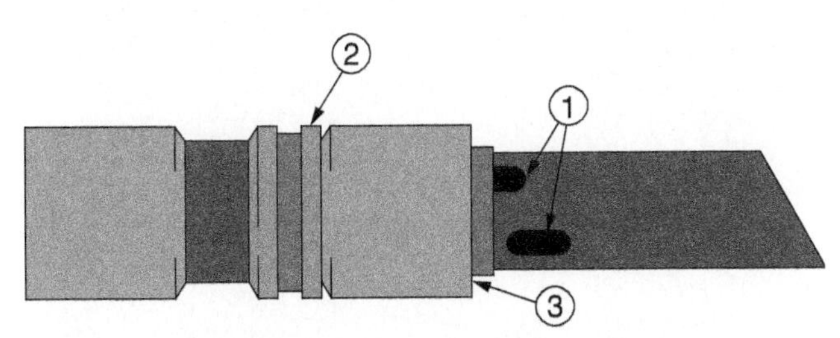

USED WITH PERMISSION OF LOKRING CORPORATION.

Figure 42 Complete first-end installation.

Figure 43 Copper crimped fittings.

Crimped Fittings for Stainless Steel

Although crimped fittings (specifically, the type that relies on an O-ring seal inside the fitting) are largely associated with copper tubing and its many applications, there are products available for other types of tubing. For example, Victaulic® offers a press-to-fit solution designed for joining Sch 10 stainless steel tubing up to 2" diameter. Several different seal materials are also available depending upon the pipe contents and application requirements. The fittings shown here are rated for pressures from a full vacuum up to 500 psi and for temperatures up to 300°F with the appropriate seal material.

Figure Credit: Courtesy of Victaulic Company

Step 4 Inspect the fitting to ensure it is clean and free of damage. In most cases, no lubricant is applied to the seal other than water.

Step 5 Refer to *Table 6*. Note that this table is an example only and should *not* be consulted for field work. It is important to follow the assembly instructions for the product in use. Locate the appropriate insertion depth for the tubing size.

Step 6 Measure from the end of the tubing and mark the appropriate insertion depth with a permanent marker.

Step 7 Prepare the crimping tool. Unplug or remove the battery from the tool while installing the appropriate jaw set. Follow the tool manufacturer's instructions for tool preparation and use.

Step 8 Insert the tubing into the fitting to the marked depth using a twisting motion.

CAUTION

If the tubing is difficult to insert into the fitting, it is likely due to deformed tubing or burrs remaining on the tube end. Ensure that all burrs are removed and discard any tubing that it is deformed to avoid a failed joint.

Step 9 Open the jaws of the crimping tool and place them around the joint.

Step 10 Ensure that the crimping tool is perpendicular to the tubing, then squeeze and hold the trigger (*Figure 44*) until the crimping cycle is complete.

Table 6 Crimped Fitting Insertion Chart

Nominal Tubing Size (ID)	Insertion Depth
1/2"	11/16"
5/8"	3/4"
3/4"	7/8"
1"	7/8"
1 1/4"	1"
1 1/2"	1 3/8"
2"	1 1/2"

Figure 44 Electric crimping tool.

Step 11 Inspect the joint. Ensure that the tubing has not moved away from the fitting by observing the insertion depth mark.

Step 12 Once both sides of all joints have been crimped, test the system under pressure for leaks per the job specifications or the fitting manufacturer's guidance.

5.0.0 Section Review

1. One unique advantage of the Lokring® joining system is that _____.
   a. no tool is needed to complete the assembly
   b. the fittings do not require a seal
   c. only a small amount of heat is needed to seal the joints
   d. all the fittings are constructed of stainless steel

2. Standard crimped fittings for copper tubing are designed for operating pressures up to _____.
   a. 125 psi
   b. 150 psi
   c. 200 psi
   d. 275 psi

Review Questions

1. Copper tubing can be connected by _____.
   a. flared or compression joints
   b. rod welding
   c. stick welding
   d. oxyacetylene welding

2. The color code for Type L copper tubing is _____.
   a. red
   b. gold
   c. blue
   d. green

3. One reason for using flared joints rather than soldered joints is that they _____.
   a. are cheaper
   b. can be disassembled and reassembled easily
   c. are stronger
   d. are faster to put together

4. The difference between Type K and Type L copper tubing is: _____.
   a. Type K has thinner walls
   b. Type L has thinner walls
   c. Type K is drawn tubing
   d. Type L is a different metal

5. Long flare nuts are preferred over short flare nuts for _____.
   a. potable water
   b. high vibration applications
   c. carbon steel tubing
   d. aluminum tubing

6. The ring that is compressed in a compression fitting is called a _____.
   a. nut
   b. flange
   c. ferrule
   d. washer

7. The main difference between soldering and brazing is the _____.
   a. type of tubing
   b. temperature of the heating procedure
   c. type of filler used
   d. depth of the cup

8. When assembling long runs of tubing, the tubing *must* be _____.
   a. properly supported
   b. made out of shorter pieces
   c. no longer than 20'
   d. allowed to hang loose

9. Soldered joints depend on _____.
   a. heat above 800 degrees
   b. capillary action
   c. quenching
   d. shallow cups

10. Soldering flux cleans and protects the tubing from _____.
    a. penetration
    b. oxidation
    c. adhesion
    d. osmosis

11. Wetting is the process that allows solder to flow into the joint by decreasing _____.
    a. penetration
    b. gravitation
    c. surface tension
    d. oxidation

12. The *best* fluxes for joining copper tubing and copper fittings are _____.
    a. corrosive fluxes
    b. inflammable
    c. lead based
    d. non-corrosive

13. The BAg series of filler metals contain _____.
    a. silver
    b. phosphorus
    c. lead
    d. aluminum

14. BAg-1 and BAg-2 can be hazardous to breath when heated because they also contain _____.
    a. mercury
    b. cadmium
    c. oxygen
    d. iron

15. To calculate the tubing needed for a brazed or soldered pipe, the face-to-face measurement should be added to the _____

    a. tangent
    b. cup depth engagement
    c. solder thickness
    d. length of the fitting

16. Bending pipe allows the process to _____.

    a. cost less
    b. heat up more
    c. be stopped
    d. flow more freely

17. The formula used to determine the bend allowance is _____.

    a. L = R – D × 0.01745
    b. L = R × D × 0.01745
    c. L = R × sinD
    d. L = R × tanD

18. The bend allowance for a 45-degree bend on a 24" radius is _____.

    a. $18\frac{7}{8}$"
    b. $24\frac{7}{8}$"
    c. $36\frac{1}{2}$"
    d. $48\frac{3}{16}$"

19. The point where two straight lines meet to form an angle is the _____.

    a. tangent
    b. arc
    c. vortex
    d. vertex

20. The inside radius of a pipe bend is calculated by taking the center line radius and subtracting _____.

    a. half the OD
    b. twice the OD
    c. the sine of the degree of bend
    d. the OD

21. Glass-lined pipe is used in _____.

    a. foundries
    b. chemical and food processing industries
    c. power plants
    d. mining

22. When working with glass-lined pipe, *never* _____.

    a. use a torque wrench
    b. weld to the outside
    c. use a gasket
    d. empty the pipe

23. With high-pressure hydraulic systems, the connection hoses and fittings *must* be rated for a *minimum* of _____.

    a. 50 degrees Fahrenheit
    b. 180 degrees Fahrenheit
    c. 10,000 psi
    d. 20,000 psi

24. When preparing a Lokring® joint, sand _____.

    a. the inside of the fitting
    b. around the circumference of the pipe
    c. parallel to the axis of the pipe
    d. the edge of the pipe

25. A common material used as a seal inside a crimped copper fitting is _____.

    a. cork
    b. EPDM
    c. brass
    d. twisted fibrous packing

Trade Terms Introduced in This Module

Acetylene: A gas composed of two parts carbon and two parts hydrogen, commonly used in combination with oxygen to cut, weld, and braze steel.

Capillary action: The flow of liquid (solder) into a small space between two surfaces.

Compression coupling: A mechanical fitting that is compressed onto a pipe, tube, or hose.

Cup depth engagement: The distance the tubing penetrates the fitting.

Ferrules: Bushings placed over a tube to tighten it.

Flaring: Increasing the diameter at the end of a pipe or tube.

Inside diameter (ID): Measurement of the inside diameter of the pipe.

Nonferrous metal: A metal that contains no iron and is therefore nonmagnetic.

Outside diameter (OD): Measurement of the outside diameter of the pipe.

Oxidation: A chemical reaction that increases the oxygen content of a compound.

Reamer: A tool used to enlarge, shape, smooth, or otherwise finish a hole.

Solder: An alloy, such as of zinc and copper, or of tin and lead, that is used to join metallic surfaces when melted.

Soldering flux: A chemical substance that aids the flow of solder. Flux removes and prevents the formation of oxides on the pieces to be joined by soldering.

Wetting: Spreading liquid filler metal or flux on a solid base metal.

Additional Resources

This module presents thorough resources for task training. The following reference material is recommended for further study.

Benfield Conduit Bending Manual. 3rd Edition. Jack Benfield. 1993. Cleveland, OH. Penton Media, Inc.

Design Manual for Lined Piping Systems. 2008. Marion, NC. CRANE ChemPharma Flow Solutions.

LP-101 or LP-105 Installation Procedure for Lokring Type 316/316 L Fittings. Lokring Corporation, 396 Hatch Drive, Foster City, CA 94404.

Pipe, Fittings, Valves, Supports, and Fasteners. 2000. United Association of Journeymen.

Figure Credits

Van Sant Enterprises/Trick-Tools.com, Figure 26

LOKRING® Technology Corporation, Figures 32–33, 35-42

Courtesy of Tri Tool, Inc., Figure 34

NIBCO INC., Figure 43

Courtesy of RIDGID Tool Company, Figure 44

Section Review Answer Key

Section 1.0.0

Answer	Section Reference	Objective
1. a	1.1.0	1a
2. a	1.2.1	1b
3. a	1.3.0	1c
4. c	1.4.0	1d

Section 2.0.0

Answer	Section Reference	Objective
1. c	2.1.0	2a
2. c	2.2.0	2b

Section 3.0.0

Answer	Section Reference	Objective
1. b	3.1.1	3a
2. c	3.2.0	3b
3. d	3.3.2	3c

Section 4.0.0

Answer	Section Reference	Objective
1. a	4.1.0	4a
2. c	4.2.0	4b

Section 5.0.0

Answer	Section Reference	Objective
1. b	5.1.0	5a
2. c	5.2.0	5b

NCCER CURRICULA — USER UPDATE

NCCER makes every effort to keep its textbooks up-to-date and free of technical errors. We appreciate your help in this process. If you find an error, a typographical mistake, or an inaccuracy in NCCER's curricula, please fill out this form (or a photocopy), or complete the online form at **www.nccer.org/olf**. Be sure to include the exact module ID number, page number, a detailed description, and your recommended correction. Your input will be brought to the attention of the Authoring Team. Thank you for your assistance.

Instructors – If you have an idea for improving this textbook, or have found that additional materials were necessary to teach this module effectively, please let us know so that we may present your suggestions to the Authoring Team.

NCCER Product Development and Revision

13614 Progress Blvd., Alachua, FL 32615

Email: curriculum@nccer.org
Online: www.nccer.org/olf

❑ Trainee Guide ❑ Lesson Plans ❑ Exam ❑ PowerPoints Other ____________

Craft / Level: Copyright Date:

Module ID Number / Title:

Section Number(s):

Description:

Recommended Correction:

Your Name:

Address:

Email: Phone:

This page is intentionally left blank.

Hot Taps

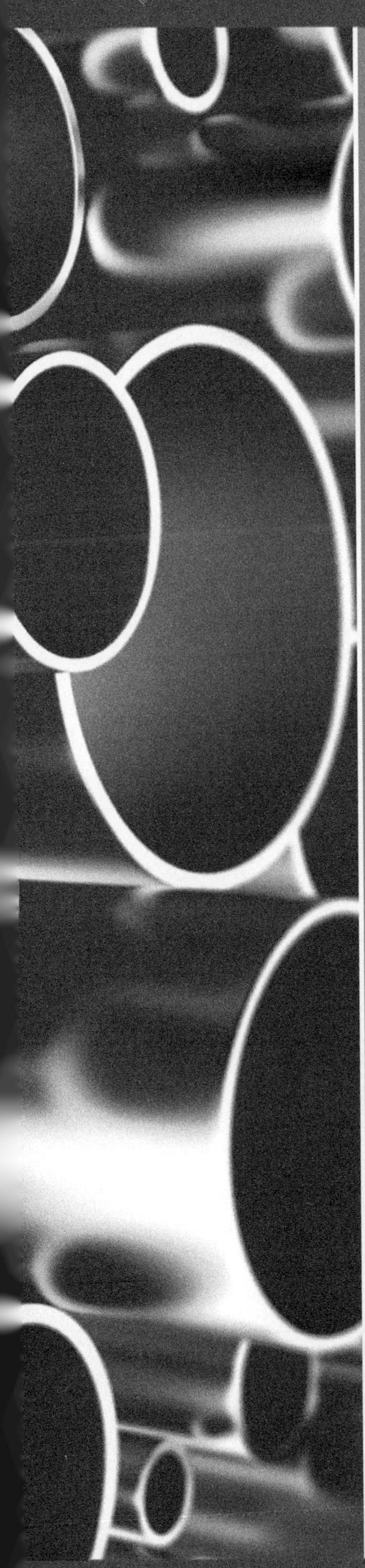

Overview

When it is necessary to connect to pipes that cannot be shut down or emptied, fluid pressure must be contained to prevent leaks. The method for doing this is referred to as hot tapping, and its procedures will vary according to what is being conveyed through the line. In some situations, it may be possible to temporarily stop the flow while connections are made; this is where line stop plugs, pipe freezing, and pipe plugging become important. Hot tapping is not a regular part of a pipefitter's career, but it is important to understand the environmental factors associated with it as well as ways to safely assist any contractors called on to perform it.

Module 08407

Trainees with successful module completions may be eligible for credentialing through the NCCER Registry. To learn more, go to **www.nccer.org** or contact us at 1.888.622.3720. Our website, **www.nccer.org**, has information on the latest product releases and training.

Your feedback is welcome. You may email your comments to **curriculum@nccer.org**, send general comments and inquiries to **info@nccer.org**, or fill in the User Update form at the back of this module.

This information is general in nature and intended for training purposes only. Actual performance of activities described in this manual requires compliance with all applicable operating, service, maintenance, and safety procedures under the direction of qualified personnel. References in this manual to patented or proprietary devices do not constitute a recommendation of their use.

08407
Hot Taps

Objective

Successful completion of this module prepares you to do the following:

1. Explain how to hot tap existing piping.
   a. State potential hazards that must be addressed before a hot tap is made.
   b. Explain how to install the required hot tap fittings.
   c. Explain how to operate hot tap machines.
   d. Identify and describe various line stop plugs.

Performance Tasks

Under supervision, you should be able to do the following:

1. Identify mechanical joint stops and fittings.
2. Identify bolt-weld stops and fittings.
3. Identify split tee fittings.

Trade Terms

Full port
Head clearance
Hot tap
Line stop
Travel distance

Industry Recognized Credentials

If you are training through an NCCER-accredited sponsor, you may be eligible for credentials from NCCER's Registry. The ID number for this module is 08407. Note that this module may have been used in other NCCER curricula and may apply to other level completions. Contact NCCER's Registry at 1.888.622.3720 or go to **www.nccer.org** for more information.

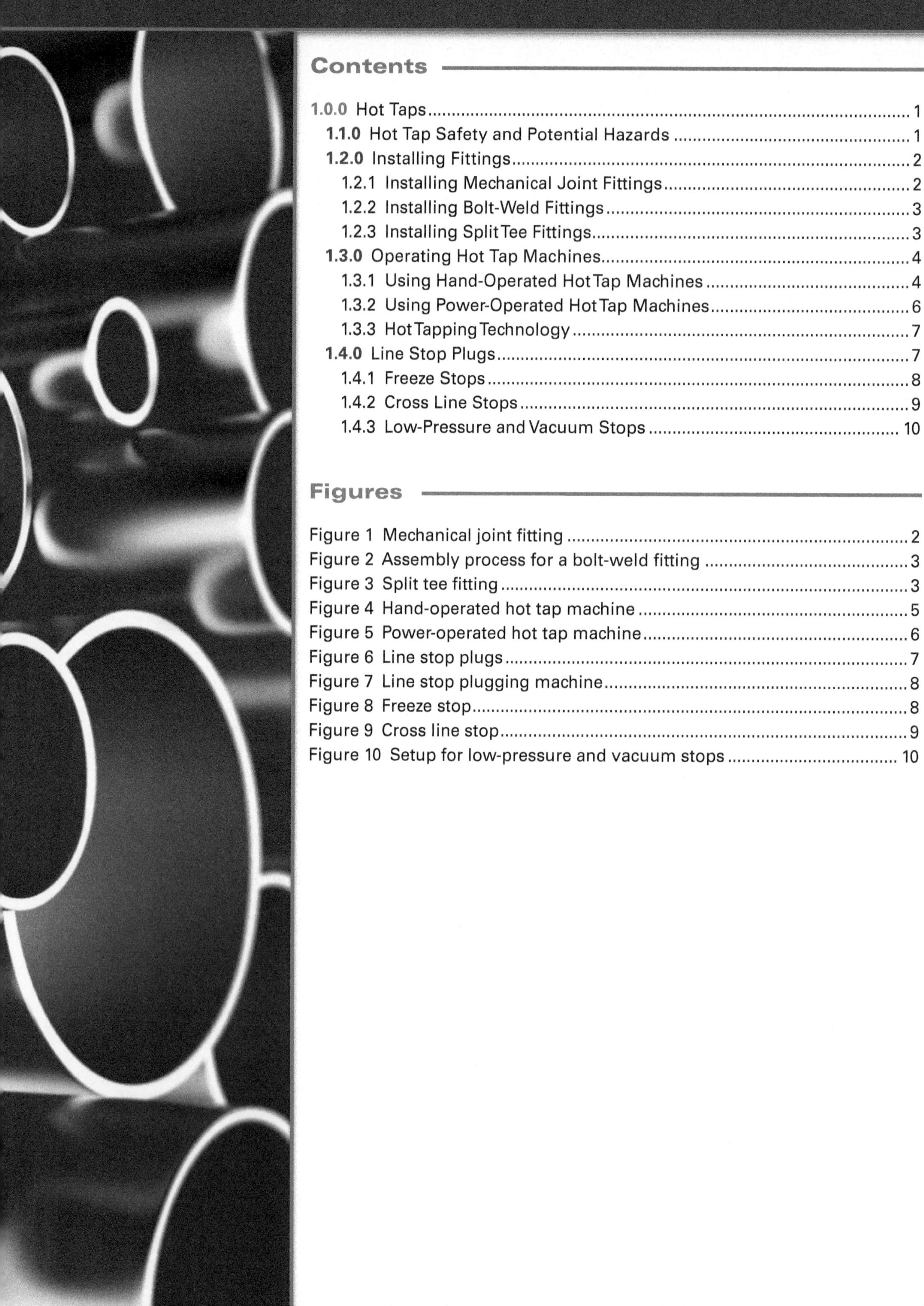

Contents

Figures

SECTION ONE

1.0.0 HOT TAPS

Objective

Explain how to hot tap existing piping.

a. State potential hazards that must be addressed before a hot tap is made.
b. Explain how to install the required hot tap fittings.
c. Explain how to operate hot tap machines.
d. Identify and describe various line stop plugs.

Performance Tasks

1. Identify mechanical joint stops and fittings.
2. Identify bolt-weld stops and fittings.
3. Identify split tee fittings.

Trade Terms

Full port: The maximum internal opening for flow through a valve that matches the ID of the pipe used.

Head clearance: The amount of space needed to install a hot tap machine on a pipe.

Hot tap: To make a safe entry into a pipe or vessel operating at a pressure or vacuum under controlled conditions without losing product.

Line stop: A device used to temporarily contain the flow or pressure of a product inside a pipe while a branch connection is being made.

Travel distance: The distance that the cutting bit moves from the top of the tapping valve to the pipe to be cut.

Hot tap machines are drilling and tapping machines used when a branch connection must be added to an existing pipeline, tank, or vessel that cannot be shut down. A fitting is installed on the pipe or vessel to secure the tap. Usually, the fitting is welded, but in cases in which it cannot be welded, it is bolted into place. After the fitting is installed, it should be tested at the pressures specified by the hot tap manufacturer and system specifications.

Hot tap machines can be used on plastics and different types of metals including carbon steel, cast iron, ductile iron, and certain alloys. Hot tapping procedures vary according to the type of pipe on which the hot tap is to be made. The installation of a fitting and line stop, as well as the ability to make hot taps, are important to advanced pipefitting.

Many pipefitters may never perform a hot tap because outside contractors are usually called in for this purpose. The training that is required for a particular machine, as well as the expense of the machine itself, makes contracting a better choice than in-house tapping. Insurance considerations may make contracting even more attractive. Whether contracting or in-house tapping are preferred at a job site, pipefitters must understand parameters of the task and ways to control the hazards involved.

1.1.0 Hot Tap Safety and Potential Hazards

Hot tap machines contain pressure inside an existing pipeline until a branch connection has been completed. Each machine has a specific temperature and pressure rating that must not be exceeded for safety reasons. These ratings depend on the operating pressure and temperature rating of the pipeline and the branch being installed.

All hot tap machines have a device for drilling into the pipeline. The drill may use flat drills, high-speed drills, or hole saws. Each hole saw is equipped with a pilot drill or bit to start the hole in the pipeline and to keep the cutter square during the operation.

When selecting a hot tap machine, consider:

- The size of the hot tap to be made
- The pressure rating of the valve flange
- The size and thickness of the pipe or vessel
- The material the pipe or vessel is made of
- The pressure and temperature inside the pipe or vessel
- The product inside the pipe or vessel
- The angle of the tap
- Head clearance for the hot tap machine
- Travel distance of the drill bit

The requirement for head clearance for a hot tap on pipe as large as 10" in diameter may be 14' or more. Using a hole saw, the drill bit must travel through the valve and through the wall of the pipe without cutting the far wall.

WARNING!

Improper use of hot tap machines can cause serious personal injury.

For the safe use of hot tap machines:

- Obtain all necessary permits and manufacturer's safety guidelines.
- Choose the proper machine for the job.
- Never use a tool for anything other than its intended use.
- Wear safety glasses or goggles.
- Wear gloves when using tools that require them.
- Wear a welding shield during welding operations.
- Observe welding codes whenever welding is being done.
- Never weld a fitting on a tank in an area where vapors may exist.
- Determine the material flowing through a pipeline before servicing the line.
- Barricade areas according to company policies.
- Follow manufacturer's instructions; special training may be necessary.

1.2.0 Installing Fittings

Fittings are used where single branch pipelines are to be tapped and they must be installed before a hot tap can be made. The basic types are mechanical joint, bolt-weld, and split tee. Proper installation procedures and the observance of welding codes are both important when installing fittings.

Tapping valves are installed between the hot tap machine and the fitting on the pipe. The valve must be a **full port** gate valve so the hot tap cutter will clear all sides of the valve when installed. After installing the valve to the fitting, the valve must be pressure-tested according to plant specifications.

1.2.1 Installing Mechanical Joint Fittings

Mechanical joint fittings support the piping and come with six pieces, forming two halves that are placed around the pipeline where the hot tap is to be made. The fitting has two halves of the body that are bolted together on the pipe, and the glands that complete the fitting are also made in two halves that bolt together. There are specific temperature limitations for mechanical joint fittings, due to the material of the gasket. *Figure 1* shows a mechanical joint fitting.

To install a mechanical joint fitting:

Step 1 Use a nonflammable solvent to remove any excess dirt from the pipeline and mechanical joint fitting.

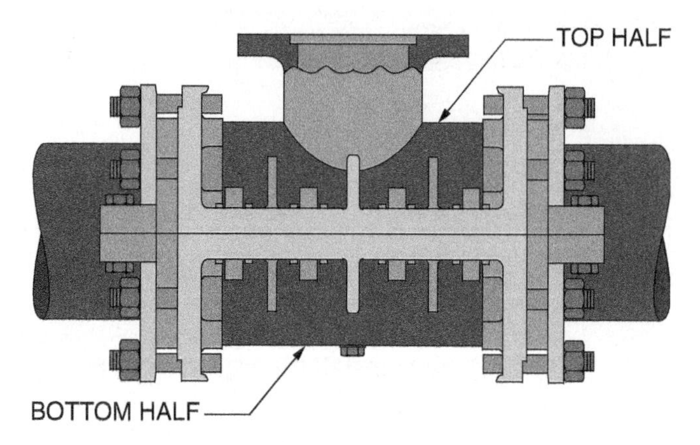

Figure 1 Mechanical joint fitting.

Step 2 Ensure that the pipe is not egg-shaped in case one uses a stopper.

Step 3 Place the top half of the mechanical joint fitting on top of the pipeline where the hot tap is to be drilled.

Step 4 Place the bottom half of the mechanical joint fitting under the pipeline.

Step 5 Fasten the top and bottom halves of the mechanical joint fittings together, using the bolts that are provided with the fitting, leaving about 1/4" between the top and bottom mechanical joint flanges.

Step 6 Slip an oval, rubber-side gasket into the groove between the flanges so that the gasket extends about 1" beyond the end of the mechanical joint fitting.

Step 7 Use a socket wrench to tighten the flange bolts.

Step 8 Use a knife to cut off the ends of the gasket, leaving about 1/8" of the gasket protruding into the socket.

Step 9 Center the mechanical joint fitting on the pipe with the outlet in the proper position.

Step 10 Place end gaskets around the pipe with the tapered end of the gaskets facing into the socket.

Step 11 Press the end gaskets into position.

Step 12 Assemble the glands on the pipe.

Step 13 Insert the bolts.

Step 14 Center the glands against the end gaskets.

Step 15 Tighten the flange bolts evenly to compress the end gaskets against the pipe, using a torque wrench.

> **WARNING!** Follow the appropriate torquing procedures for the type of mechanical joint being installed. Failure to torque the bolts properly could result in serious injury when pressure is applied to the pipe and mechanical joint fitting.

Step 16 Install the appropriate valve on the mechanical joint fitting.

Step 17 Conduct the appropriate pressure test for the mechanical joint fitting and valve assembly.

1.2.2 Installing Bolt-Weld Fittings

Bolt-weld fittings have three major parts: the top half of the fitting, the bottom half of the fitting, and the seal. Bolt-weld fittings are used in a pipeline to facilitate repairs, modifications, or additions and are also used with hot tap machines to make branch connections. They can be used with stopples with or without a temporary bypass. *Figure 2* shows the assembly process for a bolt-weld fitting.

To install a bolt-weld fitting:

Step 1 Use a nonflammable solvent to remove any excess dirt from the pipeline and bolt-weld fitting.

Step 2 Place the seal on the pipeline.

Step 3 Place the top half of the fitting on top of the seal.

Step 4 Place the bottom half of the fitting under the pipeline.

Step 5 Tighten the socket head cap screws to hold the bottom half in place.

Step 6 Fillet-weld the top and bottom halves together and to the pipe if specified.

> **WARNING!** To avoid personal injury, follow the proper welding procedures and codes for the given pipeline.

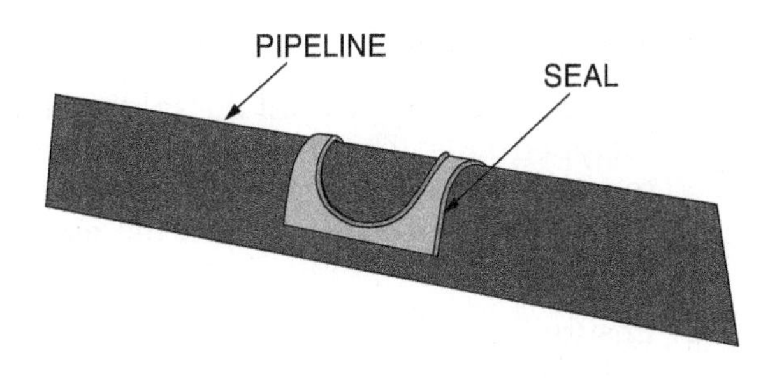

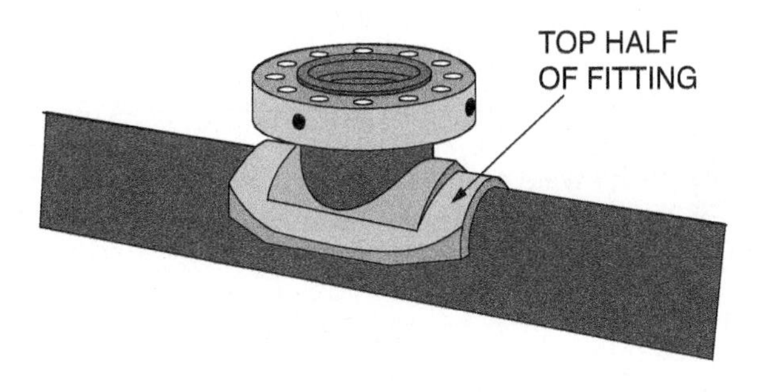

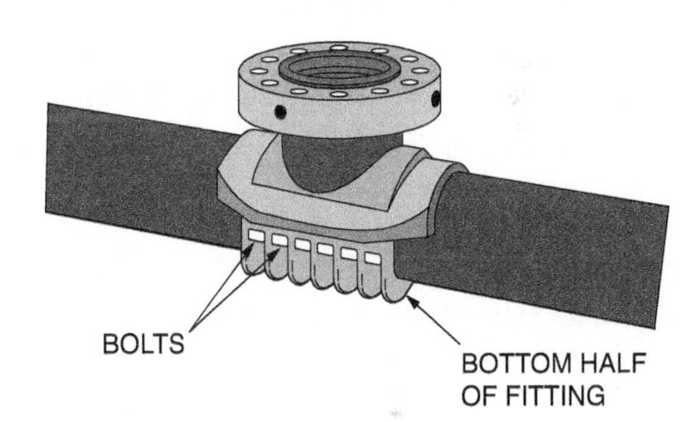

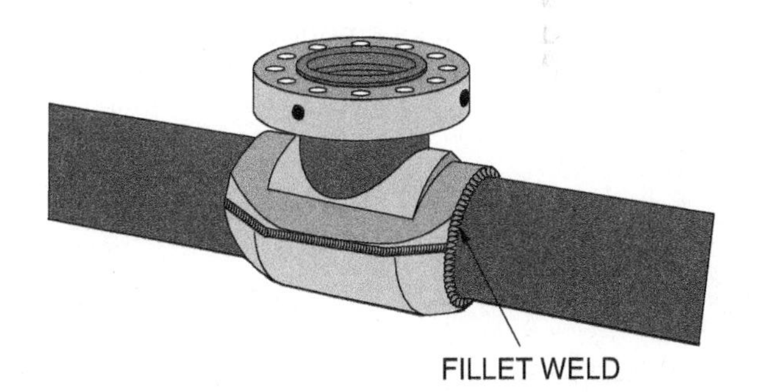

Figure 2 Assembly process for a bolt-weld fitting.

1.2.3 Installing Split Tee Fittings

A split tee fitting (*Figure 3*) is welded to the pipe and used to hold the tapping valve and machine during the tapping procedure. Split tee fittings are installed to aid in branch line taps, plugging taps, and new construction where branch connections require reinforcement around the entire circumference of the pipe. When welding on a process line, continuous flow is required in the line to avoid gas pockets.

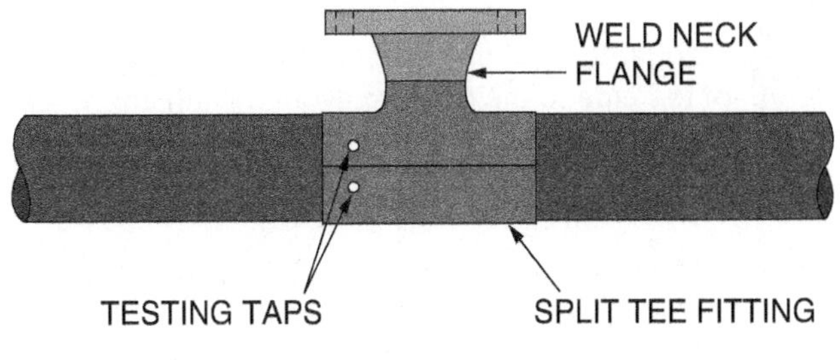

Figure 3 Split tee fitting.

To install a split tee fitting:

Step 1 Ensure that each piece of the split tee has a tapped hole. This is to provide a port to insert a nipple into, so that each weld may be pressure-tested when the tee fitting is installed.

Step 2 Ensure that the pipe has been ultrasound tested for thickness.

Step 3 Begin welding only when all required conditions are met and verified.

> **WARNING!**
> Never weld onto a stagnant pipeline. The flow through the line is needed to displace the heat from welding.

Step 4 Determine the temperature, pressure, and thickness of the pipe.

Step 5 Place the split tee fitting on the pipeline where the connection is to be made.

Step 6 Tack-weld the split tee fitting along its seams to the pipeline.

> **CAUTION**
> Follow the proper welding procedures and codes for the type of material being used. Protect the threads of the tapped hole when welding.

Step 7 Grind the tacks smooth using a 14", half-round bastard file.

> **CAUTION**
> Be careful not to file into the base metal of the pipe.

Step 8 Fillet-weld the split tee fitting along its seams to the pipeline.

> **WARNING!**
> Stagger the welding process from one part of the pipe to the other to avoid having an excessive amount of heat in one area.

Step 9 Test the split tee through the tapped holes, using the proper procedures.

Step 10 Determine if you are going to seal-weld or insert screw plugs into the testing taps.

Step 11 Seal the testing taps.

1.3.0 Operating Hot Tap Machines

The two basic types of hot tap machines are hand-operated and power-operated; special training is required before operating either type.

All hot tap machines consist of the same basic components; these include:

- A drive mechanism that is either pneumatically, hydraulically, or manually powered is used to drive the hole saw
- A drive shaft that connects the drive mechanism to the cutter
- A machine adapter that is used to attach the hot tap machine to the tapping valve
- A cutter that is used to cut a hole in the pipe wall and remove a piece of pipe (known as the *coupon*) from the pipe wall

When performing a hot tap, always remove and save the pipe coupon so that it can be inspected by the safety engineer. Do not allow the coupon to be lost inside the pipeline. Hole saws are designed with a U-shaped wire that is retracted in the saw until the saw cuts through the pipe wall. After passing through the pipe wall, the U-shaped wire protrudes from the hole saw to hold the coupon so that it can be removed.

1.3.1 Using Hand-Operated Hot Tap Machines

Hand-operated hot tap machines are normally used in highly explosive areas because the operator can control the speed of the tap. They are operated with a ratchet mechanism or a wrench to advance the drill or cutter. A measuring rod is used to ensure that the proper travel distance of the cutter is achieved when it is lowered. *Figure 4* shows a hand-operated hot tap machine.

To use a hand-operated hot tap machine:

> **WARNING!**
> This procedure is provided to teach you how hot tap machines operate. Never attempt to operate a hot tap machine unless you have received specialized training from the equipment manufacturer.

Step 1 Inspect the valve and fitting to be sure no foreign material is present.

Step 2 Check the valve to be sure it opens freely.

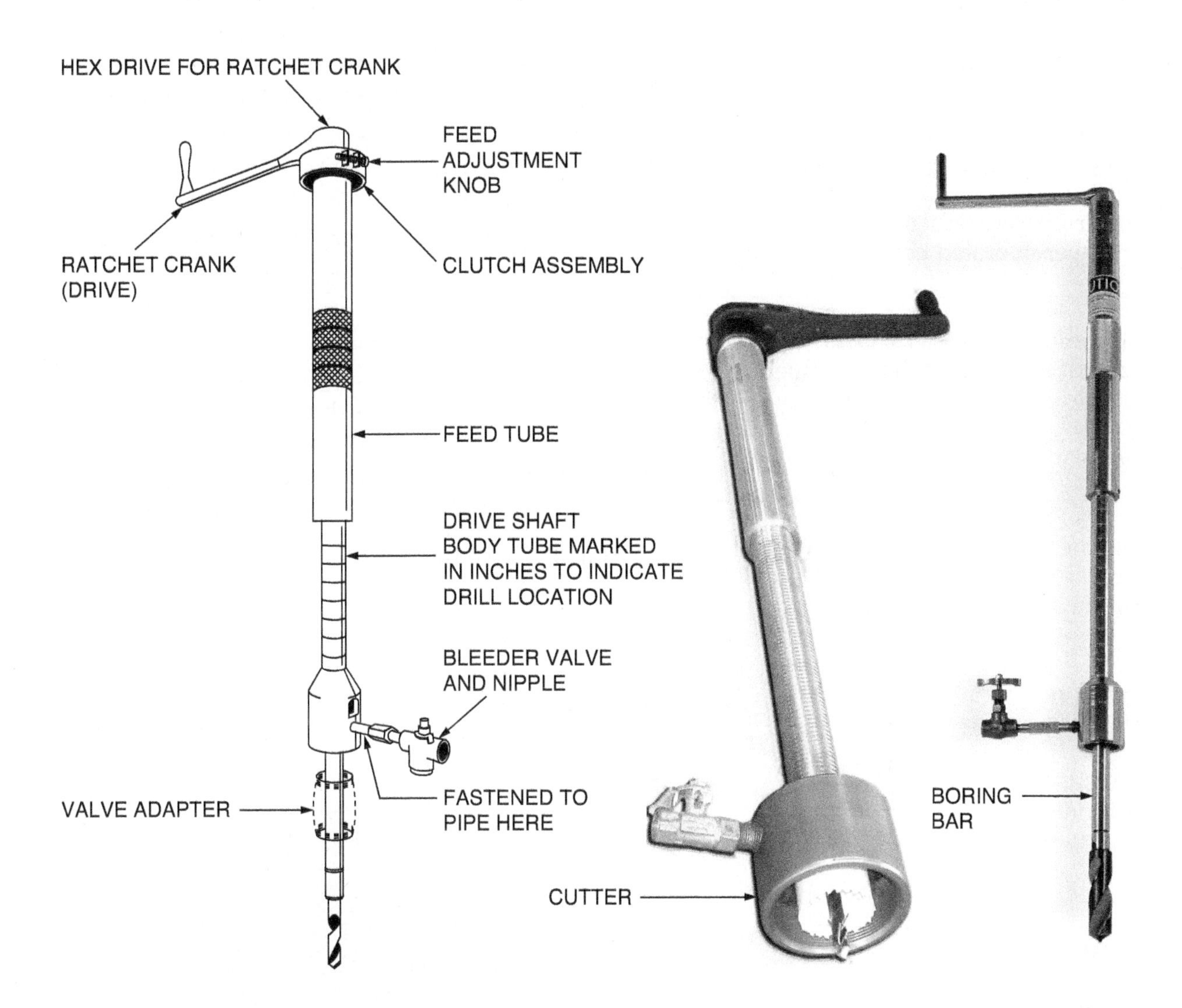

NOTE: Valve adapters are used to adapt machine to various size valves.

Figure 4 Hand-operated hot tap machine.

Step 3 Check the hot tap machine manufacturer's chart to determine the proper machine travel.

Step 4 Attach the appropriate tool to the boring bar.

Step 5 Bolt the flanges of the hot tap machine and the valve to connect the hot tap machine and the valve.

Step 6 Turn the ratchet clockwise to lower the boring bar and make the hot tap.

Step 7 Turn the ratchet counterclockwise to raise the boring bar.

Step 8 Close the valve.

Step 9 Remove the bolts from the hot tap machine and valve flanges to disconnect the hot tap machine from the valve.

1.3.2 Using Power-Operated Hot Tap Machines

Power-operated hot tap machines function like hand-operated hot taps except that the drill is hydraulically or pneumatically powered. *Figure 5* shows a power-operated hot tap machine.

To use a power-operated hot tap machine:

> **WARNING!**
> This procedure is provided to teach you how hot tap machines operate. Never attempt to operate a hot tap machine unless you have received specialized training from the equipment manufacturer.

Step 1 Bolt the flanges of the hot tap machine and valve to connect the hot tap machine and the valve.

Step 2 Activate the appropriate hot tap machine switch to lower the boring bar into place.

Step 3 Check the alignment of the boring bar in relation to the pipeline and adjust the boring bar as needed.

Step 4 Activate the appropriate hot tap machine switch to raise the boring bar.

Step 5 Attach the appropriate cutter/reamer tool to the boring bar.

Step 6 Activate the appropriate hot tap machine switch to lower the boring bar and begin the cutting/reaming process.

Step 7 Activate the appropriate hot tap machine switch to raise the cutter/reamer tool.

Step 8 Close the valve.

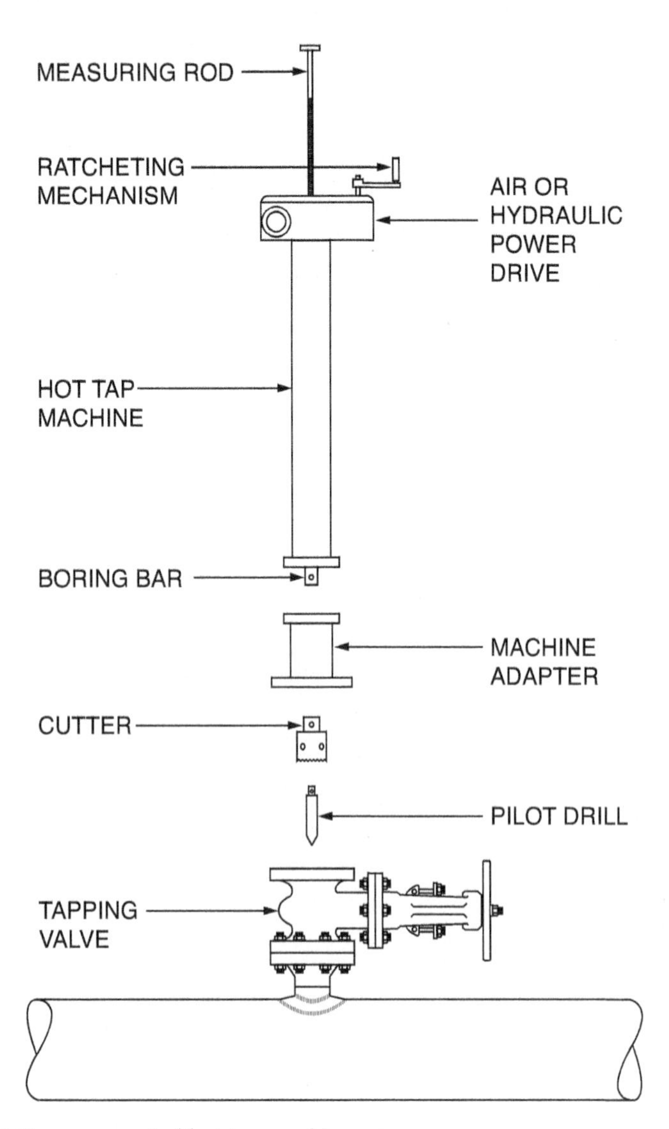

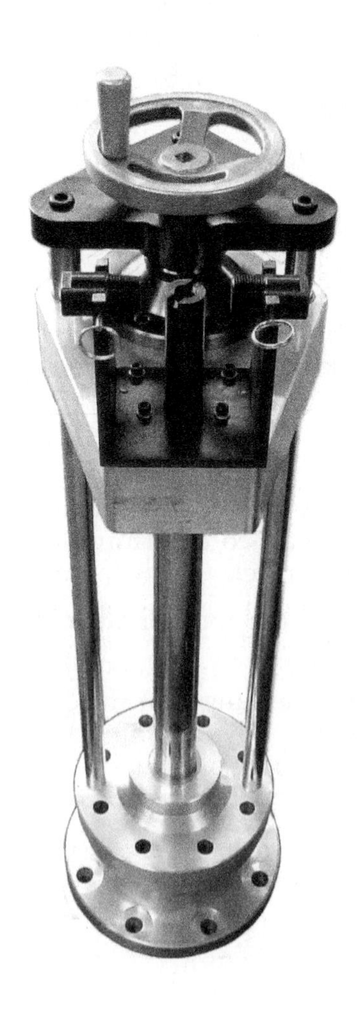

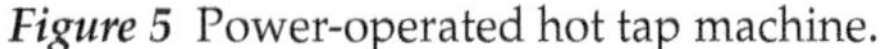

Figure 5 Power-operated hot tap machine.

Step 9 Release pressure from the fluid trapped between the valve and the hot tap machine.

Step 10 Drain the contents of the valve according to plant procedures and OSHA regulations.

Step 11 Remove the bolts from the hot tap machine and valve flanges to disconnect the hot tap machine from the valve.

Step 12 Blind-flange the valve unless a spool is to be added to this point.

1.3.3 Hot Tapping Technology

The spread of plastic pipe, including its use in gas transport, has led to the development of equipment for hot tapping plastic. This equipment allows the branch to be fusion-welded to the main pipe as part of the process, speeding the hot tap procedure.

Another development is that of diverless systems for hot tapping undersea pipe. The system employs cradles to lift the pipe clear of the bottom, to keep out muck and sand. When the pipe is in place, a remote-controlled hot tap machine is deployed, allowing the pipe to be entered at depths not easily accessed by divers.

Another place where hot tapping equipment has found a use is in draining sunken tanker vessels underwater. A hot tap fitting and valve are attached to the hull and the hole is drilled. The valve is then shut off. The machine is removed, a hose is attached, and the stored fluids are pumped to tanker vessels on the surface.

1.4.0 Line Stop Plugs

Line stop plugs (*Figure 6*) serve as temporary block valves and can be installed anywhere in the system. They are used to isolate a section of line for repairs or additions without interrupting service. T.D. Williamson, Inc., likely the largest company in the field, refers to these items as *STOPPLE®* fittings. This has become a commonly-used term for all such devices. The installation and use of line stop plugs is a highly-specialized skill and should only be done by trained, certified personnel.

Before installing the plug, you must know:

- The size of the hot tap
- The size of the valve bore
- The pressure rating of the valve
- Whether the valve is flanged or screw-type
- Whether the valve is operating properly
- The distance from the outside face of the valve to the pipe
- The clearance on the machine
- What material the pipe is made of
- The thickness of the pipe, vessel, or tank to be tapped so you can determine if it is safe to weld or bolt a fitting to the pipe without crushing it or burning a hole in it
- The pressure inside the pipe, vessel, or tank to be tapped

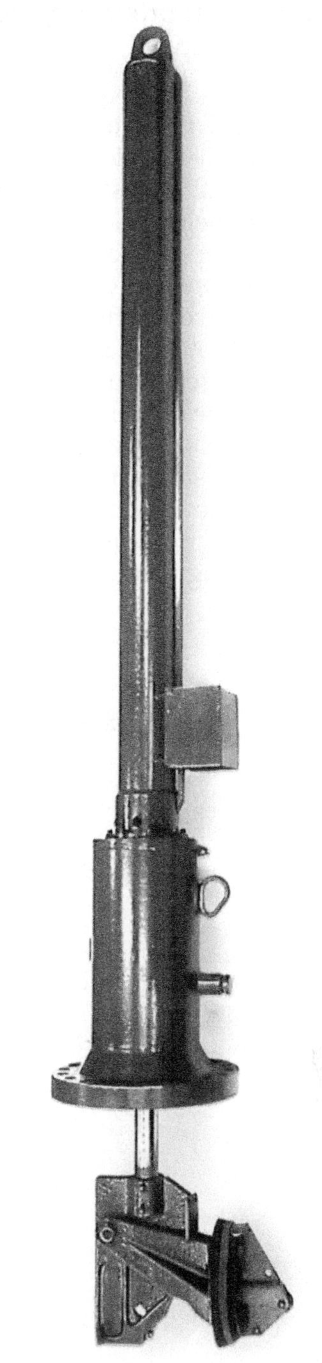

Courtesy of TD Williamson, Inc.

Figure 6 Line stop plugs.

- The temperature of the product inside the pipe, vessel, or tank to be tapped
- The rigging needs

Line stop plugs are used on low-, medium-, and high-pressure piping systems as well as on vacuum systems. Line stops use a variety of sealing elements that are selected, based on compatibility with the piping service. Line stops are used after a hot tap has been completed and the coupon has been removed. After the line stop is completed, a completion plug is installed and is locked and held in place through the use of jack screws in the custom-designed split tee fitting so that the tapping valve can be removed. Line stops can be installed in pipes ranging in size from ½" to 36" in diameter. They enable pipelines to be isolated for repairs or additions without interrupting service or losing product. *Figure 7* shows a plugging machine.

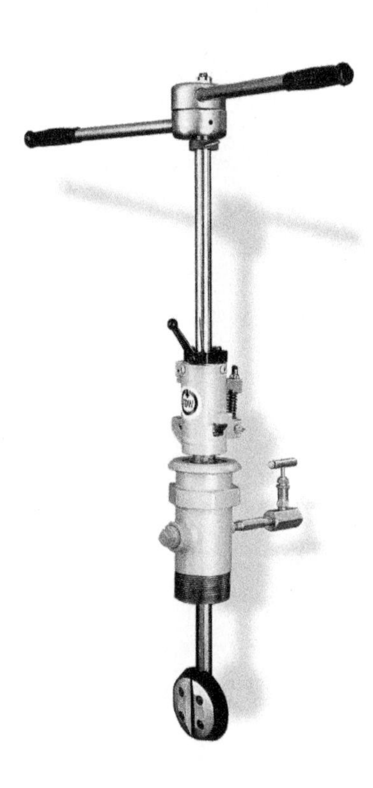

Figure 7 Line stop plugging machine.

1.4.1 Freeze Stops

Freeze stops (*Figure 8*), or cryogenic plugs, are used to temporarily plug lines containing freezable liquids under pressure. Freeze stops differ from other stops in that they use a chemical process rather than a mechanical process and therefore do not require the use of a hot tap machine. A freeze stop is the controlled formation of a solid plug inside a pipe containing flowing liquids. The pipeline is frozen so that the pipe diameter shrinks slightly and liquid crystals adhere to the inside wall, thus plugging the pipeline. This shrinkage and adherence enables the plug to maintain its position and resist high pressures that may exist in the pipeline.

To assist a technician who is performing a freeze stop:

Step 1 Tag out the area around the pipeline to be isolated in accordance with local procedures.

Step 2 Ensure that the temperature-monitoring equipment is in proper working condition.

Step 3 Ensure that the pipeline pressure is as low as possible.

Step 4 Ensure that the pipeline temperature is as low as possible.

Courtesy of TD Williamson, Inc.

Figure 8 Freeze stop.

Step 5 Ensure that the pipe section where the freeze stop is to be performed is full of a freezable liquid that can be stopped from flowing.

Step 6 Clean the pipe where the freeze stop is to be performed, being sure to remove any paint.

Step 7 Assist the technical representative in performing the freeze stop.

1.4.2 Cross Line Stops

Cross line stops are often used when there is not enough clearance for a conventional line stop of a branch line coming off a main line or header. Installing a cross line stop involves making a hot tap from the back side of the pipeline opposite the branch line. The branch is then plugged from the inside of the main line or header. A guide attached to the tap machine boring bar is lowered into place to provide the seal. This type of stopple can withstand temperatures up to 1,200°F and pressures up to 6,000 psi. *Figure 9* shows a typical cross line stop.

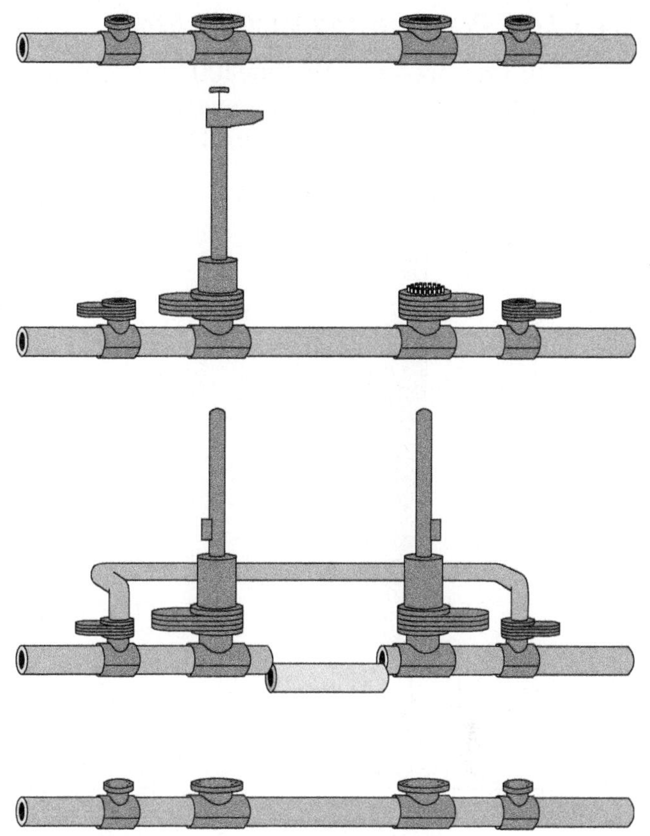

Courtesy of TD Williamson, Inc.

Figure 9 Cross line stop.

1.4.3 Low-Pressure and Vacuum Stops

Plugging low-pressure (below 100 psi) or negative pressure (vacuum) piping systems requires the installation of four split tee fittings and two equalization fittings. Two split tee fittings are installed as line stops and two are installed as bypass fittings. *Figure 10* shows a typical setup for low-pressure and vacuum stops.

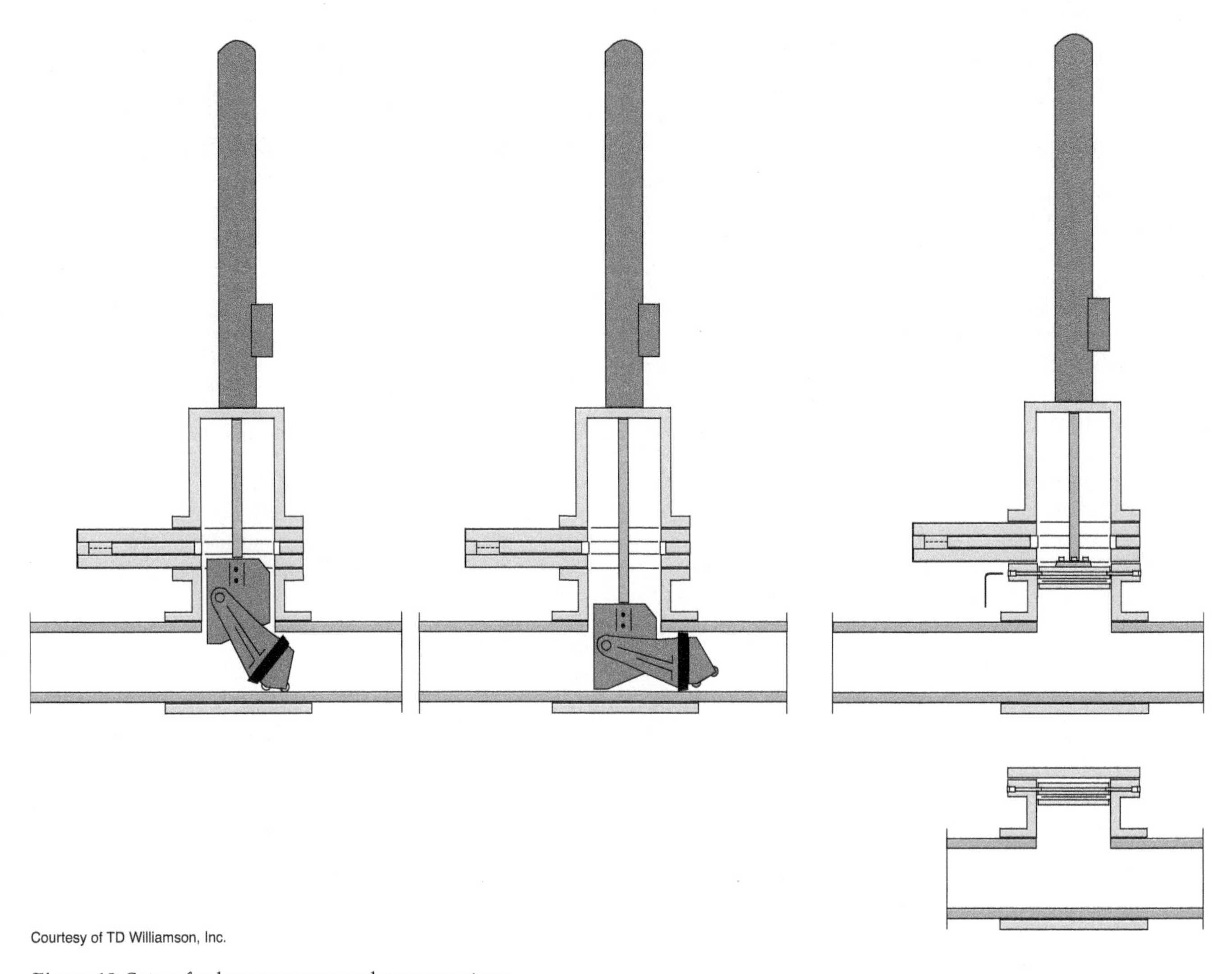

Courtesy of TD Williamson, Inc.

Figure 10 Setup for low-pressure and vacuum stops.

1.0.0 Section Review

1. Safe operation of a hot tap machine calls for pressure ratings that *must* be adhered to and attention to _____.

   a. voltage
   b. height
   c. temperature
   d. viscosity

Review Questions

1. Each hot tap machine is able to _____.
   a. contain the pressure in a pipeline
   b. operate at any pressure
   c. operate without an attendant
   d. work on any kind of pipe

2. The valve used in a hot tap *must* be a _____.
   a. butterfly valve
   b. plug valve
   c. check valve
   d. full port gate valve

3. The mechanical joint hot tap fitting has _____.
   a. no limitations
   b. specific temperature limitations
   c. only one piece
   d. one-piece glands

4. The three major parts of a bolt-weld fitting are _____.
   a. glands, gasket, and body
   b. drill, cap, and valve
   c. seal, top half, and bottom half
   d. flange, gasket, and seal

5. The testing taps on a split tee are used to _____.
   a. insert a thermowell
   b. drill holes in the pipe
   c. test the welds
   d. look at the pipe

6. The flow through a pipeline is needed to _____.
   a. carry away the coupons
   b. displace the heat from the welding
   c. keep the drill lubricated
   d. heat the pipe

7. The basic components of a hot tap machine are _____.
   a. a flange, a gasket, and a valve
   b. the drive mechanism, shaft, adapter, and cutter
   c. handle, grinder, and flange
   d. the drill, a valve, and a gasket

8. When you perform a hot tap, you *must* save the _____.
   a. shavings
   b. material from the bleed valve
   c. pipe coupon
   d. seal

9. To make the tap, hand-operated hot tap machines are turned _____.
   a. upward
   b. clockwise
   c. downward
   d. counterclockwise

10. Line stop plugs are used to _____.
    a. isolate a section of line
    b. install a branch line
    c. increase line pressure
    d. increase flow rate

Trade Terms Introduced in This Module

Full port: The maximum internal opening for flow through a valve that matches the ID of the pipe used.

Head clearance: The amount of space needed to install a hot tap machine on a pipe.

Hot tap: To make a safe entry into a pipe or vessel operating at a pressure or vacuum under controlled conditions without losing product.

Line stop: A device used to temporarily contain the flow or pressure of a product inside a pipe while a branch connection is being made.

Travel distance: The distance that the cutting bit moves from the top of the tapping valve to the pipe to be cut.

Cornerstone of Craftsmanship

Jody P. Suchanek

Bechtel Corporation
Deputy Workforce Development Manager/
NCCER Primary Administrator

How did you choose a career in the industry?

I decided after high school that college just was not for me. I had many friends that were much older than me that had already been making a good living for them and their families in the construction industry. So, I decided that would be the road I would take.

Who inspired you to enter the industry?

I was inspired by friends of mine and friends of the family that had been in the industry before me. I was also inspired by great men that helped begin to mold me and help me in my career choices.

What types of training have you been through?

I have completed four semesters of training with the local pipefitting union hall.

How important are education and training in construction? Education and training are the most important parts of our industry. If we as craftsmen fail to educate and train the upcoming generations, then eventually no one will be able to fill the shoes of those who've gone before them.

How important are NCCER credentials to your career?

NCCER credentials have given me a baseline of certifications that have been accepted as the industry standard. They have allowed me to build my career, and they've opened many doors and opportunities for me.

How has training in construction impacted your life?

Being part of a training team for the past 10 years has given me the most satisfaction of any part of my career. It's fulfilling to help others and enable them to be the best they could possibly be. Construction as a whole has provided a substantial income that has met the needs of my family completely. The relationships you build in this industry are lifelong.

What kinds of work have you done in your career?

My first three months in the industry were as part of a labor group. I then spent about 5 months with a scaffolding crew. I became a structural ironworker, maintenance technician, lumber industry mechanical technician, small equipment operator, rigger, pipefitter, foreman, general foreman, full-time instructor/evaluator, workforce development supervisor, and now a workforce development deputy manager and NCCER primary administrator.

What do you enjoy most about your job?

The most enjoyable part of my job is equipping people to be the safest, best quality minded, and productive personnel in our industry. In equipping them with industry skills and also giving them life skills, we give them the best opportunity to succeed at both.

What factors have contributed most to your success?

The most contributing factor in my career has been with the more experienced craftsmen taking the time to instill in me the grit, knowledge, and desire to be the best craftsman and leader that I could be and to give those around me the same opportunities. I have had the opportunity to encourage and enable others to excel and become great employees. Building leaders that will lead others to become the same caliber of craftsmen and leaders should be our ultimate goal.

Would you suggest construction as a career to others?

Yes, because it has provided for me and my family and will always provide for those who are willing to learn, work, and put forth the effort to stay busy.

What advice would you give to those new to the field?

I would encourage each person to look, listen, and learn. Always be aware of what is going on around you. Listen wholeheartedly to those who would like to see you succeed and are successful in their careers. Learn all that you can from every individual that influences your life, in and out of the industry, the good and the bad. Take all that into consideration and become your own individual and be the best craftsman you can. Finally, pass it on to the generation to come!

How do you define craftsmanship?

Craftsmanship is the ability to provide a product of building this industry in the safest, "do it right the first time" quality, and most proficient manner. Craftsmanship is also working together with a group of people to accomplish the goal of building the world as we know it.

Additional Resources

This module presents thorough resources for task training. The following reference material is recommended for further study.

The Pipe Fitters Blue Book. 2010. W. V. Graves. Webster, TX: Graves Publishing Co.

Figure Credits

T.D. Williamson, Inc., Module opener, Figures 4 (image on the right), 6-10
IFT Products, www.TappingMachines.com, Figures 4 (image in the middle), 5

Section Review Answer Key

Section 1.0.0

Answer	Section Reference	Objective
1. c	1.1.0	1a

NCCER CURRICULA — USER UPDATE

NCCER makes every effort to keep its textbooks up-to-date and free of technical errors. We appreciate your help in this process. If you find an error, a typographical mistake, or an inaccuracy in NCCER's curricula, please fill out this form (or a photocopy), or complete the online form at **www.nccer.org/olf**. Be sure to include the exact module ID number, page number, a detailed description, and your recommended correction. Your input will be brought to the attention of the Authoring Team. Thank you for your assistance.

Instructors – If you have an idea for improving this textbook, or have found that additional materials were necessary to teach this module effectively, please let us know so that we may present your suggestions to the Authoring Team.

NCCER Product Development and Revision

13614 Progress Blvd., Alachua, FL 32615

Email: curriculum@nccer.org
Online: www.nccer.org/olf

❑ Trainee Guide ❑ Lesson Plans ❑ Exam ❑ PowerPoints Other ____________

Craft / Level: Copyright Date:

Module ID Number / Title:

Section Number(s):

Description:

Recommended Correction:

Your Name:

Address:

Email: Phone:

This page is intentionally left blank.

Maintaining Valves

Overview

Understanding the function and assembly of valves is essential to the pipefitter's career. While most valves are replaced rather than maintained, it is important to understand the procedures for both. Knowledge on valve maintenance contributes to troubleshooting issues within a pipe run, with the function of valves as a pivotal point.

Module 08408

Trainees with successful module completions may be eligible for credentialing through the NCCER Registry. To learn more, go to **www.nccer.org** or contact us at 1.888.622.3720. Our website, **www.nccer.org**, has information on the latest product releases and training.

Your feedback is welcome. You may email your comments to **curriculum@nccer.org**, send general comments and inquiries to **info@nccer.org**, or fill in the User Update form at the back of this module.

This information is general in nature and intended for training purposes only. Actual performance of activities described in this manual requires compliance with all applicable operating, service, maintenance, and safety procedures under the direction of qualified personnel. References in this manual to patented or proprietary devices do not constitute a recommendation of their use.

08408
MAINTAINING VALVES

Objectives

Successful completion of this module prepares you to do the following:

1. Explain how to remove and install valves and bonnet gaskets.
   a. Explain the basics of removing and installing threaded and flanged valves.
   b. Explain how to replace bonnet gaskets.
2. Identify and describe how to replace valve-stem O-rings.
   a. Identify common O-ring materials.
   b. Explain how to replace valve-stem O-rings.
3. Identify and describe how to replace valve-stem packing.
   a. Identify and describe common packing materials.
   b. Explain how to repack valves.

Performance Tasks

Under supervision, you should be able to do the following:

1. Demonstrate how to remove and install threaded valves.
2. Remove and install flanged valves.
3. Replace bonnet gaskets.
4. Replace valve stem O-rings.
5. Demonstrate repacking a valve.

Trade Terms

Body
Bonnet
Direct pressure
Disc
Packing
Port
Reverse pressure
Seat
Stem
Trim

Industry Recognized Credentials

If you are training through an NCCER-accredited sponsor, you may be eligible for credentials from NCCER's Registry. The ID number for this module is 08408. Note that this module may have been used in other NCCER curricula and may apply to other level completions. Contact NCCER's Registry at 1.888.622.3720 or go to **www.nccer.org** for more information.

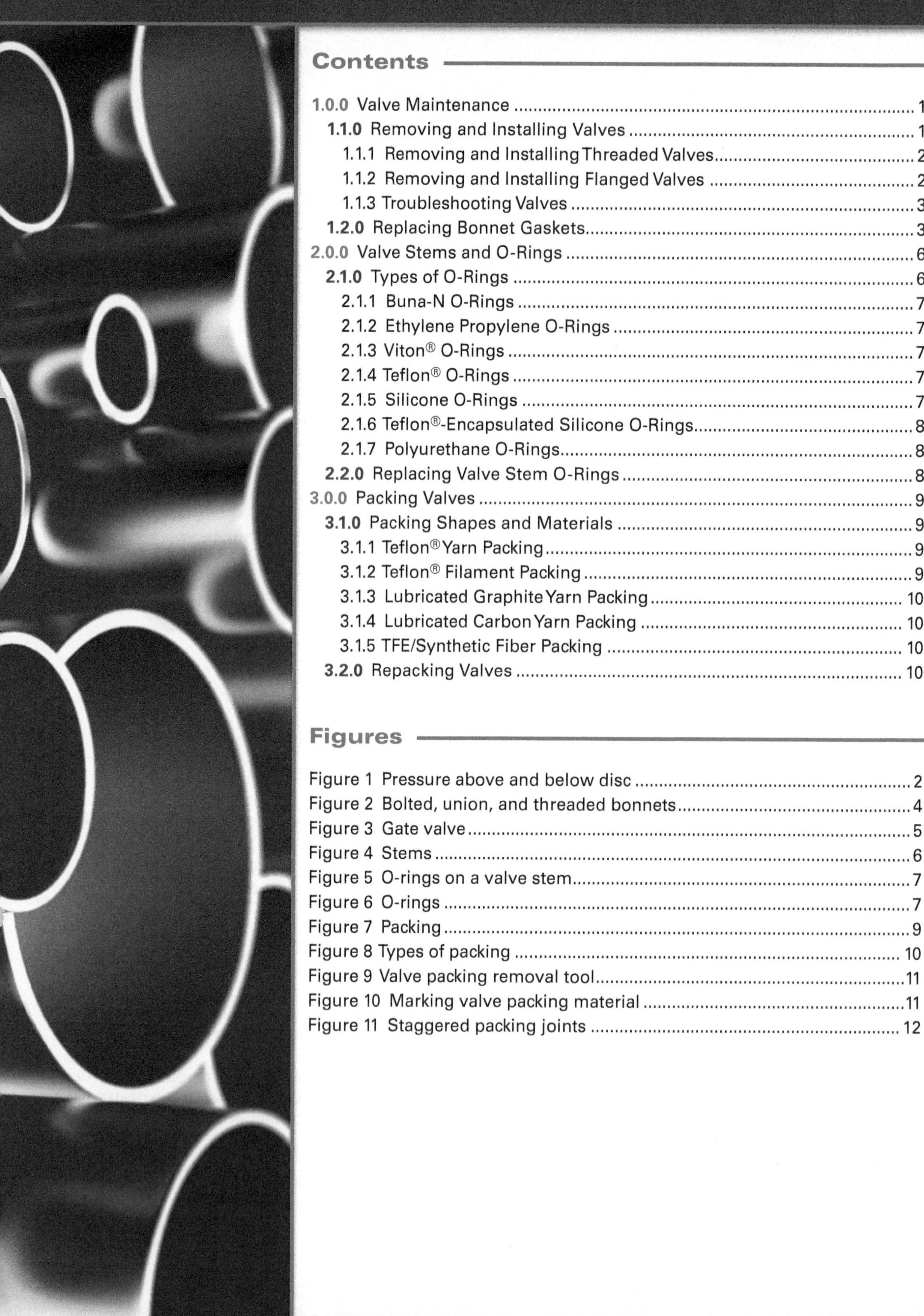

Contents

Figures

SECTION ONE

1.0.0 VALVE MAINTENANCE

Objective

Explain how to remove and install valves and bonnet gaskets.

a. Explain the basics of removing and installing threaded and flanged valves.
b. Explain how to replace bonnet gaskets.

Performance Tasks

1. Demonstrate how to remove and install threaded valves.
2. Remove and install flanged valves.
3. Replace bonnet gaskets.

Trade Terms

Body: The main part of the valve. It contains the disc, seat, and valve ports. The body of the valve is directly connected to the piping by threaded, welded, mechanically-joined, or flanged ends.

Bonnet: The part of the valve that contains the trim. The bonnet is located above the body.

Direct pressure: Flow pushing the sealing element of the valve into the seat and improving closure.

Disc: The moving part of a valve that directly affects the flow through the valve.

Packing: The material between the valve stem and bonnet that provides a leakproof seal and prevents material from leaking up around the valve stem.

Port: The internal opening for flow through a valve.

Reverse pressure: Flow pushing the closure element out of the seat.

Seat: The non-moving part of the valve on which the disc rests to form a seal and close off the valve.

Stem: The part of the valve that connects the disc to the valve operator. A stem can have linear, rotary, or helical movement.

Trim: The internal parts of a valve that receive the most wear and can be replaced. The trim includes the stem disc, seat ring, disc holder or guide, wedge, and bushings.

Valves regulate and direct the flow of material within a piping system. Therefore, the overall efficiency of any system is closely linked to its valves. Because of their importance, valves must be installed correctly and maintained regularly. It is important to closely follow the manufacturer's maintenance procedures and make sure the valve is not under warranty from the manufacturer. If it is and maintenance is performed, the warranty may be voided.

1.1.0 Removing and Installing Valves

The safe removal and installation of valves is highly important to the efficient operation of a piping system. Because of regular maintenance procedures, valves must be easily accessible when possible. When removing valves from a piping system, be sure to follow all plant procedures to lock out and relieve pressure from the valves. If working at a job site where the pipefitter is not permitted to perform lockout procedures, ensure that it has been locked out by the appropriate personnel before removing it.

The positioning of a valve in a piping system is extremely important. Valves that are installed with the stem in the upright position tend to work best. The stem can be rotated down to the horizontal position, but it should not point downward. If the valve is installed with the stem in the downward position, the bonnet acts like a trap for sediment, which may cut and damage the stem. Also, if water is trapped in the bonnet in cold weather, the body of the valve may freeze and crack.

It is important to consider the direction of flow through the valves when installing them. Butterfly valves, safety valves, pressure-relief valves, and certain others will have arrows stamped on the side indicating the direction of flow, or they will have a port labeled as the inlet or the outlet (unless they are bi-directional). When a valve is not marked, the pipefitter must determine which side of the disc will receive the most pressure.

The trim of the valve includes the stem disc, seat ring, disc holder or guide, wedge, and bushings; this is the part of the valve that receives the most wear and tear. Most gate valves can have the pressure on either side, as can some butterfly valves. Globe valves should be installed so that the pressure is below the disc unless pressure above the disc is required in the job specifications. *Figure 1* shows pressure above and below the disc.

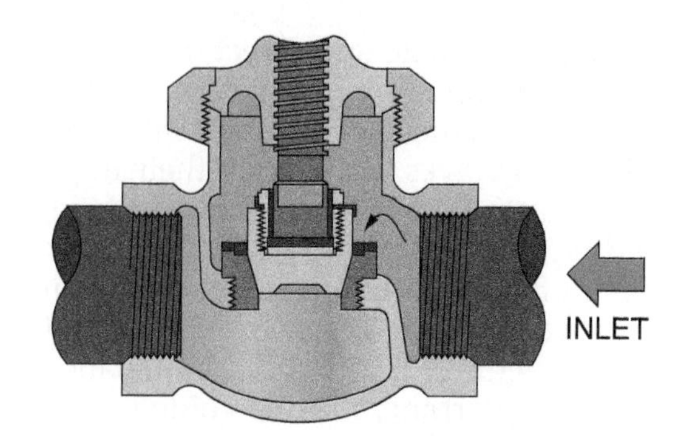

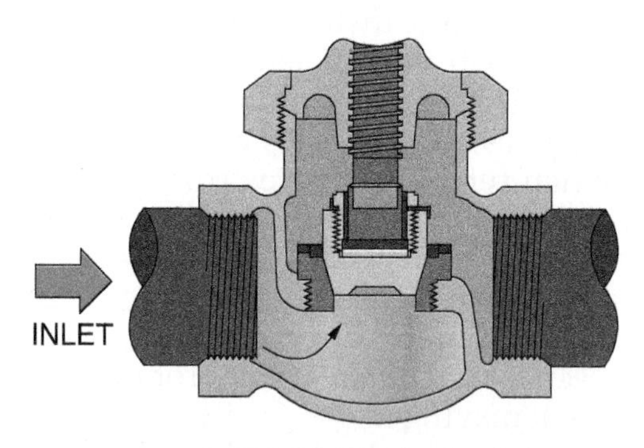

Figure 1 Pressure above and below disc.

Tapered plug valves, including multi-port plug valves, are usually installed with the flow pushing in the direction through the plug into the **seat** at any port. This is called **direct pressure**. When the plug is pushed away from the seat, the flow direction is called **reverse pressure**.

Eccentric plug valves used in air, gas, or clean liquid service are best mounted with the shaft horizontal and in a direct pressure orientation. This means that flow will enter at the end farthest from the seat, and it will push the plug into the seat. If the eccentric plug is on the discharge side of a pump, the plug should be in reverse pressure orientation. If eccentric plugs are used with material such as pulp or other suspended solids, the flow must be able to carry the solids through without clogging the valve. For that reason, in horizontal pipelines, the plug should open upward. That is, the plug should move away from the seat to the top of the pipe channel. In vertical pipelines the seat end (usually marked on the body or the flange) is installed at the top, so the plug won't catch or jam on the solids. On DeZURIK® eccentric plug valves, the word "Seat" is cast on the seat end of the body. On Val-Matic's Cam-Centric® valves, the words "Seat End" are on the flange.

Diaphragm valves will usually work in any position in vertical pipe runs because the mechanism is isolated from the fluid. The weep hole on the side of the barrel should face downward in horizontal runs, so that any leak in the diaphragm will be revealed. Butterfly and gate valves usually have a preferred orientation, such as placing the seat ring carrier of a butterfly valve upstream. In all cases, read and follow the manufacturer's instructions, as well as the job specifications, so that the system does not have to be reworked.

1.1.1 Removing and Installing Threaded Valves

The threaded joint is a very common method of joining pipe to smaller-sized valves up to about 2". In order to have a strong, leakproof, threaded joint, the threads must be clean and smoothly cut. They must have the correct pitch, lead, taper, and form, and the threads must be correctly sized. If these conditions are not met, the seal leaks.

Special precautions must be taken when installing valves with threaded ends to avoid damaging the valve. When installing valves with threaded ends:

- Never place the valve in any type of vise. Always secure the pipe in a vise and screw the valve onto the pipe.
- Always grip the valve with the pipe wrench on the body of the valve.
- Use a strap wrench on valves with brass bodies to avoid marring the valve.
- Lubricate the male threads of the pipe. Do not lubricate the female threads inside the valve, and do not lubricate the thread nearest the end of the pipe.
- Do not run the male end of the pipe all the way into the valve. This damages the seat and causes the valve to leak. Always leave three threads showing outside the valve.

1.1.2 Removing and Installing Flanged Valves

Each type of flange is available in different face styles, and each flange face style uses a certain gasket. A set of two flanges of the same face style should always be used with the proper gasket. Typically, two different face styles would never be used at the same joint, but this may be necessary when connecting to manufactured equipment. If it is necessary to connect a flat-face flange to a raised-face flange, be sure to confirm that the flat-face flange is not cast iron because the pressure of the bolts can crack or break the flange. If the flat-face flange is cast iron, change the raised-face flange to a flat-face flange with engineering approval and use a full-face gasket between them.

Flange bolts must be tightened carefully to avoid warping the flange. If the bolts are not tightened correctly, the joint may leak or crack. Pressure must be applied evenly, and in the correct amount, to make a good seal. A torque wrench can be used to apply equal pressure at all points around the flange. With the aid of a torque wrench, a pipefitter can properly adjust and tighten a flanged joint. Each piping system may use different torque settings, depending on the sizes of the flanges and bolts, the pressure or temperature of the system, and the type of gasket used. Pipefitters may consult the supervisor or piping drawings for the proper torque setting.

1.1.3 Troubleshooting Valves

When valve malfunctioning is known or suspected, operate it a few times for verification. If the actuator is very difficult to operate, there are several possible reasons:

- The packing may be too tightly compressed. If the contents are not too dangerous, loosen the packing nut until it leaks slightly, then tighten until the leak stops.
- Dirt may be obstructing operation of the valve. This is frequently a problem in ball valves because of the way they open and close. The usual solution is to flush the valve out.
- In the case of gate valves and globe values (to a lesser extent), if the valve will not open, there may be pressure above the seat. If there is an equalizer on the valve, that might solve the problem.

1.2.0 Replacing Bonnet Gaskets

The types of bonnets that contain gaskets or seals are the bolted, union, or threaded bonnets (*Figure* 2). Bolted bonnets adhere to the body with several bolts around the circumference of the bonnet. Valves with bolted bonnets are commonly used for line sizes larger than 1¼" and they are designed for use in process plants at all pressures and temperatures. When a leak occurs between the bonnet and the body of the valve, it is time to replace the bonnet gasket.

Threaded bonnets have a male threaded piece that screws inside the body of the valve, with the stem passing through the center of the male threaded top. A union bonnet has a female threaded ring or nut screwed onto the outside of the body of the valve. A bolted bonnet may involve bolts screwing into threaded holes in the body or flanges, with holes all the way through and nuts providing the threads to tighten the flanges together.

Follow these steps to replace bolted bonnet gaskets:

Step 1 Isolate the system according to plant procedures.

Step 2 Perform the standard lockout/tagout procedure. Follow your company's line break procedure.

Step 3 Remove the valve from the system so that testing can be performed after replacing the gasket.

Step 4 Secure the valve in a vise with the valve ends between the jaws of the vise.

CAUTION

Use soft covers in the vise jaws to protect the valve from damage.

Step 5 Turn the handwheel to open the gate valve (*Figure 3*). This raises the disc from the valve seat into the bonnet area.

Step 6 Remove the bonnet bolts from the bonnet.

Step 7 Remove the bonnet from the valve body.

Step 8 Remove the gasket and all gasket material remaining on the bonnet using a putty knife, gasket scraper, or single-edge razor blade. When removing the old gasket, try to keep it in one or two pieces so that it can be used as a pattern for a new gasket.

Step 9 Use the manufacturer's specifications to select the proper material to make a new gasket. Note that certain types of gaskets cannot be made on site and must be ordered from the manufacturer.

Step 10 Place the old gasket on top of the gasket material and trace the outline using a pencil. If the old gasket is completely destroyed, use the bonnet as a pattern for the new gasket.

Step 11 Cut out the gasket using a gasket cutter.

Step 12 Lubricate the gasket if specifications direct you to do so.

Step 13 Place the gasket on the valve, making sure it is positioned properly.

Step 14 Place the bonnet on the valve body.

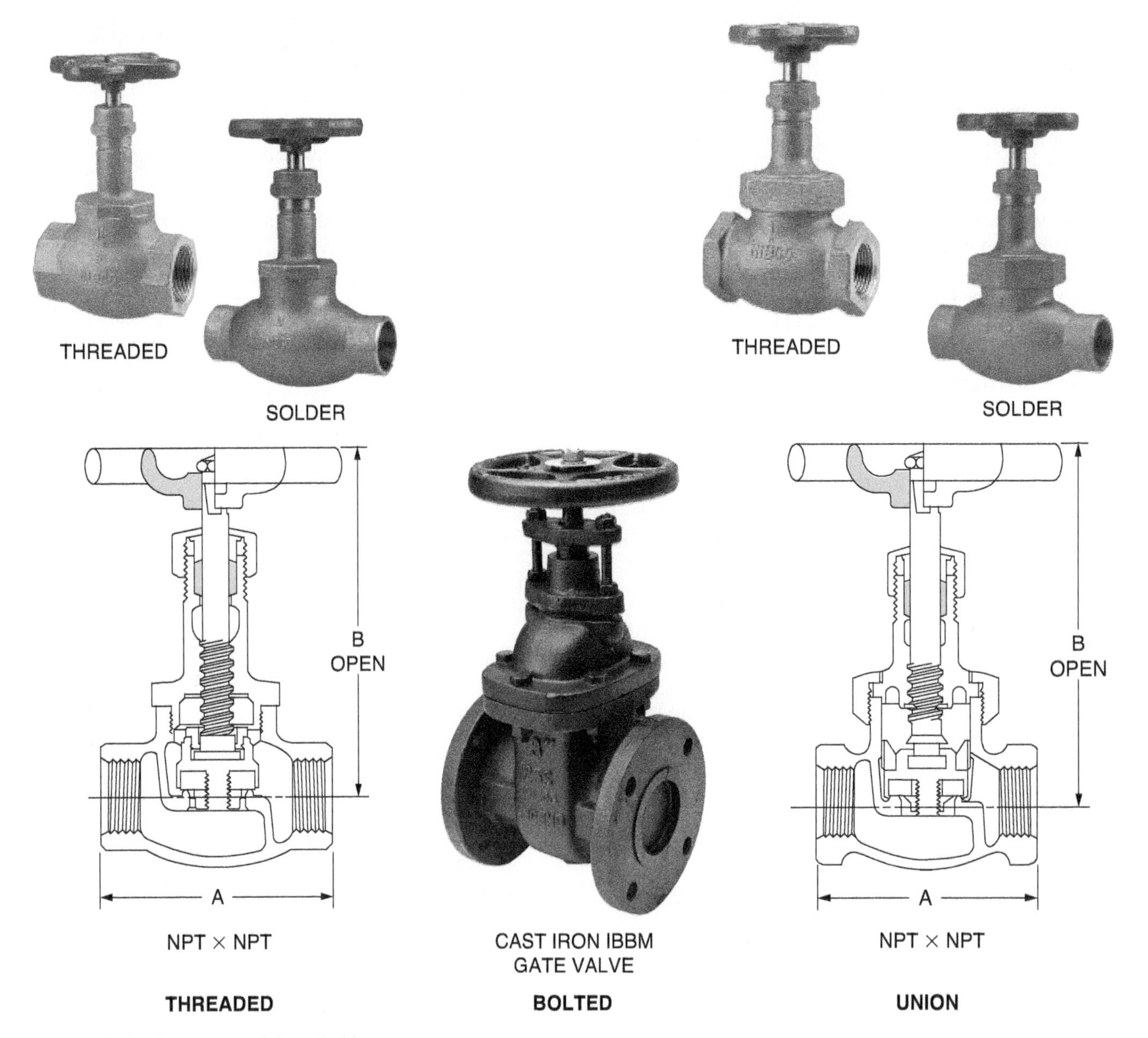

Figure 2 Bolted, union, and threaded bonnets.

Step 15 Clean and lubricate the bonnet bolt threads.

Step 16 Place the bonnet bolts in the proper position.

Step 17 Hand-tighten the bonnet bolts.

Step 18 Tighten the bonnet bolts using the crossover method. Follow the manufacturer's specifications and use a torque wrench to tighten the bolts.

Step 19 Replace the handwheel and tighten it according to specifications.

Step 20 Test the valve according to the specified test procedure.

Step 21 Return the valve to service.

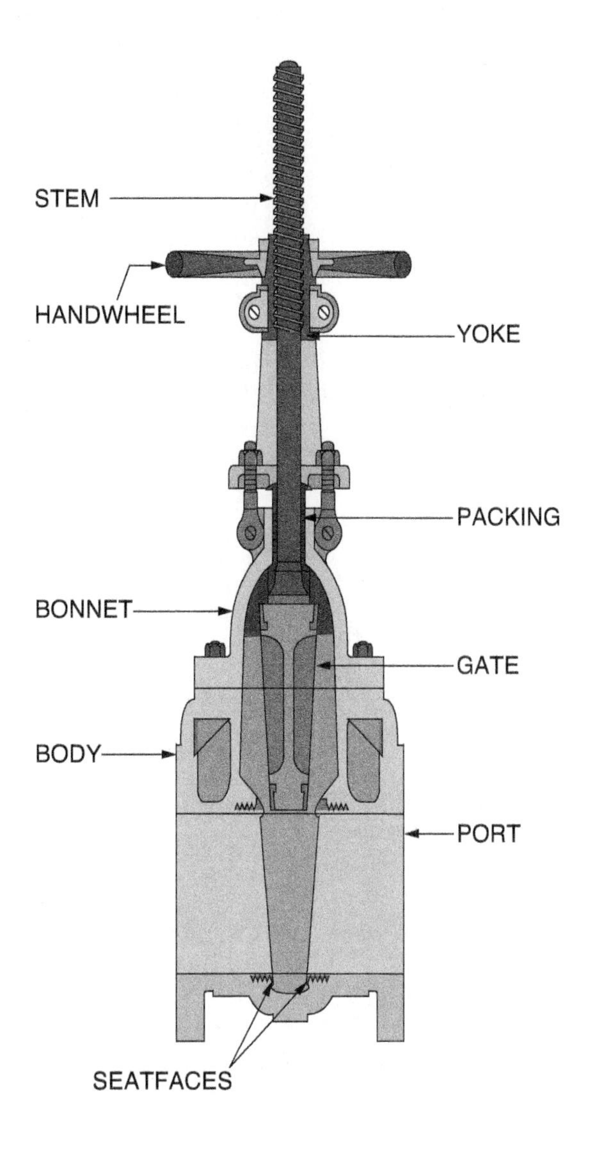

Figure 3 Gate valve.

1.0.0 Section Review

1. The stem disc, seat ring, wedge, and bushings are all part of the valve's _____.
   a. stem
   b. bonnet
   c. trim
   d. disc

2. What type of bonnet has a female threaded ring or nut screwed onto the outside of the body of the valve?
   a. Threaded
   b. Union
   c. Bolted
   d. Hinged

Section Two

2.0.0 Valve Stems and O-Rings

Objective

Identify and describe how to replace valve-stem O-rings.

a. Identify common O-ring materials.
b. Explain how to replace valve-stem O-rings.

Performance Task

4. Replace valve stem O-rings.

The valve stem provides a link between the handwheel and the gate. Many times, the pipefitter must attach the handwheel to the stem after the valve is installed. Some handwheels bolt onto the stem while others slip over a keyway cut into the stem and are then secured to the stem by a nut or machine screw. Handwheels vary according to the type of stem they fit. Three general types of stems include the rising stem, the non-rising stem, and the outside screw-and-yoke (OS&Y) stem. *Figure 4* shows types of stems.

When a rising stem valve is opened, both the handwheel and the stem rise. The height of the stem gives an approximation of how far the valve is open. These stems can be installed only where there is sufficient clearance for the handwheel to rise. Rising stems come in contact with the fluid in the valve and are used only with fluids that do not harm the threads, such as hydrocarbons, water, and steam.

Non-rising stems are suitable in spaces where space is limited. As the name implies, neither the stem nor the handwheel rises when the valve is opened. A spindle inside the valve body turns when the handwheel is turned and raises the stem so stem wear is held to a minimum. This type of stem also comes in contact with the fluid in the valve.

The OS&Y stem is suitable for corrosive fluids because the stem does not come in contact with the fluid in the line. As the handwheel is turned, the stem moves up through the handwheel. The height of the stem indicates how far the valve is open.

Some valve stems have O-rings that provide a leakproof seal for the lubrication around the stem. These O-rings fit inside specially machined grooves. *Figure 5* shows O-rings on a valve stem.

2.1.0 Types of O-Rings

O-rings (*Figure 6*) are used to seal against conditions ranging from strong vacuum to high pressure. O-rings are made from a variety of materials for different applications. Like gaskets, O-rings are often used to seal a mechanical connection between two parts of an instrument. O-rings that are used in instruments are often made of rubber or synthetic material. Occasionally, high-temperature or pressure applications may require the use of a metal O-ring. Among the characteristics useful in O-rings is low compression set, or the ability to return to their original shape and size when they are released from pressure.

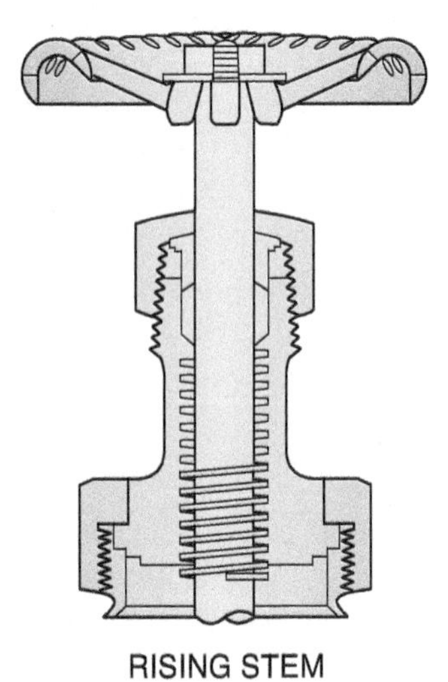

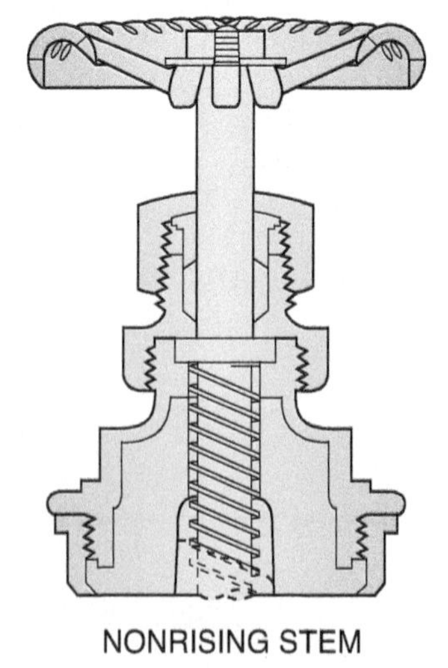

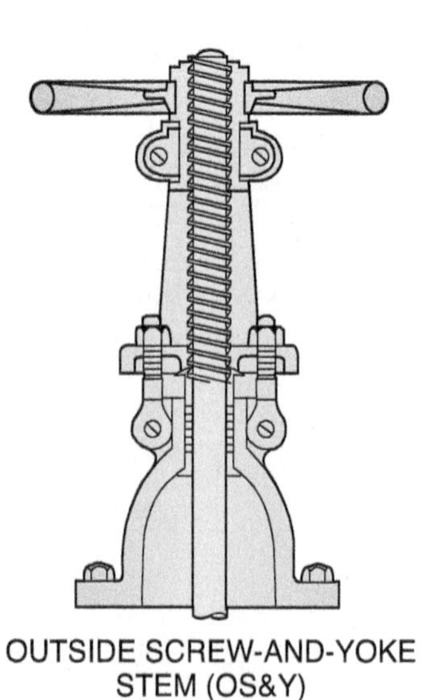

Figure 4 Stems.

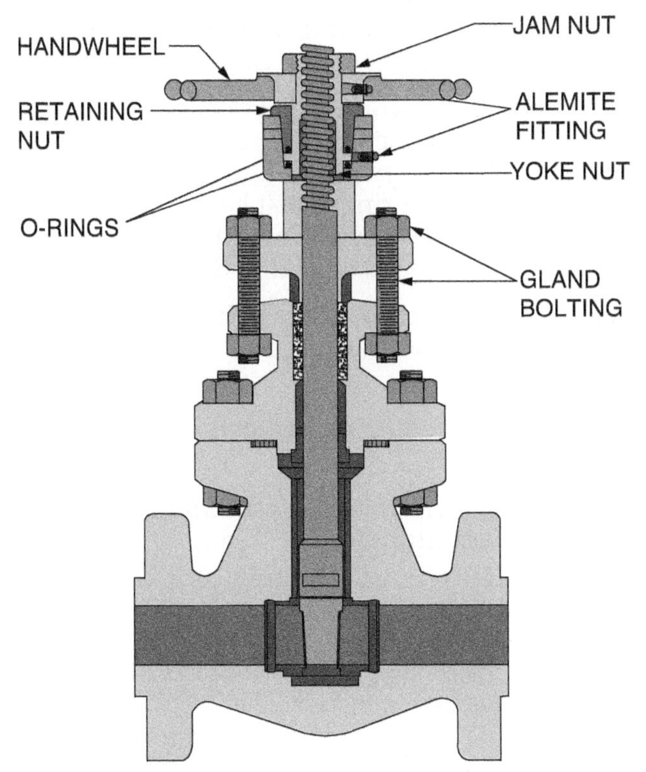

Figure 5 O-rings on a valve stem.

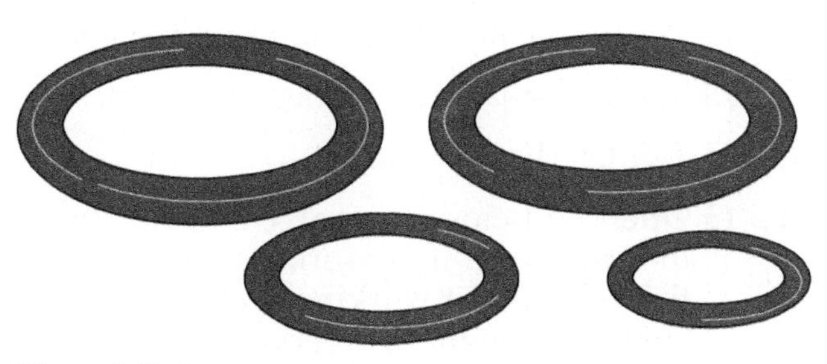

Figure 6 O-rings.

Some of the more common O-rings are made from Buna-N (Nitrile), ethylene propylene, Viton®, Teflon®, silicone, Teflon®-encapsulated silicone, and polyurethane. O-rings are used in both static and dynamic seals. A static seal is not subjected to flow but may have system pressure. A dynamic seal has both flow and pressure.

The sizes of O-rings are set by the *Aerospace Standard AS568A*, published by the Society of Automotive Engineers (SAE). Sizes are designated by the numbers following the dash, such as *AS568*-**006** and *AS568*-**216**.

2.1.1 Buna-N O-Rings

Buna-N O-rings are widely used. They are made of an elastomeric-sealing material. They are used with a variety of petroleum and silicone fluids, hydraulic and non-aromatic fuels, and solvents. Buna-N O-rings are not compatible with phosphate esters, ketones, brake fluids, strong acids, or ozone and have an approximate temperature range of –65°F to 275°F. Buna-N O-rings do not weather well, especially in direct sunlight.

2.1.2 Ethylene Propylene O-Rings

Ethylene propylene O-rings resist automotive brake fluids, hot water, steam (to approximately 400°F), silicone fluids, dilute acids, and phosphate esters. They are not compatible with petroleum fluids or diester lubricants. Ethylene O-rings have a good compression set, high abrasion resistance, and are weather resistant. They have an approximate temperature range of –70°F to 250°F.

2.1.3 Viton® O-Rings

Viton® O-rings offer excellent resistance to petroleum products, diester lubricants, silicone fluids, phosphate esters, solvents, and acids (except fuming nitric acid), but they should never be used with acetates, methyl alcohol, ketones, brake fluids, hot water, or steam. They have a low compression set and low gas permeability. Viton® O-rings are often used for hard vacuum service. The approximate temperature range of Viton® O-rings is –31°F to 400°F.

2.1.4 Teflon® O-Rings

Teflon® O-rings lubricate well and have excellent chemical and temperature resistance. They make fine static seals but need mechanical loading when used as dynamic seals. They have an approximate temperature range of –300°F to 500°F.

> **NOTE**
>
> All O-rings require lubrication. Keep in mind, however, that the lubricant and the O-ring material must be compatible. If they are not properly matched, the seal could be damaged. Because new lubricants are constantly introduced, it is important to check the manufacturer's requirements to ensure a match between the O-ring and the lubricant.

2.1.5 Silicone O-Rings

Silicone O-rings are used where long-term exposure to dry heat is expected. Due to poor abrasion resistance, they perform best in static sealing applications. Silicone O-rings resist brake fluids and high aniline point oil. They are not recommended for use with ketones and most petroleum oils. They have an approximate temperature range of –80° to 400°F.

2.1.6 Teflon®-Encapsulated Silicone O-Rings

Teflon®-encapsulated silicone O-rings are used in most of the same applications as silicone O-rings. The Teflon® coating makes them resistant to most solvents and chemicals. They have an extremely low coefficient of friction and a low compression set. They are primarily used as seals in static applications and have an approximate temperature range of –75°F to 400°F.

2.1.7 Polyurethane O-Rings

Polyurethane O-rings are abrasion resistant; they have high tensile strength and excellent tear strength. Polyurethane is the toughest of the elastomers. Polyurethane O-rings can be used with petroleum fluids, ozone, and solvents (except ketones). Polyurethane O-rings are not compatible with hot water, brake fluids, acids, or high temperatures. They have a poor compression set with an approximate temperature range of –40°F to 200°F.

2.2.0 Replacing Valve Stem O-Rings

To replace valve stem O-rings:

Step 1 Remove the jam nut from the valve stem.

Step 2 Remove the handwheel from the valve stem.

Step 3 Remove the retaining nut using a box wrench.

Step 4 Screw the yoke nut out of the top of the bonnet.

Step 5 Remove and inspect the O-rings, looking for any damage caused by the valve stem.

Step 6 Smooth the surface on the valve stem using a fine-grit emery cloth.

Step 7 Select the correct O-ring replacements. The manufacturer's specifications specify the proper material and size O-rings to use. The O-rings must be the correct size and made of the proper material to provide an adequate seal in the valve.

Step 8 Lubricate the O-rings lightly, according to specifications, to prepare them for installation.

Step 9 Slide the O-rings into the appropriate grooves.

Step 10 Screw the yoke nut onto the valve stem very carefully.

CAUTION

To avoid breaking the seal, do not pinch or break the O-rings, and do not dent or scratch the valve stem with the bonnet.

Step 11 Screw the retaining nut onto the yoke nut.

Step 12 Return the handwheel to the valve stem.

Step 13 Return the jam nut to the valve stem.

Step 14 Tighten the jam nut.

Step 15 Open and close the valve several times to make sure it is working properly. If the valve is not working properly, repeat Steps 1 through 15; otherwise, proceed to Step 16.

Step 16 Return the valve to service.

2.0.0 Section Review

1. Which type of O-ring lubricates well and has excellent chemical and temperature resistance?
   a. Buna-N
   b. Ethylene
   c. Viton®
   d. Teflon®

2. What is the last step in replacing a valve stem O-ring, before testing it and returning it to service?
   a. Tightening the jam nut
   b. Screwing the retaining nut onto the yoke nut
   c. Screwing the yoke nut onto the valve stem
   d. Returning the jam nut to the valve stem

SECTION THREE

3.0.0 PACKING VALVES

Objective

Identify and describe how to replace valve-stem packing.

a. Identify and describe common packing materials.
b. Explain how to repack valves.

Performance Task

5. Demonstrate repacking a valve.

There are four areas within a valve that must provide a complete seal if the valve is to function properly. These areas are:

- The closing mechanism inside the valve
- The end connections
- The area where the valve bonnet meets the body
- The valve stem

Achieving a good seal at the stem is somewhat difficult because of the movement of the stem. Therefore, various types of packing are used to seal it. Packing prevents fluid from leaking up through the stem, and it retains the pressure of the fluid within the valve. The packing material fills the stuffing box, which is the space between the valve stem and the bonnet (*Figure 7*). A gland presses the packing against the stem within the stuffing box. The stuffing box usually requires occasional tightening, especially if the valve has not been used for a while.

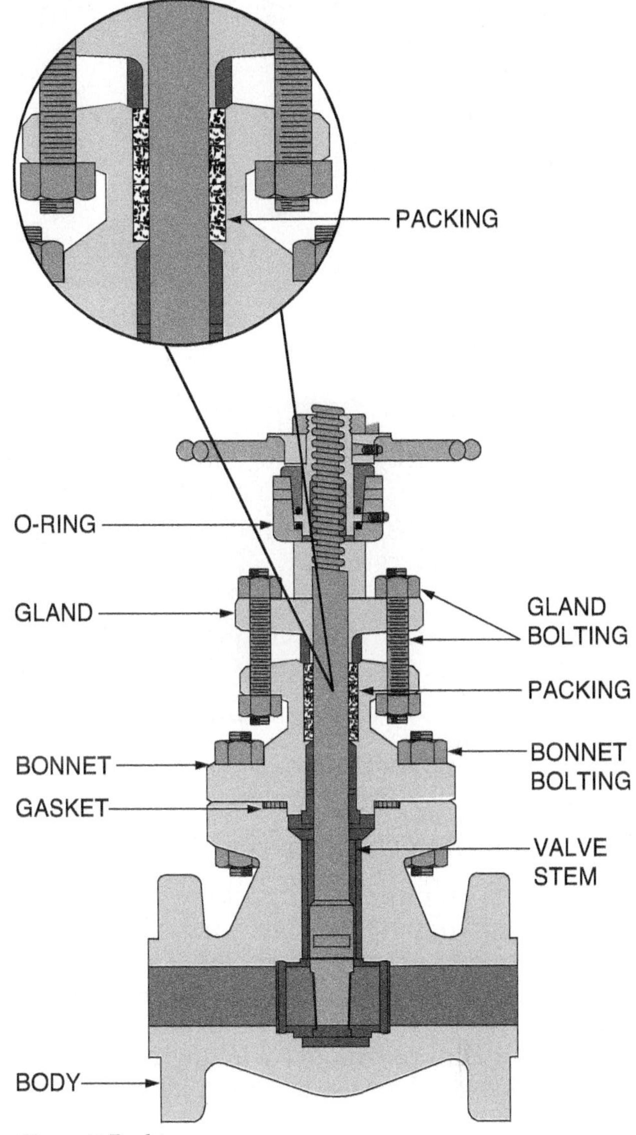

Figure 7 Packing.

3.1.0 Packing Shapes and Materials

The most common types of packing are those that are solid or that have braided or granulated fibers. Packing may be square-shaped, wedge-shaped, or ring-shaped. Commonly used types of packing include Teflon® yarn and filament, lubricated graphite yarn, lubricated carbon yarn (graphite impregnated), and tetrafluoroethylene (TFE/synthetic fiber). Types of packing are shown in *Figure 8*.

Though most packing materials will normally be specified, the following factors must be considered:

- Material flowing through the line
- Operating pressures
- Operating temperatures
- Minimum temperature of the piping system
- Composition of the valve stem

3.1.1 Teflon® Yarn Packing

Teflon® yarn packing is a cross-braided yarn impregnated with Teflon®. Teflon®-impregnated packing resists concentrated acids such as sulfuric acid and nitric acid, sodium hydroxide, gases, alkalis, and most solvents. It is good for use in applications of up to approximately 550°F.

3.1.2 Teflon® Filament Packing

Teflon® filament (cord) packing is a braided packing made from TFE filament. It is impregnated with Teflon® and an inert softener lubricant, or sometimes it is impregnated with graphite. It is often used on rotating pumps, mixers, agitators, kettles, and other equipment.

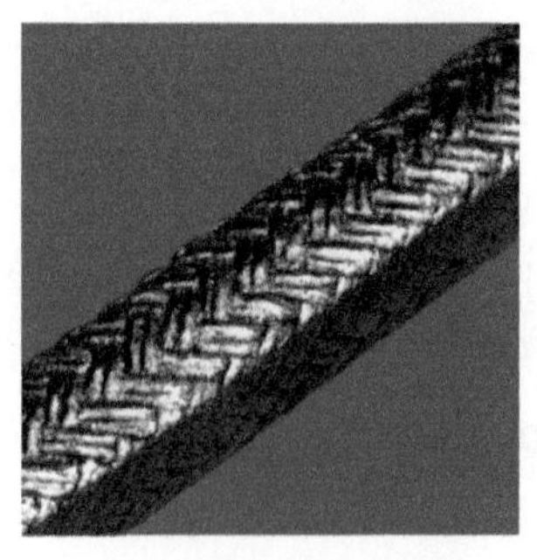

(A) INCONEL® AND CARBON WIRE YARN

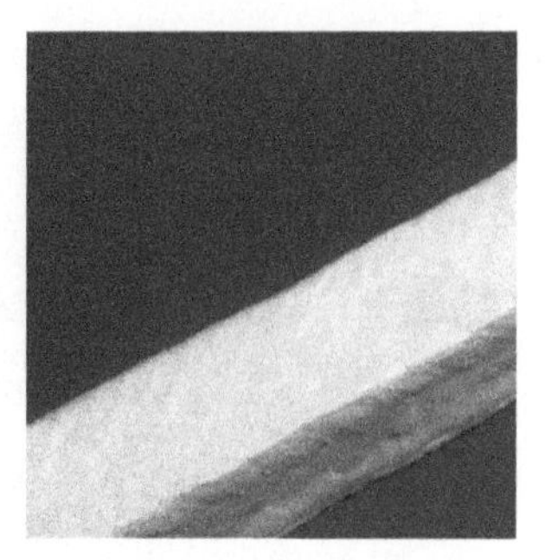

(B) TEFLON® YARN

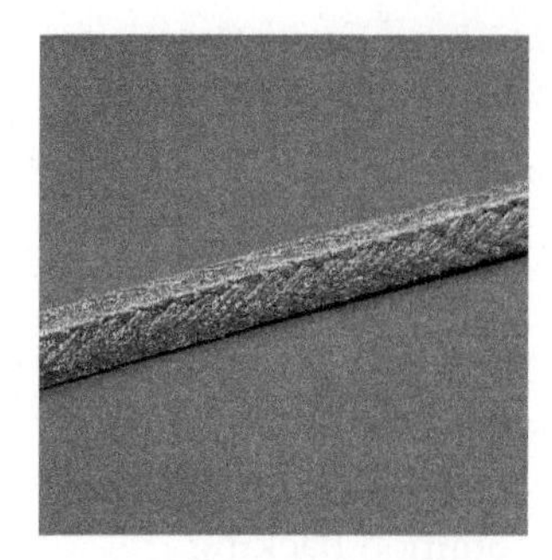

(C) GRAPHITE YARN

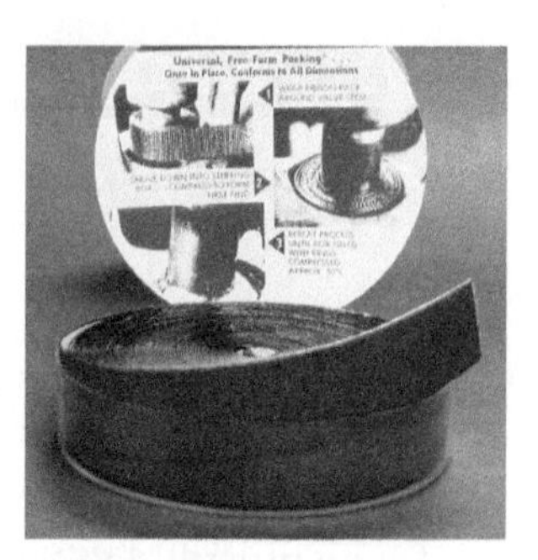

(D) GRAPHITE RIBBON

(E) PACKING IN PLACE

(F) SYNTHETIC FIBER

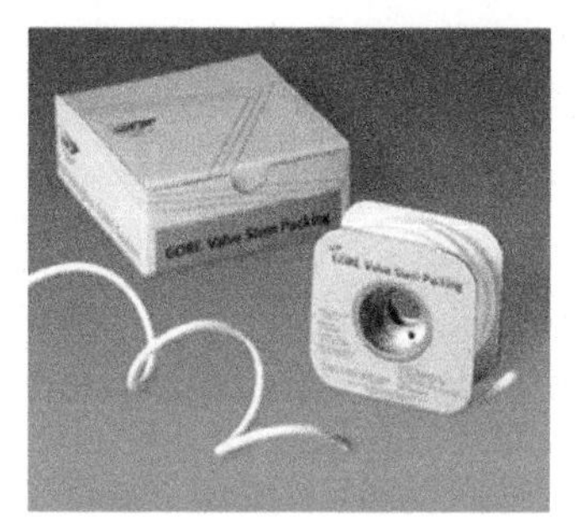

(G) EXPANDED PTFE

Figure 8 Types of packing.

3.1.3 Lubricated Graphite Yarn Packing

Lubricated graphite yarn packing is an intertwined braid of pure graphite yarn impregnated with inorganic graphite particles. The graphite particles dissipate heat. Lubricated graphite yarn packing also contains a special lubricant that provides a film to prevent wicking and reduce friction. It is good for use in high-temperature applications.

3.1.4 Lubricated Carbon Yarn Packing

Lubricated carbon yarn packing is made from an intertwined braid of carbon fibers impregnated with graphite particles and lubricants to fill voids and block leakage. Lubricated carbon yarn packing is used in systems containing water, steam, and solutions of acids and alkalis. It is considered suitable for steam applications with temperatures up to approximately 1,200°F, and up to 600°F where oxygen is present. It is reactive to oxygen atmospheres.

3.1.5 TFE/Synthetic Fiber Packing

TFE/synthetic fiber packing is made from braided yarn fibers saturated and sealed with TFE particles before being woven into a multi-lock braided packing. TFE/synthetic fiber packing protects against a variety of chemical actions. It is used in applications where caustics, mild acids, gases, and many chemicals and solvents are present. It is often used in general service for rotating and reciprocating pumps, agitators, and valves. Some variations also use TFE yarns (sometimes combined with Inconel) and a filling of graphite or other material, held in place with the yarn exterior.

3.2.0 Repacking Valves

If a valve is leaking through the stem, tightening the valve usually stops the leak. If it doesn't, the packing may need to be checked. Always store packing materials in a clean, dry place, leaving them in their original wrappings. Never allow dirt or abrasive materials to come in contact with the packing material.

The precise procedure for replacing packing varies according to the type of valve and the type of packing. The following is a general procedure for repacking a valve:

Step 1 Isolate the system according to plant procedures. The first requirement is that the valve must be fully backseatable and must be backseated.

WARNING!

Never repack a valve with the valve under pressure unless specifically told to do so by your supervisor. The circumstances in which you would do this will not be a judgment call of yours. The first requirement is that the valve be completely backseated; that is, the valve is made so that it can be opened all the way to seal the stem and packing completely from the pressure of the fluid. This is still a dangerous operation, especially with high temperature or reactive materials.

Step 2 Perform the standard lockout/tagout procedure.

Step 3 Isolate the valve and remove it from the system.

Step 4 Secure the valve in a vise, with the valve ends between the jaws of the vise.

CAUTION

Use soft covers in the vise jaws to protect the valve from damage.

Step 5 Turn the handwheel to open the valve completely.

Step 6 Remove the handwheel nut from the valve stem.

Step 7 Remove the handwheel from the valve stem.

Step 8 Remove the gland bolts.

Step 9 Slide the gland flange off the valve body and over the stem.

Step 10 Slide the packing gland off the valve stem, exposing the valve packing area.

Step 11 Remove the valve packing material, using a valve packing removal tool (*Figure 9*).

Figure 9 Valve packing removal tool.

Step 12 Clean the stem, packing gland, and stuffing box using a fine-grit emery cloth, making sure all traces of the old packing material are removed.

Step 13 Select new packing material recommended by the manufacturer or plant specifications.

Step 14 Wrap a length of packing around the valve stem.

Step 15 Mark two, 45-degree lines where the packing material overlaps. *Figure 10* shows marking the valve packing material.

Step 16 Remove the packing from the stem.

Step 17 Use a knife to cut the packing along the 45-degree marks on the packing material.

CAUTION

Do not cut the packing on the stem.

Step 18 Place the cut piece of packing around the stem to check the fit. The packing material should fit snugly around it.

Step 19 Remove the packing from the stem. This piece is the pattern for cutting more pieces of packing.

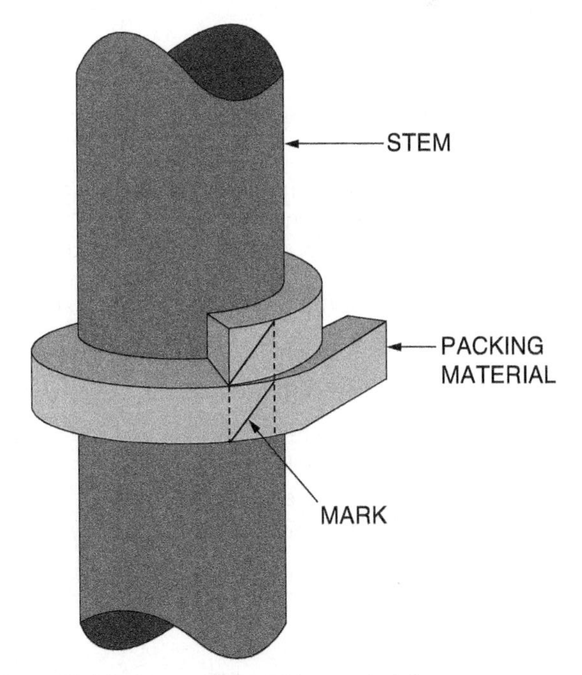

Figure 10 Marking valve packing material.

Step 20 Cut several pieces of the packing material using the pattern.

Step 21 Push the first piece of packing into the stuffing box, using the packing gland to push it down.

Step 22 Place another piece of packing in the stuffing box using the packing gland to push it down. Rotate the alignment of the packing joints 180 degrees with each layer so they do not line up (*Figure 11*).

Step 23 Continue stuffing the packing, one piece at a time, until the stuffing box is full.

Step 24 Slide the packing gland over the stem.

Step 25 Slide the gland flange over the stem and onto the valve body.

Step 26 Replace and hand tighten the gland bolts.

Step 27 Replace the handwheel on the stem.

Step 28 Replace and tighten the handwheel nut.

Step 29 Clean and lubricate the gland bolt threads.

Step 30 Insert the gland bolts into the glands.

Step 31 Tighten all bolts using the crossover method. Use a torque wrench and follow the valve manufacturer's specifications for the proper torque.

Step 32 Return the valve to service.

Step 33 Place a few drops of oil on the stem depending on the specifications.

Step 34 Turn the handwheel a few times to ensure that it is turning properly. If it is not, repeat Steps 6 through 32; otherwise, proceed to Step 35.

Step 35 Test the valve if required.

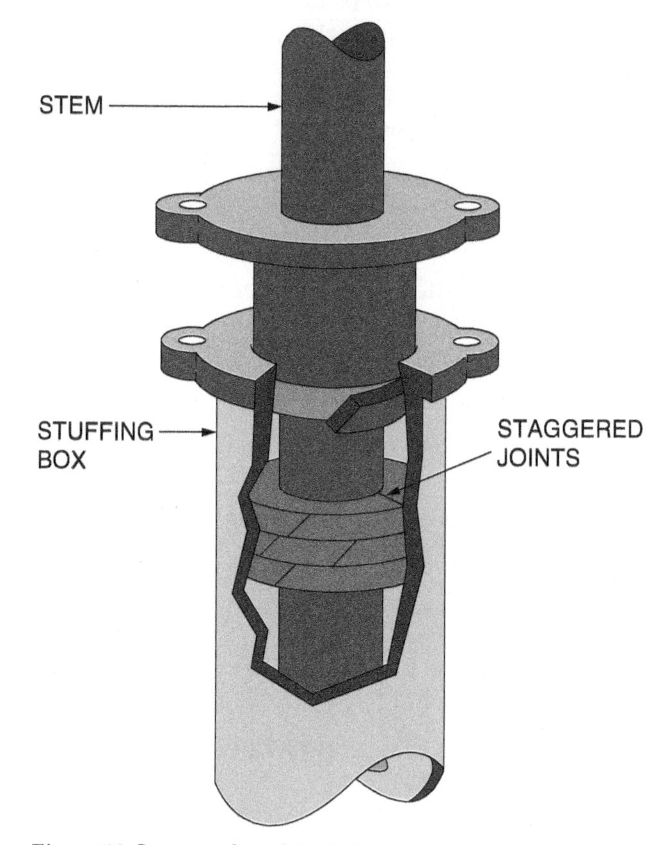

Figure 11 Staggered packing joints.

3.0.0 Section Review

1. Packing may be solid, or it may have braided or ____.
   a. asbestos fibers
   b. diametric fibers
   c. granulated fibers
   d. cabled fibers

2. What is the first step in repacking a valve?
   a. Isolate the system
   b. Perform the lockout
   c. Secure the valve in a vise
   d. Slide the packing gland off the valve stem

Review Questions

1. When the plug of a valve is pushed away from the seat, this results in _____.
   a. direct pressure
   b. a reverse pressure
   c. a Venturi port
   d. a full port

2. Which part of the valve receives the *most* wear and tear?
   a. Trim
   b. Port
   c. Stem
   d. O-ring

3. When installing a threaded valve, you will *most* often lubricate the _____.
   a. threads of the pipe
   b. threads inside the valve
   c. inside of the pipe
   d. innermost thread of the pipe in the valve

4. The first step of installing a threaded valve is to _____.
   a. isolate that part of the system
   b. cut the pipe on the downstream side
   c. open the drain valves
   d. make sure the pressure is on at that valve

5. When installing flanged valves, avoid unequal pressure on the flanges by using the crossover method of tightening flange bolts and _____.
   a. torquing bolts to the full pressure the first time around
   b. increasing torque by 25 percent each time around until the required torque has been reached
   c. increasing torque by 50 percent each time around until the required torque has been reached
   d. placing the torque wrench on the bolt, not the nut

6. What type of stem should be used with corrosive fluids?
   a. The OS&Y
   b. Silicone O-ring
   c. Rising stem
   d. Non-rising stem

7. What type of O-ring does *not* weather well?
   a. Polyurethane
   b. Silicone
   c. Teflon®
   d. Buna-N

8. Valves with bolted bonnets are commonly used for line sizes larger than _____.
   a. 1⁄8"
   b. 1⁄4"
   c. 1⁄2"
   d. 1 1⁄4"

9. All of the following *must* be considered when choosing packing for a valve *except* _____.
   a. material flowing through the line
   b. operating pressure
   c. whether the valve is threaded
   d. the composition of the valve stem

10. The ends of the packing should be _____.
    a. cut square
    b. staggered
    c. curved
    d. straight up and down

11. When replacing valve stem O-rings, the first step is to remove the _____.
    a. retaining nut
    b. handwheel
    c. O-rings
    d. jam nut

12. The challenge in producing a good seal at the stem of a valve is related to _____.
    a. the size of the stem in relation to the valve
    b. the shape of the stem, which requires a special instrument to maneuver
    c. the movement of the stem
    d. displacement of the steam due to temperature variants

13. In areas where space is limited, what type of valve stem is workable?
    a. Rising stem
    b. Non-rising stem
    c. OS&Y
    d. Hollow

14. Ethylene propylene O-Rings are the ideal choice for systems involving petroleum fluids.
    a. True
    b. False

15. If dry heat, over long periods of time, is likely, what kind of O-rings should be used?
    a. Silicone
    b. Ethylene propylene
    c. Teflon®
    d. Buna-N

Trade Terms Introduced in This Module

Body: The main part of the valve. It contains the disc, seat, and valve ports. The body of the valve is directly connected to the piping by threaded, welded, mechanically-joined, or flanged ends.

Bonnet: The part of the valve that contains the trim. The bonnet is located above the body.

Direct pressure: Flow pushing the sealing element of the valve into the seat and improving closure.

Disc: The moving part of a valve that directly affects the flow through the valve.

Packing: The material between the valve stem and bonnet that provides a leakproof seal and prevents material from leaking up around the valve stem.

Port: The internal opening for flow through a valve.

Reverse pressure: Flow pushing the closure element out of the seat.

Seat: The nonmoving part of the valve on which the disc rests to form a seal and close off the valve.

Stem: The part of the valve that connects the disc to the valve operator. A stem can have linear, rotary, or helical movement.

Trim: The internal parts of a valve that receive the most wear and can be replaced. The trim includes the stem disc, seat ring, disc holder or guide, wedge, and bushings.

Additional Resources

This module presents thorough resources for task training. The following reference material is recommended for further study.

American: The Right Way products listing. American Cast Iron Pipe Company. Available at: **https://american-usa.com**.

Choosing the Right Valve. New York, NY: Crane Company.

Piping Pointers; Application and Maintenance of Valves and Piping Equipment. New York, NY: Crane Company.

Safety Relief Valve Replacement, Maintenance, and Installation. 2000. The Air Conditioning, Heating and Refrigeration NEWS. Available at: **https://www.achrnews.com/articles/94640-safety-relief-valve-replacement-maintenance-and-installation**.

Velan Products. 2018. Velan. Available at: **http://www.velan.com/products/index.htm**.

Figure Credits

NIBCO INC., Figures 2, 6

Sealing Equipment Products Co., Inc. (SEPCO®), Figure 8 (A-E)

A.W. Chesterton, Figure 8 (F)

GORE™ Valve Stem Packing ©W.L. Gore & Associates, Figure 8 (G)

Section Review Answer Key

Section 1.0.0

Answer	Section Reference	Objective
1. c	1.1.0	1a
2. b	1.2.0	1b

Section 2.0.0

Answer	Section Reference	Objective
1. d	2.1.4	2a
2. a	2.2.0	2b

Section 3.0.0

Answer	Section Reference	Objective
1. c	3.1.0	3a
2. a	3.2.0	3b

This page is intentionally left blank.

NCCER CURRICULA — USER UPDATE

NCCER makes every effort to keep its textbooks up-to-date and free of technical errors. We appreciate your help in this process. If you find an error, a typographical mistake, or an inaccuracy in NCCER's curricula, please fill out this form (or a photocopy), or complete the online form at **www.nccer.org/olf**. Be sure to include the exact module ID number, page number, a detailed description, and your recommended correction. Your input will be brought to the attention of the Authoring Team. Thank you for your assistance.

Instructors – If you have an idea for improving this textbook, or have found that additional materials were necessary to teach this module effectively, please let us know so that we may present your suggestions to the Authoring Team.

NCCER Product Development and Revision

13614 Progress Blvd., Alachua, FL 32615

Email: curriculum@nccer.org
Online: www.nccer.org/olf

❑ Trainee Guide ❑ Lesson Plans ❑ Exam ❑ PowerPoints Other ________________

Craft / Level: Copyright Date:

Module ID Number / Title:

Section Number(s):

Description:

Recommended Correction:

Your Name:

Address:

Email: Phone:

This page is intentionally left blank.

Fundamentals of Crew Leadership

Overview

When a crew is assembled to complete a job, one person is appointed the leader. This person is usually an experienced craft professional who has demonstrated leadership qualities. While having natural leadership qualities helps in becoming an effective leader, it is more true that "leaders are made, not born." Whether you are a crew leader or want to become one, this module will help you learn more about the requirements and skills needed to succeed.

Module 46101

Trainees with successful module completions may be eligible for credentialing through the NCCER Registry. To learn more, go to **www.nccer.org** or contact us at 1.888.622.3720. Our website, **www.nccer.org**, has information on the latest product releases and training.

Your feedback is welcome. You may email your comments to **curriculum@nccer.org**, send general comments and inquiries to **info@nccer.org**, or fill in the User Update form at the back of this module.

This information is general in nature and intended for training purposes only. Actual performance of activities described in this manual requires compliance with all applicable operating, service, maintenance, and safety procedures under the direction of qualified personnel. References in this manual to patented or proprietary devices don't constitute a recommendation of their use.

46101

FUNDAMENTALS OF CREW LEADERSHIP

Objectives

When you have completed this module, you will be able to do the following:

1. Describe current issues and organizational structures in industry today.
   a. Describe the leadership issues facing the construction industry.
   b. Explain how gender and cultural issues affect the construction industry.
   c. Explain the organization of construction businesses and the need for policies and procedures.
2. Explain how to incorporate leadership skills into work habits, including communications, motivation, team-building, problem-solving, and decision-making skills.
   a. Describe the role of a leader on a construction crew.
   b. Explain the importance of written and oral communication skills.
   c. Describe methods for motivating team members.
   d. Explain the importance of teamwork to a construction project.
   e. Identify effective problem-solving and decision-making methods.
3. Identify a crew leader's typical safety responsibilities with respect to common safety issues, including awareness of safety regulations and the cost of accidents.
   a. Explain how a strong safety program can enhance a company's success.
   b. Explain the purpose of OSHA and describe the role of OSHA in administering worker safety.
   c. Describe the role of employers in establishing and administering safety programs.
   d. Explain how crew leaders are involved in administering safety policies and procedures.
4. Demonstrate a basic understanding of the planning process, scheduling, and cost and resource control.
   a. Describe how construction contracts are structured.
   b. Describe the project planning and scheduling processes.
   c. Explain how to implement cost controls on a construction project.
   d. Explain the crew leader's role in controlling project resources and productivity.

Performance Tasks

Under the supervision of your instructor, you should be able to do the following:

1. Develop and present a look-ahead schedule.
2. Develop an estimate for a given work activity.

Trade Terms

- Autonomy
- Bias
- Cloud-based applications
- Craft professionals
- Crew leader
- Critical path
- Demographics
- Ethics
- Infer
- Intangible
- Job description
- Job diary
- Legend
- Lethargy
- Letter of instruction (LOI)
- Local area networks (LAN)
- Lockout/tagout (LOTO)
- Look-ahead schedule
- Negligence
- Organizational chart
- Paraphrase
- Pragmatic
- Proactive
- Project manager
- Return on investment (ROI)
- Safety data sheets (SDS)
- Sexual harassment
- Smartphone
- Superintendent
- Synergy
- Textspeak
- Wide area networks (WAN)
- Wi-Fi
- Work breakdown structure (WBS)

Industry Recognized Credentials

If you are training through an NCCER-accredited sponsor, you may be eligible for credentials from NCCER's Registry. The ID number for this module is 46101. Note that this module may have been used in other NCCER curricula and may apply to other level completions. Contact NCCER's Registry at 888.622.3720 or go to **www.nccer.org** for more information.

Contents

Figures and Tables

SECTION ONE

1.0.0 BUSINESS STRUCTURES AND ISSUES IN THE INDUSTRY

Objective

Describe current issues and organizational structures in industry today.

a. Describe the leadership issues facing the construction industry.
b. Explain how gender and cultural issues affect the construction industry.
c. Explain the organization of construction businesses and the need for policies and procedures.

Trade Terms

Cloud-based applications: Mobile and desktop digital programs that can connect to files and data located in distributed storage locations on the Internet ("the Cloud"). Such applications make it possible for authorized users to create, edit, and distribute content from any location where Internet access is possible.

Craft professionals: Workers who are properly trained and work in a particular construction trade or craft.

Crew leader: The immediate supervisor of a crew or team of craft professionals and other assigned persons.

Demographics: Social characteristics and other factors, such as language, economics, education, culture, and age, that define a statistical group of individuals. An individual can be a member of more than one demographic.

Job description: A description of the scope and responsibilities of a worker's job so that the individual and others understand what the job entails.

Letter of instruction (LOI): A written communication from a supervisor to a subordinate that informs the latter of some inadequacy in the individual's performance and provides a list of actions that the individual must satisfactorily complete to remediate the problem. This is usually a key step in a series of disciplinary actions.

Local area networks (LAN): Communication networks that link computers, printers, and servers within a small, defined location, such as a building or office, via hard-wired or wireless connections.

Organizational chart: A diagram that shows how the various management and operational responsibilities relate to each other within an organization. Named positions appear in ranked levels, top to bottom, from those functional units with the most and broadest authority to those with the least, with lines connecting positions indicating chains of authority and other relationships.

Paraphrase: Rewording a written or verbal statement in one's own words in such a way that the intent of the original statement is retained.

Pragmatic: Sensible, practical, and realistic.

Sexual harassment: Any unwelcome verbal or nonverbal form of communication or action construed by an individual to be of a sexual or gender-related nature.

Smartphone: The most common type of cellular telephone that combines many other digital functions within a single mobile device. Most important of these, besides the wireless telephone, are high-resolution still and video cameras, text and email messaging, and GPS-enabled features.

Synergy: Any type of cooperation between organizations, individuals, or other entities where the combined effect is greater than the sum of the individual efforts.

Wide area networks (WAN): Dedicated communication networks of computers and related hardware that serve a given geographic area, such as a work site, campus, city, or a larger but distinct area. Connectivity is by wired and wireless means, and may use the Internet as well.

Wi-Fi: The technology allowing communications via radio signals over a LAN or WAN equipped with a wireless access point or the Internet. (Wi-Fi stands for wireless fidelity.) Many types of mobile, portable, and desktop devices can communicate via Wi-Fi connections.

Today's managers, supervisors, and lower-level managers face challenges different from those of previous generations. To be a **crew leader** today, it is essential to be well prepared. Crew leaders must understand how to use various types of new technology. In addition, they must have the knowledge and skills needed to manage, train, and communicate with a culturally-diverse workforce whose attitudes toward work and job expectations may differ from those of earlier generations and cultures.

A summary of the changes in the workforce, in the work environment, and of industry needs include the following:

- A shrinking workforce
- The growth of construction, communication, and scheduling technology
- Changes in the attitudes and values of **craft professionals**
- The rapidly-changing **demographics** of gender, gender-identity, and foreign-born workers
- Increased emphasis on workplace safety and health
- Greater need for more education and training

1.1.0 Leadership Issues and Training Strategies

Effective craft training programs are necessary if the industry is to meet the forecasted worker demands. Many skilled, knowledgeable craft professionals, crew leaders, and managers have reached retirement age. In 2015, the generation of workers called *"Baby Boomers,"* who were born between 1946 and 1964, represented 50 percent of the workforce. Their departure creates a large demand for craftworkers across the industry. The US Department of Labor (DOL) concludes that the best way for industry to reduce shortages of skilled workers is to create more education and training opportunities. The DOL suggests that companies and community groups form partnerships and create apprenticeship programs. Such programs could provide younger workers, including women and minorities, with the opportunity to develop job skills by giving them hands-on experience.

When training workers, it is important to understand that people learn in different ways. Some people learn by doing, some people learn by watching or reading, and others need step-by-step instructions as they are shown the process. Most people learn best through a combination of styles. While you may have the tendency to teach in the style that you learn best, you must always stay in tune with what kind of learner you are teaching. Have you ever tried to teach somebody and failed, and then another person successfully teaches the same thing in a different way? A person who acts as a mentor or trainer needs to be able to determine what kind of learner they are addressing, and teach according to those needs.

"The mediocre teacher tells.
The good teacher explains.
The superior teacher demonstrates.
The great teacher inspires."
—William Arthur Ward (1921)

The need for training isn't limited to craft professionals. There must be supervisory training to ensure there are qualified leaders in the industry to supervise the craft professionals.

1.1.1 Motivation

As a supervisor or crew leader, it is important to understand what motivates your crew. Money is often considered a good motivator, but it is sometimes only a temporary solution. Once a person has reached a level of financial security, other factors come into play. Studies show that environment and conditions motivate many people. For those people, a great workplace may mean more to them than better pay.

If you give someone a raise, they tend to work harder for a period of time. Then the satisfaction wanes and they may want another raise. A sense of accomplishment is what motivates most people. That is why setting and working toward recognizable goals tends to make employees more productive. A person with a feeling of involvement or a sense of achievement is likely to be better motivated and help to motivate others.

1.1.2 Understanding Workers

Many older workers grew up in an environment where they learned to work hard with the same employer until retirement. They expected to stay with a company for a long time, and the structure of the companies created a family-type environment.

Times have changed. Younger workers have grown up in a highly mobile society and expect frequent rewards and rapid advancement. Some might perceive this generation of workers as lazy, narcissistic, and unmotivated, but in reality, they simply have a different perspective on life, work, and priorities. For such workers, it may be better to give them small projects or break up large projects into smaller pieces so that they are rewarded more often by successfully achieving short-term goals. The following strategies are important for keeping young workers motivated and engaged:

- *Goal setting* – Set short-term and long-term goals, including tasks to be done and expected time frames. Things can happen to change, delay, or upset the short-term goals. This is one reason to set long-term goals as well. Don't set workers up for failure, as this leads to frustration, and frustration can lead to reduced productivity.
- *Feedback* – Timely feedback is important. For example, telling someone they did a good job last year, or criticizing them for a job they did a month ago, is meaningless. Simple recognition isn't always enough. Some type of reward should accompany positive feedback, even if it is simply recognizing the employee in a public way. Constructive criticism or reprimands should always be given in private. You can also provide some positive action, such as one-on-one training, to correct a problem.

1.1.3 Craft Training

Craft training is often informal, taking place on the job site, outside of a traditional training classroom. According to the American Society for Training and Development (ASTD), a qualified co-worker or a supervisor conducts craft training generally through on-the-job instruction. The Society of Human Resources Management (SHRM) offers the following tips to supervisors in charge of training their employees:

- *Help crew members establish career goals* – Once career goals are established, you can readily identify the training required to meet the goals.
- *Determine what kind of training to give* – Training can be on the job under the supervision of a co-worker. It can be one-on-one with the supervisor. It can involve cross training to teach a new trade or skill, or it can involve delegating new or additional responsibilities.
- *Determine the trainee's preferred method of learning* – Some people learn best by watching, others from verbal instructions, and others by doing. These are categorized as visual learners, auditory learners, and tactile learners, respectively. When training more than one person at a time, try to use a mix of all three methods.

Communication is a critical component of training employees. SHRM advises that supervisors do the following when training their employees:

- Explain the task, why it is important, and how to do it. Confirm that the worker trainees understand these three areas by asking questions. Allow them to ask questions as well.
- Demonstrate the task. Break the task down into manageable parts and cover one part at a time.
- Ask your trainees to do the task while you observe them. Try not to interrupt them while they are doing the task unless they are doing something that is unsafe or potentially harmful.
- Give the trainees feedback. Be specific about what they did and mention any areas where they need to improve.

1.1.4 Supervisory Training

Because of the need for skilled craft professionals and qualified supervisory personnel, some companies offer training to their employees through in-house classes, or by subsidizing outside training programs. However, for a variety of reasons, many contractors don't offer training at all. Some common reasons include the following:

- Lack of money to train
- Lack of time to train
- Lack of knowledge about the benefits of training programs
- High rate of employee turnover
- Workforce is too small
- Past training involvement was ineffective
- The company hires only trained workers
- Lack of interest from workers

For craft professionals to move up into supervisory and managerial positions, they must continue their education and training. Those who are willing to acquire and develop new skills have the best chance of finding stable employment. This makes it critical for them to take advantage of training opportunities, and for companies to incorporate training into their business culture.

If your company has recognized the need for training, your participation in a leadership training program such as this will begin to fill the gap between craft and supervisory training.

1.1.5 Impact of Technology

Many industries, including the construction industry, have embraced technology to remain competitive. Benefits of technology include increased productivity and speed, improved quality of documents, greater access to common data, and better financial controls and communication. As technology becomes a greater part of supervision, crew leaders must be able to use it properly.

Cellphones (in particular the **smartphone**) have made it easy to keep in touch through numerous forms of media. As of 2016, 95 percent of all Americans own a cellphone of some kind

Did You Know?

Learning Strategies

There is a lot more to training than simply telling a group of people what you know about a subject or talking your way through a series of slides. Studies have shown that the information and understanding that learners retain depends on how they receive the information and use it. These learning processes are as follows, listed from least effective to most effective:

1. Reading about a process
2. Hearing a description of a process
3. Observing a process
4. Observing and hearing a description of a process
5. Observing, hearing, and responding to instruction
6. Observing, hearing, and doing a process

Although some companies may attempt to assign percentage effectiveness to these instructional approaches, the results are very arbitrary and inconclusive. It is more important to understand which are the most effective methods for teaching and training.

and 77 percent own a smartphone, according to the Pew Research Center. They are particularly useful communication tools for contractors or crew leaders who are on a job site, away from their offices, or constantly on the go. Workers use smartphones at any time for phone calls, emails, text messages, and voicemail, as well as to share photos and videos. The hundreds of thousands of mobile computing apps available allow smartphones to perform numerous other functions on the go. However, the number of accidents due to inattention while focusing on a smartphone or other mobile device is also rapidly rising. Always check the company's policy regarding cell phone use on the job.

As a crew leader, you should be aware that smartphones and tablets (*Figure 1*) allow supervisors to plan their calendars, schedule meetings, manage projects, and access their company email from remote locations. These devices are far more powerful than the computers that took us to the moon, and they can hold years of information from various projects. **Cloud-based applications** now permit remote access to and updating of files, plans, and data from any place with wireless connectivity. In fact, it is becoming common for work sites to set up **local area networks (LAN)**, **wide area networks (WAN)**, and dedicated **Wi-Fi** services to support mobile communications among workers on the job.

In all forms of electronic communication (verbal, written, and visual), it is important to keep messages brief, factual, and legal. Text-based communications can be easily misunderstood because there are no visual or auditory cues to indicate the sender's intent. In other words, it is more difficult to tell if someone is just joking via email because you can't see the sender's expression or hear their tone of voice.

1.2.0 Gender and Cultural Issues

In the past several years, the construction industry in the United States has experienced a shift in worker expectations and diversity. These two issues are converging at a rapid pace.

The generation of learners is also a factor in the learning process and in the workplace. The various generations include Baby Boomers, Generation X, Millennials (Generation Y), and Generation Z. The ranges of birth years for these generations are as follows (note that there may be

Figure 1 Digital tablets are a convenient management and scheduling tool for supervisors.

some overlap, and opinions on the generational names as well as the range of years for each generation may vary):

- *Baby Boomers* – born 1946 through 1964
- *Generation X* – born 1965 through 1979
- *Millennials* (*Generation Y*) – born 1980 through 2000
- *Generation Z* – born 2000 through 2015

Each generation has been studied to some degree to determine their different interpretations of and approaches to learning and training. Family norms, religious values and morals, educational methods, music, movies, politics, and global events are all influential factors that define a generation. Remember that individuals within a generation can vary widely in their expectations and behaviors.

This trend, combined with industry diversity initiatives, has created a climate in which companies recognize the need to embrace a diverse workforce that crosses generational, gender, and ethnic boundaries (*Figure* 2). To do this effectively, they are using their own resources, as well as relying on associations with the government and trade organizations.

All current research indicates that industry will be more dependent on the critical skills of a diverse workforce—but a workforce that must be both culturally and ethnically fused. Across the United States, construction and other industries are aggressively seeking to bring new workers into their ranks, including women, racial, and ethnic minorities. Social and political issues are no longer the main factors driving workplace diversity, but by consumers and citizens who need more hospitals, malls, bridges, power plants, refineries, and many other commercial and residential structures. The construction industry needs more workers.

Figure 2 The modern workforce is diverse.

There are some potential issues relating to a diverse workforce that may be encountered on the job site. These issues include different communication styles of men and women, language barriers associated with cultural differences, sexual harassment, and gender or racial discrimination.

1.2.1 Communication Styles of Men and Women

As more and more women enter construction workforce, it becomes increasingly important to break down communication barriers between men and women and to understand differences in behaviors so that men and women can work together more effectively. The Jamestown, Area Labor Management Committee (JALMC) in New York offers the following explanations and tips (*emphasized text* is from *Exchange and Deception: A Feminist Perspective*, ed. by Caroline Gerschlager, Monika Mokre, 2002. Springer Science & Business Media):

- *Women tend to ask more questions than men do* – Men are more likely to proceed with a job and figure it out as they go along, while women are more likely to ask questions first.
- *Men tend to offer solutions before empathy; women tend to do the opposite* – Both men and women should say what they want up front, whether it's the solution to a problem, or simply a sympathetic ear. That way, both genders will feel understood and supported.
- *Women are more likely to ask for help when they need it* – Women are generally more pragmatic when it comes to completing a task. If they need help, they will ask for it. Men are more likely to attempt to complete a task by themselves, even when they need assistance.
- *Men tend to communicate more competitively, and women tend to communicate more cooperatively* – Both parties need to hear one another out without interruption.

This doesn't mean that any one method is better or worse than the other; it simply means that men and women inherently use different approaches to achieve the same result. It's either genetic or cultural, but awareness can overcome these tendencies.

1.2.2 Language Barriers

Language barriers are a real workplace challenge for crew leaders. Millions of American workers speak languages other than English. Bilingual job sites are increasingly common. As the makeup of the immigrant population continues to change with the influx of refugees from around the world, the number of non-English speakers in the American workforce is rising and diversifying dramatically.

In addition, there are many different dialects of the English language in America, which can present some communication challenges. For example, some workers may speak a dialect commonly called *Ebonics* (referred to by linguists as *African-American Vernacular English*). The dialect called *Rural White Southern English* can also be difficult to understand by those not born in the American South. Many find the American New England accent and dialect difficult to comprehend. Some sources list as many as 16 distinct American dialects.

Companies have the following options to overcome the language challenge, mainly for foreign or English-as-a-second-language (ESL) workers:

- Offer English classes either at the work site or through school districts and community colleges.
- Offer incentives for workers to learn English.

Communication with ESL workers, and varying dialects in general, becomes even more critical as the workforce grows more diverse. The following tips will help when communicating across language barriers:

- Be patient. Give workers time to process the information in a way they can comprehend.
- Avoid humor. Humor is easily misunderstood. The worker may misinterpret what you say as a joke at the worker's expense.
- Don't assume workers are unintelligent simply because they don't understand what you are saying. Explaining something in multiple ways, and having trainees **paraphrase** what they heard, is an excellent way to ensure mutual understanding and prevent miscommunication.
- If a worker is not fluent in English, ask the worker to demonstrate his or her understanding through other means.
- Speak slowly and clearly, and avoid the tendency to raise your voice.
- Use face-to-face communication whenever possible. Over-the-phone communication is often more difficult when a language barrier is involved.
- Use pictures or sketches to get your point across.

1.2.3 Cultural Differences

As workers from a multitude of backgrounds and cultures work together, there are bound to be differences and conflicts in the workplace. To overcome cultural conflicts, the SHRM suggests the following approach to resolving cultural conflicts between individuals:

- *Define the problem from both points of view* – How does each person involved view the conflict? What does each person think is wrong? This involves moving beyond traditional thought processes to consider alternate ways of thinking.
- *Uncover cultural interpretations* – What assumptions may the parties involved be making based on cultural programming? This is particularly true for certain gestures, symbols, and words that mean different things in different cultures. By doing this, the supervisor may realize what motivated an employee to act in a specific way.
- *Create cultural* **synergy** – Devise a solution that works for both parties involved. The purpose is to recognize and respect other's cultural values, and work out mutually acceptable alternatives.

1.2.4 Sexual Harassment

In today's business world, men and women are working side-by-side in careers of all kinds, creating increased opportunity for sexual harassment to occur. Sexual harassment can be defined as any unwelcome behavior that makes someone feel uncomfortable in the workplace by focusing attention on their gender or gender identity. Activities that might qualify as sexual harassment include, but are not limited to, the following:

- Telling an offensive, sexually-oriented joke
- Displaying a poster of a man or woman in a revealing swimsuit
- Wearing a patch or article of clothing blatantly promoting or degrading a specific gender
- Making verbal or physical advances (*Figure 3*)
- Speaking abusively about a specific gender

Historically, many have thought of sexual harassment as an act perpetrated by men against women, especially those in subordinate positions. However, the nature of sexual harassment cases over the years have shown that the perpetrator can be of any gender.

Figure 3 Any unwelcome physical contact can be considered harassment.

Sexual harassment can occur in various circumstances, including the following:

- The victim and harasser may be either male or female. The victim does not have to be of the opposite gender.
- The harasser can be the victim's supervisor, an agent of the employer, a supervisor in another area, a co-worker, a subordinate, or a nonemployee.
- Legally, the victim doesn't have to be the person harassed, but could be anyone offended by the relevant conduct.
- Unlawful sexual harassment may occur without economic injury to or the discharge of the victim.
- The harasser's conduct must be unwelcome.

The Equal Employment Opportunity Commission (EEOC) enforces sexual harassment laws within industries. When investigating allegations of sexual harassment, the EEOC looks at the whole record, including the circumstances and the context in which the alleged incidents occurred. A decision on the allegations is made from the facts on a case-by-case basis. Supervision can hold the crew leader responsible who is aware of sexual harassment and does nothing to stop it. The crew leader therefore should not only take action to stop sexual harassment, but should serve as a good example for the rest of the crew.

Prevention is the best tool to eliminate sexual harassment in the workplace. The EEOC encourages employers to take steps to prevent sexual harassment from occurring. Employers should clearly communicate to employees that they and their company won't tolerate sexual harassment. They do so by developing a policy on sexual harassment, establishing an effective complaint or grievance process, and taking immediate and appropriate action when an employee complains.

Controversial or sexually-related remarks, jokes, and swearing are not only offensive to co-workers, but also tarnish a worker's character. Crew leaders need to emphasize that abrasive or crude behavior may affect opportunities for advancement. If disciplinary action becomes necessary, written company policies should spell out the type of actions warranted by such behavior. A typical approach is a three-step process in which the perpetrator is first given a verbal reprimand. In the event of further violations, a written reprimand, often including a **letter of instruction (LOI)** and a warning are given. Dismissal typically accompanies any subsequent violations.

1.2.5 Gender and Minority Discrimination

Employers are giving more attention to fair recruitment, equal pay for equal work, and promotions for women and minorities in the workplace. Consequently, many businesses are analyzing their practices for equity, including the treatment of employees, the organization's hiring and promotional practices, and compensation.

The construction industry, which was once male-dominated, is moving away from this image and actively recruiting and training women, younger workers, people from other cultures, and disabled workers. This means that organizations hire the best person for the job, without regard for race, sex, religion, age, etc.

Did You Know?

Respectful Workplace Training

Some companies employ a tool called *sensitivity training* in cases where individuals or groups have trouble adapting to a multi-cultural, multi-gender workforce. Sensitivity training is a psychological technique using group discussion, role playing, and other methods to allow participants to develop an awareness of themselves and how they interact with others. Critics of earlier forms of sensitivity training noted that the methods used were not very different from brainwashing. Modern forms of cultural-awareness and gender-issue training are more ethical, and go by the term *respectful workplace training*.

To prevent discrimination cases, employers must have valid job-related criteria for hiring, compensation, and promotion. They must apply these measures consistently for every applicant interview, employee performance appraisal, and hiring or promotion decision. Therefore, employers must train all workers responsible for recruitment, selection, supervision of employees, and evaluating job performance on how to use the job-related criteria legally and effectively.

1.3.0 Business Organization

An organization is the relationship among the people within the company or project. The crew leader needs to be aware of two types of organizations: formal organizations and informal organizations.

A formal organization exists when the people within a work group led by someone direct their activities toward achieving a common goal. An example of a formal organization is a work crew consisting of four carpenters and two laborers led by a crew leader, all working together to accomplish a project.

An organizational chart typically illustrates a formal organization. It outlines all the positions that make up an organization and shows how those positions are related. Some organizational charts even identify the person within each position and the person to whom that person reports, as well as the people that the person supervises. *Figure 4* and *Figure 5* show examples of what an organization chart might look like for a construction company and an industrial company. Note that each organizational position represents an opportunity for advancement in the construction industry that a crew leader can eventually achieve.

An informal organization allows for communication among its members so they can perform as a group. It also establishes patterns of behavior that help them to work as a group, such as agreeing to use a specific training or certification program.

An example of an informal organization is a trade association such as Associated Builders and Contractors (ABC), Associated General Contractors (AGC), and the National Association of Women in Construction (NAWIC). Those, along with the thousands of other trade associations in the United States, provide forums in which members with common concerns can share information, work on issues, and develop standards for their industry.

Both formal and informal organizations establish the foundation for how communication flows. The formal structure is the means used to delegate authority and responsibility and to exchange information. The informal structure provides means for the exchange of information.

Members in an organization perform best when they can do the following:

- Know the job and how it will be done
- Communicate effectively with others in the group
- Understand their role in the organization
- Recognize who has the authority and responsibility

"Leadership is not about titles, positions, or flow charts. It is about one life influencing another."
—John C. Maxwell

1.3.1 Division of Responsibility

The conduct of a business involves certain functions. In a small organization, one or two people may share responsibilities. However, in a larger organization with many different and complex activities, distinct activity groups may lie under the responsibility of separate department managers. In either case, the following major department functions exist in most companies:

- *Executive* – This office represents top management. It is responsible for the success of the company through short-range and long-range planning.
- *Human Resources (HR)* – This office is responsible for recruiting and screening prospective employees, managing employee benefits programs, advising management on pay and benefits, and developing and enforcing procedures related to hiring practices.
- *Accounting* – This office is responsible for all recordkeeping and financial transactions, including payroll, taxes, insurance, and audits.
- *Contract Administration* – This office prepares and executes contractual documents with owners, subcontractors, and suppliers.
- *Purchasing* – This office obtains material prices and then issues purchase orders. The purchasing office also obtains rental and leasing rates on equipment and tools.
- *Estimating* – This office is responsible for recording the quantity of material on the jobs, the takeoff, pricing labor and material, analyzing subcontractor bids, and bidding on projects.
- *Operations* – This office plans, controls, and supervises all project-related activities.

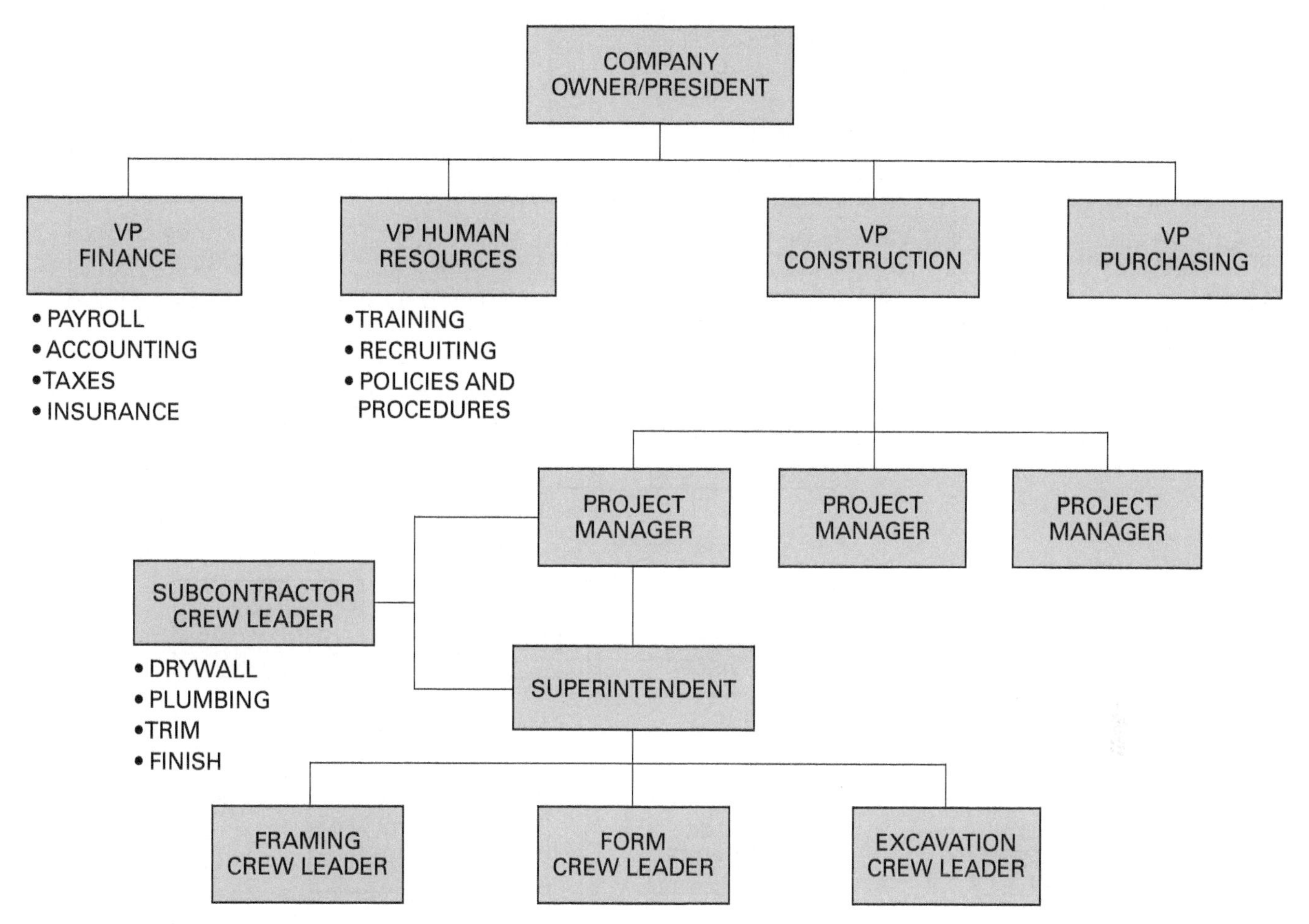

Figure 4 Sample organization chart for a construction company.

Other divisions of responsibility a company may create involve architectural and engineering design functions. These divisions usually become separate departments.

1.3.2 Responsibility, Authority, and Accountability

As an organization grows, the manager must ask others to perform many duties so that the manager can concentrate on management tasks. Managers typically assign (delegate) activities to their subordinates. When delegating activities, the crew leader assigns others the responsibility to perform the designated tasks.

Responsibility is the obligation to perform the duties. Along with responsibility comes authority. *Authority* is the right to act or make decisions in carrying out an assignment. The type and amount of authority a supervisor or worker has depends on the employee's company. There must be a balance between the authority and responsibility workers have so that they can carry out their assigned tasks. In addition, the crew leader must delegate sufficient authority to make a worker accountable to the crew leader for the results.

Accountability is holding an employee responsible for completing the assigned activities. Even though the crew leader may delegate authority and responsibility to crew members, the overall responsibility—the accountability—for the tasks assigned to the crew always rests with the crew leader.

1.3.3 Job Descriptions

Many companies furnish each employee with a written **job description** that explains the job in detail. Job descriptions set a standard for the employee. They make judging performance easier, clarify the tasks each person should handle, and simplify the training of new employees.

Each new employee should understand all the duties and responsibilities of the job after reviewing the job description. Thus, the time is shorter for the employee to make the transition from being a new and uninformed employee to a more experienced member of a crew.

A job description need not be long, but it should be detailed enough to ensure there is no misunderstanding of the duties and responsibilities of the position. The job description should

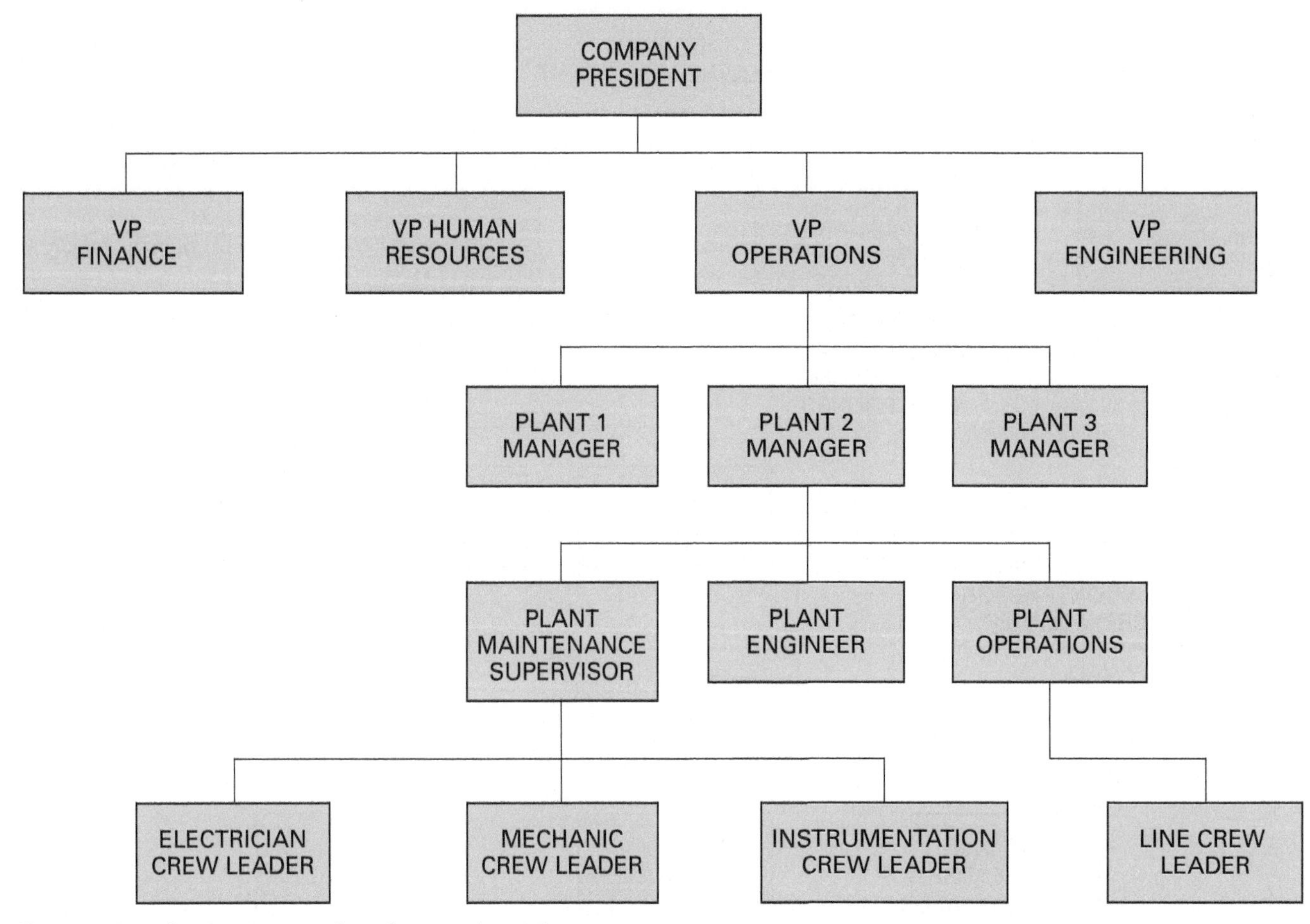

Figure 5 Sample organization chart for an industrial company.

contain all the information necessary to evaluate the employee's performance and hold the employee accountable.

A job description should contain, at minimum, the following:

- Job title
- General description of the position
- Minimum qualifications for the job
- Specific duties and responsibilities
- The supervisor to whom the position reports
- Other requirements, such as qualifications, certifications, and licenses

Figure 6 is an example of a job description.

1.3.4 Policies and Procedures

Most companies have formal policies and procedures established to help crew leaders carry out their duties. A policy is a general statement establishing guidelines for a specific activity. Examples include policies on vacations, breaks, workplace safety, and checking out tools. Procedures are formal instructions to carry out and meet policies. For example, a procedure written to implement a policy on workplace safety would include guidelines expected of all employees for reporting accidents and to follow general safety practices.

A crew leader must be familiar with the company policies and procedures, especially regarding safety practices. When OSHA inspectors visit a jobsite, they often question employees and crew leaders about the company policies related to safety. If they are investigating an accident, they will want to verify that the responsible crew leader knew the applicable company policy and followed it.

Position:
Crew Leader

General Summary:
First line of supervision on a construction crew installing concrete formwork.

Reports To:
Job Superintendent

Physical and Mental Responsibilities:
- Ability to stand for long periods
- Ability to solve basic math and geometry problems

Duties and Responsibilities:
- Oversee crew
- Provide instruction and training in construction tasks as needed
- Make sure proper materials and tools are on the site to accomplish tasks
- Keep project on schedule
- Enforce safety policies and procedures

Knowledge, Skills, and Experience Required:
- Extensive travel throughout the Eastern United States, home base in Atlanta
- Ability to operate a backhoe and trencher
- Valid commercial driver's license with no DUI violations
- Ability to work under deadlines with the knowledge and ability to foresee problem areas and develop a plan of action to solve the situation

Figure 6 A sample job description.

Did You Know?

The ADA

The Americans with Disabilities Act (ADA) is a law that prohibits employers from discriminating against disabled persons in their hiring practices. While the ADA doesn't require written job descriptions, investigators use existing job descriptions as evidence in determining whether discrimination occurred. This means that job descriptions must specifically define the job duties and the abilities needed to perform them. In making a determination of discrimination under the ADA, investigators give consideration to whether a reasonable accommodation could have qualified the disabled person to fill a job.

Additional Resources

Construction Workforce Development Professional, NCCER. 2016. New York, NY: Pearson Education, Inc.

Mentoring for Craft Professionals, NCCER. 2016. New York, NY: Pearson Education, Inc.

Generational Cohorts and their Attitudes Toward Work Related Issues in Central Kentucky, Frank Fletcher, et al. 2009. Midway College, Midway, KY. **www.kentucky.com**

The Young Person's Guide to Wisdom, Power, and Life Success: Making Smart Choices. Brian Gahran, PhD. 2014. San Diego, CA: Young Persons Press. **www.WPGBlog.com**

The following websites offer resources for products and training:

Aging Workforce News, **www.agingworkforcenews.com**

American Society for Training and Development (ASTD), **www.astd.org**

Equal Employment Opportunity Commission (EEOC), **www.eeoc.gov**

National Association of Women in Construction (NAWIC), **www.nawic.org**

Society for Human Resources Management (SHRM), **www.shrm.org**

United States Census Bureau, **www.census.gov**

United States Department of Labor, **www.dol.gov**

Wi-Fi® is a registered trademark of the Wi-Fi Alliance, **www.wi-fi.org**

1.0.0 Section Review

1. According to the US Department of Labor, the best way for the construction industry to reduce skilled-worker shortages is to _____.
   a. create training opportunities
   b. avoid discrimination lawsuits
   c. update the skills of older workers who are retiring at a later age than they previously did
   d. implement better policies and procedures

2. Which of the following is *not* helpful when dealing with language diversity in the workplace?
   a. Speaking slowly and clearly in an even tone
   b. Using humor when communicating
   c. Using sketches or diagrams to explain what needs to be done.
   d. Being patient and giving bilingual workers time to process your instructions.

3. Members tend to function best within an organization when they _____.
   a. are allowed to select their own uniform for each project
   b. understand their role within the organization
   c. don't disagree with the statements of other workers or supervisors
   d. are able to work without supervision

SECTION TWO

2.0.0 LEADERSHIP SKILLS

Objective

Explain how to incorporate leadership skills into work habits, including communications, motivation, team-building, problem-solving, and decision-making skills.

a. Describe the role of a leader on a construction crew.
b. Explain the importance of written and oral communication skills.
c. Describe methods for motivating team members.
d. Explain the importance of teamwork to a construction project.
e. Identify effective problem-solving and decision-making methods.

Trade Terms

Autonomy: The condition of having complete control over one's actions, and being free from the control of another. To be independent.

Bias: A preconceived inclination against or in favor of something.

Ethics: The moral principles that guide an individual's or organization's actions when dealing with others. It also refers to the study of moral principles.

Infer: To reach a conclusion using a method of reasoning that starts with an assumption and considers a set of logically-related events, conditions, or statements.

Legend: In maps, plans, and diagrams, an explanatory table defining all symbolic information contained in the document.

Lockout/tagout (LOTO): A system of safety procedures for securing electrical and mechanical equipment during repairs or construction, consisting of warning tags and physical locking or restraining devices applied to controls to prevent the accidental or purposeful operation of the equipment, endangering workers, equipment, or facilities.

Proactive: To anticipate and take action in the present to deal with potential future events or outcomes based on what one knows about current events or conditions.

Project manager: In construction, the individual who has overall responsibility for one or more construction projects; also called the general superintendent.

Superintendent: In construction, the individual who is the on-site supervisor in charge of a given construction project.

Textspeak: A form of written language characteristic of text messaging on mobile devices and text-based social media, which usually consists of acronyms, abbreviations, and minimal punctuation.

It is important to define some of the supervisory positions discussed throughout this module. You are already familiar with the roles of the craft professional and crew leader. A **superintendent** is essentially an on-site supervisor who is responsible for one or more crew leaders or front-line supervisors. A **project manager** or general superintendent may be responsible for managing one or more projects. This training will concentrate primarily on the supervisory role of the crew leader.

Craftworkers and crew leaders differ in that the crew leader manages the activities that the craft professionals perform. To manage a crew of craft professionals, a crew leader must have first-hand knowledge and experience in their activities. Additionally, the crew leader must be able to act directly in organizing and directing the activities of the various crew members.

This section explains the importance of developing leadership skills as a new crew leader. It will cover effective ways to communicate with co-workers and employees at all levels, build teams, motivate crew members, make decisions, and resolve problems.

Crew leaders are generally promoted up from a work crew. A worker's ability to accomplish tasks, get along with others, meet schedules, and stay within the budget have a significant influence on the selection process. The crew leader must lead the team to work safely and provide a quality product.

Making the transition from crew member to a crew leader can be difficult, especially when the new position involves overseeing a group of former peers. For example, some of the crew may try to take advantage of their friendship by seeking special favors. They may also want to be privy to supervisory information that is normally closely held. When you become a crew leader, you are no longer responsible for your work alone. Crew leaders are accountable for the work of an entire crew of people with varying skill levels, personalities, work styles, and cultural and educational backgrounds.

Crew leaders must learn to put personal relationships aside and work for the common goals of the entire crew. The crew leader can overcome these problems by working with the crew to set mutual performance goals and by freely communicating with them within permitted limits. Use their knowledge and strengths along with your own so that they feel like they are key players on the team.

As employees move from being a craftworker into the role of a crew leader and above, they will begin to spend more hours supervising the work of others (supervisory work) than practicing their own craft skills (technical work). *Figure 7* illustrates the relative amounts of time craft professionals, crew leaders, superintendents, and project managers spend on technical and supervisory work as their management responsibilities increase.

There are many ways to define a leader. One simple definition of a leader is a person who influences other people to achieve a goal. Some people may have innate leadership qualities that developed during their upbringing, or they may have worked to develop the traits that motivate others to follow and perform. Research shows that people who possess such talents are likely to succeed as leaders.

"Leadership is the art of getting someone else to do something you want done because he wants to do it."
—Dwight D. Eisenhower

2.1.0 The Qualities and Role of a Leader

Leadership traits are similar to the skills that a crew leader needs to be effective. Although the characteristics of leadership are many, there are some definite commonalities among effective leaders.

First and foremost, effective leaders lead by example. They work and live by the standards that they establish for their crew members or followers, making sure they set a positive example.

Effective leaders also tend to have a high level of drive and determination, as well as a persistent attitude. When faced with obstacles, effective leaders don't get discouraged. Instead, they identify the potential problems, make plans to overcome them, and work toward achieving the intended goal. In the event of failure, effective leaders learn from their mistakes and apply that knowledge to future situations. They also learn from their successes.

Effective leaders are typically good communicators who clearly lay out the goals of a project to their crew members. Accomplishing this may require that the leader overcome issues such as language barriers, gender bias, or differences in personalities to ensure that each member of the crew understands the established goals of the project.

Effective leaders can motivate their crew members to work to their full potential and become useful members of the team. Crew leaders try to develop crew-member skills and encourage them to improve and learn so they can contribute more to the team effort. Effective leaders strive for excellence from themselves and their team, so they work hard to provide the skills and leadership necessary to do so.

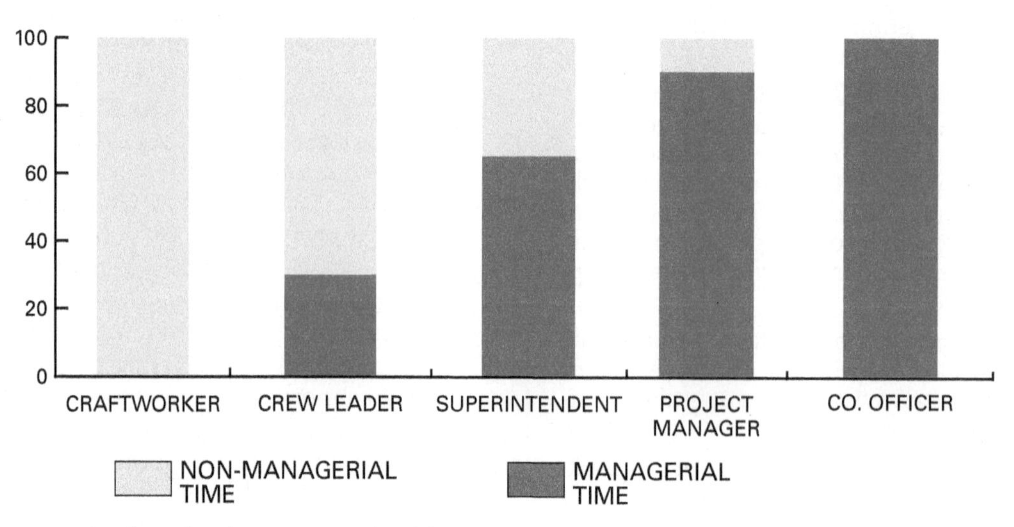

Figure 7 Percentages of technical and supervisory work by role.

In addition, effective leaders must possess organizational skills. They know what needs to be accomplished, and they use their resources to make it happen. Because they can't do it alone, leaders require the help of their team members to share in the workload. Effective leaders delegate work to their crew members, and they implement company policies and procedures to complete the work safely, effectively, and efficiently.

Finally, effective leaders have the experience, authority, and self-confidence that allows them to make decisions and solve problems. To accomplish their goals, leaders must be able to calculate risks, take in and interpret information, assess courses of action, make decisions, and assume the responsibility for those decisions.

2.1.1 Functions of a Leader

The functions of a leader will vary with the environment, the group of individuals they lead, and the tasks they must perform. However, there are certain functions common to all situations that the leader must fulfill. Some of the major functions are as follows:

- Accept responsibility for the successes and failures of the group's performance
- Be sensitive to the differences of a diverse workforce
- Ensure that all group members understand and abide by company policies and procedures
- Give group members the confidence to make decisions and take responsibility for their work
- Maintain a cohesive group by resolving tensions and differences among its members and between the group and those outside the group
- Organize, plan, staff, direct, and control work
- Represent the group

2.1.2 Leadership Traits

There are many traits and skills that help create an effective leader. Some of these include the following:

- Ability to advocate an idea
- Ability to motivate
- Ability to plan and organize
- Ability to teach others
- Enthusiasm
- Fairness
- Good communication skills
- Initiative
- Loyalty to their company and crew
- Willingness to learn from others

2.1.3 Expected Leadership Behavior

Followers have expectations of their leaders. They look to their leaders to do the following:

- Abide by company policies and procedures
- Be a loyal member of the team
- Communicate effectively
- Have the necessary technical knowledge
- Lead by example
- Make decisions and assume responsibility
- Plan and organize the work
- Suggest and direct
- Trust the team members

2.1.4 Leadership Styles

Through history, there have been many terms used to describe different leadership styles and the amount of **autonomy** they permit. The amount of crew autonomy directly relates to the leadership style used by the crew leader. *Figure 8* illustrates the linear relationship between leader authority and worker autonomy.

There are three main styles of leadership. At one extreme is the *controller style* of leadership, where the crew leader makes all of the decisions independently, without seeking the opinions of crew members. (Very little autonomy exists for workers under this type of leader.) At the other extreme is the *advisor style*, where the crew leader empowers the employees to make decisions (high autonomy). In between these extremes is the *directive style*, where the crew leader seeks crew member opinions and makes the appropriate decisions based on their input.

The following are some characteristics of each of the three leadership styles:

Controller style (high authority, low autonomy):

- Expect crew members to work without questioning procedures
- Seldom seek advice from crew members
- Insist on solving problems alone
- Seldom permit crew members to assist each other
- Praise and criticize on a personal basis
- Have no sincere interest in creatively improving methods of operation or production

Directive style (equal authority and autonomy):

- Discuss problems with their crew members
- Listen to suggestions from crew members
- Explain and instruct
- Give crew members a feeling of accomplishment by commending them when they do a job well

- Are friendly and available to discuss personal and job-related problems

Advisor style (low authority, high autonomy):

- Believe no supervision is best
- Rarely give orders
- Worry about whether they are liked by their crew members

Effective leadership takes many forms. The correct style for a particular situation or operation depends on the nature of the crew as well as the work it has to accomplish. For example, if the crew does not have enough experience for the job ahead, then a controller style may be appropriate. The controller style of leadership is also effective when jobs involve repetitive operations that require little decision-making.

However, if a worker's attitude is an issue, a directive style may be appropriate. In this case, providing the missing motivational factors may increase performance and result in the improvement of the worker's attitude. The directive style of leadership is also used when the work is of a creative nature, because brainstorming and exchanging ideas with such crew members can be beneficial.

The advisor style is effective with an experienced crew on a well-defined project. The company must give a crew leader sufficient authority to do the job. This authority must be commensurate with responsibility, and it must be made known to crew members when they are hired so that they understand who is in charge.

A crew leader must have an expert knowledge of the activities to be supervised in order to be effective. This is important because the crew members need to know that they have someone to turn to when they have a question or a problem, when they need some guidance, or when modifications or changes are warranted by the job.

Respect is probably the most useful element of authority. Respect usually derives from being fair to employees by listening to their complaints and suggestions, and by using incentives and rewards appropriately to motivate crew members. In addition, crew leaders who have a positive attitude and a favorable personality tend to gain the respect of their crew members as well as their peers. Along with respect comes a positive attitude from the crew members.

2.1.5 Ethics in Leadership

Crew leaders should maintain the highest standards of honesty and legality. Every day, the crew leader must make decisions that may have fair-

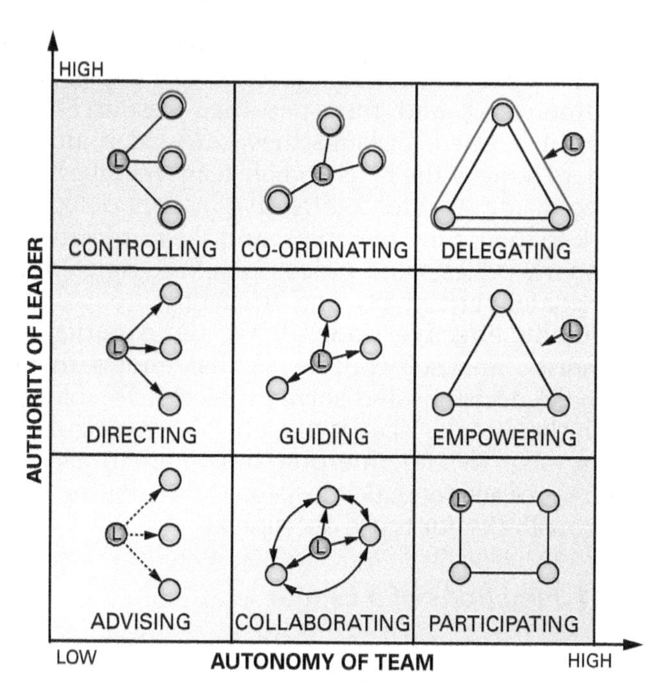

Figure 8 The authority-autonomy leadership style matrix.

ness and moral implications. When you make a dishonorable or unethical decision, it not only reflects on you, but also impacts other workers, peers, and the company as a whole.

There are three basic types of ethics:

- *Business or legal ethics* – Business, or legal, ethics concerns adhering to all laws and published regulations related to business relationships or activities.
- *Professional/balanced ethics* – Professional, or balanced, ethics relates to carrying out all activities in such a manner as to be honest and fair to everyone under one's authority.
- *Situational ethics* – Situational ethics pertains to specific activities or events that may initially appear to be a gray area. For example, you may ask yourself, "How will I feel about myself if my actions were going to be published in the newspaper or if I need to justify my actions to my family, friends, and colleagues? Would I still do the same thing?"

You will often find yourself in a situation where you will need to assess the ethical consequences of an impending decision. For example, if one of your crew members is showing symptoms of heat exhaustion, should you keep him working just because the superintendent says the project is behind schedule? If you are the only one aware that your crew did not properly erect the reinforcing steel, should you stop the pour and correct the situation? If supervision ever asks a crew leader to carry through on an unethical decision, it is up to that individual to inform the next-higher level of authority of the unethical nature of the issue. It is a sign of good character to refuse to carry out an unethical act.

"The right way is not always the popular and easy way. Standing for right when it is unpopular is a true test of character."
—Margaret Chase Smith

> **NOTE**
> Anyone who is aware of or knowingly a party to an illegal activity can face legal consequences simply due to inactivity, even if the activities are instigated by one's supervisor. However, a person of good character will never consider weighing the legal consequences against the ethics of one's actions.

2.2.0 Communication

Successful crew leaders learn to communicate effectively with people at all levels of the organization. In doing so, they develop an understanding of human behavior and acquire communication skills that enable them to understand and influence others.

There are many definitions of communication. Communication is the act of accurately and effectively conveying to or exchanging facts, feelings, and opinions with another person. Simply stated, communication is the process of exchanging information and ideas. Just as there are many definitions of communication, it also comes in many forms, including verbal, written, and nonverbal.

2.2.1 The Communication Process and Verbal Communication

There are two basic steps to clear communication, as illustrated in *Figure 9*. First, a sender sends a message (either verbal or written) to a receiver. When the receiver gets the message, he or she figures out what it means by listening or reading carefully. If anything is unclear, the receiver gives the sender feedback by asking the sender for more information.

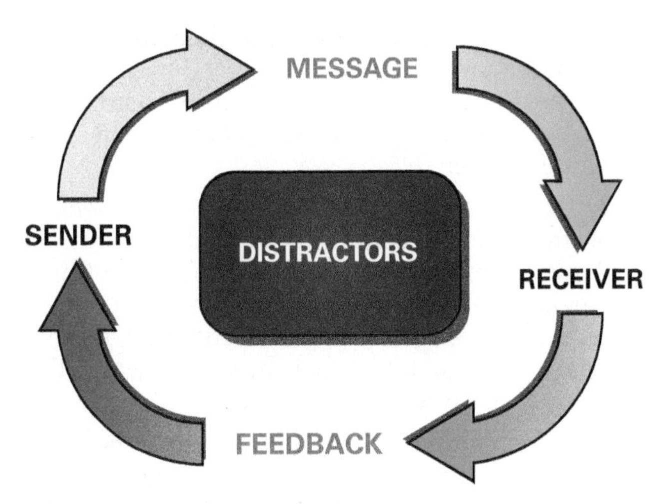

Figure 9 The communication process.

This process is called *two-way communication*, and it is the most effective way to make sure that everyone understands what's going on. This process sounds simple, so you may ask why good communication is so hard to achieve? When we try to communicate, a lot of things can get in the way. These communication obstacles and distractors are called *noise*.

The communication process illustrated in *Figure 9* consists of the following major components:

- *The Sender* – The sender is the person who creates and transmits the message. In verbal communication, the sender speaks the message aloud to the person(s) receiving the message. The sender must be sure to speak in a clear and concise manner easily understood by others. This isn't a natural skill; it takes practice. Some basic speaking tips are as follows:
  - Avoid talking with anything in your mouth (food, gum, etc.).
  - Avoid swearing, or other crude language, and acronyms.
  - Don't speak too quickly or too slowly. In extreme cases, people tend to focus on the rate of speech rather than the words themselves.

> **Did You Know?**
> **Supervisor's Communication Breakdown**
> No, this isn't about failure of communication, but how a supervisor's daily communications break down into the various types, and how much time the supervisor spends doing each. Research shows that about 80 percent of the typical supervisor's day is spent communicating through writing, speaking, listening, or using body language. Of that time, studies suggest that approximately 20 percent of communication is in the written form, and 80 percent involves speaking or listening.

- Pronounce words carefully to prevent misunderstandings.
- Speak pleasantly and with enthusiasm. Avoid speaking in a harsh voice or in a monotone.

- *The Message* – The message is what the sender is attempting to communicate to the receiver(s). A message can be a set of directions, an opinion, or dealing with a personnel matter (praise, reprimand, etc.). Whatever its function, a message is an idea or fact that the sender wants the receiver to know.

  Before speaking, determine what must be communicated, and then take the time to organize what to say. Ensure that the message is logical and complete. Take the time to clarify your thoughts to avoid confusing the receiver. This also permits the sender to get to the point quickly.

  In delivering the message, the sender should assess the audience. It is important not to talk down to them. Remember that everyone, whether in a senior or junior position, deserves respect and courtesy. Therefore, the sender should use words and phrases that the receiver can understand and avoid technical language or slang if the receiver is unfamiliar with such terms. In addition, the sender should use short sentences, which gives the audience time to understand and digest one point or fact at a time.

- *The Receiver* – The receiver is the person who takes in the message. For the verbal communication process to be successful, it is important that the receiver understands the message as the speaker intended. Therefore, the receiver must avoid things that interfere with the delivery of the message. There are many barriers to effective listening, particularly on a busy construction job site. Some of these obstacles include the following:

  - Noise, visitors, cell phones, or other distractions
  - Preoccupation, being under pressure, or daydreaming
  - Reacting emotionally to what is being communicated
  - Thinking about how to respond instead of listening
  - Giving an answer before the message is complete
  - Personal bias/prejudice against the sender or the sender's communication style
  - Finishing the sender's sentence

  Some tips for overcoming these barriers include the following:

  - Take steps to minimize or remove distractions; learn to tune out your surroundings
  - Listen for key points
  - Take notes
  - Be aware of your personal biases, and try to stay open-minded; focus on the message, not the speaker
  - Allow yourself time to process your thoughts before responding
  - Let the sender communicate the message without interruption

  There are many ways for a receiver to show that he or she is actively listening to the message. The receiver can even accomplish this without saying a word. Examples include maintaining eye contact (*Figure 10*), nodding your head, and taking notes. Feedback can also provide an important type of response to a message.

- *Feedback* – Feedback is the receiver's communication back to the sender in response to the message. Feedback is a very important part of the communication process because it shows the sender how the receiver interpreted the message and that the receiver understood it as intended. In other words, feedback is a checkpoint to make sure the receiver and sender are on the same page.

  The receiver can paraphrase the message as a form of feedback to the sender. When paraphrasing, you use your own words to repeat the message. That way, you can show the sender that you interpreted the message correctly and could explain it to others if needed.

Figure 10 Eye contact is important when communicating face-to-face.

When providing feedback, the receiver can also request clarification or additional information, generally by asking questions.

One opportunity a crew leader can take to provide feedback is in the performance of crew evaluations. Many companies use formal evaluations on a yearly basis to assess workers' performance for pay increases and eligibility for advancement. These evaluations should not come as a once-a-year surprise to workers. An effective crew leader provides frequent performance feedback, and should ultimately summarize these communications in the annual performance evaluation. It is also important to stress the importance of self-evaluation with your crew.

2.2.2 Written or Visual Communications

Some communication must be written or visual. Written or visual communication includes messages or information documented on paper, or transmitted electronically using words, sketches, or other types of images.

Many messages on a job are in text form. Examples include weekly reports, requests for changes, purchase orders, and correspondence on a specific subject. These items are in writing because they can form a permanent record for business and historical purposes. Some communications on the job must be in the form of images. People are visual beings, and sketches, diagrams, graphs, and photos or videos can more effectively communicate some things. Examples include the plans or drawings used on a job, or a video of a work site accident scene.

"One picture is worth a thousand words."
—San Antonio Express-News (1918)

When writing or creating a visual message, it is best to assess the receiver (the reader) or the audience before beginning. The receiver must be able to read the message and understand the content; otherwise, the communication process will be unsuccessful. Therefore, the sender (the writer) should consider the actual meaning of words or diagrams and how others might interpret them. In addition, the writer should make sure that handwriting, if applicable, is legible.

Here are some basic tips for writing:

- Avoid emotion-packed words or phrases.
- Avoid making judgments unless asked to do so.
- Avoid using unfamiliar acronyms and textspeak (*Figure 11*).

Figure 11 Avoid textspeak for on-the-job written communications.

- Avoid using technical language or jargon unless the writer knows the receiver is familiar with the terms used.
- Be positive whenever possible.
- Be prepared to provide a verbal or visual explanation, if needed.
- Make sure that the document is legible.
- Present the information in a logical manner.
- Proofread your work; check for spelling, typographical, and grammatical errors—especially if relying on a spell-checker or an autocorrect typing feature.
- Provide an adequate level of detail.
- State the purpose of the message clearly.
- Stick to the facts.

Entire books are available that explain how to create effective diagrams and graphs. There are many factors that contribute to clear, concise, unambiguous images. The following are some basic tips for creating effective visual communications:

- Avoid making complex graphics; simplicity is better.
- Be prepared to provide a written or verbal explanation of the diagram, if needed.
- Ensure that the diagram is large enough to be useful.
- Graphs are better for showing numerical relationships than columns of numbers.
- Present the information in a logical order.
- Provide a legend if you include symbolic information.
- Provide an adequate level of detail and accuracy.

2.2.3 Nonverbal Communication

Unlike verbal or written communication, nonverbal communication doesn't involve spoken or written words, or images. Rather, nonverbal communication refers to things the speaker and the receiver see and hear in each other while communicating face to face. Examples include facial expressions, body movements, hand gestures, tone of voice, and eye contact.

Nonverbal communication can provide an external signal of an individual's inner emotions. It occurs at the same time as verbal communication. In most cases, the transmission of the sender and receiver nonverbal cues is totally unconscious.

People observe physical, nonverbal cues when communicating, making them just as important as words. Often, nonverbal signals influence people more than spoken words. Therefore, it's important to be conscious of nonverbal cues to avoid miscommunication based on your posture or expression. After all, these things may have nothing to do with the communication exchange; instead, they may be carrying over from something else going on in your day.

2.2.4 Communication Issues

It is important to note that everyone communicates differently; that is what makes us unique as individuals. As the diversity of the workforce changes, communication becomes even more challenging because the audience may include individuals from many different backgrounds. Therefore, it is necessary to assess the audience to determine how to communicate effectively with each individual.

The key to effective communication is to acknowledge that people are different, and to be able to adjust the communication style to meet the needs of the audience or the person on the receiving end of your message. This involves relaying the message in the simplest way possible, and avoiding the use of words that people may find confusing. Be aware of how you use technical language, slang, jargon, and words that have multiple meanings. Present the information in a clear, concise manner. Avoid rambling and always speak clearly, using good grammar.

> **NOTE**
>
> When working with bilingual workers, it is sometimes helpful to learn words and phrases commonly used in their language(s). They will respect you more for attempting to learn their language as they learn yours, and the effort will generate a closer comradery.

In addition, be prepared to communicate the message in multiple ways or adjust your level of detail or terminology to ensure that everyone understands the meaning as intended. For instance, a visual learner may need a map to comprehend directions. It may be necessary to overcome language barriers on the job site by using graphics or visual aids to relay the message. *Figure 12* shows how to tailor the message to the audience.

2.3.0 Motivation

The ability to motivate others is a key skill that leaders must develop. To motivate is to influence others to put forth more effort to accomplish something. For example, a crew member who skips breaks and lunch to complete a job on time is thought to be highly motivated, but crew members who do the bare minimum or just enough to keep their jobs are considered unmotivated.

A leader can **infer** an employee's motivation by observing various factors in the employee's performance. Examples of factors that may indicate a crew's motivation include the level or rate of unexcused absenteeism, the percentage of employee turnover, and the number of complaints, as well as the quality and quantity of work produced.

Different things motivate different people in different ways. Consequently, there is no single best approach to motivating crew members. It is important to recognize that what motivates one crew member may not motivate another. In addition, what works to motivate a crew member once may not motivate that same person again in the future.

Often, the needs and factors that motivate individuals are the same as those that create job satisfaction. These include the following:

- Accomplishment
- Change
- Job importance
- Opportunity for advancement
- Personal growth
- Recognition and praise
- Rewards

A crew leader's ability to satisfy these needs increases the likelihood of high morale within a crew. Morale refers to an individual's emotional outlook toward work and the level of satisfaction gained while performing the jobs assigned. High morale means that employees will be motivated to work hard, and they will have a positive attitude about coming to work and doing their jobs.

2.3.1 Accomplishment

Accomplishment refers to a worker's need to set challenging goals and achieve them. There is nothing quite like the feeling of achieving a goal, particularly a goal one never expected to accomplish in the first place.

Participant Exercise A, Part I

Read the following verbal conversations, and identify any problems:

Conversation I:

Judy: Hey, José...
José: What's up?
Judy: Has the site been prepared for the job trailer yet?
José: Job trailer?
Judy: The job trailer—it's coming in today. What time will the job site be prepared?
José: The trailer will be here about 1:00 PM.
Judy: The job site! What time will the job site be prepared?

____

____

____

____

____

____

Conversation II:

Jimar: Hey, Mike, I need your help.
Mike: What is it?
Jimar: You and Miguel go over and help Al's crew finish laying out the site.
Mike: Why me? I can't work with Miguel. He can't understand a word I say.
Jimar: Al's crew needs some help, and you and Miguel are the most qualified to do the job.
Mike: I told you, Jimar, I can't work with Miguel.

____

____

____

____

____

____

Conversation III:

Hiro: Hey, Jill.
Jill: Sir?
Hiro: Have you received the latest DOL, EEO requirement to be sure the OFCP administrator finds our records up to date when he reviews them in August?
Jill: DOL, EEO, and OFCP?
Hiro: Oh, and don't forget the MSHA, OSHA, and EPA reports are due this afternoon.
Jill: MSHA, OSHA, and EPA?

____

____

____

____

____

____

____

____

____

Conversation IV:

Lakeisha: Good morning, Roberto, would you do me a favor?
Roberto: Okay, Lakeisha. What is it?
Lakeisha: I was reading the concrete inspection report and found the concrete in Bays 4A, 3B, 6C, and 5D didn't meet the 3,000 psi strength requirements. Also, the concrete inspector on the job told me the two batches that came in today had to be refused because they didn't meet the slump requirements as noted on page 16 of the spec. I need to know if any placement problems happened on those bays, how long the ready mix trucks were waiting today, and what we plan to do to stop these problems in the future.

____

____

____

____

____

Participant Exercise A, Part II

Read the following written memos, and identify any problems:

Memo I:

Let's start with the transformer vault $285.00 due. For what you ask? Answer: practically nothing I admit, but here is the story. Paul the superintendent decided it was not the way good ole Comm Ed wanted it, we took out the ladder and part of the grading (as Paul instructed us to do) we brought it back here to change it. When Comm Ed the architect or DOE found out that everything would still work the way it was, Paul instructed us to reinstall the work. That is the whole story there is please add the $285.00 to my next payout.

____

____

____

____

____

____

____

____

Memo II:

Let's take rooms C 307-C-312 and C-313 we made the light track supports and took them to the job to erect them when we tried to put them in we found direct work in the way, my men spent all day trying to find out what to do so ask your Superintendent (Frank) he will verify seven hours pay for these men as he went back and forth while my men waited. Now the Architect has changed the system of hanging and has the gall to say that he has made my work easier, I can't see how. Anyway, we want an extra two (2) men for seven (7) hours for April 21 at $55.00 per hour or $385.00 on April 28th DOE Reference 197 finally resolved this problem. We will have no additional charges on DOE Reference 197, please note.

____

____

____

____

____

____

Crew leaders can help their crew members attain a sense of accomplishment by encouraging them to develop performance plans, such as goals for the year, the attaining of which the crew leader will consider later in performance evaluations. In addition, crew leaders can provide the support and tools (such as training and coaching) necessary to help their crew members achieve these goals.

2.3.2 Change

Change refers to an employee's need to have variety in work assignments. Change can keep things interesting or challenging, and prevent the boredom that results from doing the same task day after day with no variety. However, frequent or significant changes in work can actually have a negative impact on morale, since people often prefer some consistency and predictability in their lives.

2.3.3 Job Importance

Job importance refers to an employee's need to feel that his or her skills and abilities are valued and make a difference. Employees who don't feel valued tend to have performance and attendance issues. Crew leaders should attempt to make every crew member feel like an important part of the team, as if the job wouldn't be possible without their help.

2.3.4 Opportunity for Advancement

Opportunity for advancement refers to an employee's need to gain additional responsibility and develop new skills and abilities. It is important that employees know that they aren't limited to their current jobs. Let them know that they have a chance to grow with the company and to be promoted as recognition for excelling in their work when such opportunities occur.

VERBAL INSTRUCTIONS (EXPERIENCED CREW)	VERBAL INSTRUCTIONS (INEXPERIENCED CREW)	WRITTEN INSTRUCTIONS	VISUAL INSTRUCTIONS (DIAGRAM/MAP)
"Please drive to the supply shop to pick up our order."	"Please drive to the supply shop. Turn right here and left at Route 1. It's at 75th Street and Route 1. Tell them the company name and that you're there to pick up our order."	1. Turn right at exit. 2. Drive 2 miles to Route 1. Turn LEFT. 3. Drive 1 mile (pass the tire shop) to 75th Street. 4. Look for supply store on right...	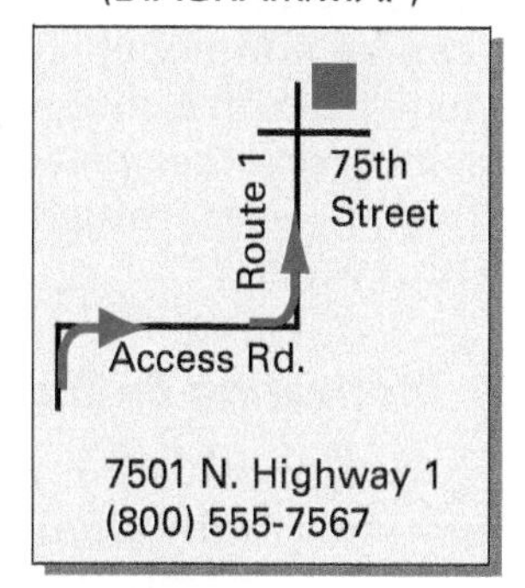

Different people learn in different ways. Be sure to communicate so you can be understood.

Figure 12 Tailor your message.

Effective leaders encourage each of their crew members to work to his or her full potential. In addition, they share information and skills with their employees to help them advance within the organization.

2.3.5 Recognition

Recognition and praise refer to the need to have good work appreciated, applauded, and acknowledged by others. You can accomplish this by simply thanking employees for helping on a project, or it can entail more formal praise, such as an award for Employee of the Month, or tangible or monetary awards. Some tips for giving recognition and praise include the following:

- Be available on the job site so that you have the opportunity to witness good work.
- Look for and recognize good work, and look for ways to praise it.
- Give recognition and praise only when truly deserved; people can quickly recognize insincere or artificial praise, and you will lose respect as a result.
- Acknowledge satisfactory performance, and encourage improvement by showing confidence in the ability of the crew members to do above-average work.

2.3.6 Personal Growth

Personal growth refers to an employee's need to learn new skills, enhance abilities, and grow as a person. It can be very rewarding to master a new competency on the job. Personal growth prevents the boredom associated with contemplation of working the same job indefinitely with no increase of technical knowledge, skills, or responsibility.

Crew leaders should encourage the personal growth of their employees as well as themselves. Learning should be a two-way street on the job site; crew leaders should teach their crew members and learn from them as well. In addition, crew members should be encouraged to learn from each other

Personal growth also takes place outside the workplace. Crew leaders should encourage their employees to acquire formal education applicable to their vocation as well as in subjects that will expand their understanding of the larger world. Encourage them to develop outside interests and ways to express their innate creativity.

2.3.7 Rewards

Rewards are additional compensation for hard work. Rewards can include an increase in a crew member's wages, or go beyond that to include bo-

Did You Know?

The Importance of Continuing Education

Education doesn't stop the day a person receives a diploma or certificate. It is a lifelong activity. Employers have long recognized and promoted continuing education as a factor in advancement, but it is essential to simply remaining in place as well. Regardless of what you do, new construction materials, methods, standards, and processes are constantly emerging. Those who don't make the effort to keep up will fall behind.

Also, studies have shown that a worker's lifetime income improved by 7–10 percent per year of community college. Consider continuing and vocational technical education (CTE and VTE) through local colleges.

nuses or other incentives. They can be monetary in nature (salary raises, holiday bonuses, etc.), or they can be nonmonetary, such as free merchandise (shirts, coffee mugs, jackets, merchant gift cards, etc.), or other prizes. Attendance at costly certification or training courses can be another form of reward.

2.3.8 Motivating Employees

To increase motivation in the workplace, crew leaders must individualize how they motivate different crew members. It is important that crew leaders get to know their crew members and determine what motivates them as individuals. Once again, as diversity increases in the workforce, this becomes even more challenging; therefore, effective communication skills are essential. Some tips for motivating employees include the following:

- Keep jobs challenging and interesting.
- Communicate your expectations. People need clear goals in order to feel a sense of accomplishment when they achieve them.
- Involve the employees. Feeling that their opinions are valued leads to pride in ownership and active participation.
- Provide sufficient training. Give employees the skills and abilities they need to be motivated to perform.
- Mentor the employees. Coaching and supporting employees boosts their self-esteem, their self-confidence, and ultimately their motivation.
- Lead by example. Employees will be far more motivated to follow someone who is willing to do the same things they are required to do.
- Treat employees well. Be considerate, kind, caring, and respectful; treat employees the way that you want to be treated.
- Avoid using scare tactics. Approaching your leadership responsibilities by threatening employees with negative consequences can result in higher employee turnover instead of motivation.
- Reward your crew for doing their best by giving them easier tasks from time to time. It is tempting to give your best employees the hardest or dirtiest jobs because you know they will do the jobs correctly.
- Reward employees for a job well done.
- Maintain a sense of humor, especially toward your own failings as a human being. No one is perfect. Your employees will appreciate that and be more motivated to be an encouragement to you.

2.4.0 Team Building

Organizations are making the shift from the traditional boss-worker mentality to one that promotes teamwork. The manager becomes the team leader, and the workers become team members. They all work together to achieve the common goals of the team.

There are several benefits associated with teamwork. These include the ability to complete complex projects more quickly and effectively, higher employee satisfaction, and reduced turnover.

2.4.1 Successful Teams

Successful teams consist of individuals who are willing to share their time and talents to reach a common goal—the goal of the team. Members of successful teams possess an "Us" or "We" attitude rather than an "I" and "You" attitude; they consider what's best for the team and put their egos aside.

Some characteristics of successful teams include the following:

- Everyone participates and every team member counts.
- All team members understand the goals of the team's work and are committed to achieve those goals.
- There is a sense of mutual trust and interdependence.
- The organization gives team members the confidence and means to succeed.
- They communicate.
- They are creative and willing to take risks.
- The team leader has strong people skills and is committed to the team.

2.4.2 Building Successful Teams

To be successful in the team leadership role, the crew leader should contribute to a positive attitude within the team. There are several ways in which the team leader can accomplish this. First, he or she can work with the team members to create a vision or purpose of what the team is to achieve. It is important that every team member is committed to the purpose of the team, and the team leader is instrumental in making this happen.

Within the construction industry, the company typically assigns a crew to a crew leader. However, it can be beneficial for the team leader to be involved in selecting the team members. The willingness of people to work on the team and the resources that they bring should be the key factors for selecting them for the team.

Participant Exercise B

You are the crew leader of a masonry crew. Sam Williams is the person whom the company holds responsible for ensuring that equipment is operable and distributed tothe jobs in a timely manner.

Occasionally, disagreements with Sam have resulted in tools and equipment arriving late. Sam, who has been with the company 15 years, resents having been placed in the job and feels that he outranks all the crew leaders.

Sam figured it was about time he talked with someone about the abuse certain tools and other items of equipment were receiving on some of the jobs. Saws were coming back with guards broken and blades chewed up, bits were being sheared in half, motor housings were bent or cracked, and a large number of tools were being returned covered with mud. Sam was out on your job when he observed a mason carrying a portable saw by the cord. As he watched, he saw the mason bump the swinging saw into a steel column. When the man arrived at his workstation, he dropped the saw into the mud.

You are the worker's crew leader. Sam approached as you were coming out of the work trailer. He described the incident. He insisted, as crew leader, you are responsible for both the work of the crew and how its members use company property. Sam concluded, "You'd better take care of this issue as soon as possible! The company is sick and tired of having your people mess up all the tools!"

You are aware that some members of your crew have been mistreating the company equipment.

1. How would you respond to Sam's accusations?

____

____

____

____

____

2. What action would you take regarding the misuse of the tools?

____

____

____

____

____

3. How can you motivate the crew to take better care of their tools? Explain.

____

____

____

____

____

When forming a new team, team leaders should do the following:

- Explain the purpose of the team. Team members need to know what they will be doing, how long they will be doing it (if they are temporary or permanent), and why they are needed.
- Help the team establish goals or targets. Teams need a purpose, and they need to know what it is they are responsible for accomplishing.
- Define team member roles and expectations. Team members need to know how they fit into the team and what they can expect to accomplish as members of the team.
- Plan to transfer a sense of responsibility to the team as appropriate. Teams should feel responsible for their assigned tasks. However, never forget that management will still hold the crew leader responsible for the crew's work.

"My model for business is The Beatles. They were four guys who kept each other's kind of negative tendencies in check. They balanced each other and the total was greater than the sum of the parts. That's how I see business: Great things in business are never done by one person, they're done by a team of people.

—Steve Jobs

2.4.3 Delegating

Once the various activities that make up the job have been determined, the crew leader must identify the person or persons who will be responsible for completing each activity. This requires that the crew leader be aware of the skills and abilities of the people on the crew. Then, the crew leader must put this knowledge to work in matching the crew's skills and abilities to accomplish the specific tasks needed to complete the job.

After matching crew members to specific activities, the crew leader must then delegate the assignments to the responsible person(s). Generally, when delegating responsibilities, the crew leader verbally communicates directly with the person who will perform or complete the activity.

When delegating work, remember to do the following:

- Delegate work to a crew member who can do the job properly. If it becomes evident that the worker doesn't perform to the desired standard, either teach the crew member to do the work correctly or turn it over to someone else who can (without making a public spectacle of the transfer).
- Make sure crew members understand what to do and the level of responsibility. Be clear about the desired results, specify the boundaries and deadlines for accomplishing the results, and note the available resources.
- Identify the standards and methods of measurement for progress and accomplishment, along with the consequences of not achieving the desired results. Discuss the task with the crew member and check for understanding by asking questions. Allow the crew member to contribute feedback or make suggestions about how to perform the task in a safe and quality manner.
- Give the crew member the time and freedom to get started without feeling the pressure of too much supervision. When making the work assignment, be sure to tell the crew member how much time there is to complete it, and confirm that this time is consistent with the job schedule.
- Examine and evaluate the result once a task is complete. Then, give the crew member some feedback as to how well the worker did the task. Get the crew member's comments. The information obtained from this is valuable and will enable the crew leader to know what kind of work to assign that crew member in the future. It will also provide a means of measuring the crew leader's own effectiveness in delegating work.

Be aware that there may be times when someone else in the company will issue written or verbal instructions to a crew member or to the crew without going through the crew leader. This kind of situation requires an extra measure of maturity and discretion on the part of the crew leader to first understand the circumstances, then establish an understanding with the responsible individual, if possible, to avoid work orders that circumvent the crew leader in the future.

2.4.4 Implementing Policies and Procedures

Every company establishes policies and procedures that crew leaders are expected to implement, and employees are expected to follow. Company policies and procedures are essentially guidelines for how the organization does business. They can also reflect organizational philosophies, such as putting safety first or making the customer the top priority. Examples of policies and procedures include safety guidelines, credit standards, and billing processes.

The following tips can help you effectively implement policies and procedures:

- Learn the purpose of each policy. This will help you follow the policy and apply it appropriately and fairly.
- If you're not sure how to apply a company policy or procedure, check the company manual or ask your supervisor.

NOTE

Try to obtain a supervisor's policy interpretation in writing or print out the email response so that you can append the decision to your copy of the company manual for future reference.

- Always follow company policies and procedures. Remember that they combine what's best for the customer and the company. In addition, they provide direction on how to handle specific situations and answer questions.

Crew leaders may need to issue orders to their crew members. An order is a form of communication that initiates, changes, or stops an activity. Orders may be general or specific, written or oral, and formal or informal. The decision of how an order will be issued is up to the crew leader, but the policies and procedures of the company may govern the choice.

When issuing orders, do the following:

- Make them as specific as possible. Avoid being general or vague unless it is impossible to foresee all the circumstances that could occur in carrying out the order.
- Recognize that it isn't necessary to write orders for simple tasks unless the company requires that supervisors write all orders.
- Write orders for more complex tasks, tasks that will take considerable time to complete, or that are permanent (standing) orders.
- Consider what is being said, the audience to whom it applies, and the situation under which it will be implemented to determine the appropriate level of formality for the order.

2.5.0 Making Decisions and Solving Problems

Decision making and problem solving and are a large part of every crew leader's daily work. They are a part of life for all supervisors, especially in fast-paced, deadlineoriented industries.

2.5.1 Decision Making Versus Problem Solving

Sometimes, the difference between decision making and problem solving isn't clear. Decision making refers to simply initiating an action, stopping one, or choosing an alternative course, as appropriate for the situation. Problem solving involves recognizing the difference between the way things are and the way things should be, then taking action to move toward the desired condition. The two activities are related because, to make a decision, you may have to use problem-solving techniques, just as solving problems requires making decisions.

2.5.2 Types of Decisions

Some decisions are routine or simple, and can be made based on past experiences. An example would be deciding how to get to work. If you've worked at the same place for a long time, you are already aware of the options for traveling to work (take the bus, drive a car, carpool with a co-worker, take a taxi, etc.). Based on past experiences with the options identified, you can make a decision about how best to get to work.

Other decisions are more difficult to make. These decisions require more careful thought about how to carry out an activity by using problem-solving techniques. An example is planning a trip to a new vacation spot. If you're not sure how to get there, where to stay, what to see, etc., one option is to research the area to determine the possible routes, hotel accommodations, and attractions. Then, you can make a decision about which route to take, what hotel to choose, and what sites to visit, without the benefit of direct experience. The Internet makes these tasks much easier, but the research still requires prioritizing what is most important to you.

2.5.3 Problem Solving

The ability to solve problems is an important skill in any workplace. It's especially important for craft professionals, whose workday is often not predictable or routine. This section provides a five-step process for solving problems, which you can apply to both workplace and personal issues. Review the following steps and then see how you can be apply them to a job-related problem. Keep in mind that you can't solve a problem until everyone involved acknowledges the problem.

Step 1 *Define the problem.* This isn't as easy as it sounds. Thinking through the problem often uncovers additional problems. Also, drilling down to the facts of the problem may mean setting aside your own biases or presumptions toward the situation or the individuals involved.

Step 2 *Think about different ways to solve the problem.* There is often more than one solution to a problem, so you must think through each possible solution and pick the best one. The best solution might be taking parts of two different solutions and combining them to create a new solution.

Step 3 *Choose the solution that seems best, and make an action plan.* It is best to receive input both from those most affected by the problem, those who must correct the problem, and from those who will be most affected by any potential solution.

Step 4 *Test the solution to determine whether it actually works.* Many solutions sound great in theory but in practice don't turn out to be effective. On the other hand, you might discover from trying to apply a solution that it is acceptable with a little modification. If a solution doesn't work, think about how you could improve it, and then test your new plan.

Step 5 *Evaluate the process.* Review the steps you took to discover and implement the solution. Could you have done anything better? If the solution turns out to be satisfactory, you can add the solution to your knowledge base.

These five steps can be applied in specific situations. Read the following example situation, and apply the five-step problem-solving process to come up with a solution.

Example:

You are part of a team of workers assigned to a new shopping mall project. The project will take about 18 months to complete. The only available parking is half a mile from the job site. The crew must carry heavy toolboxes and safety equipment from their cars and trucks to the work area at the start of the day, and then carry them back at the end of their shifts. The five-step problem-solving process can be applied as follows:

Step 1 *Define the problem.* Workers are wasting time and energy hauling all their equipment to and from the work site.

Step 2 *Think about different ways to solve the problem.* Several workers have proposed solutions:

- Install lockers for tools and equipment closer to the work site.
- Have workers drive up to the work site to drop off their tools and equipment before parking.
- Bring in another construction trailer where workers can store their tools and equipment for the duration of the project.
- Provide a round-trip shuttle service to ferry workers and their tools.

NOTE

Each solution will have pros and cons, so it's important to receive input from the workers affected by the problem. For example, workers will probably object to any plan (like the drop-off plan) that leaves their tools vulnerable to theft.

Step 3 *Choose the solution that seems best, and make an action plan.* The work site superintendent doesn't want an additional trailer in your crew's area. The workers decide that the shuttle service makes the most sense. It should solve the time and energy problem, and workers can keep their tools with them. To put the plan into effect, the project supervisor arranges for a large van and driver to provide the shuttle service.

Step 4 *Test the solution to determine whether it actually works.* The solution works, but there is another problem. The workers' schedule has them all starting and leaving at the same time. There isn't enough room in the van for all the workers and their equipment. To solve this problem, the supervisor schedules trips spaced 15 minutes apart. The supervisor also adjusts worker schedules to correspond with the trips. That way, all the workers won't try to get on the shuttle at the same time.

Step 5 *Evaluate the process.* This process gave both management and workers a chance to express an opinion and discuss the various solutions. Everyone feels pleased with the process and the solution.

2.5.4 Special Leadership Problems

Because they are responsible for leading others, it is inevitable that crew leaders will encounter problems and be forced to make decisions about how to respond to the problem. Some problems will be relatively simple to resolve, like covering for a sick crew member who has taken a day off from work. Other problems will be complex and much more difficult to handle.

Some complex problems that are relatively common include the following:

- Inability to work with others
- Absenteeism and turnover
- Failure to comply with company policies and procedures

Inability to Work with Others – Crew leaders will sometimes encounter situations where an employee has a difficult time working with others on the crew. This could be a result of personality differences, gender or gender-identity prejudices, an inability to communicate, or some other cause. Whatever the reason, the crew leader must address the issue and get the crew working as a team.

The best way to determine the reason for why individuals don't get along or work well together is to talk to the parties involved. The crew leader should speak openly with the employee, as well as the other individual(s) to uncover the source of the problem and discuss its resolution.

After uncovering the reason for the conflict, the crew leader can determine how to respond. There may be a way to resolve the problem and get the workers communicating and working as a team again. On the other hand, there may be nothing the crew leader can do that will lead to a harmonious solution. In this case, the crew leader would need to either transfer one of the involved employees to another crew or have the problem crew member terminated. Resorting to this latter option should be the last measure and taken after discussing the matter with one's immediate supervisor or the company's Human Resources Department.

Absenteeism and Turnover – Absenteeism and turnover in the industry can delay jobs and cause companies to lose money. Absenteeism refers to workers missing their scheduled work time on a job. It has many causes, some of which cannot be helped. Sickness, family emergencies, and funerals are examples of unavoidable causes of worker absence. However, there are some causes of unexcused absenteeism that crew leaders can prevent.

The most effective way to control absenteeism is to make the company's policy clear to all employees. The policy should be explained to all new employees. This explanation should include the number of absences allowed and acceptable reasons for taking sick or personal days. In addition, all workers should know how to inform their crew leaders when they miss work and understand the consequences of exceeding the number of sick or personal days allowed.

Once crew leaders explain the policy on absenteeism to employees, they must be sure to implement it consistently and fairly. This makes employees more likely to follow it. However, if enforcement of the policy is inconsistent and the crew leader gives some employees exceptions, it won't be effective. Thus, the rate of absenteeism is likely to increase.

Despite having a policy on absenteeism, there will always be employees who are chronically late or miss work. In cases where an employee abuses the absenteeism policy, the crew leader should discuss the situation directly with the employee, ensure that they understand the policy, and insist that they comply with it. If the employee's behavior does not improve, disciplinary action will be in order.

Turnover refers to the rate at which workers leave a company and are replaced by others. Like absenteeism, there are some causes of turnover that cannot be prevented and others that can. For instance, an employee may find a job elsewhere earning twice as much money. However, crew leaders can prevent some employee turnover situations. They can work to ensure safe working conditions for their crew, treat their workers fairly and consistently, and help promote good working conditions. The key is communication and promoting the motivational factors discussed earlier. Crew leaders need to know the problems if they are going to be able to successfully resolve them.

Some major causes of employee turnover include the following:

- Unfair/inconsistent treatment by the immediate supervisor
- Unsafe project sites
- Lack of job security or opportunities for advancement

For the most part, the actions described for absenteeism are also effective for reducing turnover. Past studies have shown that maintaining harmonious relationships on the job site goes a long way in reducing both turnover and absenteeism. This requires effective and **proactive** leadership on the part of the crew leader.

Failure to Comply with Company Policies and Procedures – Policies are rules that define the relationship between the company, its employees, its clients, and its subcontractors. Procedures include the instructions for carrying out the policies. Some companies have policies that dictate dress codes. The dress code may be designed partly to ensure safety, and partly to define the image a company wants to project to the outside world.

Companies develop procedures to ensure that everyone who performs a task does it safely and efficiently. Many procedures directly relate to safety. A **lockout/tagout (LOTO)** procedure is an example. In this procedure, the company defines who may perform a LOTO, how to properly complete and remove a LOTO, and who has the authority to remove it. Workers who fail to follow the procedure endanger themselves, as well as their co-workers.

Companies typically have a policy on disciplinary action, which defines steps to take if an employee violates company policies or procedures. The steps range from counseling by a supervisor for the first offense, to a written warning and/or LOI, to dismissal for repeat offenses. This will vary from one company to another. For example, some companies will fire an employee for the first violation of a safety procedure with potential for loss of life or serious injury.

The crew leader has the first-line responsibility for enforcing company policies and procedures. The crew leader should take the time with a new crew member to discuss the policies and procedures and show the crew member how to access them. If a crew member shows a tendency to neglect a policy or procedure, it is up to the crew leader to counsel that individual. If the crew member continues to violate a policy or procedure, the crew leader has no choice but to refer that individual to the appropriate authority within the company for disciplinary action.

NOTE

When a crew shows a pattern of consistent policy violations, management will scrutinize the crew leader and potentially take disciplinary action if the crew leader does not handle the violations appropriately. The crew leader is responsible for all aspects of the crew, not just its work accomplishment on the job.

Participant Exercise C

Case I:

On the way over to the job trailer, you look up and see a piece of falling scrap heading for one of the laborers. Before you can say anything, the scrap material hits the ground about five feet in front of the worker. You notice the scrap is a piece of conduit. You quickly pick it up, assuring the worker you will take care of this matter.

Looking up, you see your crew on the third floor in the area from which the material fell. You decide to have a talk with them. Once on the deck, you ask the crew if any of them dropped the scrap. The men look over at Bob, one of the electricians in your crew. Bob replies, "I guess it was mine. It slipped out of my hand."

It is a known fact that the Occupational Safety and Health Administration (OSHA) regulations state that an enclosed chute of wood shall be used for material waste transportation from heights of 20 feet or more. It is also known that Bob and the laborer who was almost hit have been seen arguing lately.

1. Assuming Bob's action was deliberate, what action would you take?

____

____

____

2. Assuming the conduit accidentally slipped from Bob's hand, how can you motivate him to be more careful?

____

____

____

3. What follow-up actions, if any, should be taken relative to the laborer who was almost hit?

____

____

____

4. Should you discuss the apparent OSHA violation with the crew? Why or why not?

____

____

____

5. What acts of leadership would be effective in this case? To what leadership traits are they related?

____

____

____

Case II:

The company just appointed Antonio crew leader of a tile-setting crew. Before his promotion into management, he had been a tile setter for five years. His work had been consistently of superior quality.

Except for a little good-natured kidding, Antonio's co-workers had wished him well in his new job. During the first two weeks, most of them had been cooperative while Antonio was adjusting to his supervisory role.

At the end of the second week, a disturbing incident took place. Having just completed some of his duties, Antonio stopped by the job-site wash station. There he saw Steve and Ron, two of his old friends who were also in his crew, washing.

"Hey, Ron, Steve, you shouldn't be cleaning up this soon. It's at least another thirty minutes until quitting time," said Antonio. "Get back to your work station, and I'll forget I saw you here."

"Come off it, Antonio," said Steve. "You used to slip up here early on Fridays. Just because you have a little rank now, don't think you can get tough with us." To this Antonio replied, "Things are different now. Both of you get back to work, or I'll make trouble." Steve and Ron said nothing more, and they both returned to their work stations.

From that time on, Antonio began to have trouble as a crew leader. Steve and Ron gave him the silent treatment. Antonio's crew seemed to forget how to do the most basic activities. The amount of rework for the crew seemed to be increasing. By the end of the month, Antonio's crew was behind schedule.

1. How do you think Antonio should've handled the confrontation with Ron and Steve?

____

____

2. What do you suggest Antonio could do about the silent treatment he got from Steve and Ron?

____

____

3. If you were Antonio, what would you do to get your crew back on schedule?

____

____

4. What acts of leadership could be used to get the crew's willing cooperation?

____

____

5. To which leadership traits do they correspond?

____

____

Additional Resources

Construction Workforce Development Professional, NCCER. 2016. New York, NY: Pearson Education, Inc.

Mentoring for Craft Professionals, NCCER. 2016. New York, NY: Pearson Education, Inc.

It's Your Ship: Management Techniques from the Best Damn Ship in the Navy, Captain D. Michael Abrashoff, USN. 2012. New York City, NY: Grand Central Publishing.

Survival of the Fittest, Mark Breslin. 2005. McNally International Press.

The Definitive Book of Body Language: The Hidden Meaning Behind People's Gestures and Expressions, Barbara Pease and Allan Pease. 2006. New York City, NY: Random House / Bantam Books.

2.0.0 Section Review

1. A crew leader differs from a craftworker in that a crew leader _____.
   a. does not need direct experience in the job duties a craft professional typically performs
   b. can expect to oversee one or more workers in addition to performing typical craft duties
   c. is exclusively in charge of overseeing, since performing technical work isn't part of this role
   d. has no responsibility to be present on the job site

2. Feedback is important in verbal communication because it _____.
   a. requires the sender to repeat the message
   b. involves the receiver repeating back the message word for word
   c. informs the sender of how the message was received
   d. consists of a written analysis of the message

3. Once achieved, setting challenging goals for workers gives them a sense of _____.
   a. accomplishment
   b. entitlement
   c. persecution
   d. failure

4. Which of the following is *not* a characteristic of a successful team?
   a. Everyone participates and everyone counts.
   b. There is a sense of mutual trust.
   c. Members minimize communication with each other.
   d. The team leader is committed to the team.

5. Problem solving differs from decision making because it _____.
   a. involves finding an answer
   b. expresses an opinion
   c. involves starting or stopping an action
   d. separates facts from non-facts

SECTION THREE

3.0.0 SAFETY AND SAFETY LEADERSHIP

Objective

Identify a crew leader's typical safety responsibilities with respect to common safety issues, including awareness of safety regulations and the cost of accidents.

a. Explain how a strong safety program can enhance a company's success.
b. Explain the purpose of OSHA and describe the role of OSHA in administering worker safety.
c. Describe the role of employers in establishing and administering safety programs.
d. Explain how crew leaders are involved in administering safety policies and procedures.

Trade Terms

Intangible: Not touchable, material, or measureable; lacking a physical presence.

Lethargy: Sluggishness, slow motion, lack of activity, or a lack of enthusiasm.

Negligence: Lack of appropriate care when doing something, or the failure to do something, usually resulting in injury to an individual or damage to equipment.

Safety data sheets (SDS): Documents listing information about a material or substance that includes common and proper names, chemical composition, physical forms and properties, hazards, flammability, handling, and emergency response in accordance with national and international hazard communication standards; also called material safety data sheets (MSDS).

Businesses lose millions of dollars every year because of on-the-job accidents. Work-related injuries, sickness, and fatalities have caused untold suffering for workers and their families. Resulting project delays and budget overruns can cause huge losses for employers, and work-site accidents damage the overall morale of the crew.

Craft professionals routinely face hazards. Examples of these hazards include falls from heights, working on scaffolds, using cranes in the presence of power lines, operating heavy machinery, and working on electrically-powered or pressurized equipment. Despite these hazards, experts believe that applying preventive safety measures could drastically reduce the number of accidents.

Did You Know?

The Fatal Four

When OSHA inspects a job site, they focus on the types of safety hazards that are most likely to cause serious and fatal injuries. These hazards result in the following most common categories of injuries:

- Falls from elevations
- Struck-by hazards
- Caught-in/between hazards
- Electrical-shock hazards

As a crew leader, one of your most important tasks is to enforce the company's safety program and make sure that all workers are performing their tasks safely. To be successful, the crew leader should do the following:

- Be aware of the human and monetary costs of accidents.
- Understand all federal, state, and local governmental safety regulations applicable to your work.
- Be the most visible example of the best safe work practices.
- Be involved in training workers in safe work methods.
- Conduct training sessions.
- Get involved in safety inspections, accident investigations, and fire protection and prevention.

"Example is not the main thing in influencing others. It is the only thing."
—Albert Schweitzer

Crew leaders are in the best position to ensure that their crew members perform all jobs safely. Providing employees with a safe working environment by preventing accidents and enforcing safety standards will go a long way towards maintaining the job schedule and enabling a job's completion on time and within budget.

3.1.0 The Impact of Accidents

Each day, workers in construction and industrial occupations face the risk of falls, machinery accidents, electrocutions, and other potentially fatal hazards. The National Institute of Occupational Safety and Health (NIOSH) statistics show that roughly 1,000 construction workers are killed on the job each year, which is more than any other industry. Falls are the leading cause of deaths in the construction industry through accident or negligence, accounting for over 60 percent of the fatalities in recent years. Nearly half of the fatal falls occurred from roofs, scaffolds, or ladders. Roofers, structural metal workers, and painters are at the greatest risk of fall fatalities (*Figure 13*).

In addition to the number of fatalities that occur each year, there are a staggering number of work-related injuries. In 2015, for example, almost 200,000 job-related injuries occurred in the construction industry. NIOSH estimates that the total cost of fatal and non-fatal injuries in the construction industry represents about 15 percent of the costs for all private industry. The main causes of injuries on construction sites include falls, electrocution, fires, and mishandling of machinery or equipment. According to NIOSH, back injuries are the leading health-related problem in workplaces.

3.1.1 Cost of Accidents

Occupational accidents cost roughly $250 billion or more every year. These costs affect the individual employee, the company, and the construction industry as a whole.

Organizations encounter both direct and indirect costs associated with workplace accidents. Direct costs are the money companies must pay out to workers' compensation claims and sick pay; indirect costs are all the other tangible and intangible things and costs a company must account for as the result of a worker's injury or death. To compete and survive, companies must control these as well as all other employment-related costs. There are many costs involved with workplace accidents. A company can insure some of these costs, but not others.

Figure 13 Falls are the leading cause of deaths and injuries in construction.

Insured costs – Insured costs are those costs either paid directly or reimbursed by insurance carriers. Insured costs related to injuries or deaths include the following:

- Compensation for lost earnings (known as *worker's comp*)
- Funeral charges
- Medical and hospital costs
- Monetary awards for permanent disabilities
- Pensions for dependents
- Rehabilitation costs

Insurance premiums or charges related to property damages include the following:

- Fire or other safety-related peril
- Structural loss; material and equipment loss or damage
- Loss of business use and occupancy
- Public liability
- Replacement cost of equipment, material, and structures

Uninsured costs – The relative direct and indirect costs of accidents are comparable to the visible and hidden portions of an iceberg, as shown in *Figure 14*. The tip of the iceberg represents direct costs, which are the visible costs. Not all of these are covered by insurance. The more numerous indirect costs aren't readily measurable, but they can represent a greater financial burden than the direct costs.

Uninsured costs from injuries or deaths include the following:

- First aid expenses
- Transportation costs
- Costs of investigations
- Costs of processing reports
- Down time on the job site
- Costs to train replacement workers

Did You Know?

The Costs of Negligence

If you receive an injury as the result of a workplace accident, and a completed investigation shows that your injuries were due to your own negligence, you can't sue your employer for damages. In addition, workers' compensation insurance companies may decline to pay claims for negligent employees who are injured or killed.

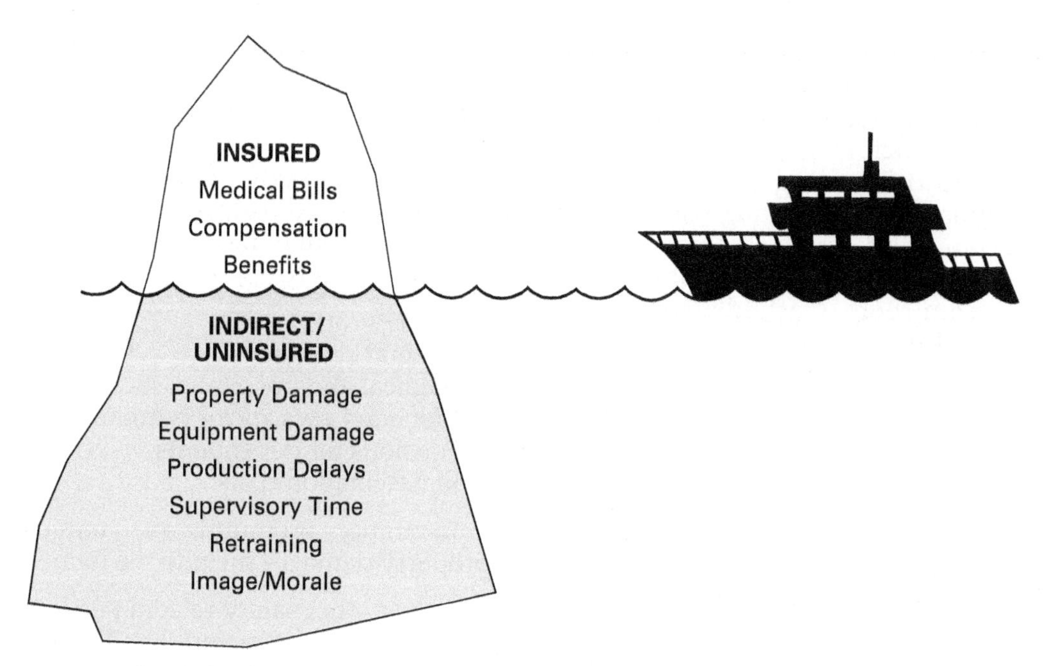

Figure 14 Costs associated with accidents.

Uninsured costs related to wage losses include the following:

- Idle time of workers whose work is interrupted
- Time spent cleaning the accident area
- Time spent repairing damaged equipment
- Time lost by workers receiving first aid
- Costs of training injured workers in a new career

Uninsured costs related to production losses include the following:

- Product spoiled by accident
- Loss of skill and experience; worker replacement
- Lowered production capacity
- Idle machine time due to lack of qualified operators

Associated costs may include the following:

- Difference between actual losses and amount recovered
- Costs of rental equipment used to replace damaged equipment
- Costs of inexperienced temp or permanent new workers used to replace injured workers
- Wages or other benefits paid to disabled workers
- Overhead costs while production is stopped
- Impact on schedule
- Loss of client bonus or payment of forfeiture for delays

Uninsured costs related to off-the-job activities include the following:

- Time spent on ensuring injured workers' welfare
- Loss of skill and experience of injured workers
- Costs of training replacement workers

Uninsured costs related to intangible factors include the following:

- Increased labor conflict
- Loss of bid opportunities because of poor safety records
- Loss of client goodwill
- Lowered employee morale
- Unfavorable public relations

3.2.0 OSHA

To reduce safety and health risks and the number of injuries and fatalities on the job, the federal government has enacted laws and regulations, including the Occupational Safety and Health Act of 1970 (OSH Act of 1970). This law created the Occupational Safety and Health Administration (OSHA), which is part of the US Department of Labor. OSHA also provides education and training for employers and workers. Through the administration of OSHA, the US Congress seeks "to assure so far as possible every working man and woman in the Nation safe and healthful working conditions and to preserve our human resources…" (OSH Act of 1970, Section 2[b]).

To promote a safe and healthy work environment, OSHA issues standards and rules for working conditions, facilities, equipment, tools, and work processes. It does extensive research into occupational accidents, illnesses, injuries, and deaths to reduce the number of occurrences and adverse effects. In addition, OSHA regulatory agencies conduct workplace inspections to ensure that companies follow the standards and rules.

To enforce OSHA regulations, the government has granted regulatory agencies the right to enter public and private properties to conduct workplace safety investigations. The agencies also have the right to take legal action if companies are not in compliance with the Act. These regulatory agencies employ OSHA Compliance Safety and Health Officers (CSHO), who are experts in the occupational safety and health field. The CSHOs are thoroughly familiar with OSHA standards and recognize safety and health hazards.

States with their own occupational safety and health programs conduct inspections by enlisting the services of qualified state CSHOs.

Companies are inspected for a multitude of reasons. They may be randomly selected, or they may be chosen as a result of employee complaints, a report of an imminent danger, or major accidents/fatalities that have occurred.

OSHA has established significant monetary fines for the violation of its regulations. *Table 1* lists the penalties as of 2016. In some cases, OSHA will hold a superintendents or crew leaders personally liable for repeat violations as well. In addition to the fines, there are possible criminal charges for willful violations resulting in death or serious injury. The attitude of the employer and their safety history can have a significant effect on the outcome of a case.

Did You Know?

State OSHA Programs

Nearly half the states in the United States and some of its territories have state-run OSHA programs. These programs are set up under federal OSHA guidelines and must establish job health and safety standards that are at least as effective as the federal standards. The states/territories have the option of adopting more stringent standards or setting standards for hazards not addressed in the federal program. Of the 22 states and territories with non-federal OSHA programs, six of them limit their coverage to state and local government employees.

3.3.0 Employer Safety Responsibilities

Each employer must set up a safety and health program to manage workplace safety and health and to reduce work-related injuries, illnesses, and fatalities. The program must be appropriate for the conditions of the workplace. It should consider the number of workers employed and the hazards they face while at work.

To be successful, the safety and health program must have management, leadership, and employee participation. In addition, training and informational meetings play an important part in effective programs. Being consistent with safety policies is the key. Regardless of the employer's responsibility, however, the individual worker is ultimately responsible for his or her own safety.

Table 1 OSHA Penalties for violations established in 2016

Type of Violation	Maximum Penalty
Serious Other-Than-Serious Posting Requirements	$12,471 per violation
Failure to Abate	$12,471 per day
Willful or Repeated Violation	$124,709 per violation

3.3.1 Safety Program

The crew leader plays a key role in the successful implementation of the safety program. The crew leader's attitude toward the program sets the standard for how crew members view safety. Therefore, the crew leader should follow all program guidelines and require crew members to do the same.

Safety programs should consist of the following:

- Safety policies and procedures
- Safety information and training
- Posting of safety notices
- Hazard identification, reporting, and assessment
- Safety record system
- Accident reporting and investigation procedures
- Appropriate discipline for not following safety procedures

3.3.2 Safety Policies and Procedures

Employers are responsible for following OSHA and state safety standards. Usually, they incorporate federal and state OSHA regulations into a safety policies and procedures manual. Employees receive such a manual when they are hired.

During orientation, appropriate company staff should guide the new employees through the general sections of the safety manual and the sections that have the greatest relevance to their job.

If the employee can't read, the employer should have someone read it to the employee and answer any questions that arise. The employee should then sign a form stating understanding of the information.

It isn't enough to tell employees about safety policies and procedures on the day they are hired and then never mention them again. Rather, crew leaders should constantly emphasize and reinforce the importance of adhering to all safety policies and procedures. In addition, employees should play an active role in determining job safety hazards and find ways to prevent and control hazards.

3.3.3 Hazard Identification and Assessment

Safety policies and procedures should be specific to the company. They should clearly present the hazards of the job and provide the means to report hazards to the proper level of management without prejudice to the individual doing the reporting. Crew leaders should also identify and assess hazards to which employees are exposed. They must also assess compliance with federal and state OSHA standards.

To identify and assess hazards, OSHA recommends that employers conduct periodic and random inspections of the workplace, monitor safety and health information logs, and evaluate new equipment, materials, and processes for potential hazards before they are used.

"*You get what you inspect, not what you expect.*"
—*Anonymous*

Crew leaders and workers play important roles in identifying and reporting hazards. It is the crew leader's responsibility to determine what working conditions are unsafe and to inform employees of hazards and their locations. In addition, they should encourage their crew members to tell them about hazardous conditions. To accomplish this, crew leaders must be present and available on the job site.

The crew leader also needs to help the employee be aware of and avoid the built-in hazards to which craft professionals are exposed. Examples include working at elevations, working in confined spaces such as tunnels and underground vaults, on caissons, in excavations with earthen walls, and other naturally-dangerous projects. In addition, the crew leader can take safety measures, such as installing protective railings to prevent workers from falling from buildings, as well as scaffolds, platforms, and shoring.

3.3.4 Safety Information and Training

The employer must provide periodic information and training to new and long-term employees. This happens as often as necessary so that all employees receive adequate training. When safety and health information changes or workplace conditions create new hazards, the company must then provide special training and informational sessions. It is important to note that the company must present safety-related information in a manner that each employee will understand.

When a crew leader assigns an inexperienced employee a new task, the crew leader must ensure that the employee can do the work in a safe manner. The crew leader can accomplish this by providing safety information or training for groups or individuals.

When assigning an inexperienced employee a new task, do the following:

- Define the task.
- Explain how to do the task safely.
- Explain what tools and equipment to use and how to use them safely.
- Identify the necessary personal protective equipment and train the employee in its use.
- Explain the nature of the hazards in the work and how to recognize them.
- Stress the importance of personal safety and the safety of others.
- Hold regular safety training sessions with the crew's input.
- Review **safety data sheets (SDS)** that may be applicable.

3.3.5 Safety Record System

OSHA regulations (29 *CFR* 1904) require that employers keep records of hazards identified and document the severity of the hazard. The information should include the likelihood of employee exposure to the hazard, the seriousness of the harm associated with the hazard, and the number of exposed employees.

In addition, the employer must document the actions taken or plans for action to control the hazards. While it is best to take corrective action immediately, it is sometimes necessary to develop a plan to set priorities and deadlines and track progress in controlling hazards.

Employers who are subject to the recordkeeping requirements of the Occupational Safety and Health Act of 1970 must maintain records of all recordable occupational injuries and illnesses. The following are some OSHA forms that should be used for this recordkeeping:

- OSHA Form 300, *Log of Work-Related Injuries and Illnesses*
- OSHA Form 300A, *Summary of Work-Related Injuries and Illnesses*
- OSHA Form 301, *Injury and Illness Incident Report*

These three OSHA forms are included in the *Appendix* at the end of this module. Note that crew leaders directly handle the OSHA Form 301.

An SDS provides both workers and emergency personnel with the proper procedures for handling or working with a substance that may be dangerous. The document will include information such as physical data (melting point, boiling point, flash point, etc.), toxicity, health effects, first aid, reactivity, storage, disposal, protective equipment required for handling, and spill/leak procedures. These sheets are of particular use if a spill, fire, or other accident occurs.

Companies not exempted by OSHA must maintain required safety logs and retain them for 5 years following the end of the calendar year to which they relate. Logs must be available (normally at the company offices) for inspection and copying by representatives of the Department of Labor, the Department of Health and Human Services, or states given jurisdiction under the Act. Employees, former employees, and their representatives may also review these logs.

3.3.6 Accident Investigation

Employees must know from their training to immediately report any unusual event, accident, or injury. Policies and definitions of what these incidents consist of should be included in safety manuals. In the event of an accident, the employer is required to investigate the cause of the accident and determine how to avoid it in the future.

According to OSHA regulations, the employer must investigate each work-related death, serious injury or illness, or incident having the potential to cause death or serious physical harm. The employer should document any findings from the investigation, as well as the action plan to

Summaries of Work-Related Injuries and Illnesses

Most companies with 11 or more employees must post an OSHA Form 300A, *Summary of Work-Related Injuries and Illnesses*, between February 1 and April 30 of each year. Employees have the right to review this form. Check your company's policies regarding this and the related OSHA forms.

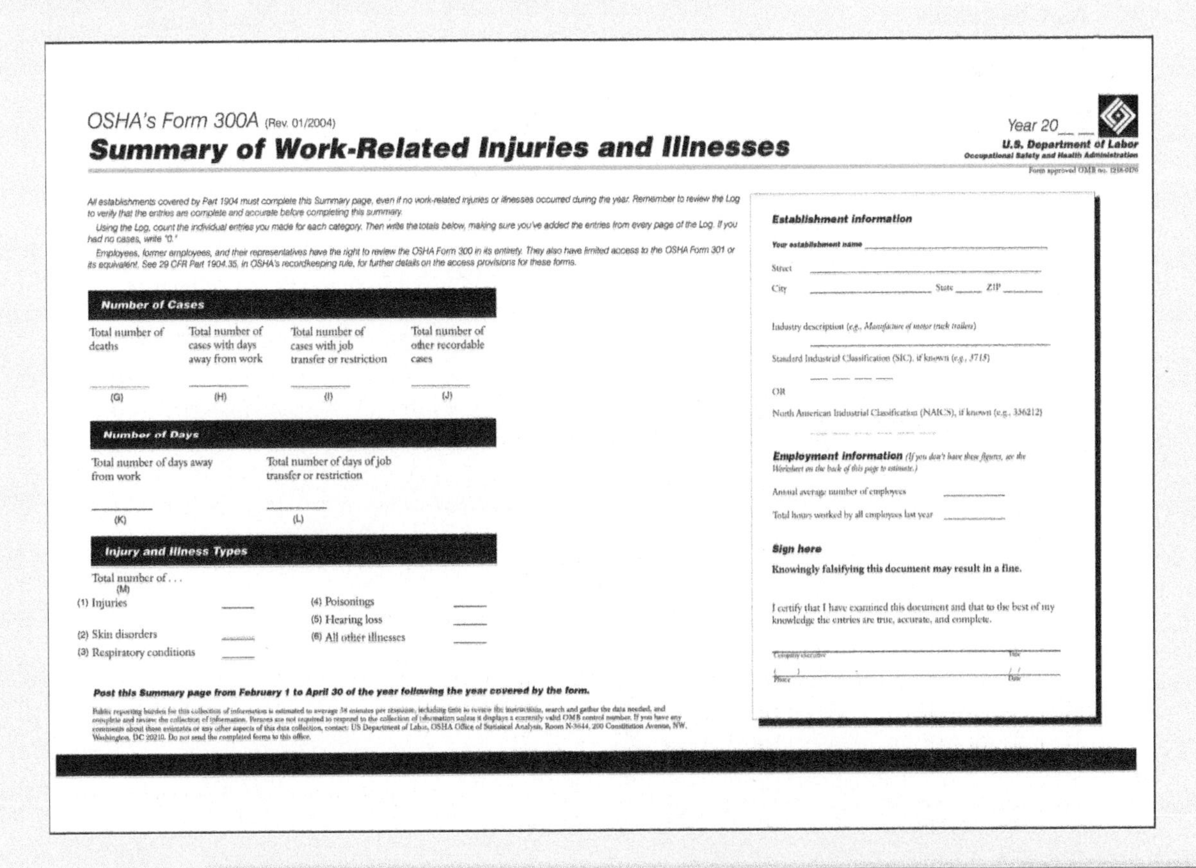

OSHA's Form 300A (Rev. 01/2004)

Summary of Work-Related Injuries and Illnesses

Year 20__ __

U.S. Department of Labor
Occupational Safety and Health Administration

Form approved OMB no. 1218-0176

All establishments covered by Part 1904 must complete this Summary page, even if no work-related injuries or illnesses occurred during the year. Remember to review the Log to verify that the entries are complete and accurate before completing this summary.

Using the Log, count the individual entries you made for each category. Then write the totals below, making sure you've added the entries from every page of the Log. If you had no cases, write "0."

Employees, former employees, and their representatives have the right to review the OSHA Form 300 in its entirety. They also have limited access to the OSHA Form 301 or its equivalent. See 29 CFR Part 1904.35, in OSHA's recordkeeping rule, for further details on the access provisions for these forms.

Number of Cases

Total number of deaths	Total number of cases with days away from work	Total number of cases with job transfer or restriction	Total number of other recordable cases
____ (G)	____ (H)	____ (I)	____ (J)

Number of Days

Total number of days away from work	Total number of days of job transfer or restriction
____ (K)	____ (L)

Injury and Illness Types

Total number of . . .
(M)

(1) Injuries	____	(4) Poisonings	____
		(5) Hearing loss	____
(2) Skin disorders	____	(6) All other illnesses	____
(3) Respiratory conditions	____		

Post this Summary page from February 1 to April 30 of the year following the year covered by the form.

Public reporting burden for this collection of information is estimated to average 58 minutes per response, including time to review the instructions, search and gather the data needed, and complete and review the collection of information. Persons are not required to respond to the collection of information unless it displays a currently valid OMB control number. If you have any comments about these estimates or any other aspects of this data collection, contact: US Department of Labor, OSHA Office of Statistical Analysis, Room N-3644, 200 Constitution Avenue, NW, Washington, DC 20210. Do not send the completed forms to this office.

Establishment information

Your establishment name ____________________

Street ____________________

City ____________ State ____ ZIP ____

Industry description (*e.g., Manufacture of motor truck trailers*)

____

Standard Industrial Classification (SIC), if known (*e.g., 3715*)

__ __ __ __

OR

North American Industrial Classification (NAICS), if known (e.g., 336212)

__ __ __ __ __ __

Employment information (*If you don't have these figures, see the Worksheet on the back of this page to estimate.*)

Annual average number of employees ____________

Total hours worked by all employees last year ____________

Sign here

Knowingly falsifying this document may result in a fine.

I certify that I have examined this document and that to the best of my knowledge the entries are true, accurate, and complete.

____________________ ____________
Company executive Title

( ) - / /
Phone Date

prevent future occurrences. The company should complete these actions immediately, with photos or video if possible. It's important that the investigation uncover the root cause of the accident to avoid similar incidents in the future. In many cases, the root cause was a flaw in the system that failed to recognize the unsafe condition or the potential for an unsafe act (*Figure 15*).

3.4.0 Leader Involvement in Safety

To be an effective, you must be actively involved in your company's safety program. Crew leader involvement includes conducting frequent safety training sessions and inspections, promoting first aid, and fire protection and prevention, preventing substance abuse on the job, and investigating accidents. Most importantly, crew leaders must practice safety at all times.

3.4.1 Safety Training Sessions

A safety training session may be a brief, informal gathering of a few employees or a formal meeting with instructional videos and talks by guest speakers. The size of the audience and the topics addressed determine the format of the meeting. You should plan to conduct small, informal safety sessions weekly.

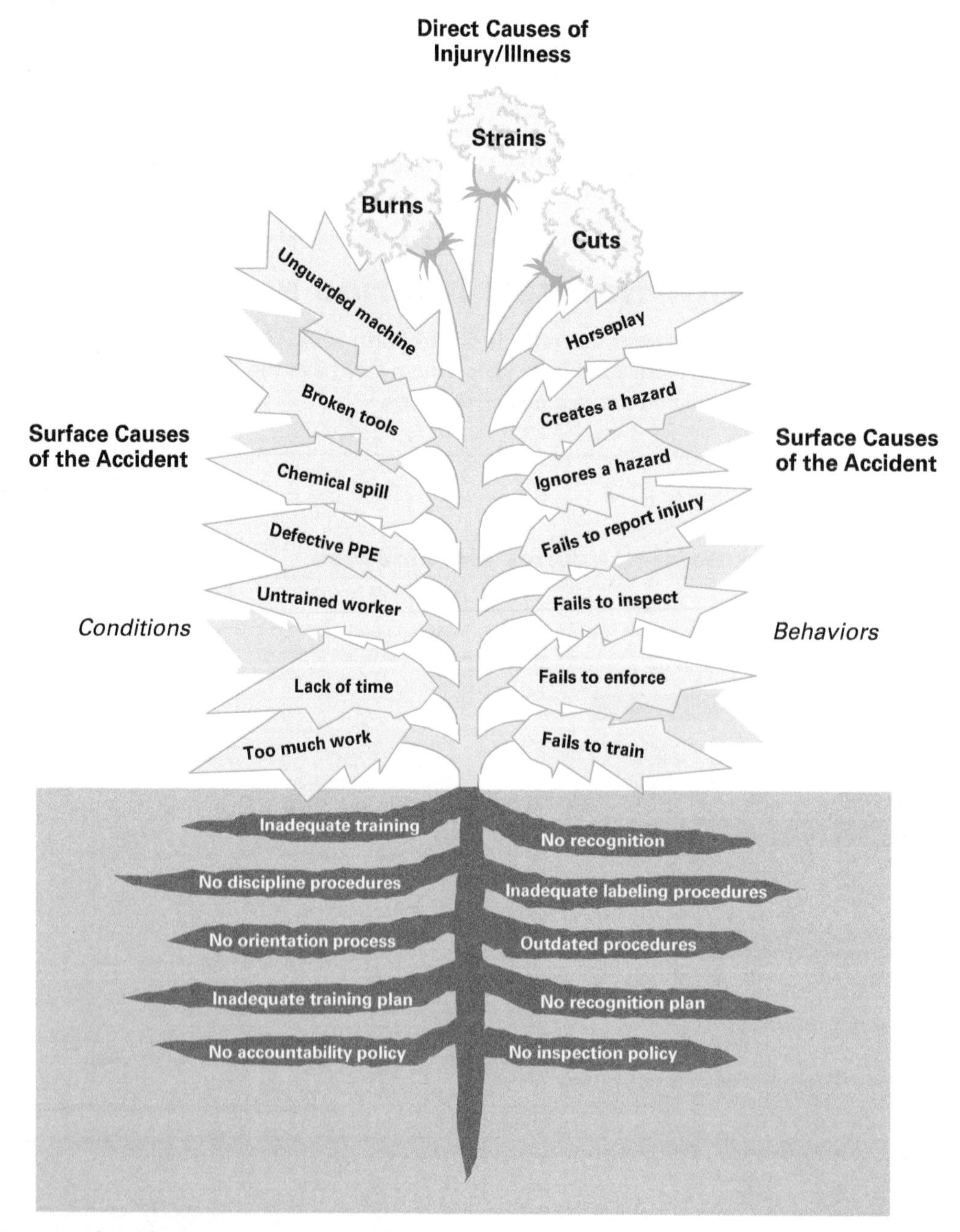

Figure 15 Root causes of accidents.

OSHA Accident Notification Requirements

There are urgent reporting requirements by the employer to OSHA for the following cases:

1. Within 8 hours: A work-related fatality.
2. Within 24 hours:
   - Work-related accident that resulted in an in-patient admission of one or more employees
   - Work-related amputation
   - Work-related loss of an eye

You should also plan safety training sessions in advance, and you should communicate the information to all affected employees. In addition, the topics covered in these training sessions should be timely and practical. Keep a log of each safety session signed by all attendees. It must be maintained as a record and available for inspection. It is advisable to attach a copy of a summary of the training session to the attendance record so you can keep track of what topics you covered and when.

3.4.2 Inspections

Crew leaders must make routine, frequent inspections to prevent accidents from happening. They must also take steps to avoid accidents. For that purpose, they need to inspect the job sites where their workers perform tasks. It's advisable to do these inspections before the start of work each day and during the day at random times.

Crew leaders must protect workers from existing or potential hazards in their work areas. Crews are sometimes required to work in areas controlled by other contractors. In these situations, the crew leader must maintain control over the safety exposure of the crew. If hazards exist, the crew leader should immediately bring the hazards to the attention of the contractor at fault, their superior, and the person responsible for the job site.

Crew leader inspections are only valuable if follow-up action corrects potential hazards. Therefore, crew leaders must be alert for unsafe and negligent acts on their work sites. When an employee performs an unsafe action, the crew leader must explain to the employee why the act was unsafe, tell that employee to not do it again, and request cooperation in promoting a safe working environment. The crew leader must document what happened and what the employee was told to do to correct the situation. It is then important that crew leaders follow up to make certain the employee is complying with the safety procedures. Never allow a safety violation to go uncorrected. There are three courses of action that you, as a crew leader, can take in an unsafe work site situation:

- Get the appropriate party to correct the problem.
- Fix the problem yourself.
- Refuse to have the crew work in the area until the responsible party corrects the problem.

"As soon as you see a mistake and don't fix it, it becomes your mistake."
—Anonymous

3.4.3 First Aid

The primary purpose of first aid is to provide immediate and temporary medical care to employees involved in accidents, as well as employees experiencing non-work-related health emergencies, such as chest pain or breathing difficulty. To meet this objective, every crew leader should be aware of the location and contents of first aid kits available on the job site. Emergency numbers should be posted in the job trailer. In addition, OSHA requires that at least one person trained in first aid be present at the job site at all times. It's also advisable, but not required, that someone on site should have cardiopulmonary resuscitation (CPR) training.

NOTE

CPR certifications must be renewed every two years.

The victim of an accident or sudden illness at a job site may be harder to aid than elsewhere since the worker may be at a remote location. The site may be far from a rescue squad, fire department, or hospital, presenting a problem in the rescue and transportation of the victim to a hospital. The worker may also have received an injury from falling rock or other materials, so immediate special rescue equipment or first-aid techniques are often needed.

Employer benefits of having personnel trained in first aid at job sites include the following:

- The immediate and proper treatment of minor injuries may prevent them from developing into more serious conditions. Thus, these precautions can eliminate or reduce medical expenses, lost work time, and sick pay.

- It may be possible to determine if the injured person requires professional medical attention.
- Trained individuals can save valuable time preparing the inured for treatment for when professional medical care arrives. This service increases the likelihood of saving a life.

The American Red Cross, Medic First Aid, and the United States Bureau of Mines provide basic and advanced first aid courses at nominal costs. These courses include both first aid and CPR. The local area offices of these organizations can provide further details regarding the training available.

3.4.4 Fire Protection and Prevention

Fires and explosions kill and injure many workers each year, so it is important that crew leaders understand and practice fire-prevention techniques as required by company policy.

The need for protection and prevention is increasing as manufacturers introduce new building materials. Some building materials are highly flammable. They produce great amounts of smoke and gases, which cause difficulties for fire fighters, and can quickly overcome anyone present. Other materials melt when they burn and may puddle over floors, preventing fire-fighting personnel from entering areas where this occurs.

OSHA has specific standards for fire safety. Employers are required to provide proper exits, fire-fighting equipment, and employee training on fire prevention and safety. For more information, consult OSHA guidelines (available at **www.osha.gov**).

3.4.5 Substance Abuse

Substance abuse is a continuing problem in the workplace. Substance abuse is the inappropriate overuse of drugs and chemicals, whether they are legal or illegal. All substance abuse results in some form of mental, sensory, or physical impairment. Some people use illegal "street drugs", such as cocaine or crystal meth. Others use legal prescription drugs incorrectly by taking too many (or too few) pills, using other people's medications, or self-medicating. Alcohol can also be abused by consuming to the point of intoxication. Other substances that are legal in some states (e.g., marijuana) can cause prolonged impairment with heavy usage.

It is essential that crew leaders enforce company policies and procedures regarding substance abuse. Crew leaders must work with management to deal with suspected drug and alcohol abuse and should not deal with these situations themselves. The Human Resources department or a designated manager usually handles these cases.

There are legal consequences of substance abuse and the associated safety implications. If you observe an employee showing impaired behavior for any reason, immediately contact your supervisor and/or Human Resources department for assistance. You protect the business, and the employee's and other workers' safety by taking these actions. It is the crew leader's responsibility to maintain safe working conditions at all times. This may include removing workers from a work site where they may be endangering themselves or others.

For example, suppose several crew members go out and smoke marijuana or drink during lunch. Then, they return to work to erect scaffolding for a concrete pour in the afternoon. If you can smell marijuana on the crew member's clothing or alcohol on their breath, you must step in and take action. Otherwise, they might cause an accident that could delay the project, or cause serious injury or death to themselves or others.

"Concern for man himself and his safety must always form the chief interest of all technical endeavors."
—Albert Einstein

It is often difficult to detect drug and alcohol abuse because the effects can be subtle. The best way is to look for identifiable effects, such as those mentioned above, or sudden changes in behavior that aren't typical of the employee. Some examples of such behaviors include the following:

- Unscheduled absences; failure to report to work on time
- Significant changes in the quality of work
- Unusual activity or **lethargy**
- Sudden and irrational temper flare-ups
- Significant changes in personal appearance, cleanliness, or health

There are other more specific signs that should arouse suspicion, especially if more than one is visible:

- Slurring of speech or an inability to communicate effectively
- Shiftiness or sneaky behavior, such as an employee disappearing to wooded areas, storage areas, or other private locations
- Wearing sunglasses indoors or on overcast days to hide dilated or constricted pupils, conditions which impair vision (*Figure 16*)

Figure 16 An impaired worker is a dangerous one.

- Wearing long-sleeved garments, particularly on hot days, to cover marks from needles used to inject drugs
- Attempting to borrow money from co-workers
- The loss of an employee's tools or company equipment

3.4.6 Job-Related Accident Investigations

Crew leaders are sometimes involved with an accident investigation. When an accident, injury, or report of work-connected illness takes place. If present on site, the crew leader should proceed immediately to the accident location to ensure that the victim receives proper first aid. The crew leader will also want to make sure that responsible individuals take other safety and operational measures to prevent another incident.

If required by company policy, the crew leader will also need to make a formal investigation and submit a report after an incident (including the completion of an *OSHA Form 301* report). An investigation looks for the causes of the accident by examining the circumstances under which it occurred and talking to the people involved. Investigations are perhaps the most useful tool in the prevention of future accidents.

The following are four major parts to an accident investigation:

- Describing the accident and related events leading to the accident
- Determining the cause(s) of the accident
- Identifying the persons or things involved and the part played by each
- Determining how to prevent reoccurrences

3.4.7 Promoting Safety

The best way for crew leaders to encourage safety is through example. Crew leaders should be aware that their behavior sets standards for their crew members. If a crew leader cuts corners on safety, then the crew members may think that it is okay to do so as well.

CAUTION

Workers often "follow the leader" when it comes to unsafe work practices. It is common for supervisory personnel to engage in unsafe practices and take more risks because they are more experienced. However, inexperienced or careless workers who take the same risks won't be as successful avoiding injury. As a leader, you must follow all safety practices to encourage your crew to do the same.

"I cannot trust a man to control others who cannot control himself."
—Robert E. Lee

The key to effectively promote safety is good communication. It is important to plan and coordinate activities and to follow through with safety programs. The most successful safety promotions occur when employees actively participate in planning and carrying out activities.

Some activities used by organizations to help motivate employees on safety and help promote safety awareness include the following:

- Safety training sessions
- Contests
- Recognition and awards
- Publicity

Safety training sessions can help keep workers focused on safety and give them the opportunity to discuss safety concerns with the crew. A previous section addressed this topic.

Did You Know?

Substance Abuse

An employee who is involved in an accident while under the influence of drugs or alcohol may be denied workers compensation insurance benefits.

Case Study

For years, a prominent safety engineer was confused as to why sheet-metal workers fractured their toes frequently. The crew leader had not performed thorough accident investigations, and the injured workers were embarrassed to admit how the accidents really occurred. Further investigation discovered they used the metal-reinforced cap on their safety shoes as a "third hand" to hold the sheet metal vertically in place when they fastened it. The rigid and heavy metal sheet was inclined to slip and fall behind the safety cap onto the toes, causing fractures. The crew leader could have prevented several injuries by performing a proper investigation after the first accident.

3.4.8 Safety Contests

Contests are a great way to promote safety in the workplace. Examples of safety-related contests include the following:

- Sponsoring housekeeping contests for the cleanest job site or work area
- Challenging employees to come up with a safety slogan for the company or department
- Having a poster contest that involves employees or their children creating safety-related posters
- Recording the number of accident-free workdays or worker-hours
- Giving safety awards (hats, T-shirts, other promotional items or prizes)

One of the positive aspects of safety contests is their ability to encourage employee participation. It is important, however, to ensure that the contest has a valid purpose. For example, workers can display the posters or slogans created in a poster contest throughout the organization as safety reminders.

CAUTION

One mistake that some companies make when offering safety contests is providing tangible or monetary awards to departments or teams specifically for the lowest number of reported accidents or accident-free work hours. While well-intentioned, this approach appeals to the tendency of people to inflate their performance to win. Consequently, history shows this has the negative effect of encouraging the under-reporting of accidents and injuries, which defeats the purposes of safety contests.

3.4.9 Incentives and Rewards

Incentives and awards serve several purposes. Among them are acknowledging and encouraging good performance, building goodwill, reminding employees of safety issues, and publicizing the importance of practicing safety standards. There are countless ways to recognize and award safety. Examples include the following:

- Supplying food at the job site when a certain goal is achieved

- Providing a reserved parking space to acknowledge someone for a special achievement
- Giving gift items such as T-shirts or gift certificates to reward employees
- Giving awards to a department or an individual (*Figure 17*)
- Sending a letter of appreciation
- Publicly honoring a department or an individual for a job well done

You can use creativity to determine how to recognize and award good safety on the work site. The only precautionary measure is that the award should be meaningful and not perceived as a bribe. It should be representative of the accomplishment.

3.4.10 Publicity

Publicizing safety is the best way to get the message out to employees. An important aspect of publicity is to keep the message accurate and current. Safety posters that hang for years on end tend to lose effectiveness. It is important to keep ideas fresh.

Examples of promotional activities include posters or banners, advertisements or information on bulletin boards, payroll mailing stuffers, and employee newsletters. In addition, the company can purchase merchandise that promotes safety, including buttons, hats, T-shirts, and mugs.

Figure 17 An example of a safety award.

Participant Exercise D

Described here are three scenarios that reflect unsafe practices by craft workers. For each of these scenarios write down how you would deal with the situation, first as the crew leader of the craft worker, and then as the leader of another crew.

1. You observe a worker wearing his hard hat backwards and his safety glasses hanging around his neck. He is using a concrete saw.

2. As you are supervising your crew on the roof deck of a building under construction, you notice that a section of guard rail has been removed. Another contractor was responsible for installing the guard rail.

3. Your crew is part of plant shutdown at a power station. You observe that a worker is welding without a welding screen in an area where there are other workers.

Additional Resources

Construction Workforce Development Professional, NCCER. 2016. New York, NY: Pearson Education, Inc.

Mentoring for Craft Professionals, NCCER. 2016. New York, NY: Pearson Education, Inc.

The following websites offer resources for products and training:

National Census of Fatal Occupational Injuries (NCFOI), **www.bls.gov**

National Institute of Occupational Safety and Health (NIOSH), **www.cdc.gov/niosh**

National Safety Council, **www.nsc.org**.

Occupational Safety and Health Administration (OSHA), **www.osha.gov**

3.0.0 Section Review

1. One of a crew leader's most important responsibilities to the employer is to _____.
   a. enforce company safety policies
   b. estimate material costs for a project
   c. make recommendations for setting up a crew
   d. provide input for fixed-price contracts

2. What amount can OSHA fine a company for willfully committing a violation of an OSHA safety standard?
   a. $1,000
   b. $7,000
   c. $12,471
   d. $124,709

3. Who is ultimately responsible for a worker's safety?
   a. The individual worker
   b. The crew leader
   c. The project superintendent
   d. The company HR department head

4. A crew leader's safety responsibilities include all of the following *except* _____.
   a. conducting safety training sessions
   b. developing a company safety program
   c. performing safety inspections
   d. participating in accident investigations

SECTION FOUR

4.0.0 PROJECT PLANNING

Objective

Demonstrate a basic understanding of the planning process, scheduling, and cost and resource control.

a. Describe how construction contracts are structured.
b. Describe the project planning and scheduling processes.
c. Explain how to implement cost controls on a construction project.
d. Explain the crew leader's role in controlling project resources and productivity.

Performance Tasks

1. Develop and present a look-ahead schedule.
2. Develop an estimate for a given work activity.

Trade Terms

Critical path: In manufacturing, construction, and other types of creative processes, the required sequence of tasks that directly controls the ultimate completion date of the project.

Job diary: A written record that a supervisor maintains periodically (usually daily) of the events, communications, observations, and decisions made during the course of a project.

Look-ahead schedule: A manual- or software-scheduling tool that looks several weeks into the future at the planned project events; used for anticipating material, labor, tool, and other resource requirements, as well as identifying potential schedule conflicts or other problems.

Return on investment (ROI): A measure of the gain or loss of money resulting from an investment; normally measured as a percentage of the original investment.

Work breakdown structure (WBS): A diagrammatic and conceptual method for subdividing a complex project, concept, or other thing into its various functional and organizational parts so that planners can analyze and plan for each part in detail.

This section describes methods of efficient project control. It examines estimating, planning and scheduling, and resource and cost control. All workers who participate in a job are responsible at some level for controlling cost and schedule performance, and for ensuring that they complete the project according to plans and specifications.

> **NOTE**
> This section mainly pertains to building-construction projects, but the project control principles described here apply generally to all types of projects.

The contractor, project manager, superintendent, and crew leader each have management responsibilities for their assigned jobs. For example, the contractor's responsibility begins with obtaining the contract, and it doesn't end until the client takes ownership of the project. The project manager is generally the person with overall responsibility for coordinating the project. Finally, the superintendent and crew leader are responsible for coordinating the work of one or more workers, one or more crews of workers within the company or, on occasion, one or more crews of subcontractors. The crew leader directs a crew in the performance of work tasks.

4.1.0 Construction Project Phases, Contracts, and Budgeting

Construction projects consist of three phases: the *development phase*, the *planning phase*, and the *construction phase*. Throughout these phases, the property owner must work directly with engineers and architects to gather data necessary for persuading a contractor to accept the job. Once a job is in progress, the crew leader fills an active role in maintaining the planned budget for the project.

4.1.1 Development Phase

A new building project begins when an owner has decided to build a new facility or add to an existing facility. The development process is the first stage of planning for a new building project. This process involves land research and feasibility studies to ensure that the project has merit. Architects or engineers develop the conceptual drawings that define the project graphically. They then provide the owner with sketches of room layouts and elevations and make suggestions about what construction materials to use.

During the development phase, architects, engineers, and/or the owner develop an estimate for the proposed project and establish a preliminary budget. Once that budget is established, the financing of the project with lending institutions begins. The development team begins preliminary reviews with government agencies. These reviews include zoning, building restrictions, landscape requirements, and environmental impact studies.

The owner must analyze the project's cost and potential **return on investment (ROI)** to ensure that its costs won't exceed its market value and that the project provides a reasonable profit during its existence. If the project passes this test, the responsible architects/engineers will proceed to the planning phase.

4.1.2 Planning Phase

When the architects/engineers begin to develop the project drawings and specifications, they consult with other design professionals such as structural, mechanical, and electrical engineers. They perform the calculations, make a detailed technical analysis, and check details of the project for accuracy.

The design professionals create drawings and specifications. They use these drawings and specifications to communicate the necessary information to the contractors, subcontractors, suppliers, and workers that contribute to a project.

During the planning phase, the owners hold many meetings (*Figure 18*) to refine estimates, adjust plans to conform to regulations, and secure a construction loan. If the project is a condominium, an office building, or a shopping center, then a marketing firm develops a marketing program. In such cases, the selling of the project often starts before actual construction begins.

Figure 18 Architects and clients meet to refine plans.

Next, the design team produce a complete set of drawings, specifications, and bid documents. Then the owner will select the method to obtain contractors. The owner may choose to negotiate with several contractors or select one through competitive bidding. Everyone concerned must also consider safety as part of the planning process. A safety crew leader may walk through the site as part of the pre-bid process.

Contracts for construction projects can take many forms. All types of contracts fall under three basic categories: *firm-fixed-price, cost reimbursable,* and *guaranteed maximum price.*

Firm-fixed-price – In this type of contract, the buyer generally provides detailed drawings and specifications, which the contractor uses to calculate the cost of materials and labor. To these costs, the contractor adds a percentage representing company overhead expenses such as office rent, insurance, and accounting/payroll costs. At the end, the contractor adds a profit factor.

When submitting the bid, the contractor will state very specifically the conditions and assumptions on which the company based the bid. These conditions and assumptions also form the basis from which parties can price allowable changes to the contract. Because contracting parties establish the price in advance, any changes in the job requirements once the job is started will impact the contractor's profit margin.

This is where the crew leader can play an important role by identifying problems that increase the amount of planned labor or materials. By passing this information up the chain of command, the crew leader allows the company to determine if the change is outside the scope of the bid. If so, they can submit a change order request to cover the added cost.

Cost reimbursable contract – In this type of contract, the buyer reimburses the contractor for labor, materials, and other costs encountered in the performance of the contract. Typically, the contractor and buyer agree in advance on hourly or daily labor rates for different categories of worker. These rates include an amount representing the contractor's overhead expense. The buyer also reimburses the contractor for the cost of materials and equipment used on the job.

The buyer and contractor also negotiate a profit margin. On this type of contract, the profit margin is likely to be lower than that of a fixed-price contract because of the significantly-reduced contractor's cost risk. The profit margin is often subject to incentive or penalty clauses that make the amount of profit awarded subject to performance by the contractor. The contract usually ties performance to project schedule milestones.

Guaranteed maximum price (GMP) contract – This form of contract, also called a *not-to-exceed contract,* is common on projects negotiated mainly with the owner. Owner's involvement in the process usually includes preconstruction, and the entire team develops the parameters that define the basis for the work.

With a GMP contract, the owner reimburses the contractor for the actual costs incurred by the contractor. The contract also includes a payment of a fixed fee up to the maximum price allowed in the contract. The contractor bears any cost overruns.

The advantages of the GMP contract vehicle may include the following:

- Reduced design time
- Allows for phased construction
- Uses a team approach to a project
- Reduction in changes related to incomplete drawings

4.1.3 Construction Phase

The designated contractor enlists the help of mechanical, electrical, elevator, and other specialty subcontractors to complete the construction phase. The contractor may perform one or more parts of the construction, and rely on subcontractors for the remainder of the work. However, the general contractor is responsible for managing all the trades necessary to complete the project. *Figure 19* shows the flow of a typical project from beginning to end.

As construction nears completion, the architect/engineer, owner, and government agencies start their final inspections and acceptance of the project. If the general contractor has managed the project, the subcontractors have performed their work, and the architects/engineers have regularly inspected the project to ensure it satisfied the local code, then the inspection process can finish up in a timely manner. This results in a satisfied client and a profitable project for all.

On the other hand, if the inspection reveals faulty workmanship, poor design, incorrect use of materials, or violation of codes, then the inspection and acceptance will become a lengthy battle and may result in a dissatisfied client and an unprofitable project.

The initial set of drawings for a construction project reflects the completed project as conceived by the architect and engineers. During construction, changes are usually necessary because of factors unforeseen during the design phase. For example, when electricians must reroute cabling or conduit, or the installed equipment location is different than shown on the original drawing, such changes must be marked on the drawings. Without this record, technicians called to perform maintenance or modify the equipment later will have trouble locating all the cabling and equipment.

Project supervision must document any changes made during construction or installation on the drawings as the changes occur. Architects usually note changes on hard-copy drawings using a colored pen or pencil, so users can readily spot the change. These marked-up versions are commonly called **redline drawings**. With mobile digital technology, architects and engineers can revise and promulgate the latest drawing versions almost instantly. After the drawings have been revised to reflect the redline changes, the final drawings are called **as-built drawings**, and are so marked. These become the drawings of record for the project.

4.1.4 Project-Delivery Systems

Project-delivery systems are processes for constructing projects, from development through construction. Project delivery systems focus on the following three primary systems, shown in *Figure 20*:

- *General contracting* – The traditional project delivery system uses a general contractor. In this type of project, the owner determines the design of the project, and then solicits proposals from general contractors. After selecting a general contractor, the owner contracts directly with that contractor, who builds the project as the prime, or controlling, contractor.
- *Design-build* – In the design-build system, a single entity manages both the design and construction of a project. Design-build delivery commonly use GMP contracts.
- *Construction management* – The construction management project delivery system uses a construction manager to facilitate the design and construction of a project. Construction managers are very involved in project control; their main concerns are controlling time, cost, and the quality of the project.

4.1.5 Cost Estimating and Budgeting

Before building a project, an estimate must be prepared. Estimating is the process of calculating the cost of a project. There are two types of costs to consider, including direct and indirect costs. Direct costs, also known as general conditions, are those that planners can clearly assigned to a

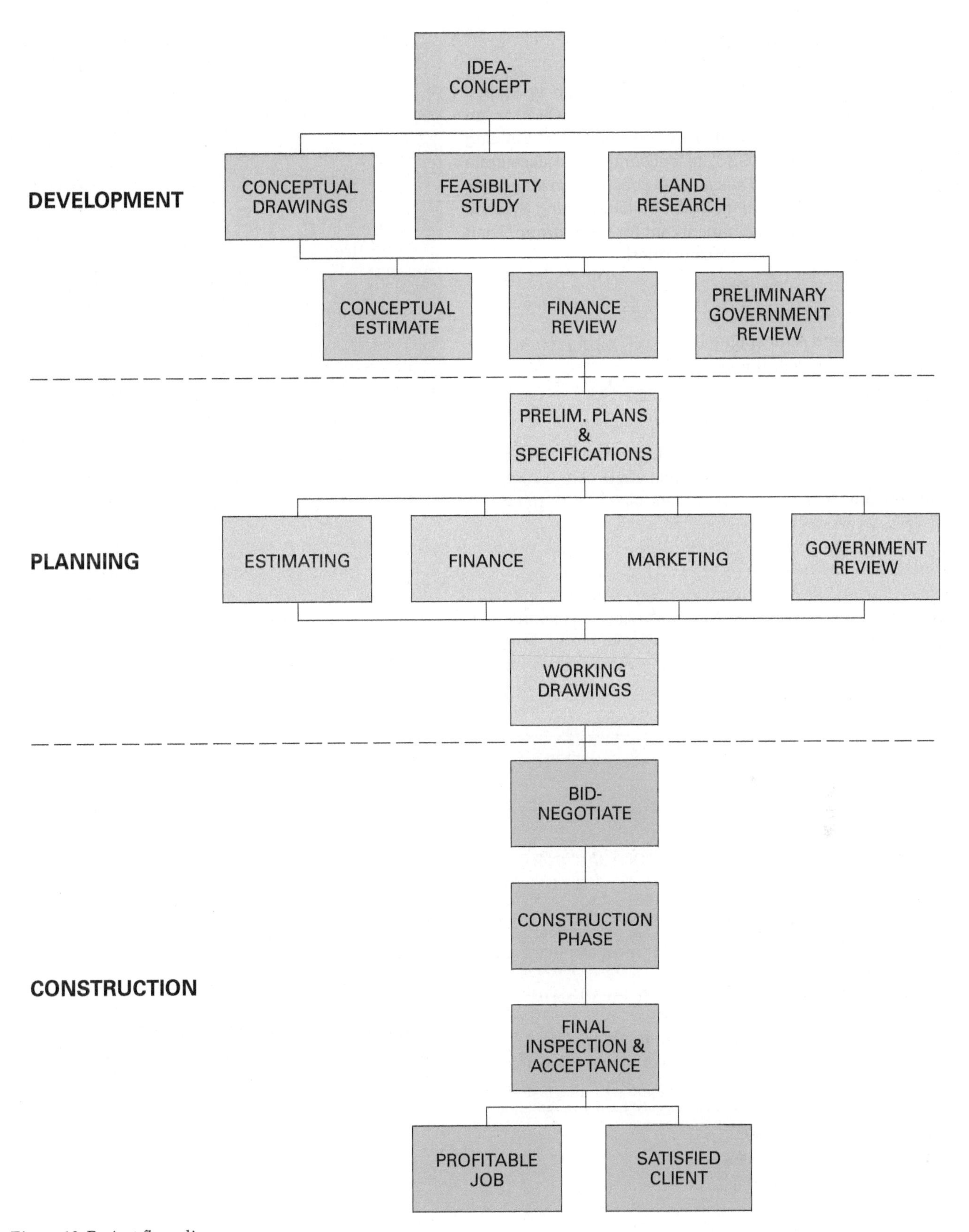

Figure 19 Project flow diagram.

GOING GREEN

LEED

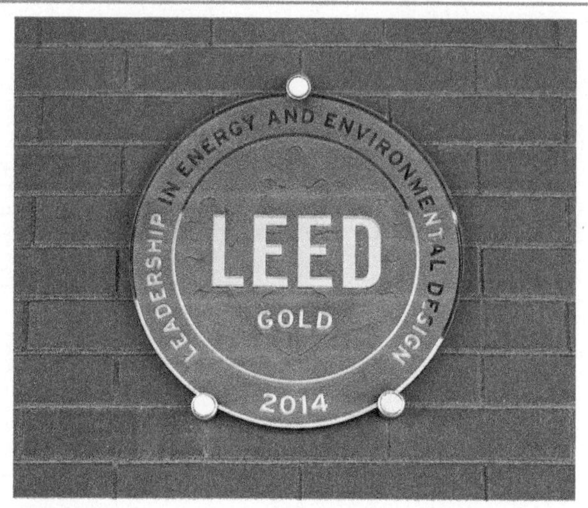

Figure Credit: © Linda Williams/Dreamstime.com

LEED stands for *Leadership in Energy and Environmental Design,* which is an initiative started by the US Green Building Council (USGBC) to encourage and accelerate the adoption of sustainable construction standards worldwide through its Green Building Rating System. USGBC is a non-government, not-for-profit group. Their rating system addresses six categories:

1. Sustainable Sites (SS)
2. Water Efficiency (WE)
3. Energy and Atmosphere (EA)
4. Materials and Resources (MR)
5. Indoor Environmental Quality (EQ)
6. Innovation in Design (ID)

Building owners are the driving force behind the LEED voluntary program. Construction crew leaders may not have input into the decision to seek LEED certification for a project, or what materials are used in the project's construction. However, crew leaders can help to minimize material waste and support recycling efforts, both of which are factors in obtaining LEED certification.

An important question to ask is whether your project is seeking LEED certification. If the project is seeking certification, the next step is to ask what your role will be in getting the certification. If you are procuring materials, what information does the project need and who should receive it? What specifications and requirements do the materials need to meet? If you are working outside the building or inside in a protected area, what do you need to do to protect the work area? How does your crew manage waste? Are there any other special requirements that will be your responsibility? Do you see any opportunities for improvement? LEED principles are described in more detail in NCCER's *Your Role in the Green Environment* (Module ID 70101-15) and *Sustainable Construction Supervisor* (Module ID 70201-11).

DELIVERY SYSTEMS

RESPONSIBLE PARTIES	GENERAL CONTRACTING	DESIGN-BUILD	CONSTRUCTION MANAGEMENT
OWNER	Designs project (or hires architect)	Hires general contractor	Hires construction management company
GENERAL CONTRACTOR	Builds project (with owner's design)	Involved in project design, builds project	Builds, may design (hired by construction management company)
CONSTRUCTION MANAGEMENT COMPANY			Hires and manages general contractor and architect

Figure 20 Project-delivery systems.

budget. Indirect costs are overhead costs shared by all projects. Planners generally calculate these costs as an overhead percentage to labor and material costs.

Direct costs include the following:

- Materials
- Labor
- Tools
- Equipment

Indirect costs refer to overhead items such as the following:

- Office rent
- Utilities
- Telecommunications
- Accounting
- Office supplies, signs

The bid price includes the estimated cost of the project as well as the profit. Profit refers to the amount of money that the contractor will make after paying all the direct and indirect costs. If the direct and indirect costs exceed the estimate for the job, the difference between the actual and estimated costs must come out of the company's profit. This reduces what the contractor makes on the job.

Profit is the fuel that powers a business. It allows the business to invest in new equipment and facilities, provide training, and to maintain a reserve fund for times when business is slow. In large companies, profitability attracts investors who provide the capital necessary for the business to grow. For these reasons, contractors can't afford to consistently lose money on projects. If they can't operate profitably, they are forced out of business. Crew leaders can help their companies remain profitable by managing budget, schedule, quality, and safety adhering to the drawings, specifications, and project schedule.

The cost estimate must consider many factors. Many companies employ professional cost estimators to do this work. They also maintain performance data for previous projects. They use this data as a guide in estimating new projects. Development of a complete estimate generally proceeds as follows:

Step 1 Using the drawings and specifications, an estimator records the quantity of the materials needed to construct the job. Construction companies call this step the *quantity takeoff*. The estimator enters the information on a hard-copy or digital takeoff sheet like the one shown in *Figure 21*.

Step 2 The estimator uses the company's productivity rates to calculate the amount of labor required to complete the project. Most companies keep records of these rates for the type and size of the jobs that they perform. The company's estimating department maintains and updates these records.

Step 3 The estimator calculates the labor hours required by dividing the estimated amount of work by the productivity rate.

- For example, if the productivity rate for concrete finishing is 40 square feet per hour, and there are 10,000 square feet of concrete to be finished, then 250 hours of concrete finishing labor is required (10,000 ÷ 4 = 250).
- The estimator multiplies this number by the hourly rate for concrete finishing to determine the cost of that labor category.
- If this work is subcontracted, then the estimator uses the subcontractor's cost estimate, raised by an overhead factor, in place of direct-labor cost.

Step 4 The estimator transfers the total material quantities from the quantity takeoff sheet to a summary or pricing sheet (*Figure 22*). The total cost of materials is calculated after obtaining material prices from local suppliers.

Step 5 Next, the estimator determines the cost of equipment needed for the project. This number could reflect rental cost or a factor applied by the company when they plan to use their own equipment.

Step 6 The estimator totals the cost of all resources on the summary sheet—materials, equipment, tools, and labor. The estimator can also calculate the material unit cost—the total cost divided by the total number of units of listed materials.

Step 7 The estimator adds the cost of taxes, bonds, insurance, subcontractor work, and other indirect costs to the direct costs of the materials, equipment, tools, and labor.

Step 8 A sum of direct and indirect costs yields the total project cost. The contractor adds the expected profit to that total.

WORKSHEET

Takeoff By: | DATE ____ | PROJECT ____ | PAGE #

Checked By: | SHEET ____ of ____ | ARCHITECT ____

REF.	DESCRIPTION	DIMENSIONS				EXTENSION	QUANTITY	UNIT	TOTAL		REMARKS
		NO	LENGTH	WIDTH	HEIGHT				QUANTITY	UNIT	

Figure 21 Quantity takeoff sheet.

By:	SUMMARY SHEET DATE ____________ PROJECT ____________ SHEET ______ of ______ TITLE: ____________ WORK ORDER # ____________													PAGE #
	DESCRIPTION	QUANTITY		MATERIAL COST		LABOR MAN HOURS FACTORS					LABOR COST		ITEM COST	
		TOTAL	UT	PER UNIT	TOTAL	CRAFT	PR UNIT	TOTAL	RATE	COST PR	PER	TOTAL	TOTAL	PER UNIT
					MATERIAL							LABOR	TOTAL	

Figure 22 Summary sheet.

> **NOTE**
>
> There are computer programs and mobile applications available to simplify the cost-estimating process. Many of them are tailored to specific industries, such as construction or manufacturing, as well as specific trades within the industries, such as plumbing, electrical, HVAC, and insulating contractors.
>
> Estimating programs are typically set up to include a takeoff form and a form for estimating labor by category. Most of these programs include a database that contains current prices for labor and materials, so they automatically price the job and produce a bid. The programs can also generate purchase orders for materials on the job.

As a crew leader, you may be required to estimate quantities of materials. You will need a set of construction drawings and specifications to estimate the amount of a certain type of material required to perform a job. You should carefully review the appropriate section of the technical specifications and page(s) of drawings to determine the types and quantities of materials required. Then enter the quantities on the worksheet. For example, you should review the specification section on finished carpentry along with the appropriate pages of drawings before taking off the linear feet of door and window trim.

If insufficient materials are available to complete the job and an estimate is required, the estimator must determine how much more construction work is necessary. Knowing this, the crew leader can then determine the materials needed. You must also reference the construction drawings in this process.

4.2.0 Planning

One definition of planning is determining the methods used and their sequence to carry out the different tasks to complete a project (*Figure 23*). Planning involves the following:

- Determining the best method for performing the job
- Identifying the responsibilities of each person on the work crew
- Determining the duration and sequence of each activity
- Identifying the tools and equipment needed to complete a job
- Ensuring that the required materials are at the work site when needed
- Making sure that heavy construction equipment is available when required

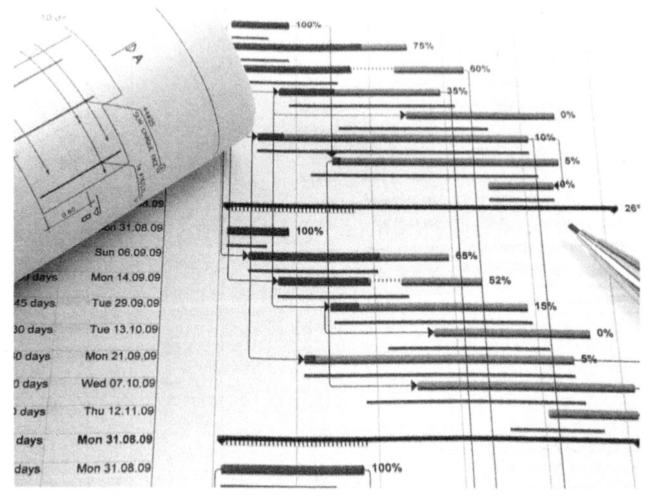

Figure 23 Proper prior planning prevents poor performance.

- Working with other contractors in such a way as to avoid interruptions and delays

Some important reasons for crew-leader planning include the following:

- Controlling the job in a safe manner so that it is built on time and within cost
- Lowering job costs through improved productivity
- Preparing for bad weather or unexpected occurrences
- Promoting and maintaining favorable employee morale
- Determining the best and safest methods for performing the job

With a plan, a crew leader can direct work efforts efficiently and can use resources such as personnel, materials, tools, equipment, work area, and work methods to their full potential.

A proactive crew leader will always have a backup plan in case circumstances prevent the original plan from working. There are many circumstances that can cause a plan to go awry, including adverse weather, equipment failure, absent crew members, and schedule slippage by other crafts.

Project planners establish time and cost limits for the project; the crew leader's planning must fit within those constraints. Therefore, it is important to consider the following factors that may affect the outcome:

- Site and local conditions, such as soil types, accessibility, or available staging areas
- Climate conditions that should be anticipated during the project
- Timing of all phases of work
- Types of materials to be installed and their availability

Participant Exercise E

Assume you are the leader of a crew building footing formwork for the construction shown in the figure below. You have used all of the materials provided for the job, yet you have not completed it. You study the drawings and see that the formwork consists of two side forms, each 12" high. The total length of footing for the entire project is 115'-0". You have completed 88'-0" to date; therefore, you have 27'-0" remaining (115' – 88' = 27').

Your task is to prepare an estimate of materials that you will need to complete the job. In this case, you will estimate only the side form materials (do not consider the miscellaneous materials here).

- Footing length to complete: 27'-0"
- Footing height: 1'-0"

Refer to the worksheet on the next page for a final tabulation of the side forms needed to complete the job.

1. Using the same footing as described in the example above, calculate the quantity (square feet) of formwork needed to finish 203 linear feet of the footing. Place this information directly on the worksheet.
2. You are the crew leader of a carpentry crew whose task is to side a warehouse with plywood sheathing. The wall height is 16', and there is a total of 480 linear feet of wall to side. You have done 360 linear feet of wall and have run out of materials. Calculate how many more feet of plywood you will need to complete the job. If you are using 4 × 8 plywood panels, how many will you need to order to cover the additional work? Write your estimate on the worksheet.

Show your calculations to the instructor.

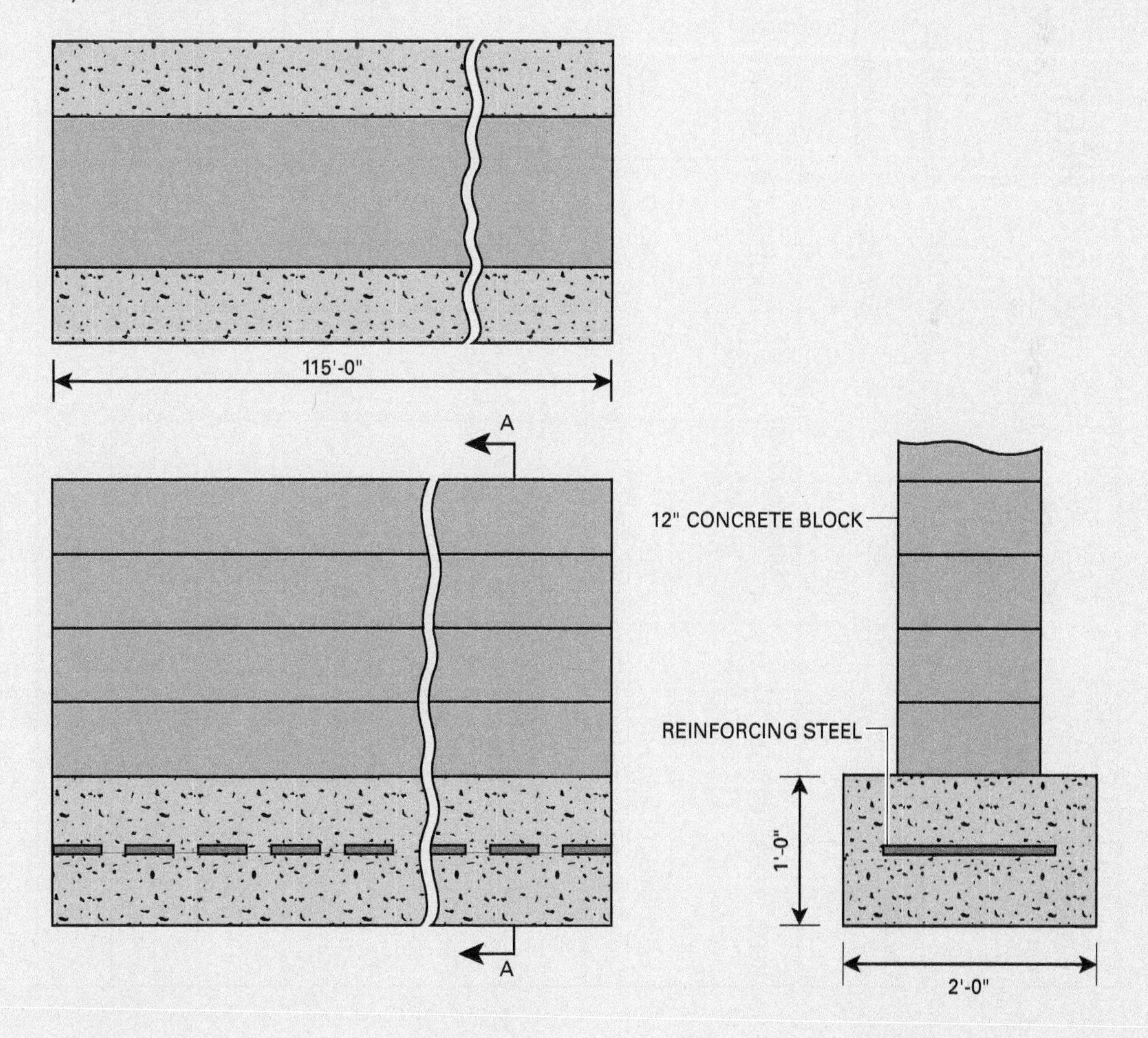

Participant Exercise E Worksheet

WORKSHEET

Takeoff By: RWH	DATE 2/1/17	PROJECT Sam's Diner	PAGE #1
Checked By:	SHEET 01 of 01	ARCHITECT 654b	

REF.	DESCRIPTION	DIMENSIONS				EXTENSION	QUANTITY	UNIT	TOTAL		REMARKS
		NO	LENGTH	WIDTH	HEIGHT				QUANTITY	UNIT	
	Footing Side Forms 2		27'0"		1'0"	2x27x1	54	SF	54	SF	

- Equipment and tools required and their availability
- Personnel requirements and availability
- Relationships with the other contractors and their representatives on the job

On a simple job, crews can handle these items almost automatically. However, larger or more complex jobs require the planner to give these factors more formal consideration and study.

4.2.1 Stages of Planning

Formal planning for a construction job occurs at specific times in the project process. The two most important stages of planning occur in the preconstruction phase and during the construction work.

Preconstruction planning – The preconstruction stage of planning occurs before the start of construction. The preconstruction planning process doesn't always involve the crew leader, but it's important to understand what it consists of.

There are two phases of preconstruction planning. The first is developing the proposal, bid, or negotiated price for the job. This is when the estimator, the project manager, and the field superintendent develop a preliminary plan for completing the work. They apply their experience and knowledge from previous projects to develop the plan. The process involves determining what methods, personnel, tools, and equipment the work will require and what level of productivity they can achieve.

The second phase of preconstruction planning occurs after the client awards the contract. This phase requires a thorough knowledge of all project drawings and specifications. During this process, planners select the actual work methods and resources needed to perform the work. Here, crew leaders might get involved, but their planning must adhere to work methods, production rates, and resources that fit within the estimate prepared during contract negotiations. If the project requires a method of construction different

Participant Exercise F

1. In your own words, define planning, and describe how a job can be done better if it is planned. Give an example.

___

___

___

___

2. Consider a job that you recently worked on to answer the following:
   a. List the material(s) used.
   b. List each member of the crew with whom you worked and what each person did.
   c. List the kinds of equipment used.

___

___

___

___

___

3. List some suggestions for how the job could have been done better, and describe how you would plan for each of the suggestions.

___

___

___

___

from what is normal, planners will usually inform the crew leader of what method to use.

Construction planning – During construction, the crew leader is directly involved in daily planning. This stage of planning consists of selecting methods for completing tasks before beginning work. Effective planning exposes likely difficulties, and enables the crew leader to minimize the unproductive use of personnel and equipment. Proper planning also provides a gauge to measure job progress.

Effective crew leaders develop a tool known as the **look-ahead schedule**. These schedules consider actual circumstances as well as projections two-to-three weeks into the future. Developing a look-ahead schedule helps ensure that all resources are available on the project when needed. Most scheduling apps and programs include this feature that automatically help you focus on this period in the job.

4.2.2 The Planning Process

The planning process consists of the following five steps:

Step 1 Establish a goal.

Step 2 Identify the completed work activities required to achieve the goal.

Step 3 Identify the required tasks to accomplish those activities.

Step 4 Communicate responsibilities.

Step 5 Follow up to verify the goal achievement.

Establishing a goal – The term *goal* has different meanings for different people. In general, a goal is a specific outcome that one works toward. For example, the project superintendent of a home construction project could establish the goal to have a house dried-in by a certain date. (The term *dried-in* means ready for the application of roofing and siding.) To meet that goal, the leader of the framing crew and the superintendent would need to agree to a goal to have the framing completed by a given date. The crew leader would then establish sub-goals (objectives) for the crew to complete each element of the framing (floors, walls, roof) by a set time. The superintendent would need to set similar goals with the crews that install sheathing, building wrap, windows, and exterior doors. However, if the framing crew doesn't meet its goal, that will delay the other crews.

Identifying the required work – The second step in planning is to identify the necessary work to achieve the goal as a series of activities in a certain sequence. You will learn how to break down a job into activities later in this section. At this point, the crew leader should know that, for each activity, one or more objectives must be set.

An objective is a statement of a condition the plan requires to exist or occur at a specific time. An objective must:

- Mean the same thing to everyone involved
- Be measurable, so that everyone knows when it has been reached
- Be achievable with the resources available
- Have everyone's full support

Examples of practical objectives include the following:

- "By 4:30 p.m. today, the crew will have completed installation of the floor joists."
- "By closing time Friday, the roof framing will be complete."

Notice that both examples meet the first three requirements of an objective. Planners assume that everyone involved in completing the task is committed to achieving the objective. The advantage in developing objectives for each work activity is that it allows the crew leader to determine if the crew is following the plan. In addition, objectives serve as sub-goals that are usually under the crew leader's control.

Some construction work activities, such as installing 12" footing forms, are done so often that they require little planning. However, other jobs, such as placing a new type of mechanical equipment, require substantial planning. This type of job demands that the crew leader set specific objectives.

Whenever faced with a new or complex activity, take the time to establish objectives that will serve as guides for accomplishing the job. You can use these guides in the current situation, as well as for similar work in the future.

Identifying the required tasks – To plan effectively, the crew leader must be able to break a work activity assignment down into smaller tasks. Large jobs include a greater number of tasks than small ones, but all jobs can be broken down into manageable components.

When breaking down an assignment into tasks, make each task identifiable and definable. A task is identifiable when one knows the types and amounts of resources it requires. A task is definable if it has a specific duration. For purposes of efficiency, the job breakdown should not be too detailed or complex, unless the job has never been done before or must be performed with strictest efficiency.

For example, a suitable breakdown for the work activity to install square vinyl floor tiles in a cafeteria might be the following:

Step 1 Prepare the floor.

Step 2 Stage the tiles.

Step 3 Spread the adhesive.

Step 4 Lay the tiles.

Step 5 Clean the tiles.

Step 6 Wax the floor.

The crew leader could create even more detail by breaking down any one of the tasks into subtasks. In this case, however, that much detail is unnecessary and wastes the crew leader's time and the project's money.

Planners can divide every work activity into three general parts:

- Preparing
- Performing
- Cleaning up

Some of the most frequent mistakes made in the planning process are forgetting to prepare and to clean up. The crew leader must not overlook preparation and cleanup.

After identifying the various tasks that make up the job and developing an objective for each task, the crew leader must determine what resources the job requires. Resources include labor, equipment, materials, and tools. In most jobs, the job estimate identifies these resources. The crew leader must make sure that these resources are available on the site when needed.

"By failing to prepare, you are preparing to fail."
—Benjamin Franklin

Communicating responsibilities – No supervisor can complete all the activities within a job alone. The crew leader must rely on other people to get everything done. Therefore, most jobs have a crew of people with various experiences and skill levels to assist in the work. The crew leader's job is to draw from this expertise to get the job done well and in a safe and timely manner.

Once the various activities that make up the job have been determined, the crew leader must identify the person or persons responsible for completing each activity. This requires that the crew leader be aware of the skills and abilities of the people on the crew. Then, the crew leader must put this knowledge to work in matching the crew's skills and abilities to specific tasks required to complete the job.

After matching crew members to specific activities, the crew leader must then assign work to the crew. The crew leader normally communicates responsibilities verbally; the crew leader often talks directly to the responsible person for the activity. There may be times when the crew leader assigns work through written instructions or indirectly through someone other than the crew leader. Either way, crew members should know what it is they are responsible for accomplishing on the job.

Following up – Once the crew leader has delegated the activities to the appropriate crew members, there must be follow-up to make sure that the crew has completed them correctly and efficiently. Task follow-up involves being present on the job site to make sure all the resources are available to complete the work, ensuring that the crew members are working on their assigned activities, answering any questions, and helping to resolve any problems that occur while the work is being done. In short, follow-up activity means that the crew leader is aware of what's going on at the job site and is doing whatever is necessary to make sure that the crew completes the work on schedule. *Figure 24* reviews the planning steps.

The crew leader should carry a small note pad or electronic device for planning and taking notes. That way, you can record thoughts about the project as they occur, and you won't forget pertinent details. The crew leader may also choose to use a manual planning form such as the one illustrated in *Figure 25*.

As the job progresses, refer to these resources to see that the tasks are being done according to plan. This is job analysis. Construction projects that don't proceed according to work plans usually end up costing more and taking longer. Therefore, it is important that crew leaders refer to the planning documents periodically.

The crew leader is involved with many activities on a day-to-day basis. Thus, it is easy to forget important events if they aren't recorded. To help keep track of events such as job changes, interruptions, and visits, the crew leader should keep a **job diary**. A job diary is a notebook in which the crew leader records activities or events that take place on the

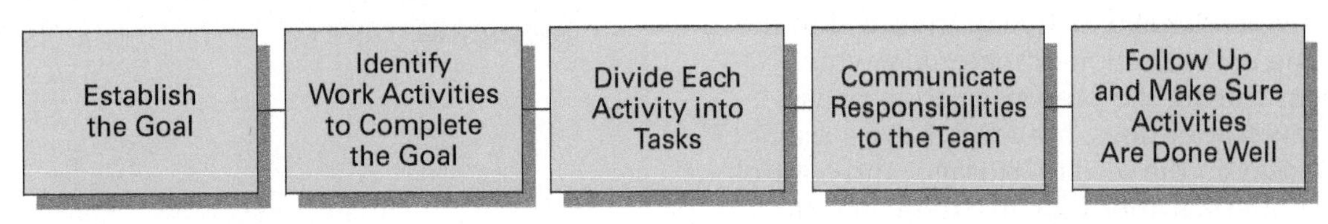

Figure 24 Steps to effective planning.

DAILY WORK PLAN

"PLAN YOUR WORK AND WORK YOUR PLAN = EFFICIENCY"

Plan of ____________ Date ________

PRIORITY	DESCRIPTION	✓ When Completed X Carried Forward

Figure 25 Planning form.

job site that may be important later. When making entries in a job diary, make sure that the information is accurate, factual, complete, consistent, organized, and up-to-date. This is especially true if documenting personnel problems. If the company requires maintaining a job diary, follow company policy in determining which events and what details you should record. However, if there is a doubt about what to include, it is better to have too much information than too little. *Figure 26* shows a sample page from a job diary.

4.2.3 Planning Resources

Once a job has been broken down into its tasks or activities, the various resources needed to perform them must be determined and accounted for. Resource planning includes the following specific considerations:

- *Safety planning* – Using the company safety manual as a guide, the crew leader must assess the safety issues associated with the job and take necessary measures to minimize any risk to the crew. This may involve working with the company or site safety officer and may require a formal hazard analysis.
- *Materials planning* – Preconstruction planning identifies the materials required for the job and lists them on the job estimate. Companies usually order the materials from suppliers who have previously provided quality materials on schedule and within estimated cost.

  The crew leader is usually not involved in the planning and selection of materials, which happens during the preconstruction phase. The crew leader does, however, have a major role to play in the receipt, storage, and control of the materials after they reach the job site.
- The crew leader is also involved in planning materials for tasks such as job-built formwork and scaffolding. In addition, the crew leader may run out of a specific material, such as fasteners, and need to order more. In such cases, be sure to consult the appropriate supervisor, since most companies have specific purchasing policies and procedures.

July 8, 2017

Weather: Hot and Humid

Project: Company XYZ Building

- The paving contractor crew arrived late (10 am).
- The owner representative inspected the footing foundation at approximately 1 pm.
- The concrete slump test did not pass. Two trucks had to be ordered to return to the plant, causing a delay.
- John Smith had an accident on the second floor. I sent him to the doctor for medical treatment. The cause of the accident is being investigated.

Figure 26 Sample page from a job diary.

- *Site planning* – There are many planning elements involved in site work. The following are some of the key elements:
  - Access roads
  - Emergency procedures
  - Material and equipment storage
  - Material staging
  - Parking
  - Sedimentation control
  - Site security
  - Storm water runoff
- *Equipment planning* – The preconstruction phase addresses much of the planning for use of construction equipment. This planning includes the types of equipment needed, the use of the equipment, and the length of time it will be on the site. The crew leader must work with the main office to make certain that the equipment reaches the job site on time. The crew leader must also ensure that crew equipment operators are properly trained.

  Coordinating the use of the equipment is also very important. Some equipment operates in combination with other equipment. For example, dump trucks are generally required when loaders and excavators are used. The crew leader should also coordinate equipment with other contractors on the job. Sharing equipment can save time and money and avoid duplication of effort.

  The crew leader must reserve time for equipment maintenance to prevent equipment failure. In the event of an equipment failure, the crew leader must know who to contact to resolve the problem. An alternate plan must be ready in case one piece of equipment breaks down, so that the other equipment doesn't sit idle. The crew leader should coordinate these contingency plans with the main office or the crew leader's immediate superior.
- *Tool planning* – A crew leader is responsible for planning tool usage for a job. This task includes the following:
  - Determining the tools required
  - Informing the workers who will provide the tools (company or worker)
  - Making sure the workers are qualified to use the tools safely and effectively
  - Determining what controls to establish for tools
- *Labor planning* – All jobs require some sort of labor because the crew leader can't complete all the work alone. When planning for labor, the crew leader must do the following:
  - Identify the skills needed to perform the work.
  - Determine the number of people having those specific skills that are required.
  - Decide who will be on the crew.

  In many companies, the project manager or job superintendent determines the size and makeup of the crew. Then, supervision expects the crew leader to accomplish the goals and objectives with the crew provided. Even though the crew leader may not have any involvement in staffing the crew, the crew leader is responsible for training the crew members to ensure that they have the skills needed to do the job. In addition, the crew leader is responsible for keeping the crew adequately staffed at all times to avoid job delays. This involves dealing with absenteeism and turnover, two common problems discussed in earlier sections.

4.2.4 Scheduling

Planning and scheduling are closely related and are both very important to a successful job. Planning identifies the required activities and how and in what order to complete them. Scheduling involves establishing start and finish times/dates for each activity.

A schedule for a project typically shows the following:

- Operations listed in sequential order
- Units of construction
- Duration of activities
- Estimated date to start and complete each activity
- Quantity of materials to be installed

There are different types of schedules used today. They include the bar chart, the network schedule, also called the **critical path** method (CPM) or precedence diagram, and the short-term, or look-ahead schedule.

The following is a summary of the steps a crew leader must complete to develop a schedule.

Step 1 Make a list of all the activities that the plan requires to complete the job, including individual work activities and special tasks, such as inspections or the delivery of materials. At this point, the crew leader should just be concerned with generating a list, not with determining how to accomplish the activities, who will perform them, how long they will take, or the necessary sequence to complete them.

Step 2 Use the list of activities created in Step 1 to reorganize the work activities into a logical sequence. When doing this, keep in mind that certain steps can't happen before the completion of others. For example, footing excavation must occur before concrete emplacement.

Step 3 Assign a duration or length of time that it will take to complete each activity and determine the start time for each. Then place each activity into a schedule format. This step is important because it helps the crew leader compare the task time estimates to the scheduled completion date or time.

The crew leader must be able to read and interpret the job schedule. On some jobs, the form provides the beginning and expected end date for each activity, along with the expected crew or worker's production rate. The crew leader can use this information to plan work more effectively, set realistic goals, and compare the starts and completions of tasks to those on the schedule.

Before starting a job, the crew leader must do the following:

- Determine the materials, tools, equipment, and labor needed to complete the job.
- Determine when the various resources are needed.
- Follow up to ensure that the resources are available on the job site when needed.

The crew leader should verify the availability of needed resources three to four working days before the start of the job. This should occur even earlier for larger jobs. Advance preparation will help avoid situations that could potentially delay starting the job or cause it to fall behind schedule.

Supervisors can use bar chart schedules, also known as *Gantt charts,* for both short-term and long-term jobs. However, they are especially helpful for jobs of short duration.

Bar charts provide management with the following:

- A visual presentation of the overall time required to complete the job using a logical method rather than a calculated guess
- A means to review the start and duration of each part of the job
- Timely coordination requirements between crafts
- Alternative sequences of performing the work

A bar chart works as a control device to see whether the job is on schedule. If the job isn't on schedule, supervision can take immediate action in the office and the field to correct the problem and increases the likelihood of completing the activity on schedule. *Figure 27* illustrates a bar or Gantt chart.

Another type of schedule is the network schedule, which shows dependent (critical path) activities and other activities completed in parallel with but not part of the critical path. In *Figure 28,* for example, reinforcing steel can't be set until the concrete forms have been built and placed. Other activities are happening in parallel, but the forms are in the critical path.

When building a house, builders can't install and finish drywall until wiring, plumbing, and HVAC ductwork have been roughed-in. Because other activities, such as painting and trim work, depend on drywall completion, the drywall work is a critical-path operation. In other words, until it is complete, workers can't start the other tasks, and the project itself will likely experience delay by the amount of delay starting any dependent

Did You Know?

Recycling Materials

Habitat for Humanity, a charitable organization that builds homes for disadvantaged families, accepts surplus building materials for use in their projects. In some cities, they have stores called ReStores, which serve as retail outlets for such materials. They may also accept materials salvaged during demolition of a structure. Companies can obtain LEED credits through practices such as salvaging building materials, segregating scrap materials for recycling, and taking steps to minimize waste.

Figure Credit: Sushil Shenoy/Virginia Tech

> **Did You Know?**
>
> ## Scheduling Apps
>
> Many companies use project scheduling software. Among the top construction project management applications at the time of publication are WorkflowMax, Procore, BuildTools, Co-Construct, and BuilderTREND. Such programs are spreadsheets on steroids, which can respond to schedule changes by altering downstream events and producing a revised schedule. Most of the newest programs are cloud-based, and users can access and update them remotely using mobile, tablet, and computer devices.

activity. Likewise, workers can't even start the drywall installation until the rough-ins are complete. Therefore, the project superintendent is likely to focus on those activities when evaluating schedule performance.

The advantage of a network schedule is that it allows project leaders to see how a schedule change with one activity is likely to affect other activities and the project in general. Planners lay out a network schedule on a timeline and usually show the estimated duration for each activity. Planners use network schedules for complex jobs that take a long time to complete. The PERT (program evaluation and review technique) schedule is a form of network schedule.

Since the crew needs to hold to the job schedule, the crew leader needs to be able to plan daily production. As discussed earlier, short-term scheduling is a method used to do this. *Figure 29* displays an example of a short-term, look-ahead schedule.

The information to support short-term scheduling comes from the estimate or cost breakdown. The schedule helps to translate estimate data and the various job plans into a day-to-day schedule of events. The short-term schedule provides the crew leader with visibility over the immediate future of the project. If actual production begins to slip behind estimated production, the schedule will warn the crew leader that a problem lies ahead and that a schedule slippage is developing.

Crew leaders should use short-term scheduling to set production goals. Generally, workers can improve production when they:

- Know the amount of work to be accomplished
- Know the time available to complete the work
- Can provide input when setting goals

Example:

A carpentry crew on a retaining wall project is about to form and pour catch basins and put up wall forms. The crew has put in several catch basins, so the crew leader is sure that they can perform the work within the estimate. However, the crew leader is concerned about their production of the wall forms. The crew will work on both the basins and the wall forms at the same time. The scheduling process in this scenario could be handled as follows:

1. The crew leader notices the following in the estimate or cost breakdown:
   a. Production factor for wall forms = 16 worker-hours (w-h) per 100 ft^2
   b. Work to be done by measurement = 800 ft^2
   c. Total time: (800 ft^2 × 16 w-h/100 ft^2) = 128 w-h
2. The carpenter crew consists of the following:
   a. One carpenter crew leader
   b. Four carpenters
   c. One laborer
3. The crew leader determines the goal for the job should be set at 128 w-h (from the cost breakdown).
4. If the crew remains the same (six workers), the work should be completed in about 21 crew-hours (128 w-h ÷ 6 workers/crew = 21.3 crew-hours).
5. The crew leader then discusses the production goal (completing 800 ft^2 in 21 crew-hours) with the crew and encourages them to work together to meet the goal of getting the forms erected within the estimated time.

In this example, the crew leader used the short-term schedule to translate production into work-hours or crew-hours and to schedule work so that the crew can accomplish it within the estimate. In addition, setting production targets provides the motivation to produce more than the estimate requires.

No matter what type of schedule is used, supervision must keep it up to date to be useful to the crew leader. Inaccurate schedules are of no value. The person responsible for scheduling in the office handles the updates. This person uses information gathered from job field reports to do the updates.

The crew leader is usually not directly involved in updating schedules. However, completing field or progress reports used by the company may be a daily responsibility to keep the schedule up to date. It's critical that the crew leader fill out any required forms or reports completely and accurately.

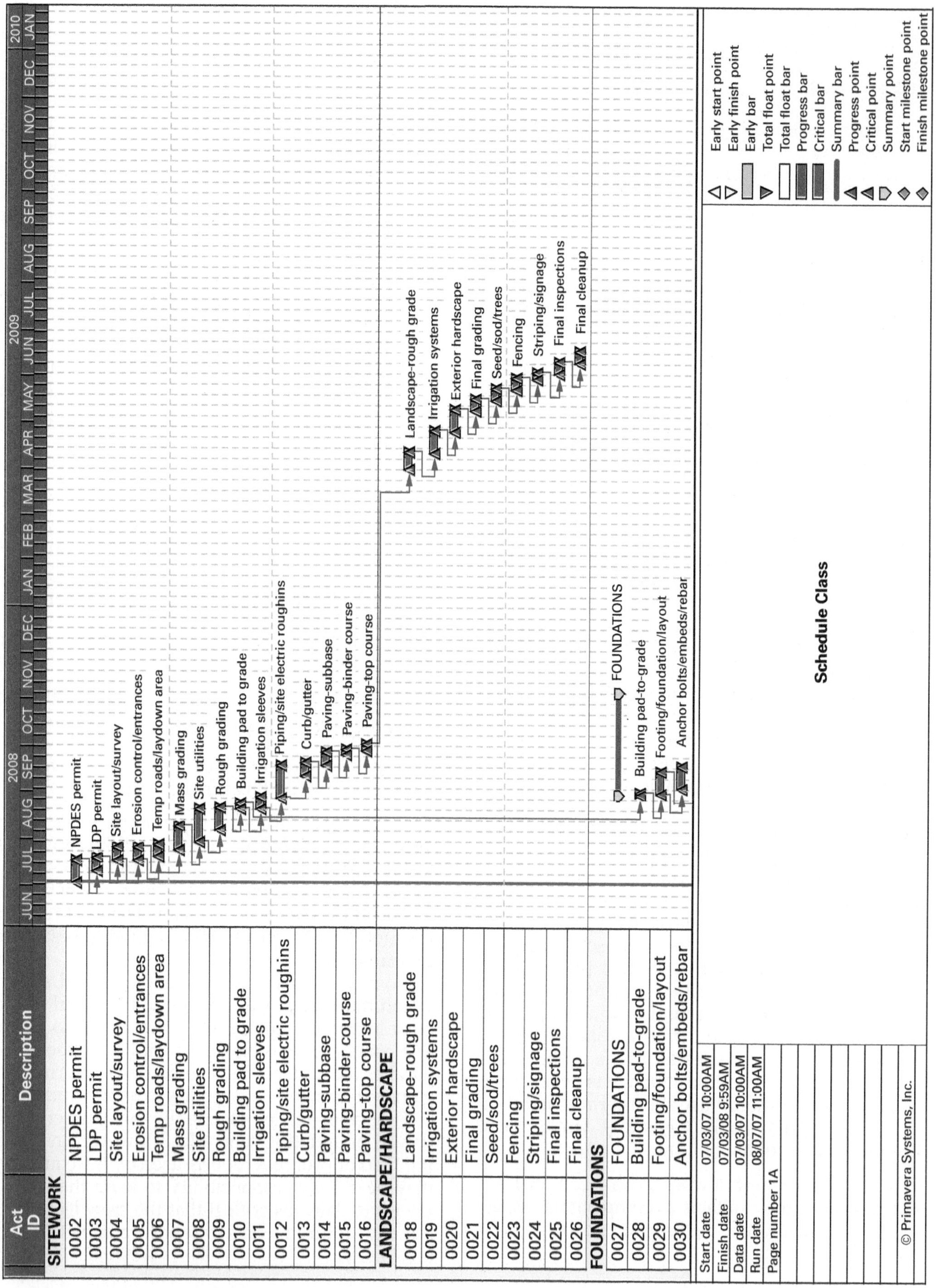

Act ID	Description
SITEWORK	
0002	NPDES permit
0003	LDP permit
0004	Site layout/survey
0005	Erosion control/entrances
0006	Temp roads/laydown area
0007	Mass grading
0008	Site utilities
0009	Rough grading
0010	Building pad to grade
0011	Irrigation sleeves
0012	Piping/site electric roughins
0013	Curb/gutter
0014	Paving-subbase
0015	Paving-binder course
0016	Paving-top course
LANDSCAPE/HARDSCAPE	
0018	Landscape-rough grade
0019	Irrigation systems
0020	Exterior hardscape
0021	Final grading
0022	Seed/sod/trees
0023	Fencing
0024	Striping/signage
0025	Final inspections
0026	Final cleanup
FOUNDATIONS	
0027	FOUNDATIONS
0028	Building pad-to-grade
0029	Footing/foundation/layout
0030	Anchor bolts/embeds/rebar

2008: JUN JUL AUG SEP OCT NOV DEC — 2009: JAN FEB MAR APR MAY JUN JUL AUG SEP OCT NOV DEC — 2010: JAN

Start date	07/03/07 10:00AM
Finish date	07/03/08 9:59AM
Data date	07/03/07 10:00AM
Run date	08/07/07 11:00AM
Page number 1A	

Schedule Class

- Early start point
- Early finish point
- Early bar
- Total float point
- Total float bar
- Progress bar
- Critical bar
- Summary bar
- Progress point
- Critical point
- Summary point
- Start milestone point
- Finish milestone point

Figure 27 Example of a bar-chart schedule.

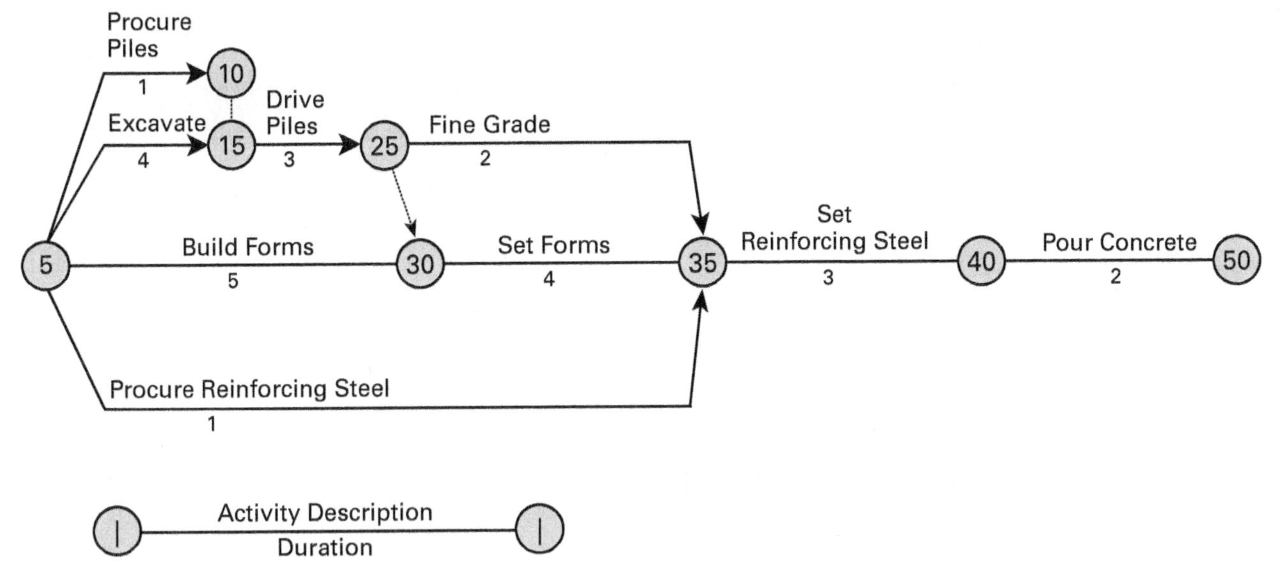

Figure 28 Example of a network schedule.

4.3.0 Cost Control

Being aware of costs and controlling them is the responsibility of every employee on the job. It's the crew leader's job to ensure that employees uphold this responsibility. Control refers to the comparison of estimated performance against actual performance and following up with any needed corrective action. Crew leaders who use cost-control practices are more valuable to the company than those who do not.

On a typical job, many activities are going on at the same time, even within a given crew. This can make it difficult to control the activities involved. The crew leader must be constantly aware of the costs of a project and effectively control the various resources used on the job.

When resources aren't controlled, the cost of the job increases. For example, a plumbing crew of four people is installing soil pipe and runs out of fittings. Three crew members wait (*Figure 30*) while one crew member goes to the plumbing-supply dealer for a part that costs only a few dollars. It takes the crew member an hour to get the part, so four worker-hours of production have been lost. In addition, the total cost of the delay must include the travel costs for retrieving the supplies.

4.3.1 Assessing Cost Performance

Cost performance on a project is determined by comparing actual costs to estimated costs. Regardless of whether the job is a contract-bid project or an in-house project, the company must first establish a budget. In the case of a contract bid, the budget is generally the cost estimate used to bid the job. For an in-house job, participants will submit labor and material forecasts, and someone in authority will authorize a project budget.

It is common to estimate cost by either breaking the job into funded tasks or by forecasting labor and materials expenditures on a timeline. Many companies create a **work breakdown structure (WBS)** for each project. Within the WBS, planners assign each major task a discrete charge number. Anyone working on that task charges that number on their time sheet, so that project managers can readily track cost performance. However, knowing how much money the company is spending doesn't necessarily determine cost performance.

Although financial reports can show that actual expenses are tracking with forecast expenses, they don't show if the work itself is occurring at the required rate. Thus, it is possible to have spent half the budget, but have less than half of the work compete. When the project is broken down into funded tasks related to schedule activities and events, there is far greater control over cost performance.

4.3.2 Field Reporting System

The total estimated cost comes from the job estimates, but managers obtain the actual cost of doing the work from an effective field-reporting system. A field-reporting system consists of a series of forms, which are completed by the crew leader and others. Each company has its own forms and methods for obtaining information. The following paragraphs describe the general information and the process of how they are used. First, records must document the number of hours each person worked on each task. This

	Calendar Dates															
	7/1	7/2	7/3	7/7	7/8	7/9	7/10	7/11	7/14	7/15	7/16	7/17	7/18	7/21	7/22	7/23
ACTIVITY DESCRIPTION	Work Days															
	1	2	3	4	5	6	7	8	9	10	11	12	13	14	15	16
Process Piles	■															
Excavate	■	■	■	■												
Build Forms	■	■	■	■	■											
Process Reinforcing Steel	■															
Drive Piles					■	■	■									
Fine Grade								■	■							
Set Forms								■	■	■	■					
Set Reinforced Steel												■	■	■		
Pour Concrete															■	■

NOTES: The project start date is July 1st, which is a Tuesday.
Time placement of activity and duration may be done anytime through shaded portion.
Bottom portion of line available to show progress as activities are completed.

ADDITIONAL TASKS:

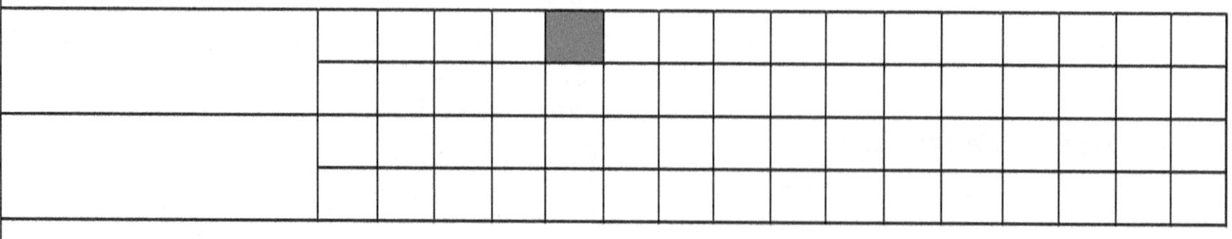

Figure 29 Short-term schedule.

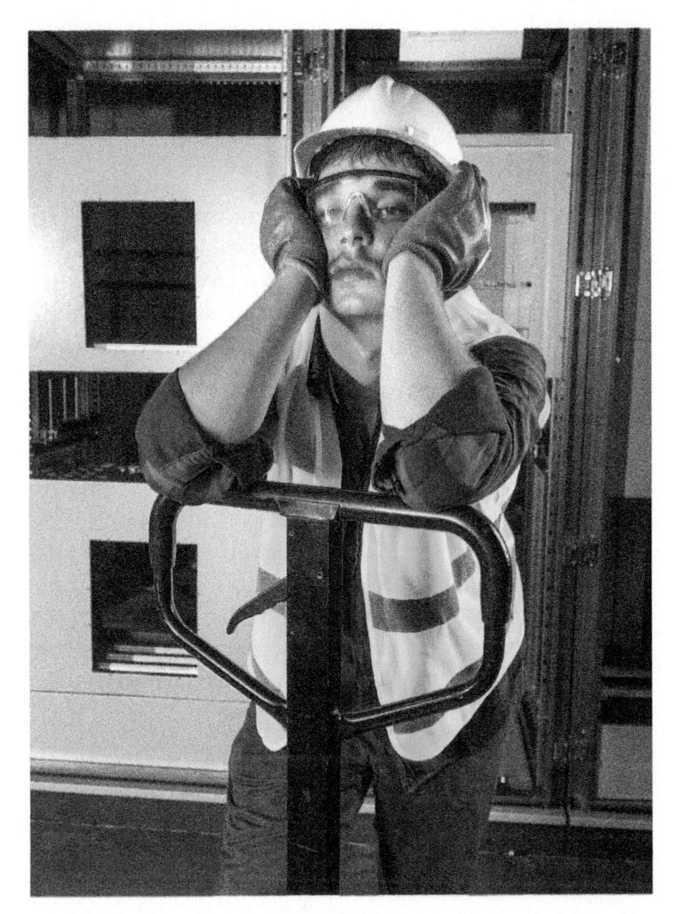

Figure 30 Idle workers cost money.

information comes from daily time cards. Once the accounting department knows how many hours each employee worked on an activity, it can calculate the total cost of the labor by multiplying the number of hours worked by the wage rate for each worker. Managers can then calculate the cost for the labor to do each task as the job progresses. They compare this cost and the estimated costs during the project and at its completion. Managers use a similar process to determine if the costs to operate equipment are comparable to the estimated equipment operating costs.

As the crew places materials, a designated person will measure the quantities from time to time, and send this information to the company office and, possibly, the crew leader. This information, along with the actual cost of the material and the number of hours it took the workers to install it, is compared to the estimated cost. If the cost is greater than the estimate, management and the crew leader must take action to reduce the cost.

For this comparison process to be of use, the information obtained from field personnel must be correct. It is important that the crew leader be accurate in reporting. This is another reason to maintain a daily job diary as discussed earlier. In the event of a legal/contractual conflict with the client, courts can use such diaries as evidence in legal proceedings, and they can be helpful in reaching a settlement.

Example:

You are running a crew of five concrete finishers for a subcontractor. When you and your crew show up to finish a slab, the general contractor (GC) informs you, "We're a day behind on setting the forms, so I need you and your crew to stand down until tomorrow." What do you do?

Solution:

You should first call your office to let them know about the delay. Then, immediately record it in your job diary. A six-man crew for one day represents 48 worker-hours. If your company charges $30 an hour, that's a potential loss of $1,440, which the company would want to recover from the GC. If there is a dispute, your entry in the job diary could result in a favorable decision for your employer.

4.3.3 Crew Leader's Role in Cost Control

The crew leader is often the company's representative in the field, where the work takes place. Therefore, the crew leader contributes a great deal to determining job costs. When supervision assigns work to a crew, the crew leader should receive a budget and schedule for completing the job. It is then up to the crew leader to make sure the crew finishes the job on time and stays within budget. The crew leader achieves this by actively managing the use of labor, materials, tools, and equipment.

"Beware of little expenses;
a small leak will sink a great ship."
—Benjamin Franklin

If the actual costs are at or below the estimated costs, the job is progressing as planned and scheduled, then the company will realize the expected profit. However, if the actual costs exceed the estimated costs, one or more problems may result in the company losing its expected profit, and maybe more. No company can remain in business if it continually loses money.

One of the factors that can increase cost is client-related changes (often called scope creep). The crew leaders must be able to assess the potential impact of such changes and, if necessary, confer with the employer to determine the course of action. If contractor-related losses are occurring, the crew leader and superintendent will need to work together to get the costs back in line.

"There is no such thing as scope creep, only scope gallop."
—Cornelius Fitchner

The following are some causes that can make actual costs exceed estimated costs, along with some potential actions the crew leader can take to bring the costs under control:

NOTE

Before starting any action, the crew leader should check with company supervision to see that the action proposed is acceptable and within the company's policies and procedures.

- *Late delivery of materials, tools, and/or equipment* – Plan ahead to ensure that job resources will be available when needed
- *Inclement weather* – Work with the superintendent and have alternate plans ready
- *Unmotivated workers* – Counsel the workers
- *Accidents* – Enforce the existing safety program (or evaluate the program for adequacy)

There are many other methods to get the job done on time if it gets off schedule. Examples include working overtime, increasing the size of the crew, prefabricating assemblies, or working staggered shifts. However, these examples may increase the cost of the job, so the crew leader should not do them without the approval of the project manager.

4.4.0 Resource Control

The crew leader's job is to complete assigned tasks safely according to the plans and specifications, on schedule, and within the scope of the estimate. To accomplish this, the crew leader must closely control how resources of materials, equipment, tools, and labor are used. The crew must minimize waste whenever possible.

Control involves measuring performance and correcting departures from plans and specifications to accomplish objectives. Control anticipates predictable deviations from plans and specifications based on experience and takes measures to prevent them from occurring.

An effective control process can be broken down into the following steps:

Step 1 Establish standards and divide them into measurable units. For example, a baseline can be created using experience gained on a typical job, where 2,000 linear feet (LF) of 1¼" copper water tube was installed in five days. Thus, dividing the total LF installed by the number of installation days gives the average installation rate. In this case, for 1¼" copper water tube, the standard rate was 400 LF/day.

Step 2 Measure performance against a standard. On another job, a crew placed 300 LF of the same tube during a given day. Thus, this actual production of 300 LF/day did not meet the standard rate of 400 LF/day.

Step 3 Adjust operations to meet the standard. In Step 2, if the plan scheduled the job for five days, the crew leader would have to take action to meet this goal. If 300 LF/day is the actual average daily production rate for that job, the crew must increase their placement rate by 100 LF/day to meet the standard. This increase may come by making the workers place piping faster, adding workers to the crew, or receiving authorization for overtime.

The crew leader's responsibility in materials control depends on the policies and procedures of the company. In general, the crew leader is responsible for ensuring on-time delivery, preventing waste, controlling delivery and storage, and preventing theft of materials.

4.4.1 Ensuring On-Time Delivery

It's essential that the materials required for each day's work be on the job site when needed. The crew leader should confirm in advance the placement of orders for all materials and that they will arrive on schedule. A week or so before the delivery date, follow up to make sure there will be no delayed deliveries or items on backorder.

If other people are responsible for providing the materials for a job, the crew leader must follow up to make sure that the materials are available when needed. Otherwise, delays occur as crew members stand around waiting for the delivery of materials.

4.4.2 Preventing Waste

Waste in construction can add up to loss of critical and costly materials and may result in job delays. The crew leader needs to ensure that every crew member knows how to use the materials efficiently. Crew leaders should monitor their crews to make certain that no materials are wasted.

An example of waste is a carpenter who saws off a piece of lumber from a full-sized piece, when workers could have found the length needed in the lumber scrap pile. Another example of waste

Did You Know?

Just in Time (JIT)

Just-in-time (JIT) delivery is a strategy in which materials arrive at the job site when needed. This means that a crew may install the materials right off the truck. This method reduces the need for on-site storage and staging areas. It also reduces the risk of loss or damage while moving products about the site. Other modern material management methods include the use of radiofrequency identification (RFID) tags that make it easy to inventory and locate material in crowded staging areas. Using an RFID inventory control system requires special scanners.

involves rework due to bad technique or a conflict in craft scheduling requiring removal and reinstallation of finished work. The waste in time spent and replacement material costs occur when redoing the task a second time.

Under LEED, waste control is very important. Companies receive LEED credits for finding ways to reduce waste and for recycling construction waste. Workers should segregate waste materials by type for recycling, if feasible (*Figure 31*).

Figure 31 Waste materials separated for recycling.

4.4.3 Verifying Material Delivery

A crew leader may be responsible for the receipt of materials delivered to the work site. When this happens, the crew leader should require a copy of the shipping invoice or similar document and check each item on the invoice against the actual materials to verify delivery of the correct amounts.

The crew leader should also check the condition of the materials to verify that nothing is defective before signing the invoice. This can be difficult and time consuming because it requires the crew to open cartons and examine their contents. However, this step cannot be overlooked, because a signed invoice indicates that the recipient accepted all listed materials and in an undamaged state. If the crew leader signs for the materials without checking them, and then finds damage, no one will be able to prove that the materials came to the site in that condition.

After checking and signing the shipping invoice, the crew leader should give the original or a copy to the superintendent or project manager. The company then files the invoice for future reference because it serves as the only record the company has to check bills received from the supplier company.

4.4.4 Controlling Delivery and Storage

Another essential element of materials control is the selection of where the materials will be stored on the job site. There are two factors in determining the appropriate storage location. The first is convenience. If possible, the materials should be stored near where the crew will use them. The time and effort saved by not having to carry the materials long distances will greatly reduce the installation costs.

Next, the materials must be stored in a secure area to avoid damage. It is important that the storage area suit the materials being stored. For instance, materials that are sensitive to temperature, such as chemicals or paints, should be stored in climate-controlled areas to prevent waste.

4.4.5 Preventing Theft and Vandalism

Theft and vandalism of construction materials increase costs. The direct and indirect costs of lost materials include obtaining replacement materials, production time lost while the needed materials are missing, and the costs of additional security precautions can add significantly to the overall cost to the company. In addition, the insurance that the contractor purchases will increase in cost as the theft and vandalism rate grows.

The best way to avoid theft and vandalism is a secure job site. At the end of each work day, store unused materials and tools in a secure location, such as a locked construction trailer. If the job site is fenced or the building has lockable accesses, the materials can be stored within. Many sites have security cameras and/or intrusion alarms to help minimize theft and vandalism.

4.4.6 Equipment Control

The crew leader may not be responsible for long-term equipment control. However, the equipment required for a specific job is often the crew leader's responsibility. The first step is to schedule when a crew member must transport the required equipment from the shop or rental yard. The crew leader is responsible for informing the shop of the location of its use and seeing a worker returns it to the shop when finished with it.

It is common for equipment to lay idle at a job site because of a lack of proper planning for the job and the equipment arrived early. For example, if wire-pulling equipment arrives at a job site before the conduit is in place, this equipment will be out of service while awaiting the conduit installation. In addition to the wasted rental cost, damage, loss, or theft can also occur with idle equipment while awaiting use.

The crew leader needs to control equipment use, ensure that the crew operates the equipment according to its instructions, and within time and cost guidelines. The crew leader must also assign or coordinate responsibility for maintaining and repairing equipment as indicated by the applicable preventive maintenance schedule. Delaying maintenance and repairs can lead to costly equipment failures. The crew leader must also ensure that the equipment operators have the necessary credentials to operate the equipment, including applicable licenses.

The crew leader is responsible for the proper operation of all other equipment resources, including cars and trucks. Reckless or unsafe operation of vehicles will likely result in damaged equipment, leading to disciplinary action in the case of the workers, fines and repair costs levied by the vehicle vendors, and a loss of confidence in the crew leader's ability to govern crew-members' actions.

The crew leader should also arrange to secure all equipment at the close of each day's work to prevent theft. If continued use of the equipment is necessary for the job, the crew leader should make sure to lock it in a safe place; otherwise, return it to the shop.

4.4.7 Tool Control

Among companies, various policies govern who provides hand and power tools to employees. Some companies provide all the tools, while others furnish only the larger power tools. The crew leader should be familiar with and enforce any company policies related to tools.

Tool control has two aspects. First, the crew leader must control the issue, use, and maintenance of all tools provided by the company. Second, the crew leader must control how crew members use the tools to do the job. This applies to tools supplied by the company as well as tools that belong to the workers.

Using the proper tools correctly saves time and energy. In addition, proper tool use reduces the chance of damage to the tool during use, as well as injury to the user and to nearby workers. This is especially true of edge tools and those with any form of fixed or rotating blade. A dull tool is far more dangerous than a sharp one.

Tools must be adequately maintained and properly stored. Making sure that tools are cleaned, dried, and lubricated prevents rust and ensures that the tools are in the proper working order. If a tool is damaged, it is essential to repair or replace it promptly. Otherwise, an accident or injury could result.

NOTE

Regardless of whether a worker or the company owns a tool, OSHA holds companies responsible for the consequences of using a tool on a job site. The company is accountable if an employee receives an injury from a defective tool. Therefore, the crew leader needs to be aware of any defects in the tools the crew members are using.

Users should take care of company-issued tools as if they were their own. Workers should not abuse tools simply because they don't their property.

One of the major sources of low productivity on a job is the time spent searching for a tool. To prevent this from occurring, supervision should establish a storage location for company-issued tools and equipment. The crew leader should make sure that crew members return all company-issued tools and equipment to this designated location after use. Similarly, workers should organize their personal toolboxes so that they can readily find the appropriate tools and return their tools to their toolboxes when they are finished using them. This is a matter of professionalism that crew leaders are in an excellent position to develop.

Studies have shown that some keys to an effective tool control system are:

- Limiting the number of people allowed access to stored tools
- Limiting the number of people held responsible for tools and holding them accountable
- Controlling the ways in which a tool can be returned to storage
- Making sure tools are available when needed

4.4.8 Labor Control

Labor typically represents more than half the cost of a project, and therefore has an enormous impact on profitability. For that reason, it's essential to manage a crew and their work environment in a way that maximizes their productivity. One of the ways to do that is to minimize the time spent on unproductive activities such as the following:

- Engaging in bull sessions
- Correcting drawing errors
- Retrieving tools, equipment, and materials
- Waiting for other workers to finish

If crew members are habitually goofing off, it is up to the crew leader to counsel those workers. The crew leader's daily diary should document the counseling. The crew leader will need to refer repeated violations to the attention of higher management as guided by company policy.

All sorts of errors will occur during a project that need correction. Some errors, such as mistakes on drawings, may be outside of the crew leader's control. However, before a drawing error results in performing an action that will need to be corrected later, a careful examination of the drawings by the crew leader may discover the error before work begins, saving time and materials. If crew members are making mistakes due to inexperience, the crew leader can help avoid these errors by providing on-the-spot training and by checking on inexperienced workers more often.

The availability and location of tools, equipment, and materials to a crew can have a profound effect on their productivity. As discussed in the previous section, the key to minimizing such problems is proactive management of these resources.

Delays caused by crews or contractors can be minimized or avoided by carefully tracking the project schedule. In doing so, crew leaders can anticipate delays that will affect the work and either take action to prevent the delay or redirect the crew to another task.

4.4.9 Production and Productivity

Production is the amount of construction materials put in place. It is the quantity of materials installed on a job, such as 1,000 linear feet of waste pipe installed for a task. On the other hand, productivity is a rate, and depends on the level of efficiency of the work. It is the amount of work done per hour or day by one worker or a crew.

Production levels are set during the estimating stage. The estimator determines the total amount of materials placed from the plans and specifications. After the job is complete, supervision can assess the actual amount of materials installed, and can compare the actual production to the estimated production.

Productivity relates to the amount of materials put in place by the crew over a certain period of time. The estimator uses company records during the estimating stage to determine how much time and labor it will take to place a certain quantity of materials. From this information, the estimator calculates the productivity necessary to complete the job on time.

Participant Exercise G

1. List the methods your company uses to minimize waste.

____

____

____

____

____

2. List the methods your company uses to control small tools on the job.

____

____

____

____

____

3. List five ways that you feel your company could control labor to maximize productivity.

____

____

____

____

____

For example, it might take a crew of two people ten days to paint 5,000 square feet. To calculate the required average productivity, divide 5,000 ft^2 by 10 days. The result is 500 ft^2/day. The crew leader can compare the daily production of any crew of two painters doing similar work with this average, as discussed previously.

Planning is essential to productivity. The crew must be available to perform the work and have all the required materials, tools, and equipment in place when the job begins.

The time on the job should be for business, not for taking care of personal problems. Anything not work-related should be handled after hours, away from the job site, if possible. Even the planning of after-work activities, arranging social functions, or running personal errands should occur after work or during breaks. Only very limited and necessary exceptions to these rules should be permitted (e.g., making or meeting medical appointments), and these should be spelled out in company or crew policies.

Organizing field work can save time. The key to effectively using time is to work smarter, not necessarily harder. Working smart can save your crew from doing unnecessary work. For example, most construction projects require that the contractor submit a set of as-built plans at the completion of the work. These plans describe the actual structure and installation of materials. The best way to prepare these plans is to mark up a set of working plans as the work is in progress as described earlier. That way, workers and supervision won't forget pertinent details and waste time trying to reconstruct how the company did the work.

"Knowledge is power is time is money."
—Robert Thier, Storm and Silence

The amount of material used should not exceed the estimated amount. If it does, either the estimator has made a mistake, undocumented changes have occurred, or rework has caused the need for additional materials. Whatever the case, the crew leader should use effective control techniques to ensure the efficient use of materials.

When bidding a job, most companies calculate the cost per worker-hour. For example, a ten-day job might convert to 160 w-h (two painters for ten days at eight hours per day). If the company charges a labor rate of $30/hour, the labor cost would be $4,800. The estimator then adds the cost of materials, equipment, and tools, along with overhead costs and a profit factor, to determine the price of the job.

After completion of a job, information gathered through field reporting allows the company office to calculate actual productivity and compare it to the estimated figures. This helps to identify productivity issues and improves the accuracy of future estimates.

The following labor-related practices can help to ensure productivity:

- Ensure that all workers have the required resources when needed.
- Ensure that all personnel know where to go and what to do after each task is completed.
- Make reassignments as needed.
- Ensure that all workers have completed their work properly.

Additional Resources

Construction Workforce Development Professional, NCCER. 2016. New York, NY: Pearson Education, Inc.

Mentoring for Craft Professionals, NCCER. 2016. New York, NY: Pearson Education, Inc.

Blueprint Reading for Construction, James A. S. Fatzinger. 2003. New York, NY: Pearson Education, Inc.

Construction Leadership from A to Z: 26 Words to Lead By, Wally Adamchik. 2011. Live Oak Book Company.

The following websites offer resources for products and training:

Architecture, Engineering, and Construction Industry (AEC), **www.aecinfo.com**

US Green Building Council (USGBC), **www.usgbc.org/leed**

4.0.0 Section Review

1. Which of these activities occurs during the development phase of a project?
   a. Architect/engineer sketches are prepared and a preliminary budget is developed.
   b. Government agencies give a final inspection of the design, check for adherence to codes, and inspect materials used.
   c. Detailed project drawings and specifications are prepared.
   d. Contracts for the project are awarded.

2. A job diary should typically record any of the following *except* _____.
   a. items such as job interruptions and visits
   b. the exact times of scheduled lunch breaks
   c. changes needed to project drawings
   d. the actual time each job task related to a particular project took to complete

3. Which of the following is a correct statement regarding project cost?
   a. Cost is handled by the accounting department and isn't a concern of the crew leader.
   b. The difference between the estimated cost and the actual cost affects a company's profit.
   c. Wasted material is factored into the estimate and is never a concern.
   d. The contractor's overhead costs aren't included in the cost estimate.

4. To prevent job delays due to late delivery of materials, the crew leader should _____.
   a. demand a discount from the supplier to compensate for the delay
   b. tell a crew member to go look for the delivery truck
   c. refuse the late delivery and re-order the materials from another supplier
   d. check with the supplier in advance of the scheduled delivery

Summary

In construction, the crew leader is the company's supervisor on the ground at the scene where the actual work takes place. The modern crew leader must be prepared to deal with a host of challenges ranging from differences in age, culture, race, gender, language, and educational backgrounds to attitudes regarding authority and personal integrity. Rather than being overwhelmed, the effective and successful crew leader will use these challenges to broaden his or her understanding of people and to grow in character.

Crew leaders must develop an understanding of the company, its policies, and how to become an effective and valuable member of the supervisory team. They must also be able to effectively communicate up and down the chain, to motivate the crew's workers, and develop a sense of ownership within the crew so that they have the confidence and will to do the jobs.

Safety is an overriding concern for the company, and crew leaders need to make safe work practices and conditions the number-one priority. Once a worker has lost an eye, a limb, or life itself, it's too late to reconsider how the task might have been done safer. Workers are a company's most valuable asset and the crew leader is in the position of being the best example of safety to the crew.

While crew leaders view their role as mainly people and work managers, these tasks gain a greater sense of importance when crew leaders understand their responsibilities to the company for whom they work. Effectively controlling worker productivity, resources, schedules, and other aspects of the project directly contribute to cost control, which reflects in greater profits for the company. Companies recognize effective crew leaders and they can expect long and successful careers as a result.

Review Questions

1. Younger-generation workers may have a better perception of success if _____.
   a. crew leaders assign their projects in smaller, well-defined units
   b. they receive closer supervision
   c. crew leaders leave them alone
   d. they are given the perception that they are in charge

2. Companies can help prevent sexual harassment in the workplace by _____.
   a. requiring employee training that avoids the potentially offensive subject of stereotypes
   b. developing a consistent policy with appropriate consequences for engaging in sexual harassment
   c. communicating to workers that, legally, the victim of sexual harassment is only the one being directly harassed
   d. establishing an effective complaint or grievance process for victims of harassment

3. Which of the following statements about a formal organization is *true*?
   a. It is a group of members created mainly to share professional information.
   b. It is a group established for developing industry standards.
   c. It is a group of individuals led by someone all working toward the same goal.
   d. It is a group that is identified by its overall function rather than by key positions.

4. Which of these departments in a large company would likely arrange equipment rentals?
   a. Accounting
   b. Estimating
   c. Human Resources
   d. Purchasing

5. When a person is promoted to crew leader, the amount of time they spend on craft work _____.
   a. increases
   b. decreases
   c. stays the same
   d. doubles

6. Which of the following is a *correct* statement regarding annual crew evaluations?
   a. Their results should never come as a surprise to the crew member.
   b. Formal evaluations should be conducted every three months.
   c. Comments on performance should be withheld until the formal evaluation.
   d. Advise the crew members to improve their performance just before the formal evaluation is scheduled.

7. As a crew leader, what should you avoid when giving recognition to crew members?
   a. Looking for good work and knowing what good work is
   b. Being available at the job site to witness good work
   c. Giving praise for poor work to make the worker feel better
   d. Encouraging work improvement through your confidence in a worker

8. Within the construction industry, who normally assigns a crew to the crew leader?
   a. The US Department of Labor
   b. The company
   c. The crew leader
   d. The crew members

9. Some decisions are difficult to make because you _____.
   a. have no experience dealing with the circumstances requiring a decision
   b. don't want to make a decision
   c. prefer to think it's someone else's responsibility to make a decision
   d. don't recognize the need for a decision

10. The first step in problem solving should be _____.
    a. choosing a solution to the problem
    b. evaluating the solution to the problem
    c. defining the problem
    d. testing the solution to the problem

11. The leading cause of fatalities in the construction industry is _____.
    a. electrocution
    b. asphyxiation
    c. struck by something
    d. falls

12. OSHA inspection of a business or job site _____.
    a. can be done only by invitation
    b. can be conducted at random
    c. is done only after an accident
    d. is conducted only if a safety violation occurs

13. To proactively identify and assess hazards, it is essential that employers _____.
    a. ensure that crew leaders are present and available on the job site
    b. conduct safety inspections of contractor offices
    c. design safety logs to document safety violations
    d. conduct accident investigations

14. Employers who are subject to OSHA recordkeeping requirements must maintain a log of recordable occupational injuries and illnesses for _____.
    a. three years
    b. five years
    c. seven years
    d. as long as the company is in business

15. Which of the following statements regarding safety training sessions is *correct*?
    a. The project manager usually holds them.
    b. They are held only for new employees.
    c. They should be conducted frequently by the crew leader.
    d. They are required only after an accident has occurred.

16. The type of contract in which the client pays the contractor for their actual labor and material expenses they incur is known as a _____.
    a. firm fixed-price contract
    b. time-spent contract
    c. performance-based contract
    d. cost-reimbursable contract

17. On a design-build project, _____.
    a. the same contractor is responsible for both design and construction
    b. the owner is responsible for providing the design
    c. the architect does the design and the general contractor builds the project
    d. a construction manager is hired to oversee the project

18. When planning labor for onsite work, the crew leader must _____.
    a. factor in the absenteeism and turnover rates
    b. factor in time for training classes
    c. select the material to be used by each crew member
    d. identify the skills needed to perform the work

19. Which of the following is *not* a source of data in the field-reporting system?
    a. The cost-estimate for the project
    b. Crew leader's daily diary
    c. Onsite inventory of placed materials
    d. Daily time cards

20. The following duties are important in work site equipment control *except* _____.
    a. scheduling the pickup and return of a piece of equipment
    b. ensuring the equipment is operated and maintained correctly
    c. monitoring safe and responsible operation of vehicles
    d. selection and purchase of equipment to be stocked at the work site equipment room

Trade Terms Introduced in This Module

Autonomy: The condition of having complete control over one's actions, and being free from the control of another. To be independent.

Bias: A preconceived inclination against or in favor of something.

Cloud-based applications: Mobile and desktop digital programs that can connect to files and data located in distributed storage locations on the Internet ("the Cloud"). Such applications make it possible for authorized users to create, edit, and distribute content from any location where Internet access is possible.

Craft professionals: Workers who are properly trained and work in a particular construction trade or craft.

Crew leader: The immediate supervisor of a crew or team of craft professionals and other assigned persons.

Critical path: In manufacturing, construction, and other types of creative processes, the required sequence of tasks that directly controls the ultimate completion date of the project.

Demographics: Social characteristics and other factors, such as language, economics, education, culture, and age, that define a statistical group of individuals. An individual can be a member of more than one demographic.

Ethics: The moral principles that guide an individual's or organization's actions when dealing with others. It also refers to the study of moral principles.

Infer: To reach a conclusion using a method of reasoning that starts with an assumption and considers a set of logically-related events, conditions, or statements.

Intangible: Not touchable, material, or measureable; lacking a physical presence.

Job description: A description of the scope and responsibilities of a worker's job so that the individual and others understand what the job entails.

Job diary: A written record that a supervisor maintains periodically (usually daily) of the events, communications, observations, and decisions made during the course of a project.

Legend: In maps, plans, and diagrams, an explanatory table defining all symbolic information contained in the document.

Lethargy: Sluggishness, slow motion, lack of activity, or a lack of enthusiasm.

Letter of instruction (LOI): A written communication from a supervisor to a subordinate that informs the latter of some inadequacy in the individual's performance and provides a list of actions that the individual must satisfactorily complete to remediate the problem. This is usually a key step in a series of disciplinary actions.

Local area networks (LAN): Communication networks that link computers, printers, and servers within a small, defined location, such as a building or office, via hard-wired or wireless connections.

Lockout/tagout (LOTO): A system of safety procedures for securing electrical and mechanical equipment during repairs or construction, consisting of warning tags and physical locking or restraining devices applied to controls to prevent the accidental or purposeful operation of the equipment, endangering workers, equipment, or facilities.

Look-ahead schedule: A manual- or software-scheduling tool that looks several weeks into the future at the planned project events; used for anticipating material, labor, tool, and other resource requirements, as well as identifying potential schedule conflicts or other problems.

Negligence: Lack of appropriate care when doing something, or the failure to do something, usually resulting in injury to an individual or damage to equipment.

Organizational chart: A diagram that shows how the various management and operational responsibilities relate to each other within an organization. Named positions appear in ranked levels, top to bottom, from those functional units with the most and broadest authority to those with the least, with lines connecting positions indicating chains of authority and other relationships.

Paraphrase: Rewording a written or verbal statement in one's own words in such a way that the intent of the original statement is retained.

Pragmatic: Sensible, practical, and realistic.

Proactive: To anticipate and take action in the present to deal with potential future events or outcomes based on what one knows about current events or conditions.

Project manager: In construction, the individual who has overall responsibility for one or more construction projects; also called the general superintendent.

Return on investment (ROI): A measure of the gain or loss of money resulting from an investment; normally measured as a percentage of the original investment.

Safety data sheets (SDS): Documents listing information about a material or substance that includes common and proper names, chemical composition, physical forms and properties, hazards, flammability, handling, and emergency response in accordance with national and international hazard communication standards; also called material safety data sheets (MSDS).

Sexual harassment: Any unwelcome verbal or nonverbal form of communication or action construed by an individual to be of a sexual or gender-related nature.

Smartphone: The most common type of cellular telephone that combines many other digital functions within a single mobile device. Most important of these, besides the wireless telephone, are high-resolution still and video cameras, text and email messaging, and GPS-enabled features.

Superintendent: In construction, the individual who is the on-site supervisor in charge of a given construction project.

Synergy: Any type of cooperation between organizations, individuals, or other entities where the combined effect is greater than the sum of the individual efforts.

Textspeak: A form of written language characteristic of text messaging on mobile devices and text-based social media, which usually consists of acronyms, abbreviations, and minimal punctuation.

Wide area networks (WAN): Dedicated communication networks of computers and related hardware that serve a given geographic area, such as a work site, campus, city, or a larger but distinct area. Connectivity is by wired and wireless means, and may use the Internet as well.

Wi-Fi: The technology allowing communications via radio signals over a LAN or WAN equipped with a wireless access point or the Internet. (Wi-Fi stands for wireless fidelity.) Many types of mobile, portable, and desktop devices can communicate via Wi-Fi connections.

Work breakdown structure (WBS): A diagrammatic and conceptual method for subdividing a complex project, concept, or other thing into its various functional and organizational parts so that planners can analyze and plan for each part in detail.

OSHA Forms for Safety Records

Employers who are subject to the recordkeeping requirements of the Occupational Safety and Health Act of 1970 must maintain records of all recordable occupational injuries and illnesses. The following are three important OSHA forms used for this recordkeeping (these forms can be accessed at **www.osha.gov**):

- OSHA Form 300, *Log of Work-Related Injuries and Illnesses*
- OSHA Form 300A, *Summary of Work-Related Injuries and Illnesses*
- OSHA Form 301, *Injury and Illness Incident Report*

OSHA's Form 300 (Rev. 01/2004)

Log of Work-Related Injuries and Illnesses

Attention: This form contains information relating to employee health and must be used in a manner that protects the confidentiality of employees to the extent possible while the information is being used for occupational safety and health purposes.

Year ________

U.S. Department of Labor
Occupational Safety and Health Administration

Form approved OMB no. 1218-0176

You must record information about every work-related injury or illness that involves loss of consciousness, restricted work activity or job transfer, days away from work, or medical treatment beyond first aid. You must also record significant work-related injuries and illnesses that are diagnosed by a physician or licensed health care professional. You must also record work-related injuries and illnesses that meet any of the specific recording criteria listed in 29 CFR 1904.8 through 1904.12. Feel free to use two lines for a single case if you need to. You must complete an injury and illness incident report (OSHA Form 301) or equivalent form for each injury or illness recorded on this form. If you're not sure whether a case is recordable, call your local OSHA office for help.

Establishment name ________________

City ________________ State ________________

Identify the person			Describe the case			Classify the case										
						CHECK ONLY ONE box for each case based on the most serious outcome for that case:			Enter the number of days the injured or ill worker was:		Check the "injury" column or choose one type of illness: (M)					
(A) Case No.	(B) Employee's Name	(C) Job Title (e.g., Welder)	(D) Date of injury or onset of illness (mo./day)	(E) Where the event occurred (e.g. Loading dock north end)	(F) Describe injury or illness, parts of body affected, and object/substance that directly injured or made person ill (e.g. Second degree burns on right forearm from acetylene torch)	Days away from work (H)	Remained at work: Job transfer or restriction (I)	Remained at work: Other recordable cases (J)	Away From Work (days) (K)	On job transfer or restriction (days) (L)	Injury (1)	Skin Disorder (2)	Respiratory Condition (3)	Poisoning (4)	Hearing Loss (5)	All other illnesses (6)
					Page totals	0	0	0	0	0	0	0	0	0	0	0
											Injury (1)	Skin Disorder (2)	Respiratory Condition (3)	Poisoning (4)	Hearing Loss (5)	All other illnesses (6)

Be sure to transfer these totals to the Summary page (Form 300A) before you post it.

Public reporting burden for this collection of information is estimated to average 14 minutes per response, including time to review the instruction, search and gather the data needed, and complete and review the collection of information. Persons are not required to respond to the collection of information unless it displays a currently valid OMB control number. If you have any comments about these estimates or any aspects of this data collection, contact: US Department of Labor, OSHA Office of Statistics, Room N-3644, 200 Constitution Ave, NW, Washington, DC 20210. Do not send the completed forms to this office.

Page 1 of 1

Figure A01 OSHA Form 300

OSHA's Form 300A (Rev. 01/2004)

Summary of Work-Related Injuries and Illnesses

Year ______

U.S. Department of Labor
Occupational Safety and Health Administration

Form approved OMB no. 1218-0176

All establishments covered by Part 1904 must complete this Summary page, even if no injuries or illnesses occurred during the year. Remember to review the Log to verify that the entries are complete and accurate before

Using the Log, count the individual entries you made for each category. Then write the totals below, making sure you've added the entries from every page of the log. If you had no cases write "0."

Employees former employees, and their representatives have the right to review the OSHA Form 300 in its entirety. They also have limited access to the OSHA Form 301 or its equivalent. See 29 CFR 1904.35, in OSHA's Recordkeeping rule, for further details on the access provisions for these forms.

Number of Cases

Total number of deaths	Total number of cases with days away from work	Total number of cases with job transfer or restriction	Total number of other recordable cases
0	0	0	0
(G)	(H)	(I)	(J)

Number of Days

Total number of days away from work	Total number of days of job transfer or restriction
0	0
(K)	(L)

Injury and Illness Types

Total number of...
(M)

(1) Injury	0	(4) Poisoning	0
(2) Skin Disorder	0	(5) Hearing Loss	0
(3) Respiratory Condition	0	(6) All Other Illnesses	0

Post this Summary page from February 1 to April 30 of the year following the year covered by the form

Public reporting burden for this collection of information is estimated to average 58 minutes per response, including time to review the instruction, search and gather the data needed, and complete and review the collection of information. Persons are not required to respond to the collection of information unless it displays a currently valid OMB control number. If you have any comments about these estimates or any aspects of this data collection, contact: US Department of Labor, OSHA Office of Statistics, Room N-3644, 200 Constitution Ave, NW, Washington, DC 20210. Do not send the completed forms to this office.

Establishment information

Your establishment name ______

Street ______

City ______ State ______ Zip ______

Industry description (e.g., Manufacture of motor truck trailers) ______

Standard Industrial Classification (SIC), if known (e.g., SIC 3715) __ __ __ __

OR North American Industrial Classification (NAICS), if known (e.g., 336212) __ __ __ __ __ __

Employment information

Annual average number of employees ______

Total hours worked by all employees last year ______

Sign here

Knowingly falsifying this document may result in a fine.

I certify that I have examined this document and that to the best of my knowledge the entries are true, accurate, and complete.

______ Company executive ______ Title

______ Phone ______ Date

Figure A02 OSHA Form 300A

U.S. Department of Labor
Occupational Safety and Health Administration

OSHA's Form 301
Injuries and Illnesses Incident Report

Attention: This form contains information relating to employee health and must be used in a manner that protects the confidentiality of employees to the extent possible while the information is being used for occupational safety and health purposes.

Form approved OMB no. 1218-0176

This *Injury and Illness Incident Report* is one of the first forms you must fill out when a recordable work-related injury or illness has occurred. Together with the *Log of Work-Related Injuries and Illnesses* and the accompanying *Summary*, these forms help the employer and OSHA develop a picture of the extent and severity of work-related incidents.

Within 7 calendar days after you receive information that a recordable work-related injury or illness has occurred, you must fill out this form or an equivalent. Some state workers' compensation, insurance, or other reports may be acceptable substitutes. To be considered an equivalent form, any substitute must contain all the information asked for on this form.

According to Public Law 91-596 and 29 CFR 1904, OSHA's recordkeeping rule, you must keep this form on file for 5 years following the year to which it pertains

If you need additional copies of this form, you may photocopy and use as many as you need.

Completed by ____________

Title ____________

Phone ____________ Date ____________

Information about the employee

1) Full Name ____________

2) Street ____________

City ____________ State ______ Zip ______

3) Date of birth ____________

4) Date hired ____________

5) ☐ Male
☐ Female

Information about the physician or other health care professional

6) Name of physician or other health care professional ____________

7) If treatment was given away from the worksite, where was it given?

Facility ____________

Street ____________

City ____________ State ______ Zip ______

8) Was employee treated in an emergency room?
☐ Yes
☐ No

9) Was employee hospitalized overnight as an in-patient?
☐ Yes
☐ No

Information about the case

10) Case number from the Log ________ *(Transfer the case number from the Log after you record the case.)*

11) Date of injury or illness ________

12) Time employee began work ________ AM/PM

13) Time of event ________ AM/PM ☐ Check if time cannot be determined

14) **What was the employee doing just before the incident occurred?** Describe the activity, as well as the tools, equipment or material the employee was using. Be specific. Examples: "climbing a ladder while carrying roofing materials"; "spraying chlorine from hand sprayer"; "daily computer key-entry."

15) **What happened?** Tell us how the injury occurred. Examples: "When ladder slipped on wet floor, worker fell 20 feet"; "Worker was sprayed with chlorine when gasket broke during replacement"; "Worker developed soreness in wrist over time."

16) **What was the injury or illness?** Tell us the part of the body that was affected and how it was affected; be more specific than "hurt", "pain", or "sore." Examples: "strained back"; "chemical burn, hand"; "carpal tunnel syndrome."

17) **What object or substance directly harmed the employee?** Examples: "concrete floor"; "chlorine"; "radial arm saw." If this question does not apply to the incident, leave it blank.

18) **If the employee died, when did death occur?** Date of death

Public reporting burden for this collection of information is estimated to average 22 minutes per response, including time for reviewing instructions, searching existing data sources, gathering and maintaining the data needed, and completing and reviewing the collection of information. Persons are not required to respond to the collection of information unless it displays a current valid OMB control number. If you have any comments about this estimate or any other aspects of this data collection, including suggestions for reducing this burden, contact: US Department of Labor, OSHA Office of Statistics, Room N-3644, 200 Constitution Ave, NW, Washington, DC 20210. Do not send the completed forms to this office.

Figure A03 OSHA Form 301

Additional Resources

This module presents thorough resources for task training. The following reference material is recommended for further study.

Construction Workforce Development Professional, NCCER. 2016. New York, NY: Pearson Education, Inc.

Mentoring for Craft Professionals, NCCER. 2016. New York, NY: Pearson Education, Inc.

Blueprint Reading for Construction, James A. S. Fatzinger. 2003. New York, NY: Pearson Education, Inc.

Construction Leadership from A to Z: 26 Words to Lead By, Wally Adamchik. 2011. Live Oak Book Company.

Generational Cohorts and their Attitudes Toward Work Related Issues in Central Kentucky, Frank Fletcher, et al. 2009. Midway College, Midway, KY. **www.kentucky.com.**

It's Your Ship: Management Techniques from the Best Damn Ship in the Navy, Captain D. Michael Abrashoff, USN. 2012. New York City, NY: Grand Central Publishing.

Survival of the Fittest, Mark Breslin. 2005. McNally International Press.

The Definitive Book of Body Language: The Hidden Meaning Behind People's Gestures and Expressions, Barbara Pease and Allan Pease. 2006. New York City, NY: Random House / Bantam Books.

The Young Person's Guide to Wisdom, Power, and Life Success: Making Smart Choices, Brian Gahran, PhD. 2014. San Diego, CA: Young Persons Press. **www.WPGBlog.com**.

The following websites offer resources for products and training:

Aging Workforce News, **www.agingworkforcenews.com**.

American Society for Training and Development (ASTD), **www.astd.org**.

Architecture, Engineering, and Construction Industry (AEC), **www.aecinfo.com**.

Equal Employment Opportunity Commission (EEOC), **www.eeoc.gov**.

National Association of Women in Construction (NAWIC), **www.nawic.org**.

National Census of Fatal Occupational Injuries (NCFOI), **www.bls.gov**.

National Institute of Occupational Safety and Health (NIOSH), **www.cdc.gov/niosh**.

National Safety Council, **www.nsc.org**.

Occupational Safety and Health Administration (OSHA), **www.osha.gov**.

Society for Human Resources Management (SHRM), **www.shrm.org**.

United States Census Bureau, **www.census.gov**.

United States Department of Labor, **www.dol.gov**.

US Green Building Council (USGBC), **www.usgbc.org/leed**.

Wi-Fi® is a registered trademark of the Wi-Fi Alliance, **www.wi-fi.org**

Figure Credits

© Photographerlondon/Dreamstime.com, Module Opener
© iStockphoto.com/shironosov, Figure 1
© iStockphoto.com/kali9, Figures 2, 10
© iStockphoto.com/nanmulti, Figure 3
© iStockphoto.com/Halfpoint, Figure 13
© iStock.com/vejaa, Figure 16
DIYawards, Figure 17
© iStockphoto.com/Tashi-Delek, Figure 18
© iStockphoto.com/kemaltaner, Figure 23
Sushil Shenoy/Virginia Tech, Figure 31
John Ambrosia, Figure 27
© iStockphoto.com/izustun, Figure 30

Section Review Answer Key

Answer	Section Reference	Objective
Section One		
1. a	1.1.0	1a
2. b	1.2.2	1b
3. b	1.3.0	1c
Section Two		
1. b	2.0.0	2a
2. c	2.2.1	2b
3. a	2.3.1	2c
4. c	2.4.1	2d
5. a	2.5.1	2e
Section Three		
1. a	3.0.0	3a
2. d	3.2.0; Table 1	3b
3. a	3.3.0	3c
4. b	3.4.0	3d
Section Four		
1. a	4.1.1	4a
2. b	4.2.2	4b
3. b	4.3.3	4c
4. d	4.4.1	4d

NCCER CURRICULA — USER UPDATE

NCCER makes every effort to keep its textbooks up-to-date and free of technical errors. We appreciate your help in this process. If you find an error, a typographical mistake, or an inaccuracy in NCCER's curricula, please fill out this form (or a photocopy), or complete the online form at **www.nccer.org/olf**. Be sure to include the exact module ID number, page number, a detailed description, and your recommended correction. Your input will be brought to the attention of the Authoring Team. Thank you for your assistance.

Instructors – If you have an idea for improving this textbook, or have found that additional materials were necessary to teach this module effectively, please let us know so that we may present your suggestions to the Authoring Team.

NCCER Product Development and Revision

13614 Progress Blvd., Alachua, FL 32615

Email: curriculum@nccer.org
Online: www.nccer.org/olf

❑ Trainee Guide ❑ Lesson Plans ❑ Exam ❑ PowerPoints Other ____________

Craft / Level: Copyright Date:

Module ID Number / Title:

Section Number(s):

Description:

Recommended Correction:

Your Name:

Address:

Email: Phone:

This page is intentionally left blank.

Glossary

Acetylene: A gas composed of two parts carbon and two parts hydrogen, commonly used in combination with oxygen to cut, weld, and braze steel.

Alloy: A metal that has had other elements added to it, which substantially changes its mechanical properties.

Autonomy: The condition of having complete control over one's actions, and being free from the control of another. To be independent.

Base line: A straight line drawn around a pipe to be used as a measuring point.

Bias: A preconceived inclination against or in favor of something.

Body: The main part of the valve. It contains the disc, seat, and valve ports. The body of the valve is directly connected to the piping by threaded, welded, mechanically joined, or flanged ends.

Bonnet: The part of the valve that contains the trim. The bonnet is located above the body.

Branch: A line that intersects with another line.

Capillary action: The flow of liquid (solder) into a small space between two surfaces.

Carbon: An element which, when combined with iron, forms various kinds of steel. In steel, it is the carbon content that affects the physical properties of the steel.

Chord: A straight line crossing a circle that does not pass through the center of the circle.

Cloud-based applications: Mobile and desktop digital programs that can connect to files and data located in distributed storage locations on the Internet ("the Cloud"). Such applications make it possible for authorized users to create, edit, and distribute content from any location where Internet access is possible.

Compression coupling: A mechanical fitting that is compressed onto a pipe, tube, or hose.

Condensate: The liquid byproduct of cooling steam.

Conductivity: A measure of the ability of a material to transmit electron flow.

Constituents: The elements and compounds, such as metal oxides, that make up a mixture or alloy.

Craft professionals: Workers who are properly trained and work in a particular construction trade or craft.

Crew leader: The immediate supervisor of a crew or team of craft professionals and other assigned persons.

Critical path: In manufacturing, construction, and other types of creative processes, the required sequence of tasks that directly controls the ultimate completion date of the project.

Critical temperature: The temperature at which iron crystals in a ferrous-based metal transform from being face-centered to body-centered. This dramatically changes the strength, hardness, and ductility of the metal.

Cup depth engagement: The distance the tubing penetrates the fitting.

Cutback: The point at which a miter fitting is to be cut.

Demographics: Social characteristics and other factors, such as language, economics, education, culture, and age, that define a statistical group of individuals. An individual can be a member of more than one demographic.

Differential pressure: The measurement of one pressure with respect to another pressure, or the difference between two pressures.

Dimension: A measurement between two points on a drawing.

Direct pressure: Flow pushing the sealing element of the valve into the seat and improving closure.

Disc: The moving part of a valve that directly affects the flow through the valve.

Drip legs: Drains for condensate in steam lines placed at low points in the line and used with steam traps.

Ethics: The moral principles that guides an individual's or organization's actions when dealing with others. Also refers to the study of moral principles.

Ferrules: Bushings placed over a tube to tighten it.

Flaring: Increasing the diameter at the end of a pipe or tube.

Flash steam: Steam formed when hot condensate is released to lower pressure and reevaporated.

Full port: The maximum internal opening for flow through a valve that matches the ID of the pipe used.

Hardenability: A characteristic of a metal that makes it able to become hard, usually through heat treatment.

Head clearance: The amount of space needed to install a hot tap machine on a pipe.

Heat-affected zone (HAZ): The area of the base metal that is not melted but whose mechanical properties have been altered by the heat of the welding.

Hot tap: To make a safe entry into a pipe or vessel operating at a pressure or vacuum under controlled conditions without losing product.

Induction heating: The heating of a conducting material by means of circulating electrical currents induced by an externally applied alternating magnetic field.

Infer: To reach a conclusion using a method of reasoning that starts with an assumption and considers a set of logically-related events, conditions, or statements.

Inside diameter (ID): Measurement of the inside diameter of the pipe.

Intangible: Not touchable, material, or measureable; lacking a physical presence.

Interpass heat: The temperature to which a metal is heated while an operation is performed on the metal.

Job description: A description of the scope and responsibilities of a worker's job so that the individual and others understand what the job entails.

Job diary: A written record that a supervisor maintains periodically (usually daily) of the events, communications, observations, and decisions made during the course of a project.

Legend: In maps, plans, and diagrams, an explanatory table defining all symbolic information contained in the document.

Lethargy: Sluggishness, slow motion, lack of activity, or a lack of enthusiasm.

Letter of instruction (LOI): A written communication from a supervisor to a subordinate that informs the latter of some inadequacy in the individual's performance, and provides a list of actions that the individual must satisfactorily complete to remediate the problem. Usually a key step in a series of disciplinary actions.

Line stop: A device used to temporarily contain the flow or pressure of a product inside a pipe while a branch connection is being made.

Local area networks (LAN): Communication networks that link computers, printers, and servers within a small, defined location, such as a building or office, via hard-wired or wireless connections.

Lockout/tagout (LOTO): A system of safety procedures for securing electrical and mechanical equipment during repairs or construction, consisting of warning tags and physical locking or restraining devices applied to controls to prevent the accidental or purposeful operation of the equipment, endangering workers, equipment, or facilities.

Look-ahead schedule: A manual- or software-scheduling tool that looks several weeks into the future at the planned project events; used for anticipating material, labor, tool, and other resource requirements, as well as identifying potential schedule conflicts or other problems.

Miter: A specified angle to which a piece of pipe is cut.

Negligence: Lack of appropriate care when doing something, or the failure to do something, usually resulting in injury to an individual or damage to equipment.

Nonferrous metal: A metal that contains no iron and is therefore nonmagnetic.

Ordinate lines: Straight lines drawn along the length of the pipe connecting the ordinate marks.

Ordinates: Divisions of segments obtained by dividing the circumference of a pipe into equal parts.

Organizational chart: A diagram that shows how the various management and operational responsibilities relate to each other within an organization. Named positions appear in ranked levels, top to bottom, from those functional units with the most and broadest authority to those with the least, with lines connecting positions indicating chains of authority and other relationships.

Outside diameter (OD): Measurement of the outside diameter of the pipe.

Oxidation: A chemical reaction that increases the oxygen content of a compound.

Packing: The material between the valve stem and bonnet that provides a leakproof seal and prevents material from leaking up around the valve stem.

Paraphrase: Rewording a written or verbal statement in one's own words in such a way that the intent of the original statement is retained.

Port: The internal opening for flow through a valve.

Post-weld heat treatment (PWHT): The temperature to which a metal is heated after an operation is performed on the metal.

Pragmatic: Sensible, practical, and realistic.

Preheat: The temperature to which a metal is heated before an operation is performed on the metal.

Preheat weld treatment (PHWT): The controlled heating of the base metal immediately before welding begins.

Proactive: To anticipate and take action in the present to deal with potential future events or outcomes based on what one knows about current events or conditions.

Project manager: In construction, the individual who has overall responsibility for one or more construction projects; also called the general superintendent.

Quenching: Rapid cooling of a hot metal using air, water, or oil.

Reamer: A tool used to enlarge, shape, smooth, or otherwise finish a hole.

Return on investment (ROI): A measure of the gain or loss of money resulting from an investment; normally measured as a percentage of the original investment.

Reverse pressure: Flow pushing the closure element out of the seat.

Saddle: A fabricated 90-degree intersection of pipe.

Safety data sheets (SDS): Documents listing information about a material or substance that includes common and proper names, chemical composition, physical forms and properties, hazards, flammability, handling, and emergency response in accordance with national and international hazard communication standards. Also called material safety data sheets (MSDS).

Saturated steam: Pure steam without water droplets that is at the boiling temperature of water for the given pressure.

Seat: The nonmoving part of the valve on which the disc rests to form a seal and close off the valve.

Segments: Parts of a circle that are defined by a chord and the curve of the circumference.

Sexual harassment: Any unwelcome verbal or nonverbal form of communication or action construed by an individual to be of a sexual or gender-related nature.

Smartphone: The most common type of cellular telephone that combines many other digital functions within a single mobile device. Most important of these, besides the wireless telephone, are high-resolution still and video cameras, text and email messaging, and GPS-enabled features.

Solder: An alloy, such as of zinc and copper, or of tin and lead, that is used to join metallic surfaces when melted.

Soldering flux: A chemical substance that aids the flow of solder. Flux removes and prevents the formation of oxides on the pieces to be joined by soldering.

Stem: The part of the valve that connects the disc to the valve operator. A stem can have linear, rotary, or helical movement.

Stress: The load imposed on an object.

Stress relieving: The even heating of a structure to a temperature below the critical temperature followed by a slow, even cooling.

Superheated steam: Saturated steam to which heat has been added to raise the working temperature.

Superintendent: In construction, the individual who is the onsite supervisor in charge of a given construction project.

Synergy: Any type of cooperation between organizations, individuals, or other entities where the combined effect is greater than the sum of the individual efforts.

Tempering: Increases the toughness of quenched steel and helps avoid breakage and failure of heat-treated steel. Also called drawing.

Textspeak: A form of written language characteristic of text messaging on mobile devices and text-based social media, which usually consists of acronyms, abbreviations, and minimal punctuation.

Title block: A section of an engineering drawing blocked off for pertinent information, such as the title, drawing number, date, scale, material, draftsperson, and tolerances.

Tracer: A steam line piped beside product piping to keep the product warm or prevent it from freezing.

Travel distance: The distance that the cutting bit moves from the top of the tapping valve to the pipe to be cut.

Trim: The internal parts of a valve that receive the most wear and can be replaced. The trim includes the stem disc, seat ring, disc holder or guide, wedge, and bushings.

Wetting: Spreading liquid filler metal or flux on a solid base metal.

Wide area networks (WAN): Dedicated communication networks of computers and related hardware that serve a given geographic area, such as a work site, campus, city, or a larger but distinct area. Connectivity is by wired and wireless means, and may use the Internet as well.

Wi-Fi: The technology allowing communications via radio signals over a LAN or WAN equipped with a wireless access point or the Internet. (Wi-Fi stands for wireless fidelity.) Many types of mobile, portable, and desktop devices can communicate via Wi-Fi connections.

Work breakdown structure (WBS): A diagrammatic and conceptual method for subdividing a complex project, concept, or other thing into its various functional and organizational parts so that planners can analyze and plan for each part in detail.

This page is intentionally left blank.

This page is intentionally left blank.

This page is intentionally left blank.

This page is intentionally left blank.

This page is intentionally left blank.

This page is intentionally left blank.

This page is intentionally left blank.

This page is intentionally left blank.

This page is intentionally left blank.

This page is intentionally left blank.